Contemporary Clinical Neuroscience

Series Editor

Mario Manto, Division of Neurosciences, Department of Neurology, CHU-Charleroi, Charleroi, Belgium, University of Mons, Mons, Belgium

Contemporary Clinical Neurosciences bridges the gap between bench research in the neurosciences and clinical neurology work by offering translational research on all aspects of the human brain and behavior with a special emphasis on the understanding, treatment, and eradication of diseases of the human nervous system. These novel, state-of-the-art research volumes present a wide array of preclinical and clinical research programs to a wide spectrum of readers representing the diversity of neuroscience as a discipline. The book series considers proposals from leading scientists and clinicians. The main audiences are basic neuroscientists (neurobiologists, neurochemists, geneticians, experts in behavioral studies, neurophysiologists, neuroanatomists), clinicians (including neurologists, psychiatrists and specialists in neuroimaging) and trainees, graduate students, and PhD students.

Volumes in the series provide in-depth books that focus on neuroimaging, ADHD (attention deficit hyperactivity disorder and other neuropsychiatric disorders, neurodegenerative diseases, G protein receptors, sleep disorders, addiction issues, cerebellar disorders, and neuroimmune diseases. The series aims to expand the topics at the frontiers between basic research and clinical applications. Each volume is available in both print and electronic form.

Parthvi Ravat • Jorge C. Kattah • Aasef G. Shaikh
Editors

Advances in Translational Neuroscience of Vestibular and Eye Movement Disorders

Second Edition

Springer

Editors
Parthvi Ravat
Department of Neurology
Flinders Medical Centre, Bedford Park
Adelaide, SA, Australia

Aasef G. Shaikh
Neurological Institute, University
Hospitals Neurology Service, Louis Stokes
Cleveland, VA Medical Center
Department of Neurology, Case Western
Reserve University
Cleveland, OH, USA

Jorge C. Kattah
Peoria, IL, USA

ISSN 2627-535X ISSN 2627-5341 (electronic)
Contemporary Clinical Neuroscience
ISBN 978-3-032-07538-3 ISBN 978-3-032-07539-0 (eBook)
https://doi.org/10.1007/978-3-032-07539-0

This Springer imprint is published by the registered company Springer Nature Switzerland AG
The registered company address is: Gewerbestrasse 11, 6330 Cham, Switzerland

Preface

The field of eye movements and vestibular control, though niche, boasts a rich tradition of celebrating careers of remarkable scientists and physicians. This book is testament of one of such gatherings, dedicated to one such luminary, Sergio Carmona, MD, whose contributions have profoundly shaped the landscape of neurotology in South America.

In honor of Dr. Carmona, colleagues and friends gathered for a scientific symposium, sharing novel research findings and didactic teachings from junior faculty. This symposium, held from May 5–7, 2023, marked a significant moment as academics eagerly returned to in-person meetings after the pandemic, celebrating science together. Dr. Sergio Carmona is a prominent figure in neurotology in Argentina, serves as the Chief of Neurotology at the Buenos Aires Institute of Neuroscience (INEBA) and as a Professor of Neurology at the National University of Rosario. His dedication to the field of eye movement and vestibular control is unparalleled, and his efforts to disseminate knowledge have reached far beyond his native Rosario and Buenos Aires, impacting the global scientific community.

Throughout his illustrious career, Dr. Carmona has organized numerous congresses and symposia, including the renowned Barany Society Congress in 2014 in Buenos Aires. As the former president of la Fundacion Iberoamericana de Neurotologia (FINO), he has tirelessly educated physicians, audiologists, and physical therapists across Latin America. His books, originally written in Spanish, have been translated into English, Italian, and German, further extending his influence. An academically active neurologist, Dr. Carmona has published three books on neurotology, covering topics such as the inner ear, imbalance, ataxia, eye movements, and neurotology. His special expertise lies in the neurotology of Fabry's disease.

On a personal note, Dr. Carmona is known for his exceptional organizational skills and his ability to host successful meetings. His expertise as a sommelier adds a unique touch to his gatherings, making them memorable for all attendees. Many of us have enjoyed traveling with Sergio and his family, always grateful for the exploratory and fun opportunities he arranged.

For the past 30 years, Sergio and his family have been close to my own family, sharing both joyful and sorrowful moments. My late wife had a special affection for the Carmonas, and I am certain she would be pleased to see this tribute to Sergio included here.

Adelaide, SA, Australia Parthvi Ravat
Peoria, IL, USA Jorge C. Kattah
Cleveland, OH, USA Aasef G. Shaikh

Contents

About the Editors

Parthvi Ravat, MD, FRACP, is a neurologist currently practising in Australia with fellowship training in epilepsy from Flinders Medical Centre, Adelaide, Australia. She has previously served as Assistant Professor at Seth GS Medical College and King Edward Memorial Hospital, Mumbai. Her clinical and research work spans across epilepsy, autoimmune encephalitis and eye movement disorders. She has authored peer reviewed publications in various international journals and leads collaborative research in the Young Epilepsy Section of the International League Against Epilepsy in her capacity as Vice Chair. Her academic interests revolve around comprehensive epilepsy care, incorporation of neurotechnology in clinical practice, eye movement abnormalities, and understanding the lived experiences of people with neurological illnesses. She has founded "Win Epilepsy", a mobile application for patient education and improving accessibility in India. Her professional trajectory across India and Australia has provided a broad, multicultural lens that informs her approach to neurological care. Beyond academics, she enjoys time with family, fitness and analysing equity markets.

Jorge C. Kattah, M.D., is Professor and Head of Neurology at the University of Illinois College of Medicine, Peoria. He obtained an MD degree from El Colegio mayor de Nuestra Señora Del Rosario, Bogotá, Colombia, followed by a residency in Neurology at Georgetown University and a fellowship in Neuroophthalmology at the Eye and Ear Hospital, Pittsburgh, Pennsylvania. He was a faculty member at Georgetown University for 20 years and moved to the University of Illinois College of Medicine in Peoria to initiate a new Neurology residency. He has been Professor and Head for 27 years. His research interest over the last 50 years includes neuroophthalmology, eye movements, vestibular and cerebellar function, the acute vestibular syndrome in peripheral and central lesions, Wernicke's encephalopathy, and paraneoplastic syndromes. He has published over 200 peer-reviewed papers. He received the 2009 Stroke Innovation Award, from the Editors of Stroke Second Prize, and the Gold Medal from La Fundacion Iberoamericana de Neurotologia in 2018. In 2024, he completed a 50th year membership of the American Academy of Neurology.

Aasef G. Shaikh, MD, PhD, is Professor of Neurology at Case Western Reserve University and holds the Penni and Stephen Weinberg Chair in Brain Health. He serves as Vice-Chair for Research in the Department of Neurology and Associate Medical Director at the Cleveland Functional Electrical Stimulation Center. He is internationally recognized for pioneering work in movement disorders, eye movements, and neuromodulation, with about 200 peer-reviewed publications, multiple edited volumes, and leadership in NIH- and VA-funded research programs exceeding $5 million. His numerous honors include the Grass Foundation–American Neurological Association Award in Neuroscience, the American Academy of Neurology Career Award, and the Jon Stolk Award for Excellence in Movement Disorders. He is Editor-in-Chief of *Dystonia* and serves on editorial and scientific advisory panels worldwide, reflecting his sustained impact on advancing translational neuroscience and brain health innovation.

Chapter 1
Vestibular Topodiagnosis Using the Vestibulospinal Pathway

Guillermo Zalazar, Carlos Martínez, and Sergio Carmona

Keywords Vestibular system · Truncal ataxia · Vestibulospinal pathway · Acute vestibular syndrome · Central vs peripheral vertigo

The vestibular system is a sensorimotor network that participates in perceptive, oculomotor, and postural functions (Fig. 1.1) [1, 2]. The lesion in this network may manifest itself as an acute vestibular syndrome [3]. Its correct diagnosis is important, as its cause may be either peripheral or central [4]. For this reason, many works have been carried out over the years with the intention of achieving an accurate diagnosis [5–11].

Since the publication of our first paper [5], which was based on the findings of Lee et al. [6], Kattah et al. [7], and, of course, of Joseph Babinski [12], a lot has been published on truncal ataxia and its use in the differential diagnosis of acute vestibular syndromes.

An important finding was that grade 3 truncal ataxia always indicates a central pathology, which is explained by the compromise of the central otolithic pathways (Fig. 1.2.) [13, 14].

In the paper published by Carlos Martínez et al. in 2024 [15], a systematic review and meta-analysis of the diagnostic accuracy of trunk and gait instability rating in acutely dizzy and ataxic patient, they found that, although this is less accurate than HINTS (Head Impulse - Nystagmus - Torsion at the Skew), when the characteristics of nystagmus are added, its accuracy increases. Its advantage is that it can be applied in emergency rooms, and it requires little training.

A series of cases of patients with acute truncal ataxia and no nystagmus in patients with acute vertigo was also published, which highlights the importance of examining

G. Zalazar
Hospital Central Ramón Carrillo, San Luis, Argentina

C. Martínez
Head of the Neurology Department, Hospital J. M. Cullen, Santa Fe, Argentina

S. Carmona (✉)
Fundación San Lucas Para La Neurociencia, Rosario, Argentina

P. Ravat et al. (eds.), *Advances in Translational Neuroscience of Vestibular and Eye Movement Disorders*, Contemporary Clinical Neuroscience,
https://doi.org/10.1007/978-3-032-07539-0_1

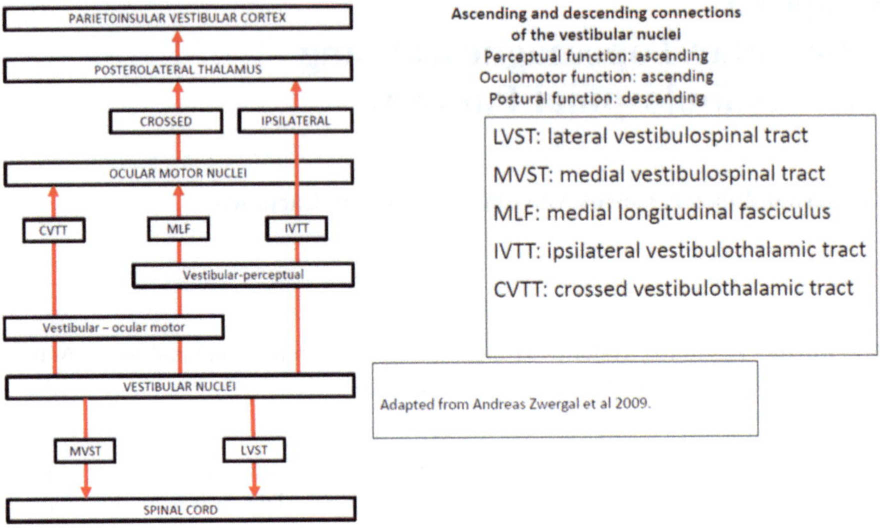

Fig. 1.1 Ascending and descending connections of the vestibular nuclei, summarizing perceptual and ocular motor functions as ascending and postural as descending

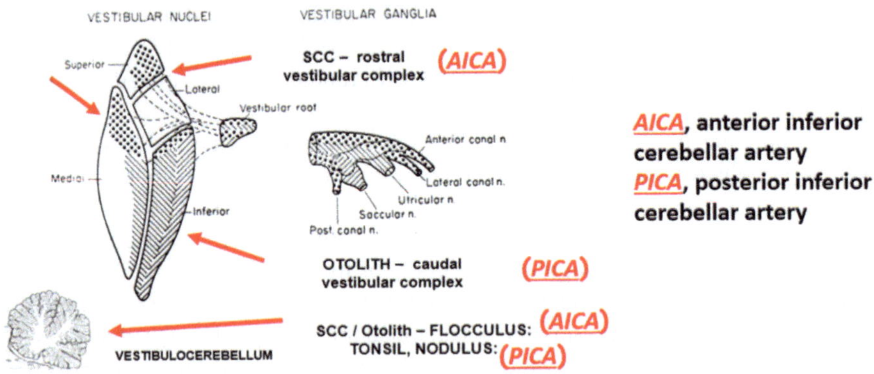

Fig. 1.2 Vascular territories of the projection sites of the vestibular nucleus complex

vestibulospinal functions in this type of patient [16]. We should not forget that, generally speaking, these people go to an emergency room, where doctors usually have little or no training in the evaluation of oculomotor signs, and this would make it very difficult for them to recognize the subtle ocular signs that patients with acute vestibular syndrome without nystagmus present. In contrast, vestibulospinal signs are easy to evaluate and interpret.

Let us not forget that several strategies to be applied in emergency rooms have been developed, and even validation trials have been carried out, such as the one carried out by Vani et al. [8], where the assessment of truncal ataxia plays an important role in differential diagnosis.

The findings of the work we carried out are not only clinically important but also allow us to understand, in part, the functioning of the vestibular system.

The paper published by Kattah et al. [14], apart from establishing that the presence of incubitus truncal ataxia has the same meaning as the presence of grade 3 truncal ataxia, includes two patients who presented Babinski asynergia but with grade 1 truncal ataxia, which indicates a discrepancy between these signs in detecting central lesions. Unfortunately, we do not have the neuroimages to determine the exact location of the ischemic lesion that causes this dissociation. Martínez, M.D., has reported the case of a male patient with ischemic infarction in the Posterior Inferior Cerebellar artery (Posterior Inferior Cerebellar artery) (PICA) territory, where the left cerebellar hemisphere was compromised, and this resulted in the appearance of Babinski asynergia, but with grade 1 truncal ataxia. This demonstrates that the dissociation of central signs exists and that it should be taken into account when examining patients with acute vestibular syndromes (personal communication). We have interpreted this finding as an indicator of spinocerebellar involvement (lack of coordination in contraction of trunk flexor muscles) and sparing of the vestibulocerebellum (which controls the extensor muscles to maintain bipedalism), or a greater vulnerability of the flexor response than the extensor response in central lesions (Fig. 1.3.) [17–19]. A lateral ocular deviation to the left, which is a sign that

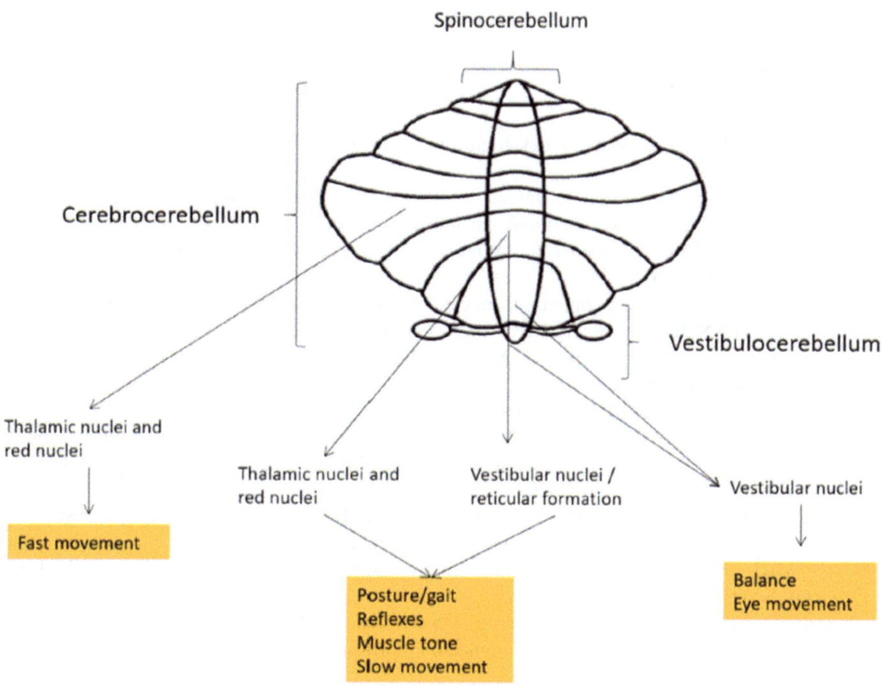

Fig. 1.3 Cerebellum and its lobules and their corresponding functions

supports the central origin of the vestibular syndrome, was also noted in this patient. This sign was also assessed following the guidelines established in the paper by Kattah et al. [20]: The instructions to the patient were (1) Without moving your head, fix on a center fixation point (which was located ~4 feet in front of the patient); (2) gently close your eyes for 3 to 5 s; and (3) open your eyes and look at the center fixation point.

The paper by Sergio Carmona et al. [16] on acute truncal ataxia without nystagmus describes patients with truncal ataxia without nystagmus, both in cerebellar lesions and in the thalamus and cerebral hemisphere. This implies that the cerebellum has sectors that may be compromised without the presence of nystagmus, only with alterations in balance—the presence of subtle ocular signs can be evidenced in almost all cases, but it is difficult to examine them. In the case of the thalamus, its lesion can cause thalamic astasia—postural instability, with retro- or lateropulsion—and we interpret it as secondary to the lesion of the ipsilateral graviceptive pathways of the vestibulothalamic tract, which sends signals to the posterolateral and central thalamic area. In patients with hemispheric lesions, they were on the right side, which may indicate a lateralization of the vestibular network, but given the small number of patients, it is necessary to evaluate these findings with a larger number of patients.

This brief review highlights the need to continue working on the vestibulospinal pathways to better understand how their alteration is manifested in patients and how they are explained from a pathophysiological point of view.

References

1. Carmona S, Kattah J. Manejo del Síndrome Vestibular Agudo. 2nd edición ed. Akadia Editorial; 2023.
2. Zwergal A, Strupp M, Brandt T, Büttner-Ennever JA. Parallel ascending vestibular pathways: anatomical localization and functional specialization. Ann N Y Acad Sci. 2009;1164(1):51–9.
3. Mahmud M, Kaski D. Vestibular neurology for the generalist. J Neurol Neurosurg Psychiatry. 2024;95(12):1196–206.
4. Edlow JA. Distinguishing peripheral from central causes of dizziness and vertigo without using HINTS or STANDING. J Emerg Med. 2024;67(6):e622–33.
5. Carmona S, Martínez C, Zalazar G, Moro M, Batuecas-Caletrio A, Luis L, Gordon C. The diagnostic accuracy of truncal ataxia and HINTS as cardinal signs for acute vestibular syndrome. Front Neurol. 2016;7:125. https://doi.org/10.3389/fneur.2016.00125.
6. Lee H, Sohn SI, Cho YW, Lee SR, Ahn BH, Park BR, et al. Cerebellar infarction presenting isolated vertigo. Frequency and vascular topographical patterns. Neurology. 2006;67:1178–83. https://doi.org/10.1212/01.wnl.0000238500.02302.b4.
7. Kattah JC, Talkad AV, Wang DZ, Hsieh YH, Newman-Toker DE. HINTS to diagnose stroke in the acute vestibular syndrome: three-step bedside oculomotor examination more sensitive than early MRI diffusion-weighted imaging. Stroke. 2009;40:3504–10. https://doi.org/10.1161/STROKEAHA.109.551234.
8. Vanni S, Pecci R, Edlow JA, Nazerian P, Santimone R, Pepe G, Moretti M, Pavellini A, Caviglioli C, Casula C, Bigiarini S, Vannucchi P, Grifoni S. Differential diagnosis of vertigo in

the emergency department: a prospective validation study of the STANDING algorithm. Front Neurol. 2017;8:590. https://doi.org/10.3389/fneur.2017.00590.

9. Gerlier C, Hoarau M, Fels A, Vitaux H, Mousset C, Farhat W, Firmin M, Pouyet V, Paoli A, Chatellier G, Ganansia O. Differentiating central from peripheral causes of acute vertigo in an emergency setting with the HINTS, STANDING, and ABCD2 tests: a diagnostic cohort study. Acad Emerg Med. 2021;28(12):1368–78. https://doi.org/10.1111/acem.14337.
10. Edlow JA, Carpenter C, Akhter M, Khoujah D, Marcolini E, Meurer WJ, Morrill D, Naples JG, Ohle R, Omron R, Sharif S, Siket M, Upadhye S, et al. Guidelines for reasonable and appropriate care in the emergency department 3 (GRACE-3): acute dizziness and vertigo in the emergency department. Acad Emerg Med. 2023;30(5):442–86. https://doi.org/10.1111/acem.14728.
11. Kuroda R, Nakada T, Ojima T, Serizawa M, Imai N, Yagi N, Tasaki A, Aoki M, Oiwa T, Ogane T, Mochizuki K, Kobari M, Miyajima H. The TriAGe+ score for vertigo or dizziness: a diagnostic model for stroke in the emergency department. J Stroke Cerebrovasc Dis. 2017;26(5):1144–53. https://doi.org/10.1016/j.jstrokecerebrovasdis.2017.01.009.
12. Clarac M, Massion J, Smith AM. History of neuroscience: Joseph Babinski (1857–1932), IBRO History of Neuroscience. http://www.ibro.info/Pub/Pub_Main_Display.asp?LC_Docs_ID=2990 (2008)
13. Stein BM, Carpenter MB. Central projections of portions of the vestibular ganglia innervating specific parts of the labyrinth in the rhesus monkey. Am J Anat. 1967;120:281–317. https://doi.org/10.1002/aja.1001200205.
14. Kattah JC, Martinez C, Zalazar G, Batuecas Á, Lemos J, Carmona S. Role of incubitus truncal ataxia, and equivalent standing grade 3 ataxia in the diagnosis of central acute vestibular syndrome. J Neurol Sci. 2022;441:120374. https://doi.org/10.1016/j.jns.2022.120374.
15. Martinez C, Wang Z, Zalazar G, Carmona S, Kattah J, Tarnutzer AA. Systematic review and meta-analysis of the diagnostic accuracy of a graded gait and truncal instability rating in acutely dizzy and ataxic patients. Cerebellum. 2024;23(6):2244–56. https://doi.org/10.1007/s12311-024-01718-6.
16. Carmona S, Martínez C, Zalazar G, Koohi N, Kaski D. Acute truncal ataxia without nystagmus. Eur J Neurol. 2024;31(4):e16185. https://doi.org/10.1111/ene.16185.
17. Manto M, Gruol DL, Schmahmann JD, Koibuchi N, Rossi F, editors. Handbook of the cerebellum and cerebellar disorders. Springer; 2013.
18. Grimaldi G, Manto M. Topography of cerebellar deficits in humans. Cerebellum. 2012;11:336–51.
19. Bodranghien F, Bastian A, Casali C, Hallett M, Louis ED, Manto M, et al. Consensus paper: revisiting the symptoms and signs of cerebellar syndrome. Cerebellum. 2016;15(3):369–91. https://doi.org/10.1007/s12311-015-0687-3.
20. Kattah JC, Badihian S, Pula JH, Tarnutzer AA, Newman-Toker DE, Zee DS. Ocular lateral deviation with brief removal of visual fixation differentiates central from peripheral vestibular syndrome. J Neurol. 2020;267(12):3763–72.

Chapter 2
The Bedside Approach to the Dizzy Patient: Hypothesis Testing Guides the Successful Evaluation

David E. Hale, Anand K. Bery, and David S. Zee

Keywords Dizziness · Vertigo · Eye movement · Vestibulo-ocular reflex · Medical diagnosis

2.1 Introduction

The past few decades have seen a dramatic increase in the way we use the laboratory to probe the function of various structures within the labyrinth in patients with dizziness, vertigo, and imbalance [1–4]. Vestibular evoked myogenic potentials (VEMPs), both cervical and ocular, evaluate the function of the saccule and utricle, respectively. Video goggles are used for head impulse testing, to evaluate the function of each of the semicircular canals, and for measurement of eye alignment and ocular counterroll, to evaluate the function of the utricle. New imaging techniques allow one to evaluate both structures within the labyrinth and the structures within the brain that mediate and modulate vestibulo-ocular and vestibulo-spinal reflexes. Despite these remarkable technological advances, there remains a pressing need to refine the bedside evaluation of vestibular patients to allow clinicians to choose the best and most efficient diagnostic roadmap to reach the correct diagnosis. Laboratory tests are expensive and often nondiagnostic except in the context of the patient's history and examination findings. Hence, focused history-taking and skillful bedside testing become more important than ever. Furthermore, there are so many

D. E. Hale
Department of Neurology, The Johns Hopkins Hospital, Baltimore, MD, USA
e-mail: dhale12@jh.edu

A. K. Bery
Department of Otolaryngology, Massachusetts Eye and Ear/Harvard Medical School, Boston, MA, USA
e-mail: abery@meei.harvard.edu

D. S. Zee (✉)
Departments of Neurology, Ophthalmology, Otolaryngology-Head and Neck Surgery, and Neuroscience, The Johns Hopkins Hospital, Baltimore, MD, USA
e-mail: dzee@jhu.edu

P. Ravat et al. (eds.), *Advances in Translational Neuroscience of Vestibular and Eye Movement Disorders*, Contemporary Clinical Neuroscience,
https://doi.org/10.1007/978-3-032-07539-0_2

bedside tests that the time it would take to use all of them on every patient is prohibitive. We therefore must ask when and which bedside tests are most useful. Here, we develop a simplified scheme, emphasizing the bedside evaluation of the vestibular patient and reviewing bedside testing techniques and their interpretation. In this chapter, we will concentrate on using abnormalities of eye movements as a window to the evaluation of patients with vestibular disorders. But, as in all neurological diagnoses, any abnormal finding must be put into the context of the results of the general neurological examination. In the particular case of the acutely vertiginous patient, examination of gait and balance, as well as of hearing, is essential [5, 6].

First, we briefly review the diagnostic process, which is driven by the patient's history and then guides a focused examination. We emphasize that the questions we ask and the tests we do must be driven by *hypotheses.* In other words, we are describing the clinical application to the bedside of the *scientific method* (Box 2.1).

Box 2.1: The Diagnostic Process

1. The HISTORY (the patient's account of their symptoms) and the EXAMINATION (clinical and laboratory) of the dizzy patient are to answer the key questions:
   (a) WHAT structures (ANATOMY) within the brain or inner ear are affected?
   (b) WHY (PHYSIOLOGY) are they malfunctioning?
   (c) WHAT are the possible causes (the list [differential diagnosis] of disease possibilities)?
   (d) WHAT is the true cause?
   (e) HOW can we treat the disease and ameliorate symptoms?
2. To narrow the list of possible causes, we choose examination techniques, both bedside and laboratory-based, to help include or exclude possible causes.
3. The bedside examination must be HYPOTHESIS-DRIVEN:
   (a) DIRECTED by the HISTORY.
   (b) PHYSIOLOGICALLY and ANATOMICALLY based.
   (c) FOCUSED and TAILORED toward a LOCALIZATION, and then toward an evolving but narrowing DIFFERENTIAL DIAGNOSIS.
   (d) USED TO GUIDE LABORATORY TESTING to finalize the correct diagnosis, i.e., to prove or disprove HYPOTHESES.

2.2 Understanding Normal Vestibular Function

To interpret the symptoms of a patient with a suspected vestibular disorder, we must first understand how the normal vestibular system functions. The prime goals of the vestibular system are balance, spatial orientation, and steady gaze while the head (or body) is in motion. Our labyrinthine sensors—the semicircular canals, which sense angular (rotational) acceleration, and the otoliths, which sense linear (translational) acceleration or the pull of gravity—are the first line of defense against any perturbations of the body that might impair vision or balance. They do this through two phylogenetically ancient reflexes, the vestibulo-ocular (VOR) and vestibulo-spinal (VSR) reflexes. Of course, vision and proprioception are also used to thwart disturbances that might throw one off balance, but they are relatively sluggish compared to the almost instantaneous (milliseconds) response of the vestibular reflexes, making the VOR and the VSR the first line of defense.

Patients with vestibular dysfunction often describe symptoms of *oscillopsia*, or *double vision* (diplopia). Oscillopsia refers to the illusory movement of a stationary object. Oscillopsia is a symptom, and not a sign or a diagnosis. When oscillopsia occurs with the head at rest (so-called "sitting oscillopsia"), it usually reflects the patient's own perception of the motion of images on the retina produced by uncalled-for eye movements such as spontaneous nystagmus. Oscillopsia might be experienced as a back-and-forth or drifting of the visual world, sometimes in one direction with a reset in the other (if there is a jerk nystagmus). One patient with a vertical nystagmus described her vertical oscillopsia "as if the vertical hold on my television screen is not working since the picture on the television screen keeps drifting up, but never seems to go anywhere". When oscillopsia occurs primarily with the head moving, it usually means the VOR cannot properly steady gaze when the head is moving, usually because of an abnormality in amplitude or direction of compensatory slow phases. Oscillopsia when the head is moving is often called "walking" oscillopsia. Such patients might, for example, describe difficulty reading road signs when riding in a car on a bumpy road [credit for coining the terms "sitting" vs. "walking" oscillopsia goes to Dr. David Newman-Toker].

2.3 Diagnostic Process: Begin with a Detailed History

When taking the history from a patient, one develops a decision tree to arrive at a likely diagnosis(es) and then uses the examination to test the hypotheses that emerged from the history. Many algorithms have been developed to aid this diagnostic process based on various combinations of the history and examination findings for acute [7–11] and episodic vertigo [12–15], respectively. Here, we propose a relatively simple algorithm that will suffice for most of the common vestibular disorders (Fig. 2.1).

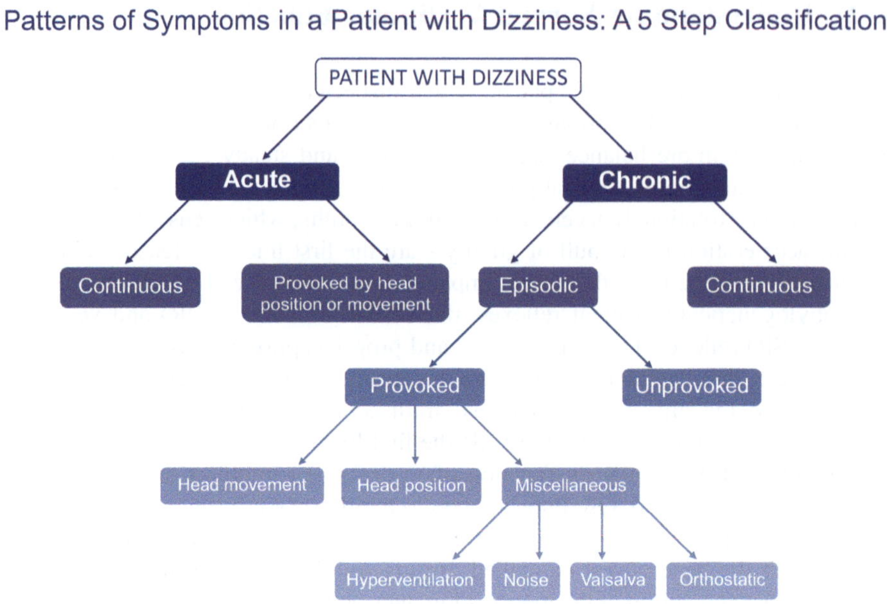

Fig. 2.1 A flowchart to evaluate the dizzy patient. The first important historical branch point is whether the symptoms are acute or chronic. If acute, are they continuous (which might suggest a diagnosis such as vestibular neuritis or a posterior fossa stroke), or are they provoked by specific factors such as a change in head position relative to gravity (which may suggest, for example, benign paroxysmal positional vertigo (BPPV))? Similarly, in chronic situations, an understanding of whether the symptoms are episodic (and provoked or spontaneous) or continuous helps to refine the differential diagnosis. As one example, chronic, episodic, and unprovoked dizziness might suggest recurrent episodes of vestibular migraine

Benign paroxysmal positional vertigo (BPPV) will be discussed in other chapters in this volume, but we emphasize here how important it is to ask about recent circumstances that might have been the precipitating event. These include prolonged recumbency or inversion of the head/body, or exposure to vibration, gravity, or linear acceleration. Examples include being in an unusually long, sustained recumbent position (such as at the hairdresser's, dentist, or in an MRI gantry) or upside-down in yoga classes. Exposure to vibration, such as with dental drills or hand-held massage guns, may also be an inciting factor. Always ask about unusual exposure to linear acceleration, such as on amusement park rides, or aerobic exercises with jumping up and down.

We recommend two essential questions to rule in or out BPPV. (1) Is the dizziness provoked by lying down or turning over in bed? If yes, this supports BPPV. (2) Is the dizziness sustained (rather than lasting seconds)? If no, this supports BPPV [15].

There is a plethora of physical therapy maneuvers to treat BPPV; by our own count, well over 30. For most clinical situations, it will suffice for the clinician to know just a few maneuvers for each canal type. To pinpoint the offending semicircular canal, the test maneuvers must be performed according to protocol, with strict attention to the orientation of the head as it is moved from one position to the other [16]. Furthermore, inherent variability in the anatomical orientations of the cupula within the ampulla and the location of the semicircular canals within the head can make precise localization difficult [17]. Computer simulations, which allow one to visualize the effects of changes in head orientation on the location and movement of the otolithic debris in the canals or on the cupula, can aid interpretation and guide treatment, but even so, the patterns of induced nystagmus do not always fit into a simple classification [18].

Finally, we will not be discussing common medical causes of dizziness, such as metabolic disturbances, hypotension, and especially orthostatic hypotension, hyperventilation/anxiety, and changes in medications, but they should always be considered in every patient with dizziness or imbalance.

2.4 Next, Let the History Guide the Exam

As indicated above in Fig. 2.2, there is a cornucopia of bedside tests, including the various subclasses of eye movements—saccades, pursuit, vestibular, and vergence—and one must cull the ones that best test your hypotheses. The tests you choose are based on two fundamental principles.

Principal 1 Your choice of tests depends upon the clinical situation: urgency and time available. For example, the acute, severely ill patient with new onset vertigo seen in the emergency department requires a rapid triage decision: Is the patient having a stroke? In this case, one focuses on the five tests of the "HINTS-Plus" battery (Head Impulse Test, Gaze nystagmus, Skew, Imbalance (Ataxia), Hearing) [19]. On the other hand, the chronic, mildly ill patients with persistent undiagnosed symptoms of a possible vestibular disturbance may require an exhaustive history and detailed examination. In other words, pull out all the tricks.

Principal 2 Focus your bedside test battery according to the diagnostic question that is raised by your hypothesis about where and what the problem is. In Fig. 2.3, we summarize the choice of tests depending on the diagnostic question.

Fig. 2.2 The full array of bedside tests available to the clinician, organized by category

-Cover testing (strabismus, phoria/tropia)

-Eye Movements

- Saccades (spontaneous, anti-saccades, target induced)
- Pursuit (horizontal, vertical)
- Convergence
- Optokinetic nystagmus (OKN – horizontal, vertical, oblique)
- Eccentric gaze-holding

-Nystagmus evaluation (including all types of oscillations)

- Effect of fixation (Frenzel lenses, penlight cover test, ophthalmoscope)
- Effects of eccentric gaze
- Effect of convergence
- Rebound nystagmus
- Eye closure

-VOR testing

- Head impulse (three planes)
- Dynamic visual acuity (DVA – three planes)
- Ocular counterroll (OCR)
- Subjective visual vertical (SVV)
- Fixation suppression of VOR

-Induced nystagmus

- Hyperventilation
- Head shaking (horizontal, vertical, roll, circular)
- Valsalva
- Noise induced
- Tragal compression, pneumatic otoscope
- Caloric nystagmus
- Positional testing
- Dix Hallpike, Supine Roll, Bow and Lean

For localizing a **spontaneous nystagmus** (central vs. peripheral)

- The 3D vector (horizontal, vertical, torsional)
- Effect of removing fixation and of eye closure
- Effect of eccentric gaze
- Effect of convergence

For detecting an **abnormal vestibulo-ocular reflex (VOR)**

- Head impulse test
- Dynamic visual acuity (DVA)

For detecting **unilateral vestibular hypofunction**

- Head impulse test
- Head shaking induced nystagmus
- Vibration induced nystagmus

For detecting **central dysfunction**

- Holding eccentric gaze
- Saccades
- Pursuit and VOR suppression
- Alternate cover for skew deviation (vertical misalignment)

For evaluating **positional symptoms**

- Supine roll
- Dix-Hallpike
- Bow and Lean

Special tests to **bring out nystagmus**

- Hyperventilation
- Valsalva
- Vibration
- Head shaking
- Noise Induced

Fig. 2.3 Bedside vestibular and ocular motor tests, based on the diagnostic question. For example, the head impulse test (HIT) and dynamic visual acuity (DVA) help identify abnormal functioning of the vestibulo-ocular reflex VOR

2.5 Six Central Issues to Address on Physical Exam

2.5.1 Issue 1: Is There a Spontaneous Nystagmus? If So, Is It Caused by a Central or Peripheral Vestibular Lesion?

The three-dimensional **vector** of a nystagmus is often a clue to its source. Based on the orientation of the semicircular canals in the head, in which the lateral canals are nearly horizontal, and the vertical canals lie at 45 degrees from the sagittal plane, one can extract what an individual semicircular canal contributes to the pattern of eye movements that are elicited when that canal is stimulated. This organization underlies Ewald's first law: compensatory eye or head movements move in a plane parallel to that of a stimulated canal [20]. The experimental results from artificial stimulation of the semicircular canals in experimental animals support this interpretation. Taken together with the parallel segregation of the pattern of perfusion and innervation in the labyrinth (superior vestibular nerve and anterior vestibular artery supply the anterior and lateral canals and utricle; and the inferior vestibular nerve and posterior vestibular artery supply the posterior canal and saccule), one can deduce that pure vertical or pure torsional nystagmus almost never comes from a peripheral lesion and a mixed horizontal-torsional nystagmus is often the sign with a complete unilateral labyrinthine lesion. Furthermore, a mixed vertical torsional nystagmus, when observed after provocative maneuvers, is often the sign of a posterior canal BPPV (upbeat torsional nystagmus) or a superior canal bone dehiscence (downbeat torsional nystagmus) (Box 2.2).

Box 2.2: Four Questions Need to Be Asked About the Nystagmus
1. What is its three-dimensional **vector**: horizontal, vertical, or torsional?
2. What is the **effect of the removal of fixation**?
3. What is the effect of **eccentric gaze**?
4. What is the effect of **convergence**?

Removal of fixation by Frenzel lenses or equivalent, or using an ophthalmoscope [21] or a penlight and at the same time covering the other eye [22], should always be a part of the vestibular examination. A spontaneous nystagmus of peripheral origin is almost always enhanced or even made evident in the case of a mild vestibular imbalance, with removal of fixation. A novel way to bring out a vestibular nystagmus is to look for a horizontal nystagmus appearing during vertical pursuit [23]. On the other hand, if fixation does not diminish a spontaneous nystagmus, the origin is usually central. With some central lesions, however, fixation suppression of nystagmus is partially spared [24, 25]. One must also remember that the true three-dimensional vector of a nystagmus is best determined in the absence of fixation since fixation may suppress one component more easily than another. For example, in a patient with posterior canal BPPV, if the examiner does not remove visual fixation, one may only see a subtle torsional nystagmus while the vertical component is suppressed, making the diagnosis less certain.

The **effect of gaze** (e.g., changing of horizontal eye position in the orbit when there is a vertical nystagmus or changing vertical eye position when there is a horizontal nystagmus) and the effect of convergence, on a spontaneous vertical or horizontal nystagmus are critical to sorting out the origin of the nystagmus. A downbeating nystagmus is often accentuated or only brought out on lateral gaze or with convergence. An upbeat nystagmus may change direction to downbeating with convergence. A horizontal nystagmus may increase its intensity or be brought out when looking in the direction of the quick phase (Alexander's law). This is a constant feature of nystagmus from peripheral lesions. If a horizontal nystagmus increases in intensity when looking in the direction of the slow phase, the origin is central and reflects an unstable neural integrator [26]. If a horizontal nystagmus changes direction depending on gaze direction (e.g., left-beating on left gaze, and right-beating on right gaze), this is usually a central sign (and a key part of the HINTS algorithm in pointing toward a stroke in a patient with the acute vestibular syndrome) [7, 27]. Convergence usually suppresses so-called congenital or *infantile nystagmus,* which appears at birth or in the first few months of life. Likewise, a horizontal infantile nystagmus usually remains purely horizontal on far up and far down gaze, while a horizontal vestibular nystagmus, obeying Ewald's first law, may develop a torsional component on far up and far down gaze [28]. Finally, note that a full deviation of the eyes under closed lids (inferred from corrections back to the center fixation target) points to a central lesion, usually in the dorsal-lateral medulla as part of Wallenberg's syndrome [29].

2.5.2 Issue 2: Is the VOR Abnormal?

There are two principal ways to evaluate the VOR at the bedside: **dynamic visual acuity** (DVA, changes in visual acuity with head movement) and the **head impulse test** (HIT). Recall that the raison d'être of the VOR is to enable us to see clearly when our head is moving. Even with the head still, when one uses an oscillating visual stimulus changing direction at a frequency of two cycles per second or more, the visual tracking system (pursuit) is unable to keep up with the target. If the target contains a visual discrimination task, the ability to see it will be decreased. Thus, in the absence of a VOR, when the head is moving more than two cycles per second, even a normal tracking system will not ensure clear vision. So, decreased visual acuity on this task reveals that the VOR is impaired. This test is easily performed at the bedside with a Snellen visual acuity card. Normal individuals show at most a one-line loss of visual acuity during head oscillation, while patients with no VOR can show a five-line loss. Two practical clinical points. First, the loss of DVA is greater with horizontal or vertical head oscillations than with torsional (roll, ear-to-shoulder) head oscillations because with the latter, the image of the target does not move far from the fovea. Second, and especially with a chronic loss of vestibular function, patients may learn to generate compensatory saccades *during* head motion in lieu of their absent VOR so that the bedside head impulse test may appear normal. However, DVA is still often abnormal in those patients and may be an easier way to reveal an underlying impaired VOR.

The head impulse test (HIT) is a particularly effective bedside test, and once the technique is mastered [30, 31], it becomes a cornerstone of the bedside vestibular examination (along with Frenzel lenses and positional testing). The key finding on head impulse testing is the presence or absence of corrective saccades to compensate for an abnormal VOR (in amplitude and/or direction). The presence of corrective saccades with impulses in one direction is the hallmark of a peripheral unilateral lesion, though occasionally this is the case with central causes (see section on anterior inferior cerebellar artery (AICA) (ischemia, below). Corrective saccades are relatively easy to discern at the bedside in acute lesions, but less easy to discern in long-standing defects. In chronic situations, the patient has had time to compensate by using short-latency covert saccades embedded in the VOR response while the head is still moving. Cross-coupling of the VOR in which the slow-phase response has a component directed around an axis away from that around which the head is rotating (e.g., a vertical eye movement in response to a horizontal head rotation) is often detected by the appearance of a corrective saccade that is orthogonal to the expected slow-phase response. While the responses of patients with peripheral lesions may show small off-axis components, for example, with selective posterior canal hypofunction or with superior division vestibular neuritis [32, 33], they are much harder to discern at the bedside than when the abnormal cross-axis component has a central origin [34]. When corrective saccades are in the opposite direction to the compensatory slow phase, the VOR gain is usually too high, which is also a sign of a central, usually cerebellar disturbance [34].

2.5.3 Issue 3: Is There Unilateral Hypofunction?

As described above, the **head impulse test** is a sensitive way to appreciate a unilateral loss of vestibular function. It is important to remember that an abnormal HIT may be a sign of ischemia in the distribution of the anterior inferior cerebellar artery (AICA) for at least three reasons: involvement of the labyrinth, involvement of the vestibular nuclei, and involvement of the cerebellum. With AICA ischemia, the HIT may show a more marked and more enduring deficit than with a peripheral cause because central structures are involved. In this case, the loss of the convergence on neurons in the vestibular nuclei of reciprocal, excitation-inhibition information from both labyrinths, and the loss of central adaptive mechanisms make the deficit more obvious and enduring. When providing virtual care (telemedicine), one usually needs to ask the patient to perform active head impulses, given that the examiner is not on site. With acute lesions, corrective saccades may still be seen, but in patients with more chronic lesions, in whom adaptation has taken place, the corrective saccades may be hard to see because they are triggered early and become covert and appear while the head is still moving [35]. Covert saccades that are not discernible at the bedside can be detected using the video head impulse test.

Nystagmus induced after **head shaking (HSN)** can be another sign of unilateral peripheral vestibular loss [36]. After a sustained period of 2–3 Hz back-and-forth horizontal head oscillation, there may be decaying horizontal nystagmus with slow phases directed toward the affected ear. After vertical head shaking, there may be a decaying horizontal nystagmus with slow phases directed toward the intact ear. The nystagmus reflects the decay of activity in a central *velocity storage mechanism* that normally acts to improve upon the inherently poor response of the peripheral labyrinth to sustained low-frequency stimuli. The action of this central mechanism is reflected in a prolongation of the time constant of the vestibulo-ocular reflex as measured during rotational chair testing, either with velocity steps or sinusoidal stimuli of different frequencies. With a unilateral loss of function, during horizontal head shaking, there is an asymmetrical (left versus right) charging of the velocity storage mechanism, and when the head stops shaking, the activity within the velocity storage mechanism slowly decays, producing the post head shaking-induced nystagmus with slow phases directed toward the paretic labyrinth. With central disorders after horizontal head shaking, there may be a vertical, usually downbeat nystagmus, or a horizontal nystagmus directed oppositely to any spontaneous nystagmus, or a post head-shaking nystagmus that quickly reverses direction [37, 38].

Circular head shaking is a simple way for a person sitting still in a chair to elicit a postrotatory nystagmus from both vertical canals in one labyrinth. In this test, the head rapidly traverses a "circle" at a constant velocity (such that the nose traces out a clock) for ten cycles or so, which is then followed by a post head-shaking torsional nystagmus. This pattern of nystagmus reflects activation of the anterior and posterior semicircular canals on one side during rotation and then activation of the vertical canals in the other labyrinth when the head stops [39]. It is a maneuver that is as if a subject were sitting on a vestibular chair that was being rotated around its earth vertical axis, but with the subject's nose pointing up to the sky. The direction of chair rotation determines the pattern of canal stimulation. If one suspects a bilateral loss of function of the vertical canals, an absence of the postrotatory response with circular head shaking is an easy way to reveal the deficit. This test is complementary to vertical canal head impulse testing since circular head shaking is generated actively, while the HIT is induced passively by the examiner.

Vibration of the skull, usually on the mastoid process, also induces nystagmus (skull vibration-induced nystagmus [SVIN]) when there is a unilateral loss of function. The vibration stimulus is transmitted to both labyrinths through the skull so that normal individuals have little or no vibration-induced nystagmus. In the case of a unilateral loss of function, the slow phase is usually directed toward the paretic labyrinth, though during the recovery period, the slow phase may be temporarily directed away from the paretic labyrinth. There is no after-nystagmus when vibration stops, probably because the vibration stimulus only engages type I, irregularly firing vestibular afferents, while with the head shaking nystagmus test, the type II, regularly firing vestibular afferents, are also engaged, which in turn activate the velocity storage mechanism [40, 41]. Thus, the vibration-induced activity bypasses the velocity storage mechanism, and there is no nystagmus when the vibration stimulus is removed. With central lesions, SVIN may show a vertical component much like the cross-coupling shown with horizontal head-shaking or head impulse testing when lesions are central.

Taken together, SVIN, HSN, and an abnormal HIT are three findings to look for, and the patterns they show can point to a peripheral versus a central lesion. The three tests are complementary, so it is often necessary to use all three.

2.5.4 Issue 4: How Can Examination of Saccades, Alignment, Vergence, and Pursuit Help in the Localization of Patients with Vestibular Symptoms?

It is beyond the purview of this chapter to comprehensively review the use of abnormalities of each subtype of eye movements for vestibular localization [42], but we will review a few archetypal syndromes in which specific patterns point to a localization. Saccades are particularly helpful with localization. In the absence of a limitation of the range of eye motion, as might be the case with an oculomotor palsy, slow horizontal saccades localize to pontine structures and slow vertical saccades localize to midbrain structures. In disorders of the medulla, cerebellum, or labyrinth, saccades are usually of normal speed. Inaccurate saccades, either overshooting or undershooting the target, are referred to as dysmetric saccades. Saccade dysmetria suggests dysfunction in cerebellar structures. Undershooting, hypometric saccades are a feature of lesions in the dorsal cerebellar vermis (lobules 5, 6, and 7), while overshooting, hypermetric saccades localize to lesions in the posterior fastigial nucleus (called the fastigial oculomotor region, FOR). Importantly, projections from each FOR decussate immediately and pass through the contralateral FOR on their way, via the uncinate fasciculus that runs adjacent to the superior cerebellar peduncle, to the brainstem. Thus, a *unilateral structural lesion* in the FOR will invariably involve the projections of both FOR and produce bilateral saccade hypermetria. On the other hand, a *functional lesion* of the FOR, such as caused by a change in activity of the Purkinje cells of the dorsal cerebellar vermis, which normally inhibits the underlying FOR, can produce a direction-specific dysmetria of saccades with overshooting toward the side of the impaired FOR. This is the case with Wallenberg's syndrome, a lateral medullary ischemic stroke, because the lesion interrupts the climbing fiber projections that traverse the inferior cerebellar peduncle on their way to the dorsal vermis (See Fig. 2.4). The climbing fibers normally act to inhibit the Purkinje cells in the dorsal cerebellar vermis. Therefore, a lack of inhibition at the dorsal cerebellar vermis increases the inhibitory activity of Purkinje cells, causing a *functional unilateral* FOR lesion with hypermetria toward the side of

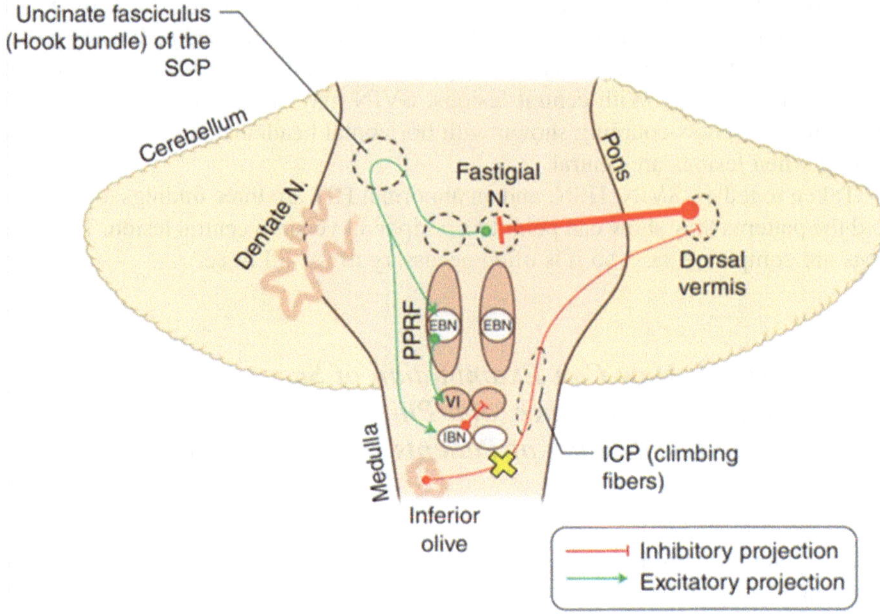

Fig. 2.4 Saccadic pathways in the cerebellum and saccadic dysmetria in Wallenberg's syndrome. (Reproduced with permission from Neuro-Ophthalmology and Neuro-Otology [2021] by Dr. Daniel Gold [43])

the lesion. On the other hand, a lesion in the dorsal cerebellar vermis that interrupts the Purkinje cell projections to the FOR leads to disinhibition of the FOR, causing saccades to undershoot toward the side of the cerebellar vermis lesion. During attempted vertical saccades, lesions of one side of the dorsal vermis or functional lesions of only one FOR cause a horizontal eye deviation away from and toward the side of the lesion, respectively. In Wallenberg's syndrome, a related characteristic sign is a deviation of the eyes under closed lids toward the side of the lesion, which is followed on eye opening by a series of hypometric saccades back toward the center eye position. The tonic deviation is referred to as ipsipulsion [29]. Finally, note that normal subjects may show a small undershoot for horizontal saccades to far eccentric positions, and a small overshoot for horizontal saccades back to the center. There may be a larger overshoot for downward saccades starting from looking up. In normal subjects, dysmetria usually lessens or disappears with repetitive testing compared to persistent dysmetria in patients with a structural cause.

The climbing fibers traveling via the inferior cerebellar peduncle (ICP) normally inhibit the Purkinje cells of the dorsal vermis. The Purkinje cells have an inhibitory effect on the ipsilateral fastigial nucleus, also called the fastigial oculomotor region (FOR). The FOR provides contralateral excitatory projections to both the excitatory burst neurons (EBN) and inhibitory burst neurons (IBN). The IBN normally inhibits the contralateral abducens nucleus to help stop saccades directed toward the side of the inhibited abducens nucleus, while the EBN normally provides excitatory projections to the ipsilateral CN VI nucleus to facilitate saccades directed toward the side of the excited abducens nucleus.

In the case of left Wallenberg's syndrome, a lesion of the left lateral medulla (indicated by the yellow 'X' in Fig. 2.4) prevents the ICP climbing fibers from providing inhibitory effects on the Purkinje cells of the dorsal vermis. In turn, there is unregulated inhibition of the left FOR. Inhibition of the left FOR causes decreased excitation of the IBN on the right, leading to less inhibition of the left CN VI nucleus. This results in saccades overshooting to the left (ipsilesional saccade hypermetria). The decreased excitation of the EBN on the right leads to less excitation of the right CN VI nucleus. This results in saccades undershooting to the right (contralesional saccade hypometria).

A "skew deviation" is a vertical ocular misalignment that occurs in the absence of peripheral ocular muscle weakness and is often highly localizing in patients with a vestibular disturbance. Skew deviation reflects an abnormality in the otolith-ocular response to a lateral head tilt. Normally, with a lateral head tilt, there is a sustained torsion of the eyes, called ocular counterroll, away from the side to which the head is tilted. In pathological circumstances, however, an eye movement response is seen similar to a pattern shown by lateral-eyed animals attempting to maintain their eyes along the horizontal meridian when tilting their head toward one ear. When lateral-eyed animals tilt their head toward one ear, the eye on the side of the downturned ear deviates upward while the eye on the opposite side deviates downward with the goal of maintaining eye alignment along the horizon. In the case of humans with abnormalities in otolith pathways, an ocular tilt reaction (OTR) emerges, which is a response to an abnormal perception that the gravity vector is no longer upright [44]. Classically, the OTR consists of a head tilt, a skew deviation, and an ocular counterroll, though not all three components are always present. In the OTR, the eye contralateral to the head tilt is deviated upward, the eye ipsilateral to the head tilt is deviated downward, and both eyes are counter-rolled toward the side of the head tilt. Patients subjectively experience vertical diplopia while the examiner finds a vertical misalignment of the eyes using alternate cover or Maddox rod testing. Patients can also experience a shift in the subjective visual vertical (SVV). The interruption of otolith-ocular pathways in the caudal vestibular nuclei in medulla, as with Wallenberg's syndrome, produces a skew deviation with the lower eye ipsilateral to the side of the lesion, while interruption of otolith-ocular pathways after they cross in the pons and ascend in the medical longitudinal fasciculus (MLF), as with internuclear ophthalmoplegia (INO), produces a skew deviation with the higher eye ipsilateral to the side of the lesion. This is known as the "high-high, low-low" rule, popularized by Dr. Steve Galetta. With cerebellar lesions, there may be a characteristic pattern of skew deviation in which the *abducting* eye is always higher when the patient moves gaze to an eccentric horizontal position on either side, referred to as an alternating skew deviation [45]. Occasionally, the *adducting* eye is higher, and, in this case, the pattern of misalignment must be distinguished from a bilateral superior oblique muscle palsy.

Not only does evaluation of vergence help identify potential abnormalities, including convergence and divergence insufficiency, but it also can play a role in the modulation of nystagmus, as was discussed above. Convergence testing is also helpful in patients presenting with thiamine (vitamin B1) deficiency in Wernicke's encephalopathy (WE). While patients with WE classically present with the triad of confusion, ophthalmoparesis, and ataxia, they can also present acutely with spontaneous upbeat nystagmus. With convergence, the upbeat nystagmus may be markedly attenuated or reverse direction to downbeat nystagmus. This is hypothesized to be due to an abnormal processing of linear acceleration information from the otolith pathway and is relatively specific to lesions involving the medulla, including WE [46, 47].

Pursuit movements keep a slowly moving target stable on the fovea. With damage to the pursuit system, small catch-up saccades are needed to maintain fixation on the target, producing the saccadic appearance of abnormal pursuit. A classic example of severely impaired pursuit occurs with damage to the middle cerebellar peduncle (MCP), which receives both ipsilateral and contralateral input from the pons, leading to bilateral impairment [48]. Lesions of the flocculus and paraflocculus (tonsil), in addition to causing downbeat and gaze-evoked nystagmus, can have prominently impaired horizontal and vertical pursuit, which contrasts with lesions of the nodulus and uvula, which classically have normal horizontal but impaired vertical pursuit [49, 50].

These classic examples show how the various classes of eye movements can be used to further localize the deficit in a patient with vestibular symptoms.

2.5.5 Issue 5: Positional Testing

Reviews of how to perform positional testing in patients with positional vertigo are abundant, and there is extensive literature on the many ways to treat BPPV. The importance of following strict standardized guidelines in performing positional testing and treatment maneuvers was emphasized earlier in this chapter. While most patients with positional vertigo have a peripheral cause, one must always be vigilant for a central cause when the pattern of positional nystagmus does not conform to the appearance expected for a peripheral lesion. Central positional nystagmus often lacks latency, is without a crescendo-decrescendo pattern, and is sustained. For example, a positional sustained downbeat nystagmus can be the only sign of a central disorder. Likewise, when patients do not respond to treatment maneuvers for BPPV and have persistent positionally-induced symptoms and nystagmus, one must image the patient with a contrast-enhanced MRI. Patients with occult lesions in the cerebellar nodulus may present solely with a positional *apogeotropic* nystagmus, making a careful inspection of the region of the inferior cerebellar vermis and fourth ventricle on their MRI an absolute requirement. Similarly, patients with lesions of the flocculus/paraflocculus can show a positional *geotropic* nystagmus. Typically, lesions of the flocculus/paraflocculus also produce abnormalities in other classes of eye movements (saccadic pursuit, gaze-evoked nystagmus), reminding the examiner that the patient's positional nystagmus should be interpreted in the context of the entire ocular motor examination [51]. While many patients with positional vertigo have a peripheral cause, vigilance is essential in patients with atypical clinical findings or a prolonged course without resolution.

2.5.6 Issue 6: Other Provocative Bedside Tests

We have already discussed skull vibration-induced nystagmus and head shaking-induced nystagmus. The Valsalva maneuver and hyperventilation testing are also useful as screening tests for an occult vestibular imbalance in patients who have dizziness but with few other objective findings of a vestibular disorder [52–54]. Hyperventilation-induced nystagmus can be a sign in patients with various peripheral lesions (e.g., inflammatory or compressive lesions on the VIII nerve, such as cholesteatoma, acoustic nerve tumors, or microvascular compression) or central lesions (demyelinating lesions or cranio-cervical junction abnormalities). Hyperventilation can produce a "recovery" nystagmus, thought to occur by causing a respiratory alkalosis and, in turn, lowering the amount of free calcium available, which can improve conduction along myelinated fibers. This may make any previous adaptation to a loss of function inappropriate so that the acute "recovery" of function can lead to an excitatory nystagmus with the slow phase directed away from the lesion. Valsalva maneuvers can produce nystagmus with labyrinthine fistulas, such as superior canal dehiscence or central problems, such as craniocervical junction abnormalities with tumors or the Chiari malformation. Noise-induced vertigo and nystagmus (Tullio's phenomenon) are characteristic features of superior canal dehiscence.

2.6 Conclusion

Here, we reiterate the propositions about hypothesis testing that were laid out in the introduction and Fig. 2.1. The bedside evaluation best proceeds by constantly asking hypothesis-driven questions, beginning with an understanding of the patient's symptoms. The examiner then hypothesizes the anatomical and physiological substrate for the patient's symptoms. Finally, bedside testing techniques are used to prove or disprove this hypothesis. The choice of examination technique is based on the specific question to be answered. The differential diagnosis evolves and narrows as the examiner proceeds along this path and eventually guides the laboratory testing necessary to arrive at the final correct diagnosis.

References

1. Lennox-Bowley A, Dasgupta S. Modernising vestibular assessment. J Laryngol Otol. 2024;138 (S2):S3–7.
2. Yang Y, Tian J, Otero-Millan J, Kheradmand A. Video ocular counter roll: a bedside test of otolith-ocular function. Ann Clin Transl Neurol. 2023;10(12):2426–9.
3. Yacovino DA, Zanotti E, Cherchi M. The spectrum of acute vestibular neuropathy through modern vestibular testing: a descriptive analysis. Clin Neurophysiol Pract. 2021;6:137–45.
4. Halmagyi GM, Curthoys IS. Vestibular contributions to the Romberg test: testing semicircular canal and otolith function. Eur J Neurol. 2021;28(9):3211–9.
5. Carmona S, Martinez C, Zalazar G, Koohi N, Kaski D. Acute truncal ataxia without nystagmus. Eur J Neurol. 2024;31(4):e16185.
6. Martinez C, Wang Z, Zalazar G, Carmona S, Kattah J, Tarnutzer AA. Systematic review and meta-analysis of the diagnostic accuracy of a graded gait and truncal instability rating in acutely dizzy and ataxic patients. Cerebellum. 2024;23(6):2244–56.
7. Kattah JC, Talkad AV, Wang DZ, Hsieh YH, Newman-Toker DE. HINTS to diagnose stroke in the acute vestibular syndrome: three-step bedside oculomotor examination more sensitive than early MRI diffusion-weighted imaging. Stroke. 2009;40(11):3504–10.
8. Newman-Toker DE, Edlow JA. TiTrATE: a novel, evidence-based approach to diagnosing acute dizziness and vertigo. Neurol Clin. 2015;33(3):577–99.
9. Vanni S, Pecci R, Casati C, Moroni F, Risso M, Ottaviani M, et al. STANDING, a four-step bedside algorithm for differential diagnosis of acute vertigo in the emergency department. Acta Otorhinolaryngol Ital. 2014;34(6):419–26.
10. Edlow JA, Gurley KL, Newman-Toker DE. A new diagnostic approach to the adult patient with acute dizziness. J Emerg Med. 2018;54(4):469–83.
11. Strobl R, Grozinger M, Zwergal A, Huppert D, Filippopulos F, Grill E. A set of eight key questions helps to classify common vestibular disorders-results from the DizzyReg patient registry. Front Neurol. 2021;12:670944.
12. Kim HJ, Song JM, Zhong L, Yang X, Kim JS. Questionnaire-based diagnosis of benign paroxysmal positional vertigo. Neurology. 2020;94(9):e942–e9.
13. Wan Y, Li Y, Zhou L, Ma R, Sun J. Significance of subtype-determining questionnaire in efficient diagnosis and treatment of BPPV. Acta Otolaryngol. 2023;143(2):106–12.
14. Imai T, Takeda N, Ikezono T, Shigeno K, Asai M, Watanabe Y, et al. Classification, diagnostic criteria and management of benign paroxysmal positional vertigo. Auris Nasus Larynx. 2017;44 (1):1–6.
15. Lindell E, Finizia C, Johansson M, Karlsson T, Nilson J, Magnusson M. Asking about dizziness when turning in bed predicts examination findings for benign paroxysmal positional vertigo. J Vestib Res. 2018;28(3–4):339–47.
16. Young L, Badihian S, Zee DS, Newman-Toker DE, Kerber KA, Bhandari A, et al. Benign paroxysmal positional vertigo: a case report in which four wrongs made a right. Clin Case Rep. 2024;12(11):e9529.
17. Nuti D, Zee DS, Mandala M. Benign paroxysmal positional vertigo: what we do and do not know. Semin Neurol. 2020;40(1):49–58.
18. Ludwig D, Schubert MC. Case report: keep your eyes open! Nystagmus guides atypical BPPV. Front Rehabil Sci. 2024;5:1384151.
19. Newman-Toker DE, Kerber KA, Hsieh YH, Pula JH, Omron R, Saber Tehrani AS, et al. HINTS outperforms ABCD2 to screen for stroke in acute continuous vertigo and dizziness. Acad Emerg Med. 2013;20(10):986–96.
20. Suzuki JI, Cohen B. Head, eye, body and limb movements from Semicircular Canal nerves. Exp Neurol. 1964;10:393–405.
21. Zee DS. Ophthalmoscopy in examination of patients with vestibular disorders. Ann Neurol. 1978;3(4):373–4.

22. Newman-Toker DE, Sharma P, Chowdhury M, Clemons TM, Zee DS, Della Santina CC. Penlight-cover test: a new bedside method to unmask nystagmus. J Neurol Neurosurg Psychiatry. 2009;80(8):900–3.
23. Korda A, Zee DS, Wyss T, Zamaro E, Caversaccio MD, Wagner F, et al. Impaired fixation suppression of horizontal vestibular nystagmus during smooth pursuit: pathophysiology and clinical implications. Eur J Neurol. 2021;28(8):2614–21.
24. Bery AK, Wang CF, Gold DR, Chang TP. The fixation suppression test can uncover vertical nystagmus of central origin in some patients with dizziness. Neurol Sci. 2021;42(12):5343–52.
25. Mantokoudis G, Wyss T, Zamaro E, Korda A, Wagner F, Sauter TC, et al. Stroke prediction based on the spontaneous nystagmus suppression test in dizzy patients: a diagnostic accuracy study. Neurology. 2021;97(1):e42–51.
26. Choi JH, Oh EH, Kim HS, Park JY, Lee SM, Choi SY, et al. Spontaneous nystagmus violating the Alexander's law: neural substrates and mechanisms. Cerebellum. 2025;24(2):34.
27. Mantokoudis G, Korda A, Zee DS, Zamaro E, Sauter TC, Wagner F, et al. Bruns' nystagmus revisited: a sign of stroke in patients with the acute vestibular syndrome. Eur J Neurol. 2021;28 (9):2971–9.
28. Fetter M, Hain TC, Zee DS. Influence of eye and head position on the vestibulo-ocular reflex. Exp Brain Res. 1986;64(1):208–16.
29. Kattah JC, Badihian S, Pula JH, Tarnutzer AA, Newman-Toker DE, Zee DS. Ocular lateral deviation with brief removal of visual fixation differentiates central from peripheral vestibular syndrome. J Neurol. 2020;267(12):3763–72.
30. Tarnutzer AA, Gold D, Wang Z, Robinson KA, Kattah JC, Mantokoudis G, et al. Impact of clinician training background and stroke location on bedside diagnostic test accuracy in the acute vestibular syndrome – a meta-analysis. Ann Neurol. 2023;94(2):295–308.
31. Korda A, Sauter TC, Caversaccio MD, Mantokoudis G. Quantifying a learning curve for video head impulse test: pitfalls and pearls. Front Neurol. 2020;11:615651.
32. d'Onofrio F, Neri L, Neri G. Vertical saccades during horizontal head impulses: a sign of posterior semicircular canal hypofunction. Acta Otorhinolaryngol Ital. 2024;44(3):198–203.
33. D'Onofrio F. Vertical eye movements during horizontal head impulse test: a new clinical sign of superior vestibular neuritis. Acta Otorhinolaryngol Ital. 2013;33(6):418–24.
34. Choi JY, Kim HJ, Kim JS. Recent advances in head impulse test findings in central vestibular disorders. Neurology. 2018;90(13):602–12.
35. Green KE, Pogson JM, Otero-Millan J, Gold DR, Tevzadze N, Saber Tehrani AS, et al. Opinion and special articles: remote evaluation of acute vertigo: strategies and technological considerations. Neurology. 2021;96(1):34–8.
36. Hain TC, Fetter M, Zee DS. Head-shaking nystagmus in patients with unilateral peripheral vestibular lesions. Am J Otolaryngol. 1987;8(1):36–47.
37. Yang TH, Lee JH, Oh SY, Kang JJ, Kim JS, Dieterich M. Clinical implications of head-shaking nystagmus in central and peripheral vestibular disorders: is perverted head-shaking nystagmus specific for central vestibular pathology? Eur J Neurol. 2020;27(7):1296–303.
38. Huh YE, Kim JS. Patterns of spontaneous and head-shaking nystagmus in cerebellar infarction: imaging correlations. Brain. 2011;134(Pt 12):3662–71.
39. Haslwanter T, Minor LB. Nystagmus induced by circular head shaking in normal human subjects. Exp Brain Res. 1999;124(1):25–32.
40. Shemesh AA, Kattah JC, Zee DS, Zuma EMF, Otero-Millan J. Amplification of vibration induced nystagmus in patients with peripheral vestibular loss by head tilt. Front Neurol. 2024;15:1420699.
41. Curthoys IS, Zee DS, Dumas G, Pastras CJ, Dlugaiczyk J. Skull vibration induced nystagmus, velocity storage and self-stability. Front Neurol. 2025;16
42. Kheradmand A, Zee DS. Clinical bedside examination. In: Bronstein A, editor. Oxford textbook of vertigo and imbalance. 2nd ed. Oxford University Press; 2025.
43. Gold D. Neuro-ophthalmology and neuro-otology: a case-based guide for clinicians and scientists. Cham: Springer; 2021.

44. Halmagyi GM, Gresty MA, Gibson WP. Ocular tilt reaction with peripheral vestibular lesion. Ann Neurol. 1979;6(1):80–3.
45. Zee DS. Considerations on the mechanisms of alternating skew deviation in patients with cerebellar lesions. J Vestib Res. 1996;6(6):395–401.
46. Kattah JC, Tehrani AS, du Lac S, Newman-Toker DE, Zee DS. Conversion of upbeat to downbeat nystagmus in Wernicke encephalopathy. Neurology. 2018;91(17):790–6.
47. Kattah JC, McClelland C, Zee DS. Vertical nystagmus in Wernicke's encephalopathy: pathogenesis and role of central processing of information from the otoliths. J Neurol. 2019;266 (Suppl 1):139–45.
48. Fracica E, Hale D, Gold DR. Diagnosing and localizing the acute vestibular syndrome – beyond the HINTS exam. J Neurol Sci. 2022;442:120451.
49. Zee DS, Yamazaki A, Butler PH, Gucer G. Effects of ablation of flocculus and paraflocculus of eye movements in primate. J Neurophysiol. 1981;46(4):878–99.
50. Walker MF, Tian J, Shan X, Tamargo RJ, Ying H, Zee DS. Lesions of the cerebellar nodulus and uvula impair downward pursuit. J Neurophysiol. 2008;100(4):1813–23.
51. Shemesh AA, Zee DS. Eye movement disorders and the cerebellum. J Clin Neurophysiol. 2019;36(6):405–14.
52. Minor LB, Haslwanter T, Straumann D, Zee DS. Hyperventilation-induced nystagmus in patients with vestibular schwannoma. Neurology. 1999;53(9):2158–68.
53. Walker MF, Zee DS. The effect of hyperventilation on downbeat nystagmus in cerebellar disorders. Neurology. 1999;53(7):1576–9.
54. Minor LB, Solomon D, Zinreich JS, Zee DS. Sound- and/or pressure-induced vertigo due to bone dehiscence of the superior semicircular canal. Arch Otolaryngol Head Neck Surg. 1998;124(3):249–58.

Chapter 3
Video Ocular Counter-Roll (vOCR): A Clinical Bedside Evaluation of Otolith-Ocular Function

Lien Van Laer and Amir Kheradmand

Keywords Oculography · Portable VOG · Video ocular counter roll · Torsional VOR

3.1 Introduction

The primary aim of clinical vestibular testing is to detect the extent of any vestibular impairment. Traditionally, this has been answered by only evaluating semicircular canal function through methods such as caloric and rotatory chair testing [1]. In these tests, vestibular function is evaluated as a response elicited by either thermal caloric or rotatory stimuli, which consists of a slow-phase vestibulo-ocular reflex (VOR) followed by a fast-phase correction, manifesting as nystagmus. For assessment of otolith function, vestibular evoked myogenic potentials (VEMP) have been used as a clinical test. The VEMP testing measures the electromyographic (EMG) activity evoked by vestibular stimulation typically through vibration or sound [2, 3]. This

L. Van Laer
Department of Rehabilitation Sciences and Physiotherapy/Movant, Faculty of Medicine and Health Science, University of Antwerp, Antwerp, Belgium

Laboratory of Vestibular NeuroAdaptation, Department of Otolaryngology – Head and Neck Surgery, Johns Hopkins University School of Medicine, Baltimore, MD, USA

Department of Neuroscience, The Johns Hopkins University School of Medicine, Baltimore, MD, USA

A. Kheradmand (✉)
Department of Neuroscience, The Johns Hopkins University School of Medicine, Baltimore, MD, USA

Department of Neurology, The Johns Hopkins University School of Medicine, Baltimore, MD, USA

Department of Otolaryngology-Head and Neck Surgery, The Johns Hopkins University School of Medicine, Baltimore, MD, USA
e-mail: akherad@jhu.edu

P. Ravat et al. (eds.), *Advances in Translational Neuroscience of Vestibular and Eye Movement Disorders*, Contemporary Clinical Neuroscience,
https://doi.org/10.1007/978-3-032-07539-0_3

includes cervical VEMP (cVEMP), which evaluates saccular function via the activity at the Sternocleidomastoid muscle, and ocular VEMP (oVEMP), which evaluates utricular function via the activity of ocular muscles. While these tests are valuable, they are primarily laboratory-based, requiring technical equipment and are time-consuming to implement, making them less practical for timely clinical application.

Recent advances, in lightweight video oculography (VOG) has transformed vestibular testing, enabling precise, bedside quantification and analysis. Portable VOG has been effectively incorporated into bedside vestibular assessments. The video head impulse testing (vHIT) is now widely used to evaluate the function of individual semicircular canals [4, 5]. For vHIT evaluation, VOG systems rely on pupil detection to track eye movements while an integrated accelerometer measures head motion in real time. This simultaneous measurement enables the calculation of VOR gain and detection of catch-up saccades, which serve as markers of vestibular dysfunction. While vHIT is a valuable bedside tool for semicircular canal evaluation, it does not assess otolith function. Building on the principles of otolith-ocular function, a VOG test for clinical evaluation of otolith-ocular function has been developed in recent years. This test is known as the video ocular counter-roll (vOCR) and is based on the measurement of the torsional vestibulo-ocular reflex with a lateral head tilt [6]. In this chapter, we review the fundamental principles of otolith-ocular function, discuss the technical aspects of vOCR measurement, and highlight its clinical significance in evaluating otolith function.

3.2 Ocular Counter-Roll and Otolith-Ocular Pathophysiology

Understanding how the otoliths generate VOR is a key step in evaluating the role of VOG in assessing otolith-ocular function. The VOR stabilizes gaze when the head is rotated around the horizontal (yaw), torsional (roll), or vertical (pitch) axes. By drawing parallels with semicircular canal function, it is known that the brain maintains a natural balance between vestibular inputs from both sides. When the head of a healthy individual is still, the left and right vestibular inputs have equal resting discharge rates (vestibular tone). When the head is moved in one direction, excitation occurs on the side of the movement, while inhibition occurs on the opposite side (i.e., Ewald's first law) [7]. In cases of sudden imbalance, such as in acute unilateral vestibular loss, spontaneous nystagmus emerges, characterized by a slow-phase eye movement directed toward the impaired side, followed by fast-phase corrections toward the intact side [7, 8]. Gradually, compensatory mechanisms restore equilibrium between the two sides over time. Once recovery is achieved, the vestibular deficit becomes apparent only during head movements, such as with rapid head impulses. In these cases, VOR gain is reduced on the affected side, and compensatory catch-up saccades can be brought out during head impulses.

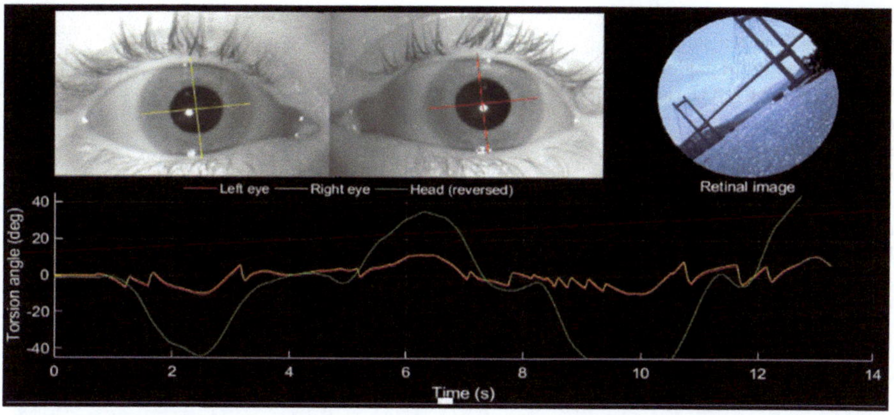

Fig. 3.1 Sensory mismatch between the head tilt and ocular counter-roll (OCR), resulting in changes in the retinal image orientation during head tilt. The green line on the graph represents the head tilt ranging between ±0 and 40°. The yellow and orange lines on the graph represent OCR in both eyes, ranging between ±0 and 10°

But how does this principle extend to otolith function? The utricular macula is oriented in the horizontal plane and responds to translational movements, including both linear translations and static tilts against the pull of gravity. Thus, functioning as sensors that detect linear acceleration, the otolith organs are stimulated by the force of gravity when the head is tilted laterally [9]. Consequently, a head tilt elicits a compensatory torsional VOR in the opposite direction, known as the ocular counter-roll (OCR). Unlike the horizontal and vertical VOR, which closely match the amount of head movement, the OCR is not equivalent to the amount of head tilt in humans. The nature of this response to head tilt also varies across species. For example, in a rabbit—a lateral-eyed animal—a head tilt elicits a nearly complete compensatory vertical eye movement. In contrast, humans with front-facing eyes exhibit only partial torsional eye movement to counteract a head tilt. As shown in Fig. 3.1, a 40° head tilt may result in approximately 10° of ocular torsion. This incomplete compensation creates a misalignment between eye and head positions, causing a shift in image orientation on the retina during head tilt. While this partial response might suggest that OCR is inefficient or undercompensatory, it may actually provide an evolutionary advantage. By preserving some disparity between eye and head position, the brain can extract crucial visual cues, helping to distinguish between self-induced motion and movement with in the environment. This advantage, however, comes at a computational cost, requiring the brain to process and correct for shifts in the retinal image to maintain spatial orientation [10].

The OCR response during a stationary or static head tilt is primarily driven by the inputs from the otolith organs. During the movement of the head in the roll plane, however, the combination of inputs from the otolith organs and semicircular canals results in a torsional nystagmus with both slow and fast-phase torsional eye movements. For example, during a right head tilt, the dynamic torsional response consists

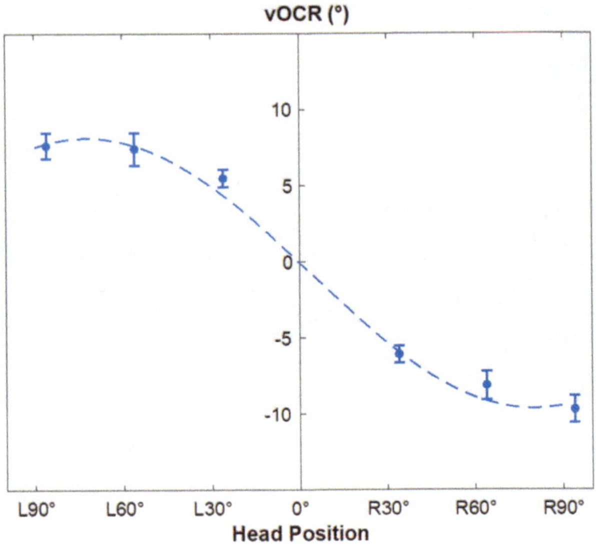

Fig. 3.2 Example vOCR recordings from ten subjects at varying body tilt angles (0° to 90°) to the left (L) and right (error bars represent standard error of mean). The vOCR response plateaus at approximately 6° with a head tilt of 60°. The error bars represent the standard error of the mean

of a series of slow-phase changes of ocular torsion to the left side, each followed by a fast-phase correction to the right side. In contrast, when the head is sustained to the right side, there is only a static OCR to the left side without any fast-phase correction. Thus, assessing otolith function requires isolating the static phase of head tilt, where OCR reflects sustained otolith-driven activity without interference from canal-driven fast-phase corrections.

Early human studies applied search coils to measure OCR across different head tilts ranging from 0° to 90°, demonstrating a sinusoidal response that plateaued at larger tilts [11–14]. For instance, at approximately 60° of head tilt, the OCR reaches its maximum response of around 6°, with no further increase beyond this point at larger tilt angles. This plateau effect has also been validated through vOCR recordings (Fig. 3.2), highlighting the inherently low gain of OCR. A smaller tilt of 30° induces approximately 5° of OCR, yielding a gain of 0.15—substantially lower than the angular VOR gain, which is typically 1.00. While vOCR has produced similar gain values to the coil recordings, it has also shown that slippage may pose a challenge when measuring ocular torsion using contact-based techniques, unlike VOG-based recordings [15].

In terms of the neural pathway underlying OCR, the otolith inputs that detect head tilt against the pull of gravity primarily originate from the utricle with projections to the vestibular nucleus that cross over and ascend within the brainstem to reach the ocular motor nuclei (Fig. 3.3a). This pathway drives the OCR through a coordinated ocular motor response. On the side of the head tilt, the superior rectus and superior oblique muscles produce elevation and intorsion of the eye, while on the opposite side, the inferior rectus and inferior oblique muscles produce depression and extorsion of the eye.

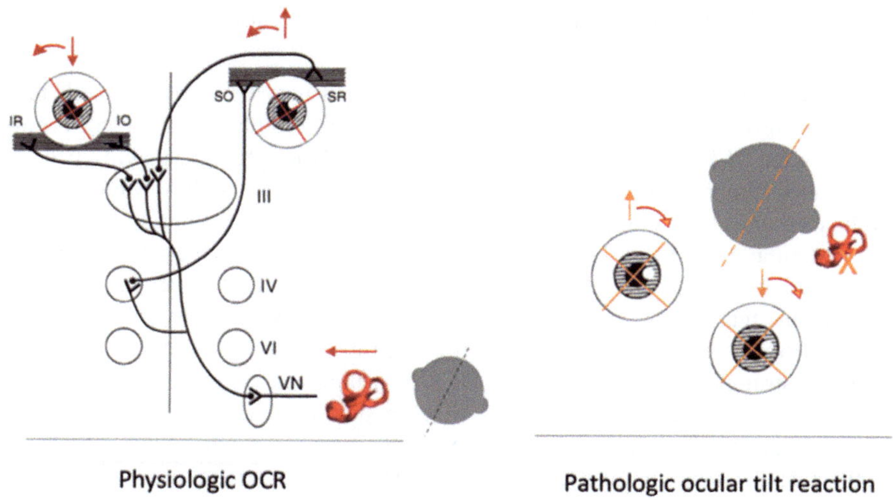

Fig. 3.3 (**a**) Physiologic OCR response to lateral head tilt is mediated by a neural pathway that originates in the labyrinth and projects to the ocular motor nuclei. A head tilt to one side stimulates the utricle on the same side. Through the otolith-ocular pathway, the superior rectus (SR) and superior oblique (SO) muscles lead to elevation and intorsion of the eye on the side of the tilt. Inferior rectus (IR) and inferior oblique (IO) muscles lead to depression and extorsion of the eye on the opposite side of the tilt. (**b**) Pathological ocular tilt reaction is characterized by ocular roll, skew deviation, and head tilt. In lesions affecting the labyrinth or lower brainstem, ocular roll deviates toward the lesion, the lower eye in skew deviation is on the lesion side, and head tilt is ipsilesional. In contrast, lesions at higher brainstem levels produce the opposite pattern. *VN* vestibular nuclei, *III, IV*, and *VI* Oculomotor nuclei, *IR* inferior rectus muscle, *IO* inferior oblique muscle, *SO* superior oblique muscle, *SR* superior rectus muscle

Vestibular pathologies can significantly impact otolith function, with acute peripheral vestibular lesions leading to a pathological ocular roll. This is often accompanied by vertical misalignment of the eyes, known as skew deviation, indicating impaired vestibular control of ocular alignment, along with a compensatory head tilt toward the side of the lesion [16]. Collectively, these signs constitute the ocular tilt reaction (OTR) [17–19]. Within the OCR pathway (Fig. 3.3), a peripheral lesion affecting the labyrinth results in an ocular roll toward the lesion side, skew deviation with the lower eye on the lesion side, and an ipsilesional head tilt. A similar pattern is observed with the lower brainstem lesions, where the pathway crosses at this level. In contrast, lesions at higher brainstem levels produce the opposite pattern—ocular roll deviates away from the lesion, the lower eye in skew deviation is on the side opposite to the lesion, and the head tilt is contralesional. Based on this distinction, The OCR during head tilt serves as a normal physiological response, whereas the OTR reflects a pathological disturbance within the otolith-ocular pathway that generates the OCR response. Similar to the resolution of spontaneous nystagmus after canal dysfunction, the OTR may diminish over time as compensation occurs.

3.3 OCR Measurement and Video Oculography

With recent advances in VOG platforms, OCR can now be measured with high accuracy, providing a valuable tool for assessment of otolith function [15]. Earlier methods for OCR measurement relied on bulky VOG technology, making bedside application impractical [20–22]. Additionally, previous studies were largely conducted in healthy volunteers and not in patient populations, limiting their utility for vestibular assessment. Unlike horizontal and vertical VOR measurement, where eye movements can be tracked using pupil detection, OCR involves torsional eye movements that require iris pattern recognition and tracking. This adds a layer of computational complexity, especially for real-time tracking, which made earlier methods struggle to achieve real-time OCR measurement.

In recent years, a newly developed VOG method has enabled accurate tracking and measurement of ocular torsion [15]. This method is binocular, operates at 100 Hz, and achieves a noise level of less than 0.1°. Additionally, it can incorporate eyelid detection to account for partial occlusion of the iris. First, the captured image of the eye is enhanced, and the iris pattern is isolated to track and compare in real time against a reference position to determine changes in torsional eye position (Fig. 3.4). Building on the principles of OCR physiology, this VOG method has been used for clinical evaluation of otolith-ocular function as the vOCR test [6, 15, 23–26].

3.4 vOCR Test

As a clinical test, vOCR can detect vestibular deficits affecting otolith function [6, 23–25]. This test consists of a simple bedside maneuver while the patient is fixating on a visual target in front of them at eye level about 1.5 meters away. The head and torso are tilted 30° en bloc by the examiner in the roll plane, which can be comfortably maintained while the patient is sitting on a chair (Fig. 3.5). With the head and torso tilted 30° to one side, maintaining straight-ahead fixation requires two additional, lateral targets alongside the central fixation point. Patients are instructed to fix gaze on the appropriate lateral target to preserve straight-ahead fixation during the tilt. All three targets are aligned at equal distances from the central target. The spacing between the targets is determined by calculating half the vertical distance from the sitting surface to the patient's eyes, i.e., multiplying by the sine of the 30° tilt, ensuring proper alignment for maintaining fixation with a 30° head and body tilt. During the maneuver, VOG goggles equipped with accelerometers simultaneously record both eye and head positions, allowing real-time tracking of head movement and ocular response. This entire procedure could be completed within a few minutes, and an automated analysis can provide instant results.

The reliability of the vOCR test has been validated through repeated measurements, with minimal differences between trials, ensuring consistent results [24]. In

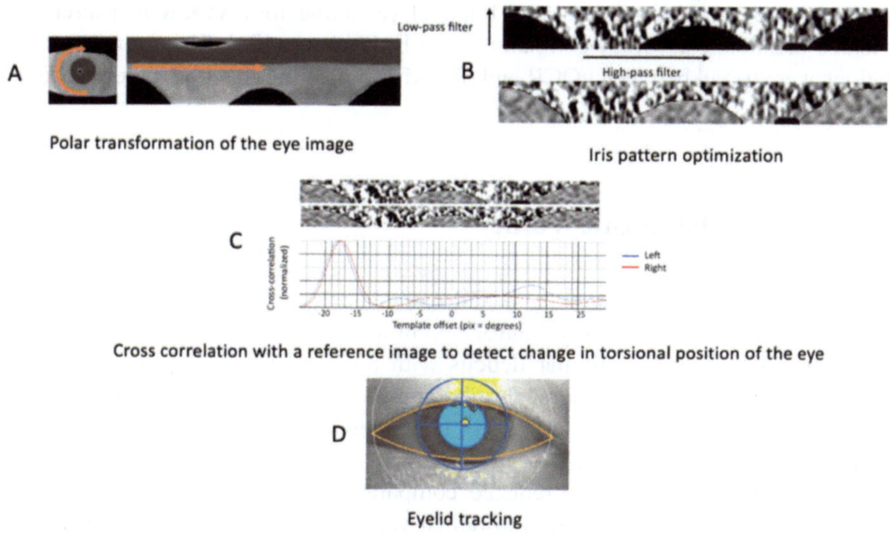

Fig. 3.4 (**a**): The eye image undergoes a polar transformation. (**b**): A low-pass filter is applied to enhance the iris's features. (**c**): Ocular torsion is measured by comparing the transformed image to a reference image. (**d**): The system also tracks eyelid position to determine the visible portions of the iris and pupil

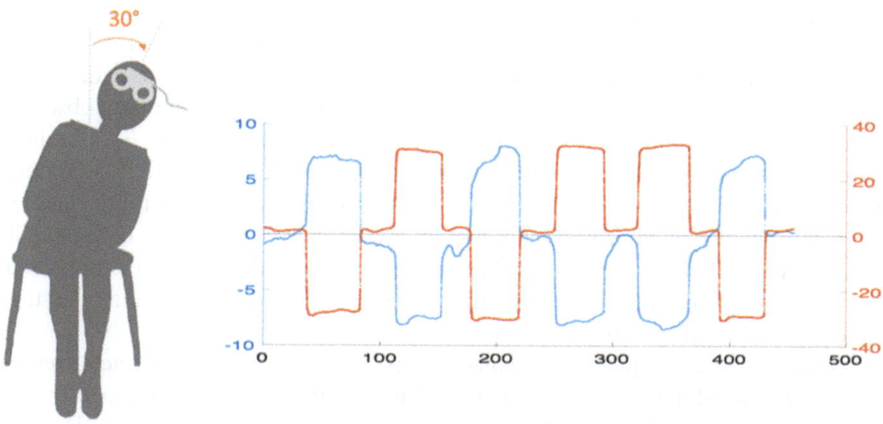

Fig. 3.5 The vOCR test consists of a simple bedside maneuver while the patient is fixating on a visual target in front of them. The head and torso are tilted 30° en bloc by the examiner in the roll plane, which can be comfortably maintained while the patient is sitting on a chair. A sample recording is shown on the graph. Note the different scales on the ordinate for the head tilt angle (orange trace/axis) and vOCR (blue trace/axis). The positive value on the graph indicates a rightward direction, and the negative value indicates a leftward direction. With 30° head tilt, there is about 7° of vOCR response

patients with chronic unilateral and bilateral vestibular loss, vOCR measurements yielded diagnostic accuracy comparable to VEMP results [24]. The strongest correlations were found between vOCR and oVEMPs, reflecting that both assess utricular function.

3.5 vOCR Diagnostic Value

As a measure of otolith function, vOCR can detect vestibular loss and distinguish vestibular patients from healthy controls. A cutoff of 4.5° vOCR at a 30° head tilt has been shown to detect vestibular deficits with 83% accuracy [25]. Measurements show that patients with bilateral vestibular loss exhibit low, symmetrical vOCR values (mean ~ 1.6° with 30° head tilt), compared with larger, symmetrical vOCR observed in healthy controls (mean: 4.7° with 30° head tilt) [24]. In unilateral vestibular loss, vOCR is also reduced compared to controls (mean ~ 2.7 ° with 30° head tilt) [24], but the pattern of deficit varies with the stage of vestibular loss [25]. During the acute stage, vOCR is lower on the lesion side, whereas in the chronic stage, vOCR values become symmetric between the healthy and affected sides [25].

The above findings highlight vOCR as a consistent and accurate measure for detecting loss of otolith function. In chronic unilateral vestibular loss, vOCR may, however, exhibit a symmetrical response [25]. Asymmetry in vOCR deficit was most pronounced in patients with acute vestibular loss, characterized by a lower response on the affected side. This asymmetry was lower in subacute patients and minimal in chronic patients, eventually approaching the level of symmetry observed in healthy controls, though vOCR values remained lower overall (Fig. 3.6). In these patients, integrating vHIT and vOCR into a single VOG battery offers a complementary diagnostic approach. The vOCR measurement can assess otolith function and determine the stage of vestibular loss by evaluating the symmetry of the vOCR deficit. The vHIT measurement, on the other hand, can provide assessment of canal function and identify the side of vestibular impairment without determining the stage of vestibular loss (Fig. 3.6c) [27–29].

As noted above, peripheral vestibular lesions can lead to an OTR characterized by ocular roll toward the lesion side, skew deviation, and head tilt. Consequently, this pathological ocular roll may influence vOCR measurements. For instance, in right-sided vestibular loss, pathological ocular roll toward the right may enhance vOCR when the head is tilted to the left. Conversely, when the head is tilted to the lesion (i.e., right) side, the vOCR is reduced due to impaired vestibular function. This pathological ocular roll gradually recovers over time following unilateral vestibular loss, reflecting a long-term adaptive process that potentially supports recovery [30]. Thus, its influence on vOCR measurements may also diminish over time.

The evolving pattern of vOCR deficit following vestibular loss highlights its clinical value in identifying the stage of vestibular loss. Unlike vHIT deficits, which remain low in the affected side and may show no correlation with recovery after

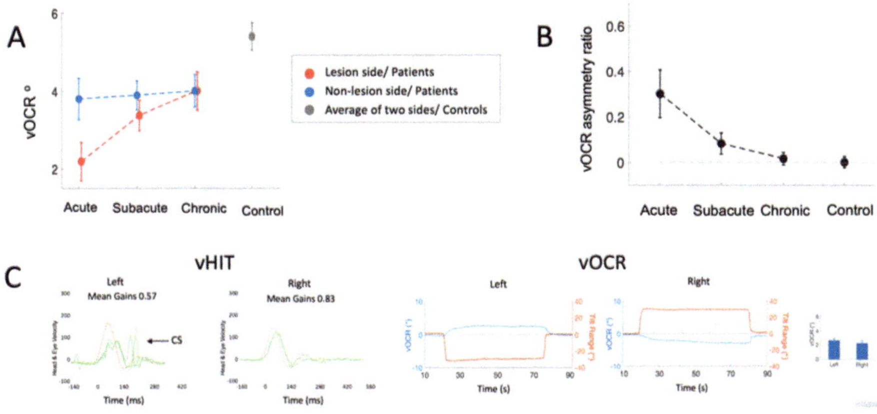

Fig. 3.6 (**a**): vOCR measurements in patients at different stages following unilateral loss of vestibular function. (**b**): Asymmetry ratio between vOCR measurements across the same patient groups. (**c**): vOCR and vHIT measurements in a patient with chronic unilateral (left-sided) vestibular loss. Negative values indicate leftward tilt, and positive values indicate rightward tilt. The bar graphs represent mean vOCR values for three tilt angles on each side. The vHIT traces display superimposed head (orange) and eye (green) velocities during lateral head impulses, with reduced gain (eye velocity/head velocity) on the left side. There are also compensatory catch-up saccades (CS) with head impulses to the left side

vestibular loss, vOCR deficits are more pronounced in the acute stage. [23, 31, 32] In chronic patients, however, vOCR can improve and become symmetric between the affected and healthy sides, while the values may typically remain lower than those observed in healthy controls (Fig. 3.6) [25]. The vOCR test can also be valuable as a clinical marker to measure how vestibular and non-vestibular neck inputs contribute to recovery at different stages following loss of vestibular function. This can be done by measuring vOCR during head-on-body tilt in comparison to the en-block whole body tilt maneuver [23]. While vOCR responses are reduced on the side of vestibular loss during both head and body tilt maneuvers, the deficit is more pronounced during whole-body tilt (Fig. 3.7). This finding indicates that neck inputs can enhance vOCR gain in early or late stages of vestibular loss and thus can facilitate compensation in otolith-ocular function following vestibular loss. Additionally, neck inputs could improve the time course of vOCR in acute vestibular loss, suggesting a compensatory role in both the static and dynamic phases of the vOCR response [23].

3.6 Conclusions

Recent advances in VOGtechnology have made it possible to assess otolith function with simple tilt maneuvers for clinical assessment of otolith function. The quick application of the vOCR test allows examination of otolith-ocular function and assessment of vestibular recovery at the bedside. Patients with unilateral vestibular

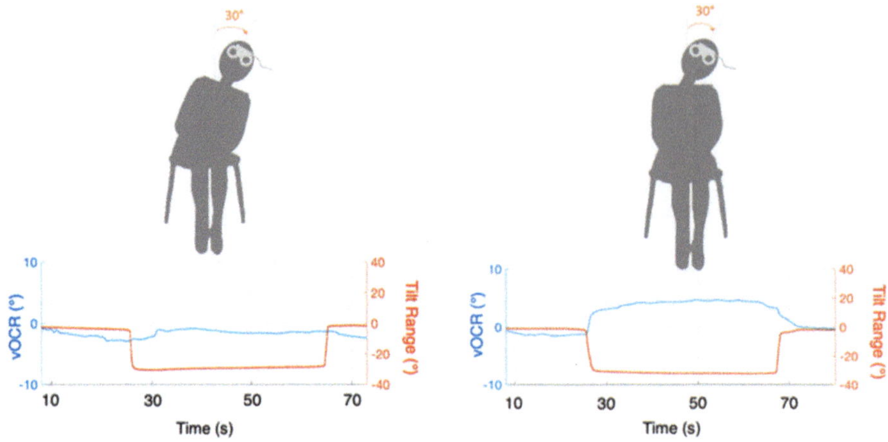

Fig. 3.7 Example of vOCR recordings from a patient with subacute vestibular loss (7 weeks), comparing two tilt conditions: en bloc head and body tilt (left) and isolated head tilt on the body (right). When the head is tilted independently of the body, the neck effect can enhance vOCR on the side of vestibular loss. The vOCR trace shows eye position (blue axis) and head position (orange axis), with negative values indicating leftward tilt and positive values indicating rightward tilt

loss demonstrate a vOCR reduction during the acute phase of vestibular loss. In chronic patients, however, vOCR may improve and become symmetric between the affected and healthy sides, while the values remain lower than healthy controls. This evolving pattern of vOCR deficit at different stages following vestibular loss can be helpful to track recovery. The vOCR test could also serve as a valuable tool for evaluating non-vestibular, neck contributions to recovery by comparing head-on-body and whole-body tilt maneuvers. Finally, when vOCR is combined with vHIT, together they provide a comprehensive VOG assessment of vestibular function, incorporating both otolith and canal function. The vOCR test can evaluate otolith-ocular function and determine the stage of vestibular loss, while the vHIT test can assess the function of semicircular canals and identify the side of the vestibular deficit.

Acknowledgments We thank all the team members who contributed to the development of the video Ocular Counter Roll test: Jorge Otero-Millan, Jing Tian, Francesco Fornasari, Shirin Sadeghpour, Yuchen Yang, Krishna Mukunda, Dale Roberts, Adrian Lasker, Ariel Winnick, Caroline Trevino, Michael Schubert, John Carey, and David Zee.

Funding This work was supported through funding from the National Institute of Health, NIH/NIDCD 1R01DC021186 and NIH/NIA K02AG083134. Lien Van Laer was awarded a Fulbright Postdoctoral Scholarship by Fulbright Belgium to support this work.

References

1. Wolfe JW, Engelken EJ, Olson JW, Kos CM. Vestibular responses to bithermal caloric and harmonic acceleration. Ann Otol Rhinol Laryngol. 1978;87(6 Pt 1):861–7. https://doi.org/10.1177/000348947808700614.
2. Dlugaiczyk J. Evidence-based diagnostic use of VEMPs: from neurophysiological principles to clinical application. HNO. 2020;68(Suppl 2):69–78. https://doi.org/10.1007/s00106-019-00767-2.
3. Colebatch JG, Halmagyi GM. Vestibular evoked potentials in human neck muscles before and after unilateral vestibular deafferentation. Neurology. 1992;42(8):1635–6. https://doi.org/10.1212/wnl.42.8.1635.
4. Halmagyi GM, Curthoys IS. A clinical sign of canal paresis. Arch Neurol. 1988;45(7):737–9. https://doi.org/10.1001/archneur.1988.00520310043015.
5. Haslwanter T, Clarke AH. Chapter 5 – eye movement measurement: electro-oculography and video-oculography. In: Vertigo and imbalance: clinical neurophysiologyof the vestibular system. Vol 9. Handbook of clinical neurophysiology. Elsevier; 2010. p. 61–79. https://doi.org/10.1016/S1567-4231(10)09005-2.
6. Yang Y, Tian J, Otero-Millan J, Kheradmand A. Video ocular counter roll: a bedside test of otolith-ocular function. Ann Clin Transl Neurol. 2023;10(12):2426–9. https://doi.org/10.1002/acn3.51921.
7. Leigh RJ, Zee DS. The neurology of eye movements. Oxford University Press; 2015. https://doi.org/10.1093/med/9780199969289.001.0001.
8. Kheradmand A, Zee DS. The bedside examination of the vestibulo-ocular reflex (VOR): an update. Rev Neurol (Paris). 2012;168(10):710–9. https://doi.org/10.1016/j.neurol.2012.07.011.
9. Hegemann SCA, Bery AK, Kheradmand A. Focused update on clinical testing of otolith organs. Audiol Res. 2024;14(4):602–10. https://doi.org/10.3390/audiolres14040051.
10. Kheradmand A, Winnick A. Perception of upright: multisensory convergence and the role of Temporo-parietal cortex. Front Neurol. 2017;8:552. https://doi.org/10.3389/fneur.2017.00552.
11. Diamond SG, Markham CH. Ocular counterrolling as an indicator of vestibular otolith function. Neurology. 1983;33(11):1460–9. https://doi.org/10.1212/wnl.33.11.1460.
12. Schmid-Priscoveanu A, Straumann D, Böhmer A, Obzina H. Vestibulo-ocular responses during static head roll and three-dimensional head impulses after vestibular neuritis. Acta Otolaryngol (Stockh). 1999;119(7):750–7. https://doi.org/10.1080/00016489950180379.
13. Takemori S, Tanaka M, Moriyama H. An analysis of ocular counter-rolling measured with search coils. Acta Otolaryngol Suppl. 1989;468:271–6. https://doi.org/10.3109/00016488909139060.
14. Collewijn H, Van der Steen J, Ferman L, Jansen TC. Human ocular counterroll: assessment of static and dynamic properties from electromagnetic scleral coil recordings. Exp Brain Res. 1985;59(1):185–96. https://doi.org/10.1007/BF00237678.
15. Otero-Millan J, Roberts DC, Lasker A, Zee DS, Kheradmand A. Knowing what the brain is seeing in three dimensions: a novel, noninvasive, sensitive, accurate, and low-noise technique for measuring ocular torsion. J Vis. 2015;15(14):11. https://doi.org/10.1167/15.14.11.
16. Sheth MS, Primiani CT, Kheradmand A, Green KE. Pearls & oy-sters: vertical diplopia and ocular torsion: peripheral vs central localization. Neurology. 2022;99(5):212–5. https://doi.org/10.1212/WNL.0000000000200835.
17. Halmagyi GM, Gresty MA, Gibson WP. Ocular tilt reaction with peripheral vestibular lesion. Ann Neurol. 1979;6(1):80–3. https://doi.org/10.1002/ana.410060122.
18. Brandt T. Vestibular disorders in (frontal) roll plane. In: Brandt T, editor. Vertigo: its multisensory syndromes. New York: Springer; 2003. p. 175–97. https://doi.org/10.1007/978-1-4757-3801-8_10.
19. Kheradmand A, Colpak AI, Zee DS. Chapter 8 – eye movements in vestibular disorders. In: Furman JM, Lempert T, editors. Handbook of clinical neurology, vol. 137. Elsevier; 2016. p. 103–17. https://doi.org/10.1016/B978-0-444-63437-5.00008-X.

20. Kingma H, Meulenbroeks A, De Jong I. Vestibular ocular reflexes in Ménière's disease patients evaluated by passive high frequency head rotation (yaw) and sidewards acceleration. Acta Otolaryngol Suppl. 2000;544:19–26. https://doi.org/10.1080/000164800750044443-1.
21. Schworm HD, Ygge J, Pansell T, Lennerstrand G. Assessment of ocular counterroll during head tilt using binocular video oculography. Invest Ophthalmol Vis Sci. 2002;43(3):662–7.
22. Zingler VC, Kryvoshey D, Schneider E, Glasauer S, Brandt T, Strupp M. A clinical test of otolith function: static ocular counterroll with passive head tilt. Neuroreport. 2006;17(6):611–5. https://doi.org/10.1097/00001756-200604240-00011.
23. Yang Y, Tian J, Otero-Millan J, Schubert MC, Kheradmand A. Video ocular counter-roll (vOCR): otolith-ocular function and compensatory effect of the neck following vestibular loss. Otolaryngol Head Neck Surg. 2023;169(3):669–78. https://doi.org/10.1002/ohn.304.
24. Otero-Millan J, Treviño C, Winnick A, Zee DS, Carey JP, Kheradmand A. The video ocular counter-roll (vOCR): a clinical test to detect loss of otolith-ocular function. Acta Otolaryngol (Stockh). 2017;137(6):593–7. https://doi.org/10.1080/00016489.2016.1269364.
25. Sadeghpour S, Fornasari F, Otero-Millan J, Carey JP, Zee DS, Kheradmand A. Evaluation of the video ocular counter-roll (vOCR) as a new clinical test of otolith function in peripheral Vestibulopathy. JAMA Otolaryngol-Head Neck Surg. 2021;147(6):518–25. https://doi.org/10.1001/jamaoto.2021.0176.
26. Kartal Özcan A, Satici S, Akbulut AA, Kiliç M, Çankaya S, Polat Z. Investigation of video ocular counter-roll findings according to head and body tilt positions in healthy subjects. Ear Hear. 2025;46(1):53–9. https://doi.org/10.1097/AUD.0000000000001563.
27. Welgampola MS, Taylor RL, Halmagyi GM. Video head impulse testing. Adv Otorhinolaryngol. 2019;82:56–66. https://doi.org/10.1159/000490272.
28. Halmagyi GM, Chen L, MacDougall HG, Weber KP, McGarvie LA, Curthoys IS. The video head impulse test. Front Neurol. 2017;8:258. https://doi.org/10.3389/fneur.2017.00258.
29. Alhabib SF, Saliba I. Video head impulse test: a review of the literature. Eur Arch Otorrinolaringol. 2017;274(3):1215–22. https://doi.org/10.1007/s00405-016-4157-4.
30. Faralli M, Ricci G, Manzari L, Zambonini G, Lapenna R, Pettorossi VE. Different time course of compensation of subjective visual vertical and ocular torsion after acute unilateral vestibular lesion. Eur Arch Otorrinolaringol. 2021;278(7):2269–76. https://doi.org/10.1007/s00405-020-06312-0.
31. Chang TP, Schubert MC. Association of the video head impulse test with improvement of dynamic balance and fall risk in patients with dizziness. JAMA Otolaryngol-Head Neck Surg. 2018;144(8):696–703. https://doi.org/10.1001/jamaoto.2018.0650.
32. Millar JL, Gimmon Y, Roberts D, Schubert MC. Improvement after vestibular rehabilitation not explained by improved passive VOR gain. Front Neurol. 2020;11:79. https://doi.org/10.3389/fneur.2020.00079.

Chapter 4
A Clinical and Instrumented Approach of Vestibulo-Spinal Function

Ricardo D'Albora and María Julia Mónaco

Keywords Vestibular system · Truncal ataxia · Vestibulo-spinal pathway · Acute vestibular syndrome · Central vertigo · Peripheral vertigo

4.1 Introduction

Assessing and isolating vestibulo-spinal function (VSF) is still a challenge, given the complexity of the system that governs *Posture*. There is no symptom and no clinical test or instrumented study that can assess VSF in a pure form. VSF is only a small component involved in *Posture Control* (PC). Of all the structures involved in postural control, two of them have vestibular origin: the vestibulo-spinal function via the vestibulo-spinal reflex (VSR) and the vestibulo-colic function via the vestibulo-colic reflex (VCR). It is also the duty of the otoneurologist to know the basic functioning of the medullary and subcortical structures that control tone and posture, as it allows him/her to correctly interpret many of the neurological alterations that affect tone and posture, to keep them in mind as differential diagnoses, and not to confuse them with alterations of the vestibular system.

4.2 Definitions

Following Bobath's concepts, we define POSTURE as the position of the body or part of the body in relation to the vector of gravity [1]. This is affected by physiological situations, such as age, sex, and environment, as well as various conditions ranging from the cerebral cortex to the osteo-musculo-articular system,

R. D'Albora (✉)
Faculty of Medicine, UDELAR, Montevideo, Uruguay

M. J. Mónaco (✉)
UMSA, Buenos Aires, Argentina

P. Ravat et al. (eds.), *Advances in Translational Neuroscience of Vestibular and Eye Movement Disorders*, Contemporary Clinical Neuroscience,
https://doi.org/10.1007/978-3-032-07539-0_4

including cognitive and psychological alterations. The function of POSTURE is to maintain permanent balance in relation to the vector of gravity, so that a series of motor responses can be generated to adapt to different situations and tasks. A good movement requires good posture. We define POSTURAL BALANCE as the state where the forces exerted on the body and their movements are zero; in other words, when there is an equitable distribution of body weight to maintain the center of gravity within well-defined limits, thus avoiding a fall [1]. This way, the body remains static or can develop movements without losing balance. POSTURAL CONTROL (PC) is the mechanism of sensory input, with complex central processing (organic-cognitive-motivational) and motor output, which has the characteristics of being predominantly unconscious and automatic, and is performed by permanent changes and adjustments of muscle tone [1, 2].

4.3 Physiological Considerations of Postural Control

To understand PC, it is mandatory to understand the phylogenetic development of the anatomo-functional structures of the Neural System (NS) [1], especially the components of the extrapyramidal systems, as this ultimately initiates, regulates, and refines (movement programming) movement [3]. The spinal cord (SC) was the first structure to participate in PC in primitive beings. In humans, as in most living organisms, the SC operates as a reflex center and a relay center [4].

From a phylogenetic point of view, SC is the oldest organ of the central nervous system that we have; at the beginning of evolution the SC only acted as a *reflex center*; in fact, those primitive beings only managed with those few and stereotyped sensorimotor reflexes, which, when working together, obtained motor responses that allowed them to defend themselves from injuries (flexor reflex, scratching reflex, cross-extension reflex); it also allowed them to have an antigravitational posture (extensor myotatic reflex, extensor thrust reflex), and finally to walk thanks to combinations of the above reflexes [4]. Although these types of reflexes are studied as segmental reflexes (horizontal portion of the SC), they are actually intersegmental, i.e., they combine more than one segment of the SC, which means that when the information reaches the dorsal root it is distributed through several segments of the SC, while the interneuron outputs are carried out in several segments before contacting the motor neuron of the anterior horn (common final pathway of movement), via the SC's own bundle (a phylogenetically more primitive bundle). Segmental reflexes can be classified according to their origin as follows: *exteroceptive* and *proprioceptive*. As living beings evolved, supraspinal nerve structures appeared, whose function was to modify the stereotyped sensorimotor integration circuits of the SC, thus shaping the segmental activity, allowing a great variety and fineness of movements; due to this, a series of bundles appeared in the structure of the SC that allowed the execution of these higher functions (descending motor bundles), and in turn ascending bundles that would allow these higher structures to control what was being carried out (ascending sensory bundles), with the SC becoming not only a

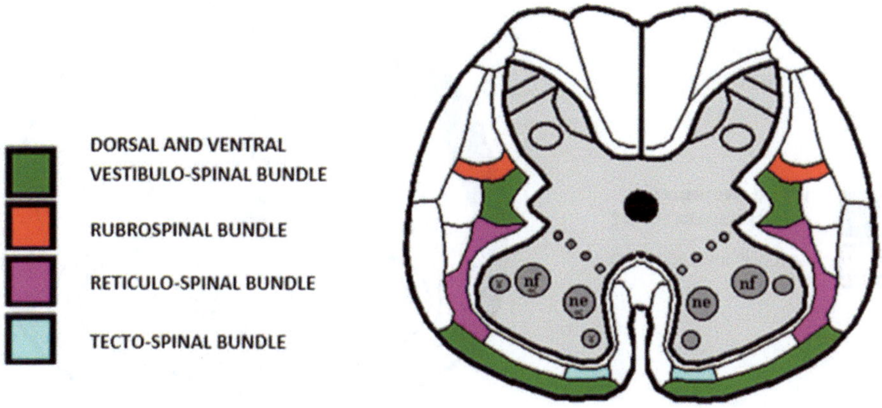

Fig. 4.1 Descending bundle involved in posture and balance

reflex center, but also a relay center to the phylogenetically more modern and higher structures [3].

The *descending bundles* originate in the brainstem nuclei and are transmitted to the SC through axons, thus forming the descending bundles, which in the SC are all motor (Fig. 4.1). All descending influences are mainly on the gamma motor neurons, requiring the integrity of the gamma loop for them to act [3, 4].

4.3.1 Descending Bundles

Vestibulo - Spinal Bundle: It originates at the level of the vestibular nuclei. The ventral or lateral bundle originates in the lateral vestibular nucleus, and the dorsal or medial bundle is formed by fibers that come from the contralateral inferior vestibular nucleus and the ipsilateral medial vestibular nucleus, so the information is direct and crossed. The lateral vestibulo-spinal bundle is involved in maculospinal reflexes. The medial vestibulo-spinal bundle is involved in canaliculo-spinal reflexes, and it maintains ipsilateral antigravity tone. At the neck level, it is an extensor of the neck, and it is responsible for all static and dynamic vestibular reflexes [3]. *Rubrospinal Bundle:* It originates at the level of the contralateral red nucleus; these are located at the level of the calotte of the cerebral peduncles; the bundles originating there decussate and descend through the trunk and then through the anterolateral cord of the SC. From a functional point of view, it can be considered as the antagonist of the vestibulo-spinal bundle, since it predominates in antigravitational tone by stimulating the flexor gamma motor neurons. The afferents to the red nucleus involved in posture and gait are: the pallidorubral, the corticorubral, and the tectorubral tracts; here we see how the gray nuclei of the base, the neurons of the cerebral cortex, and the vision, respectively, participate in the tone of the spinal musculature and the cranial pairs [3]. *Reticulo-spinal bundle:* It originates at the level of the reticular

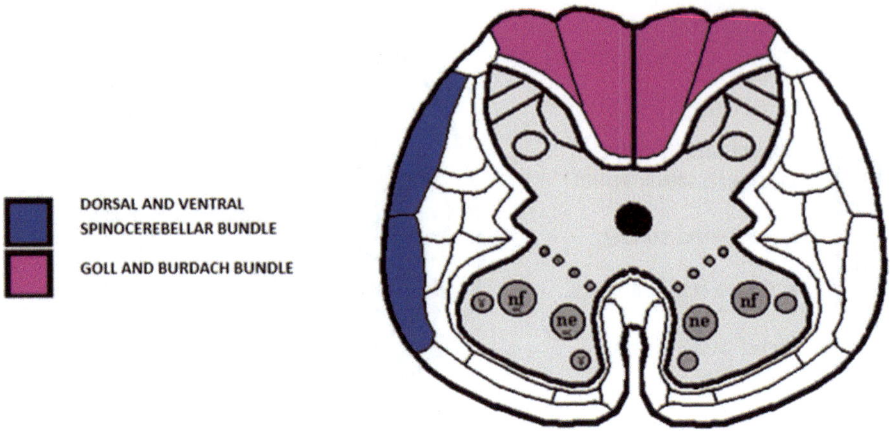

Fig. 4.2 Ascending bundles involved in posture and balance

substance; it seems that there are two distinct sectors: a superior sector whose function is to facilitate pro-G (flexor) neurons and an inferior bulbar sector of opposite motor action [3]. *Tecto-spinal bundle:* It originates at the level of the contralateral superior quadrigeminal nucleus. The fibers, on their descent, give collaterals to the reticular substance. The function is to coordinate the movements of the head and neck in relation to primary visual responses, optic straightening reflexes, as well as those of auditory origin [3].

The ascending bundles, typically studied in posture and gait, are the spinocerebellar bundles and the posterior cord bundle of the SC. Also, the spinothalamic bundles are involved, which conduct the sensation of thermoalgesia and pressure of the lower limbs. The most important ascending bundles in the assessment of balance are the spino-cerebellar bundles, which conduct unconscious deep sensitivity, and the Goll and Burdach bundles, responsible for conducting conscious deep sensitivity and epicritical tactile sensitivity [5] (Fig. 4.2).

4.3.2 Ascending Bundles -Spinocerebelar Bundles (SCB)

They originate at the level of the muscle receptors, neuromuscular spindles (NMS), and Golgi tendon organs (GTO); the impulses are conducted by the ia and ib. fibers, respectively. Once the fibers penetrate the SC, they synapse with the second neuron found in Clarke's column; from there, the fibers that originated the posterior bundle do so ascending ipsilaterally, while those that formed the anterior bundle decussate in the midline. The posterior or dorsal SCB ascends directly and enters the cerebellum through the inferior cerebellar peduncle. The anterior or ventral SCB ascends contralaterally; when it reaches the cerebral peduncle, it decussates again and enters the cerebellum through the superior cerebellar peduncle. The function of these bundles is to report in real time the state of all muscles to the cerebellum in an unconscious way. *Goll and Burdach bundles:* They originate at the level of muscle and tendon exteroceptors, as well as at the touch receptors. Impulses are conducted from the muscles and tendons. The fibers ascend through the spinocerebelar bundles in search of the second neuron located in the nuclei of Goll and Burdach. From there, the axons organize themselves and form the medial lemniscus, heading to the thalamus in search of the third neuron, and finally ascend to the primary sensory cerebral cortex. The function of these bundles is to transmit deep conscious sensation and epicritical touch. *The cerebellum* regulates movement and tone using the same strategy. It acts indirectly on the vestibular nuclei by inhibiting or suppressing ipsilateral inhibition, or through the contralateral red nucleus by stimulating or suppressing stimulation on it. No cerebellar efference synapses on motor neurons [6]. *The gray nuclei of the base* participate in muscle tone indirectly through the corticobulbar and corticospinal tracts (effectors of the modulation imparted by the gray nuclei), and directly through their action on the red nucleus and the reticular substance [7].

4.4 Pathological Considerations from the Vestibular Perspective

The lesions that alter postural control are many, and they range from musculoskeletal disorders to the cerebral cortex. The vestibular system is only one part of this complex system, which through the five inner ear sensors (three semicircular canals, the utricle and the saccule) is able to stabilize the gaze when there are rapid head turns [8], to stabilize the neck when there are head movements in order to minimize eye movements, to optimize the Vestibulo-Occulomotor Reflex (VOR) and finally to maintain the body's center of gravity within precise limits, also when there are head movements [9]. Acute peripheral vestibular lesions (posterior labyrinth and vestibular nerve) affect both VSF and vestibulo-occulomotor function (VOMF), because projections go to the oculomotor and spinal pathways [10]. Therefore, there are no pure VSF lesions in peripheral conditions; in other words, an acute VSF lesion without a condition in the VOMF is central [10]. Compensations of acute peripheral lesions can be complete or incomplete, both of the VOMF and of the VSF. Compensations of the VOMF (VOR) are usually seen, but not of the VSF (VSR) and vice versa [11]. In daily practice, we observe three possible instrumented clinical patterns of patients:

1. VOR and posturography improvement.
2. VOR improvement and no improvement in posturography.
3. Posturography improvement without VOR improvement.

4.5 Clinical Assessment of Vestibulo-Spinal Function

A correct clinical assessment of VSF requires an evaluation of the other systems involved in postural control, since there is no specific symptom of VSF. An evaluation of the visual, proprioceptive, sensory, cognitive, and osteoarticular systems is mandatory (it is not the aim of this work to assess these areas). The alterations of the VSF will present symptoms and signs [12–15]:

Symptoms: We can divide them into major and minor symptoms according to the discomfort they generate in the patient, and depending on the etiology or the degree of compensation, these may be static or dynamic. Static symptoms are those that are present even though the patient does not move, while dynamic symptoms only appear when there is body movement [12, 14, 15].

Major Symptoms: They are static (s) and dynamic (d):

1. Postural alteration (s).
2. Pulsation (s).
3. Falls (s & d).

Minor Symptoms: They are dynamic only:

1. Dynamic instability (uni or bilateral).
2. Daily life symptoms (e.g., unsteadiness in the shower, not being able to put on shoes while standing, dressing in the dark, etc.)

4.6 Clinical Signs That Assess Vestibulo-Spinal Function

Throughout history, many signs that assess VSF have been described; many of them have not passed the filter of time, and some have been preserved to date [13–15]. Like symptoms, signs can be divided into static and dynamic.

Static Signs:

1. Romberg.
2. Bárány's Test.
3. Unipodal Test.
4. Balance test under adverse situations (clinical posturography under different conditions).

Dynamic Signs:

1. Exploration of gait.
2. Unterberger-Fukuda.
3. Star gait.
4. Dynamic Index Test.
5. Miniconi Spinal Impulsive Test.

In addition to the classic signs mentioned above, other more recent signs involving the vestibulo-spinal pathway provide relevant information for the approach of patients with vestibular dysfunction. Among them, we find truncal ataxia and cephalic tilt, components of the triad: ocular tilt reaction.

4.6.1 *Truncal Ataxia*

Ataxia is defined as a disorder characterized by a decreased ability to coordinate movements [16].

To explore ataxia and for it to have vestibular value, we must ask the patient to walk; since the publication of our work on acute vestibular syndromes, [17] we have adopted the very simple and easy-to-apply classification proposed by Lee and Baloh [18], which is divided into three grades:

- Grade 1: Instability when walking, but the patient can walk without help.
- Grade 2: Instability when standing, unable to walk without assistance.
- Grade 3: The patient cannot stand or walk.

Always speaking in the context of acute vestibular syndromes, we can state that

- If the patient has vertigo and does not have ataxia, he/she does NOT have a central lesion.
- If the patient has vertigo and has grade I ataxia, there is a very low possibility that the cause is central.
- If the patient has vertigo and has grade II ataxia, it can be of peripheral or central cause, and other signs are useful.
- If the patient has vertigo and grade III ataxia, it is ALWAYS central [17, 19, 20].

The data of the last paper published by Dr. Jorge Kattah and Dr. Sergio Carmona show that finding a grade 2–3 truncal ataxia has a sensitivity of 92% (95% CI 79–100%) and a specificity of 67% (95% CI 47–86%), managing to detect most patients with central lesions, although at the cost of sacrificing specificity to some extent. But this is justified considering that the most important thing in acute vestibular syndrome is not to lose patients with central lesions. In addition, in this work it was found that the evaluation of truncal ataxia with the patient seated (incubitus ataxia) is a good substitute for standing truncal ataxia [21].

So, at both ends of the spectrum, we can be sure of the significance of the reason for consultation and can make decisions.

4.6.2 *Cephalic Tilt*

The cephalic tilt test is useful in a number of clinical situations. From the semiological point of view, we recommend doing it by first observing the patient in the primary position and then asking him/her to close their eyes [22]. With the mobile application, both situations can be photographed for a more detailed comparison, although in our experience, they are visible to the naked eye.

In general, when fixation is removed, which occurs when the patient is asked to close his/her eyes, the magnitude of the cephalic tilt increases in the case of vestibular lesion, while it disappears in the tilt due to oculomotor cause (paresis of the IV pair). In the context of an ocular tilt reaction [23] it will be towards the affected side if the condition of the vestibular graviceptic pathway is peripheral and towards the opposite side if the lesion is central, above the decussation at the level of the pons.

This test is localizing, but unfortunately, its sensitivity and specificity to distinguish between peripheral or central lesions have not been studied.

Below, we will describe the classic Bárány Test:

4.6.3 Bárány

This classic test of the deviation of the Bárány indexes [24, 25] consists of asking the seated patient to point his/her index fingers at the examiner's index fingers and asking him/her to close his/her eyes. If the patient has lower vestibulo-spinal tone on one side, he/she will deviate to that side (unilateral deviation) with both upper limbs (conjugate deviation); this has a number of advantages over other vestibulo-spinal examination strategies:

It avoids the presence of proprioception in the lower limbs, as it occurs in the Romberg test, which can be altered when deep sensibility is compromised or there is weakness in the legs.

It does not have, according to our experience, false positives as it often occurs with the Fukuda or Unterberger test.

It does not need wide spaces to be performed as in the case of the Babinsky-Weil star gait.

This test is considered positive when the patient's fingers deviate or lateropulse >1 cm towards the affected side when it is peripheral, and the deviation can be disconjugate (only one of the fingers deviates) or towards the opposite side of the lesion in central cases.

Conclusion: This test adds to the ataxia examination a side diagnosis, allowing a topodiagnosis of a genuinely vestibular lesion.

4.7 Instrumented Assessment

So far, the only instrumented and validated tool that can differentiate VSF from proprioceptive and visual function is POSTUROGRAPHY.

Posturography:

The attempt to measure balance in the human being is a very old concept, and many ideas have elapsed since the end of the nineteenth century. Different equipment has been described and used, the first based on physical and photographic methods, such as cranio-corpography; then, with the advent of pressure sensors connected to computer systems came *posturography*, based on the pressure paradigm [26] and considered the "Gold Standard." Currently, angular *posturographs* are being developed by means of gyroscopes and accelerometers, which facilitate measurement in terms of recording time, cost, and practicality for the patient and the professional [27]. *Posturography* is an OBJECTIVE method to evaluate and quantify the functional postural situation of patients. It is of little help in the etiological or specific diagnosis of diseases, for which other types of studies are available. Conceptually, *posturography* should be seen as the final common pathway or the functional result of the interactions of the complex system of balance, assessing at the same time both its afferences (proprioceptive- visual-vestibular) and its central processing, its efferences (osteo-muscular), and also the emotional and cognitive aspects, defining the role played by each one of them.

4.8 The Emergence of Evidence-Based Medicine in Diagnosis

For years, evidence-based medicine has invaded all aspects of medical practice, not only in treatment, but also in diagnosis. In the latter aspect, the need for accurate diagnostic tests that can be used in situations where practitioners have little time to check patients has been demonstrated. In the case of vestibular pathology, much has been studied on vestibulo-ocular signs in the differential diagnosis, but research on vestibulo-spinal signs is still scarce. Below, we review some studies that have evaluated some tests, taking into account the principles of evidence-based medicine:

Cohen H, Stitz, J, Sangi-Haghpeykar H, Williams S, Mulavara A, Peters B, Bloomberg, J. (2017). Tandem walking as a quick screening test for vestibular disorders: Tandem Walking for Vestibular Screening. The Laryngoscope. 128. 10.1002/lary.27022. Study design: Case control. Level of evidence 3b. The goal of this study was to test an updated version of an earlier test, tandem gait, to determine how useful it is for screening patients with vestibular disorders. The level of sensitivity and specificity of this test, using the dependent measure of interest in this study (the total number of correct tandem steps), would not help the clinician distinguish between peripheral and central vestibular disorders.

Schmidheiny A, et al. *Discriminant validity and test re-test reproducibility of a gait assessment in patients with vestibular dysfunction. BMC Ear, Nose and Throat Disorders (2015) 15:6.* Study design: Case control. Level of evidence: 3b. The purpose of the study was to investigate the discriminatory validity of a gait assessment protocol in patients with vestibular disorders and healthy participants.

The good validity and reproducibility of the measurements of the GAITRite® walkway (for temporal and spatial parameters of gait) suggest that this system could facilitate the study of gait in patients with vestibular disorders in different clinical settings and vestibular rehabilitation programs.

Cohen HS, Kimball KT. Usefulness of some current balance tests for identifying individuals with disequilibrium due to vestibular impairments. J Vestib Res. 2008;18 (5, 6):295–303. Study design: Case control. Level of evidence: 3b. The goal of this study was to determine which of several clinical balance tests best identifies patients with vestibular disorders. We compared scores of normal subjects and vestibular patients on the Berg Balance Scale, Dynamic Gait Index (DGI), Timed Up and Go (TUG), Computerized Dynamic Posturography Sensory Organization Test (SOT), and a new obstacle avoidance test: Functional Mobility Test (FMT). The main outcome measures were test sensitivity to patients and specificity to normal subjects. When adjusted by age, Berg, TUG, DGI, and FMT had moderate sensitivity and specificity. SOT had moderately high sensitivity and specificity. SOT and FMT combined had high sensitivity and moderate specificity.

Swanenburg J, Hegeemann SC, Zurbrugg A, Palla A, de Bruin ED. NeuroRehabilitation. 2014;34(4):799–807. Study design: Case control. Level of evidence: 3b. The aim of the study was to evaluate the reliability and validity of the extended test, modified version "get up and go" (ETGUG) (Wall, Bell, Campbell, & Davis, 2000) in patients with chronic bilateral vestibular loss (BVL) and healthy controls. This study has demonstrated high reliability of several of the test components, with excellent ICC (Intraclass Correlation Coefficient), small SEMs (Standard Error of the Mean), and SDC (Smallest Detectable Change) in adults with BVL. The results for the discriminative abilities were essentially conclusive, with the exception of gait initiation (Task 2). Thus, there are indications that the ETGUG can distinguish between patients with BVL and healthy individuals.

4.9 Final Thoughts

Based on our experience, we believe that the best way to isolate VSF is to combine a careful clinical examination with posturographic evaluation, based on anatomofunctional and pathological concepts. Please add the references 16 to 26 and shift references 16 and 17 to reference 27 and 28.

References

1. Soria Ayuda RE, Blas Martínez A, Jaime Sánchez A, Villarroya Bielsa E, Iglesias Triviño AR, Laborda Lalaguna G. El control postural según el concepto Bobath. Monographic article. June 18 2021. Revista Sanitaria de investigación.

2. Duclos C, Mesure S. Control postural: fisiología, conceptos principales e implicaciones para la readaptación. EMC Kinesiterapia. 2017;38(2):1–9.
3. Nova S. Sistema extrapiramidal. www.kenhub.com (2023).
4. Andrada B. Sistema motor tono muscular. In Temas de neurofisiologia en fonoaudiología. Ed Akadia BsAs. Argentina. 2020. pp. 19–22.
5. Serrano C. Medula espinal. https://www.kenhub.com/es/library/anatomia-es/medula-espinal-es (2023)
6. Zalazar Guillermo. El Cerebelo y sus funciones. Chapter 4B. Temas de neurofisiologia en fonoaudiología. Ed Akadia BsAs. Argentina; 2020. pp. 29–30.
7. Ferrri María Eugenia. Fisiología de los ganglios basales. Chapter 4ª. Temas de neurofisiologia en fonoaudiología. Ed Akadia BsAs. Argentina; 2020. pp. 23–27.
8. Zalazar Guillermo, Carmona Sergio. Chapter 11: Neurofisiología vestibular. Temas de neurofisiologia en fonoaudiología. Ed Akadia BsAs. Argentina; 2020. pp. 61–67.
9. Leigh RL, Zee DS, Ojo-cabeza M. La neurología de los movimientos oculares. 5th ed. Oxford University Press; 2015. https://doi.org/10.1093/med/9780199969289.001.0001.
10. Zalazar Guillermo, Carmona Sergio. Bases anatomicas y fisiológicas del SVA Chapter 4. Manejo del Sindrome Vestibular Agudo 2nd Ed. Ed Akadia BsAs; Argentina. 2023. pp. 53–62.
11. Martín-Bailón M, Yáñez-González R, Sánchez-Gómez H, Sánchez-Blanco C, González-Sánchez M, Martín-Sánchez V, Coscarón-Blanco E. Compensación vestibular. Revista ORL. 2020;11(1):19–28.
12. Ceballos. R. La historia clínica en el vértigo agudo Cap. 2 Manejo del Sindrome Vestibular Agudo 2nd Ed. Ed Akadia BsAs. Argentina; 2023, pp. 15–29.
13. Zuma e Maia F; Mangabeira P; Carmona S; Cal R. Chapter 3 Manejo del Sindrome Vestibular Agudo. Evaliuación clínica del paciente con vértigo. 2nd Ed. Ed. Akadia BsAs; 2023. pp. 31–52
14. Mónaco MJ. Chapter Metodos de estudio de las patologías vestibulares periféricas y su influencia en la rehabilitación vestibular. Aportes a la Audiología. Ed. Nueva Editorial Universitaria NEU. BsAs Argentina; 2021. pp. 297–328.
15. D'Albora. R. Chapter 3. Oído (2nd Part): Laberinto Posterior. Semiologia ORL y Semiología de Cabeza y Cuello. Oficina del Libro FEFMUR. 1st Ed. Montevideo Uruguay; 2004. pp. 83–151.
16. Biller J, Gruener G, Brazis P. DeMyer's the neurologic examination – a programmed text. Editorial Mc Graw Hill Education. Asociación Madrileña de Ataxias Hereditarias. 2017. https://www.asociacionmadrilenadeataxias.org/sobre-la-ataxia/. [Cited on 16/05/2020].
17. Carmona S, Martinez C, et al. The diagnostic accuracy of Truncal Ataxia and HINTS as cardinal signs for acute vestibular syndrome. Front Neurol. 2016;7:125. https://doi.org/10.3389/fneur.2016.00125.
18. Lee H, Sohn SI, Cho YW, et al. Cerebellar infarction presenting isolated vertigo: frequency and vascular topographical patterns. Neurology. 2006;67(7):1178–83.
19. Kattah JC, Talkad AV, Wang DZ, et al. HINTS to diagnose stroke in the acute vestibular syndrome: three-step bedside oculomotor examination more sensitive than early MRI diffusion-weighted imaging. Stroke. 2009;40:3504–10. Available at: https://www.ahajournals.org/doi/full/10.1161/STROKEAHA.109.551234
20. Vanni S, Pecci R, Edlow J, et al. Differential Diagnosis of Vertigo in the Emergency Department: A Prospective Validation Study of the STANDING Algorithm. Front Neurol. 2017;8:590. https://doi.org/10.3389/fneur.2017.00590.
21. Kattah JC, Martinez C, Zalazar G, Batuecas Á, Lemos J, Carmona S. Role of incubitus truncal ataxia, and equivalent standing grade 3 ataxia in the diagnosis of central acute vestibular syndrome. J Neurol Sci. 2022;441:120374. https://doi.org/10.1016/j.jns.2022.120374. Epub 2022 Aug 9
22. Carmona S, Martino N.: Differential diagnosis between fourth nerve palsy and ocular tilt reaction based on the observation of the cephalic tilt. OcularMotor Disorders, 155–159. Neuroophthalmology at the beginning of the new millennium, Editor James A. Sharpe. Medimond; 2000

23. Carmona S, Asprella Libonati G.: Neuro-Otología. Editorial Akadia, 4th ed, 2018; Vertigo, BT: Its multisensory syndromes. Springer editorial, 2nd ed, 2003.
24. Bárány R. Neue Untersuchungsmethoden, die Beziehungen zwischen Vestibularapparat, Kleinhirn, Grosshirn and Ruckenmark betreffend. Wien Med Wochenschr. 1910;60:2033.
25. Biller J, Gruener G, Brazis P. DeMyer's the neurologic examination – a programmed text. Editorial Mc Graw Hill Education; 2017.
26. Rama López J, Pérez Fernández N. Pruebas vestibulares y posturografía. RevMed Univ Navarra/vol. 2003;47(4):21–8.
27. Brock CM, et al. How to study the effect of vestibular activation on Vestibulo-spinal pathway using portable Posturography. EC Neurol. 2020;12(10):50–7.

Chapter 5
Advanced Diagnostics in Acute Vestibular Syndromes: Integrating Video-Oculography

Georgios Mantokoudis

Keywords Acute vestibular syndrome · Video-oculography · vHINTS · Artificial intelligence · Vertigo · Emergency neurology · Stroke diagnosis · Telemedicine · Machine learning

Chapter Overview

In recent years, video-oculography (VOG) has emerged as a critical tool for the objective evaluation of patients presenting with acute vestibular syndrome (AVS), particularly in emergency settings where distinguishing between peripheral vestibular disorders and central causes, such as ischemic strokes, is vital. The diagnostic challenge posed by AVS lies in its often-ambiguous clinical presentation, with overlapping symptoms between vestibular neuritis and posterior circulation strokes.

This chapter will delve into the advanced diagnostic approaches for acute vestibular syndrome (AVS), with a focus on the utilization of modern technologies such as Video-oculography (VOG), artificial intelligence (AI), and the application of tests that extend beyond the traditional HINTS (Head Impulse, Nystagmus, Test of Skew) protocol.

5.1 Introduction

Patients with dizziness who present in emergency departments (EDs) often suffer from a variety of symptoms, including nausea, vomiting, and gait instability, which can indicate underlying vestibular syndromes [1]. The most critical condition within this group is the acute vestibular syndrome (AVS), characterized by sudden-onset, continuous vertigo, nystagmus, and postural instability lasting for more than 24 hours [2]. The challenge of diagnosing AVS patients is compounded by the

G. Mantokoudis (✉)
Department of Otorhinolaryngology, Head and Neck Surgery, Inselspital, Bern University Hospital, University of Bern, Bern, Switzerland

P. Ravat et al. (eds.), *Advances in Translational Neuroscience of Vestibular and Eye Movement Disorders*, Contemporary Clinical Neuroscience,
https://doi.org/10.1007/978-3-032-07539-0_5

fact that AVS can be caused by both benign peripheral vestibulopathies and dangerous central pathologies, particularly strokes [3].

Acute vestibular syndrome (AVS) can arise from various etiologies grouped into several key categories. Vascular causes, such as ischemic strokes [4] and vertebral artery dissections [5], are particularly critical due to their potential for life-threatening outcomes. Inflammatory and infectious causes, including acute unilateral vestibulopathy (AUVP) and brainstem encephalitis [6], are common. Autoimmune diseases [7], such as autoimmune inner ear disease (AIED), paraneoplastic [8], and neoplastic conditions like vestibular schwannomas [9], further expand the range of AVS causes. Traumatic [10], degenerative, and nutritional deficiencies [11] add complexity to the diagnostic process, underscoring the need for accurate differentiation between central and peripheral origins.

Approximately 20% of patients with dizziness presenting to a tertiary hospital were found to have AVS, and nearly one-quarter of these cases were due to strokes [1]. A timely and accurate diagnosis in these patients is important. Despite extensive workups, a significant portion of patients (45%) remain undiagnosed [1], reflecting the diagnostic challenges in this patient population. Current bedside evaluations, including the widely adopted HINTS (Head Impulse, Nystagmus, Test of Skew) protocol introduced by Jorge Kattah in 2009 [12], require substantial expertise. The HINTS protocol involves a three-step clinical assessment: the head impulse test (HIT), nystagmus, and the test of skew [12, 13]. Among these, the head impulse test is the most challenging to perform but is also the most powerful tool for detecting stroke [14]. This test has proven to be more sensitive for stroke detection than early MRI in AVS.

HINTS has a sensitivity of over 91% [14]; however, the diagnostic accuracy of the HIT is highly examiner-dependent [15–17], making it crucial to have trained professionals for reliable interpretation. The bedside head impulse test is an effective diagnostic tool when performed by experienced clinicians; however, distinguishing corrective saccades during the test can be challenging, potentially leading to false-positive or false-negative results [16, 17]. Corrective saccades are often misinterpreted as nystagmus beats, and covert saccades may be missed entirely, complicating the accuracy of the assessment [16, 17]. Experts often underestimate borderline abnormal head impulses, frequently classifying them as normal [15]. This cautious approach is understandable, as a false-negative result (incorrectly identifying a normal head impulse as abnormal) could have serious consequences, potentially resulting in an undiagnosed stroke.

The first use of video-oculography (VOG) in the ED was reported in 2012 by Newman-Toker et al., demonstrating a high accuracy in diagnosing AVS patients by objectively measuring the HIT (aka vHIT or video HIT) using specially designed goggles [18].

Since then, new generations of VOG goggles have been developed, allowing the objective measurement of all three components of the HINTS protocol [19, 20]—head and eye movements—without relying on the subjective interpretation of the examiner. These devices provide objective data on vHIT and vestibulo-ocular reflex (VOR) metrics, such as gain, and saccade metrics, like saccade latency and

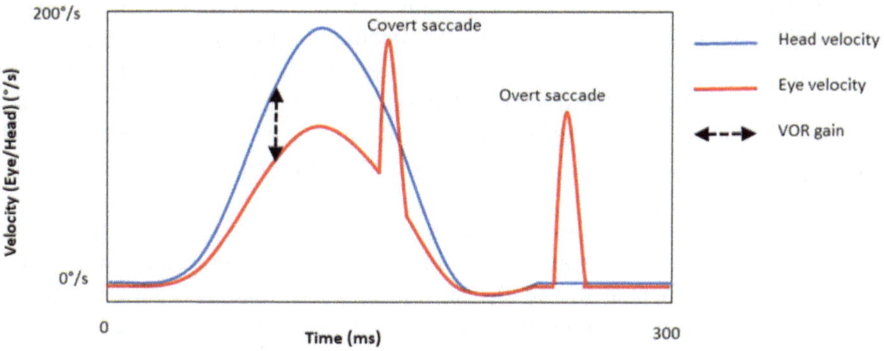

Fig. 5.1 A video head impulse recording showing head velocity traces (blue) and eye velocity traces (red). VOR gain can be calculated as the ratio between eye and head velocity. Corrective saccades can be analyzed in terms of latency, amplitude (velocity), or cumulative amplitude (sum of all saccade amplitudes)

cumulative amplitude, but the interpretation of the raw traces remains challenging [21], much like the evaluation of ECG waveforms (Fig. 5.1).

While fully automated and validated HINTS assessments using machine learning or artificial intelligence are not yet available, proof-of-concept studies have shown promising results for future applications [22]. The following sections will delve deeper into advancements in diagnostic technologies, focusing on the role of VOG in the emergency setting and its potential to improve the accuracy of AVS diagnosis.

5.2 The Video Head Impulse Test (vHIT)

In 2012, Newman-Toker et al. first introduced the use of video-oculography (VOG) in the emergency department to measure head impulses [18]. An important quantitative measure obtained from vHIT is the calculation of vestibulo-ocular reflex (VOR) gain, which represents the ratio of eye movement velocity (slow phase of the VOR) to head movement velocity during head impulses [23] (Fig. 5.1). The use of vestibulo-ocular reflex (VOR) gain measured by video head impulse testing (vHIT) has been investigated to differentiate between vestibular neuritis and stroke in patients with acute vestibular syndrome (AVS) [24]. Patients with vestibular neuritis exhibited significantly reduced VOR gain on the affected side, with a mean gain of 0.52. In contrast, patients with posterior inferior cerebellar artery (PICA) strokes showed bilaterally normal VOR gains, with a mean of 0.94, demonstrating the ability of vHIT to distinguish between these conditions based on VOR asymmetry. Anterior inferior cerebellar artery (AICA) stroke cases were more variable, showing heterogeneous gain patterns (bilateral normal, unilateral abnormal, or bilateral abnormal). The findings suggest that vHIT can be a valuable diagnostic tool for distinguishing peripheral vestibulopathies from central causes in AVS, particularly PICA strokes. The study reported an overall diagnostic accuracy of 90%, with 88% sensitivity and 92% specificity for detecting strokes when a VOR gain threshold of 0.70 was used. However, AICA strokes, which can affect several structures such as the inner ear (labyrinth), brainstem, and cerebellum, may still pose diagnostic challenges based on VOR gain alone [24]. Combining vHIT with other clinical assessments, such as skew deviation and nystagmus, can enhance diagnostic precision.

In addition, the application of vHIT by nonexperts, such as ED physicians or triage nurses, requires hands-on instructional training, as proper execution demands fast (impulse-like) manual (passive) head movements. This procedure involves a significant learning curve for the examiner to ensure accurate and reliable test results [16, 17].

Recent advancements have shown that the video head impulse test (vHIT) is a highly sensitive (91.7%) and specific (88.7%) tool for detecting vestibular strokes, providing a more reliable and examiner-independent method for distinguishing between peripheral and central causes of dizziness [25]. Another study [26] focused on the diagnostic accuracy of video head impulse testing (vHIT) as part of the VOG-assisted "HINTS"protocol (vHINTS) in differentiating between central and peripheral causes of acute vestibular syndrome (AVS). In a cohort of 46 patients, including 35 with acute unilateral vestibulopathy (AUVP) and 11 with strokes, both clinical HINTS (cHINTS) and VOG-enhanced HINTS (vHINTS) were performed, followed by MRI confirmation. The results demonstrated that vHIT alone had a sensitivity of 91% and specificity of 89% for detecting strokes. The diagnostic accuracy of HINTS was enhanced by allowing for objective, examiner-independent assessment of several eye movements: VOG-enhanced HINTS (vHINTS) outperformed clinical HINTS, achieving 100% sensitivity for detecting strokes in AVS patients, compared to 90.9% sensitivity for expert examiners performing bedside tests without VOG [19, 20]. The overall accuracy for vHINTS was 94.2%, outperforming the expert-conducted cHINTS, which had an accuracy of 88.3%.

These findings highlight the utility of VOG as an effective tool for detecting central causes in AVS, particularly strokes, providing clinicians with a valuable adjunct to traditional bedside examinations in settings with limited access to advanced imaging.

5.3 The Nystagmus Suppression Test

Historically, failure of fixation suppression (FFS) of spontaneous nystagmus has been suggested as a sign of central pathology, but its diagnostic accuracy in AVS remained uncertain. The traditional approach of evaluating nystagmus suppression at the bedside can be unreliable due to its subjective nature, leading to potential misdiagnosis. Since human beings are better at evaluating the fast phase of nystagmus (beating direction and amplitude), assessing very low or very high intensity nystagmus in darkness (no fixation) might be very challenging. Siegrist et al.

investigated the ability of clinicians to detect nystagmus fixation suppression at the bedside using a nystagmus simulation model [27]. Results showed that both novices and expert physicians had difficulty detecting small reductions in nystagmus velocity, with test sensitivity ranging from 19% to 65% for detecting differences of less than 2°/s. There was no significant difference in detection ability between experts and novices, highlighting the overall low accuracy of clinical nystagmus suppression tests, particularly for subtle changes, and underscoring the importance of using video-oculography (VOG) to improve diagnostic accuracy.

VOG offers an objective, quantifiable assessment for slow phase velocity of nystagmus, offering the potential for a more precise distinction between peripheral and central causes of dizziness. A prospective cross-sectional study evaluated the effects of visual suppression on nystagmus in 148 patients with AVS using VOG measurements [28, 29]. The nystagmus suppression occurred in both peripheral and central causes, which makes this test challenging and highly sensitive to false interpretations. The reduction in nystagmus velocity measured with VOG was significantly more pronounced in stroke patients. A reduction in nystagmus velocity of less than 2°/s under visual fixation had a diagnostic accuracy of 76.9% for stroke detection, with a sensitivity of 62.2% and specificity of 84.8% [28, 29]. Discerment of velocity changes of 2°/s is challenging and often beyond the sensitivity of a human examiner [27].

The OFI score (ocular fixation index), which represents the ratio of nystagmus suppression with visual fixation, proved variable and lacked diagnostic reliability, suggesting that a direct measurement of nystagmus velocity using VOG is a more valuable tool. The failure of fixation suppression (FFS), which is a qualitative bedside assessment, could not reliably differentiate between stroke and vestibular neuritis. Importantly, the study showed that patients with vestibular neuritis exhibited higher overall nystagmus slow phase velocity than stroke patients, further supporting the role of quantitative VOG in the diagnostic process. The study echoes findings from previous research that highlight the diagnostic limitations of bedside clinical tests in detecting subtle central pathologies without the aid of advanced tools like VOG.

Another study by Korda et al. [30, 31] explored how vertical smooth pursuit can impair the suppression of horizontal vestibular nystagmus, which is otherwise often masked by fixation. The authors demonstrated that during vertical pursuit, horizontal spontaneous nystagmus (SN) was enhanced, revealing underlying vestibular pathology that was less discernible during standard visual fixation. This suggests that vertical tracking can disengage fixation mechanisms, allowing for the detection of low-intensity nystagmus that may not be visible without video-oculography (VOG) or Frenzel lenses. The "vertical pursuit nystagmus test" has the potential as a bedside tool to uncover subtle nystagmus, providing a more accessible diagnostic method for clinicians without the need for specialized equipment. However, while this approach shows promise as a bedside tool, its diagnostic value and clinical applicability with VOG remain uncertain and require further investigation.

5.4 Gaze-Evoked Nystagmus (Nystagmus Test)

VOG and, in particular, videonystagmography (VNG) are diagnostic tools used to record and analyze various eye movements, specifically nystagmus. VNG records both the slow and fast phases of nystagmus, allowing for detailed assessment of its characteristics, including direction, intensity, and waveform. By providing a precise, objective measurement of nystagmus patterns and characteristics [32], VNG enables clinicians to distinguish between different types of vestibular and neurological disorders. VOG offers more accurate diagnoses than visual observation of nystagmus alone (VNS, videonystagmoscopy). The assessment of gaze-evoked nystagmus (GEN) is part of the HINTS protocol [12]. The presence of GEN is considered a central sign [33], but distinguishing between pathological and physiological GEN can be challenging, especially in the presence of spontaneous nystagmus (SN). It was found that physicians had greater difficulty detecting direction-changing nystagmus (also known as gaze-evoked nystagmus, GEN), whereas VOG accurately identified these nystagmus patterns [19, 20, 28, 29]. Physiologic GEN was clinically difficult to distinguish by nonexperts. From the clinical viewpoint, physiologic gaze-evoked nystagmus (aka end-point nystagmus) is a low-amplitude, low-frequency, unsustained nystagmus that appears at far lateral gaze (horizontal or upbeating), lacks a vertical component, and has minimal or absent rebound, with otherwise normal ocular motor function [33]. The criteria of physiologic GEN observed at the bedside are vague and nonquantitative.

Gaze-evoked nystagmus can be distinguished from physiologic GEN using VOG measurements and time constant calculations [28, 29].

In addition, VOG is a valuable tool for assessing GEN in patients with an acute vestibular syndrome [28, 29]. A recent study used video-oculography (VOG) to quantify GEN in 47 AVS patients, including 12 with strokes and 35 with vestibular neuritis, and compared them to 64 healthy subjects [28, 29]. The analysis aimed to differentiate pathological from physiological GEN based on slow-phase velocity (SPV), the time constant, and "null point" positioning: The "null point" in the context of gaze-evoked nystagmus (GEN) refers to the specific gaze direction where nystagmus is minimal or absent. When a patient with GEN moves his eyes to different positions, the intensity of the nystagmus can vary, often disappearing at a certain angle close to the midline (straight gaze), which is referred to as the null point (Fig. 5.2). The time constant refers to the rate at which the nystagmus decays after the eyes are held in an eccentric position (Fig. 5.2). It is a metric about the ability of the neural integrator to maintain stable eccentric gaze.

One-third of stroke patients exhibit a pathological GEN, characterized by a combination of SN in straight-ahead gaze and direction-changing GEN, consistent with a so-called Bruns' nystagmus [28, 29]. Interestingly, the "null point"—the gaze position where no nystagmus is observed—is shifted away from the midline, but the direction of the shift may vary. In some cases, the null point shifts ipsilesionally (toward the side of the lesion), while in others, it shifts contralesionally. This shift in the null point's position remains an intriguing finding, as it is still unclear whether it

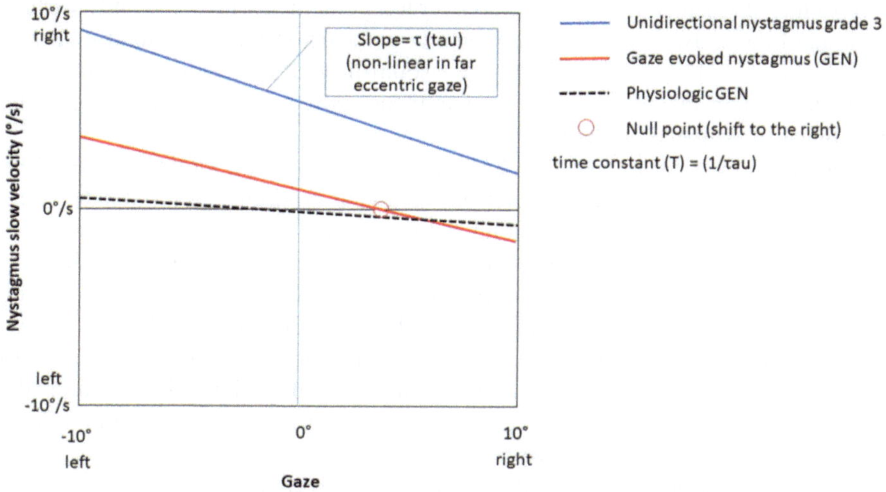

Fig. 5.2 Nystagmus intensity as a function of gaze

consistently indicates the side of the lesion. The variability suggests that multiple brainstem and cerebellar circuits contribute to gaze holding, making lesion localization based solely on the null point unreliable.

Pathological GEN can be reliably detected by VOG [28, 29]. The sensitivity for stroke detection is 33.3%, and the specificity of 100%. While the sensitivity is modest, the high specificity suggests that when GEN is present, it is a strong indicator of a central cause. Testing for GEN is part of the HINTS protocol and not meant to be applied as a standalone test. This automated method for GEN testing matched the diagnostic accuracy of expert clinicians, reinforcing the value of VOG for examiner-independent assessments.

5.5 Video Test of Skew

Video-oculography (VOG) can be used to quantitatively measure vertical ocular misalignment (skew deviation) by tracking eye movements during the cross-cover test. Various VOG methods are available for quantitatively measuring skew deviation. One approach involves using a fixation target displayed on a tablet while the subject wears VOG goggles equipped with color filters that alternately occlude each eye. The VOG system synchronizes the eye movement analysis with the timing of the occlusion, ensuring precise measurements. In this setting, eye positions are recorded while the target color alternates, allowing for precise saccade detection and skew measurement. This method was compared to the gold standard of the alternate prism cover test (APCT) performed by an orthoptist, showing a good correlation and an overall accuracy of 80.53% for VOG measurements [35]. Another approach is the manual alternate cross-cover test combined with VOG and synchronization of the software with a metronome or a manual trigger (e.g., mouse click at each change of the eye cover). A more sophisticated approach is the use of a binocular system recording both eyes simultaneously; however, such goggles are not clinically available on the market.

The study by Korda et al. [30, 31] aimed to evaluate the prevalence of skew deviation in patients presenting with acute vestibular syndrome (AVS) and to assess its diagnostic utility in distinguishing between peripheral and central causes of dizziness, particularly strokes. Skew deviation, a vertical misalignment of the eyes due to disruption of the graviceptive pathways, is traditionally considered a central sign, often included in the HINTS (Head Impulse, Nystagmus, Test of Skew) protocol for stroke diagnosis. Skew deviation appears to be present in 26% of AVS patients, including 24.5% of those with an acute unilateral vestibulopathy and 29.2% of those with stroke. Interestingly, the prevalence of skew in peripheral causes like an acute unilateral vestibulopathy is higher than previously reported, challenging the notion that skew deviation is exclusively a central sign. A vertical misalignment cut-off of 3.3° was identified, above which the likelihood of a stroke significantly increased [31, 32]. Notably, all patients with a skew deviation exceeding 3.3° were confirmed to have experienced a stroke. Clinically, the sensitivity in detecting a stroke was low (12.5%), but the specificity was high (98.1%). Similarly, the test of skew recorded with VOG showed a higher sensitivity (29.2%) but lower specificity (75.5%) compared to the clinical test [30, 31].

Despite these findings, skew deviation, whether detected clinically or via VOG, did not enhance stroke detection when used as a standalone test. The implementation of the HINTS protocol with VOG-enhanced test of skew (video test of skew) marginally increased sensitivity for stroke detection (87%) but reduced specificity (67.9%), suggesting that skew deviation adds little diagnostic value compared to more robust tests like the head impulse test and nystagmus analysis. In addition, while large skew deviations (>3.3°) are a strong indicator of stroke, the overall prevalence of skew in peripheral disorders like AUVP complicates its diagnostic utility. Skew deviation, when assessed using VOG, provides a quantitative, standardized measure, especially useful for nonexperts in emergency settings. However, its limited sensitivity and specificity suggest that it should be used in conjunction with other diagnostic tools rather than as a primary indicator of stroke.

5.6 Automated Assessment of VOG

Artificial intelligence (AI) and, in particular, machine learning (ML) can be utilized for the automated analysis of neurootological data [23, 36–41]. AI may automatically assess video head impulse test (vHIT) data to aid in the early diagnosis of stroke in AVS [19, 20, 22]. A recent study evaluated whether AI could classify stroke based on unprocessed vHIT data, comparing the results to the traditional method of calculating vestibulo-ocular reflex gain [19, 20]. Using a recurrent neural network, the AI model was trained on time-series data that captured both head and eye velocities during vHITs. The AI-based stroke classification achieved an accuracy of 87.9%, with a sensitivity of 87.7% and specificity of 88.4%. These results were comparable to stroke classification based on VOR gain, which yielded an accuracy of 91.2%, a sensitivity of 88.8%, and a specificity of 92.3%. There was no significant difference between the two methods.

A key advantage of the AI approach is its ability to analyze the entire velocity profile from vHIT, including both the slow phase of the VOR and compensatory saccades, without relying on a single VOR gain value. This comprehensive analysis allows for the identification of subtle features that may be missed in traditional assessments. AI has the potential to automate vHIT assessment in the emergency department, offering a promising tool for early stroke diagnosis in AVS patients, though larger datasets and further refinement of the neural network are needed to improve its performance.

AI can be trained to process large datasets of eye and head movement recordings after a first cleaning of raw data from artifacts like blinks and tracking errors. After data cleaning, key portions of eye movement, such as slow and quick phases, can be identified for analysis. Machine learning algorithms can then extract features, such as slow-phase velocity, saccades, and other metrics, to differentiate between physiological and pathological nystagmus or between peripheral and central causes of AVS.

A neural network, for example, can be trained using labeled data from expert clinicians to detect and classify patterns in VOG recordings, including the presence of abnormal vestibulo-ocular reflex (VOR) gains or corrective saccades. This training process involves feeding the system labeled datasets (e.g., stroke vs. nonstroke) and allowing the AI to learn the specific features that distinguish different pathologies. Over time, the AI becomes adept at making diagnostic predictions based on the extracted features, offering an automated, objective, and efficient way to assess VOG results.

Moreover, ML can integrate multiple data points from various tests (such as head impulse tests, nystagmus, and skew tests) to produce more accurate diagnostic outputs, supporting emergency physicians in distinguishing between strokes and benign peripheral vestibulopathies. The ultimate goal is to create AI-driven decision support systems that can process real-time VOG data and assist clinicians in delivering more timely and accurate diagnoses of AVS.

Federated learning is a cutting-edge approach in machine learning that allows multiple institutions to collaboratively train models while preserving data privacy. Unlike traditional machine learning models, which are often static and unable to adapt to new data, federated learning enables real-time learning across different institutions without directly sharing sensitive data. This is particularly important in healthcare, where privacy concerns limit the exchange of patient information. A notable example is the open-source platform for privacy-preserving distributed collaborative learning (DISCO: DIStributed COllaborative Learning, discolab.ai) developed by École Polytechnique Fédérale de Lausanne (EPFL), Switzerland. This open platform leverages federated learning to allow continual updates to models based on new data while ensuring that privacy is maintained. Federated learning is seen as a future cornerstone for digital health, facilitating dynamic, adaptive models that improve over time without compromising data privacy.

5.7 VOG and the Potential for Telemedicine

The use of video-oculography (VOG) in telemedicine presents a significant opportunity to enhance the diagnosis and management of patients with acute vestibular syndrome (AVS). VOG allows for the objective measurement of eye movements, enabling remote specialists to analyze data and assist in diagnostic decision-making without the need for in-person evaluations [42]. This approach is particularly beneficial in emergency settings where subspecialists may not be available. Through telemedicine, a standardized battery of VOG tests, including the HINTS examination and Dix-Hallpike maneuvers, can be administered remotely, allowing experts to review and interpret the results in real time. The ability to record and share VOG data electronically enhances diagnostic accuracy and reduces unnecessary imaging, streamlining care for patients with AVS.

5.8 Clinical Decision-Making with VOG

Triage in the emergency department often follows unclear rules that depend on factors such as the availability of specialists, accessibility to neuroimaging, and general risk stratification protocols. A more structured triage pathway (Fig. 5.3)

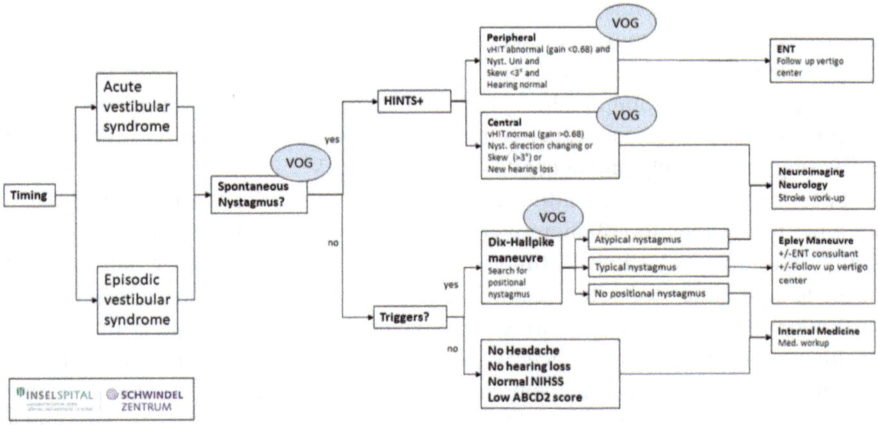

Fig. 5.3 Clinical triage pathway based on VOG assessments

could incorporate key features like the timing and triggers of symptoms, along with early video-oculography (VOG) testing, to guide clinical decision-making. The presence or absence of spontaneous or triggered nystagmus would play a pivotal role in determining the patient's pathway. For acute vestibular syndrome (AVS), spontaneous nystagmus is assessed using video-oculography (VOG). If nystagmus is present, the HINTS+ protocol (Head Impulse, Nystagmus, Test of Skew, and hearing) is employed to distinguish between central and peripheral causes.

The presence of spontaneous nystagmus with abnormal vHIT (vestibulo-ocular reflex gain <0.68) and normal hearing indicates a peripheral cause. In contrast, a normal vHIT, direction-changing nystagmus, positive skew deviation, or new hearing loss suggest a central cause, prompting further neurological evaluation and imaging (MRI, MRA) for potential stroke.

For episodic vestibular syndrome, triggered by positional changes, the Dix-Hallpike maneuver is performed to check for positional nystagmus recorded with VOG. If typical nystagmus is detected, benign paroxysmal positional vertigo (BPPV) is likely, and repositioning maneuvers are recommended. Atypical nystagmus or absence of positional nystagmus may indicate a central cause or another medical condition, requiring further investigation.

VOG plays a crucial role in the decision tree by providing objective, examiner-independent data on nystagmus and vestibulo-ocular reflex gain. It aids in distinguishing between central and peripheral causes of dizziness early in the triage process, thereby reducing diagnostic uncertainty and ensuring timely and appropriate management.

Clinical support systems based on VOG data (incorporating additional clinical data) in conjunction with machine learning might assist ED physicians in decision-making. In medical triage, machine learning models often face bias from missing data when certain tests or results are only available for specific patient groups or conditions. For example, if a particular test is only done for patients suspected of having a stroke, and not for other conditions, a decision-making model might incorrectly learn that simply having the test result means the patient is more likely to have a stroke. This is not because of the test result itself, but because the test was only done in stroke cases. This kind of bias can lead to incorrect conclusions when the model assumes the presence or absence of a test is more meaningful than it really is. To solve this, a system called MoDN (Modular Clinical Decision Support) was developed [43]. In MoDN, the system avoids this issue by treating each piece of data (symptoms, test results, etc.) independently in its own module. This way, the model does not make biased assumptions based on whether a test was done or not, but rather focuses on the actual values and information provided by each test or observation. This ensures more accurate and fair decision-making, even when some data is missing. Unlike traditional black-box models, MoDN is designed to be transparent. Healthcare professionals can see how each piece of data (e.g., a test result or symptom) contributes to the final decision. Another important application of MoDN is its ability to continuously learn from new data.

5.9 Conclusions

In summary, AVS is a critical condition that requires accurate differentiation between benign peripheral disorders and serious central causes, such as strokes. Traditional bedside protocols like HINTS are effective but rely on expert interpretation, leading to variability in diagnostic accuracy. VOG enhances diagnostic precision by objectively measuring eye movements and VOR, reducing examiner dependence. VOG improves the detection of central causes, particularly when combined with artificial intelligence. Integrating VOG into ED workflows and telemedicine might support timely, examiner-independent AVS diagnosis and improve patient outcomes, especially in resource-limited settings.

References

1. Comolli L, Korda A, Zamaro E, Wagner F, Sauter TC, Caversaccio MD, Nikles F, Jung S, Mantokoudis G. Vestibular syndromes, diagnosis and diagnostic errors in patients with dizziness presenting to the emergency department: a cross-sectional study. BMJ Open. 2023;13(3):e064057.
2. Bisdorff A, Von Brevern M, Lempert T, Newman-Toker DE. Classification of vestibular symptoms: towards an international classification of vestibular disorders. J Vestib Res. 2009;19(1–2):1–13.
3. Tarnutzer AA, Berkowitz AL, Robinson KA, Hsieh YH, Newman-Toker DE. Does my dizzy patient have a stroke? A systematic review of bedside diagnosis in acute vestibular syndrome. CMAJ. 2011;183(9):E571–92.
4. Kim JS, Newman-Toker DE, Kerber KA, Jahn K, Bertholon P, Waterston J, Lee H, Bisdorff A, Strupp M. Vascular vertigo and dizziness: diagnostic criteria. J Vestib Res. 2022;32(3):205–22.
5. Horii A, Okumura K, Kitahara T, Kubo T. Intracranial vertebral artery dissection mimicking acute peripheral vertigo. Acta Otolaryngol. 2006;126(2):170–3.
6. Smiatacz T, Kowalik MM, Hlebowicz M. Prolonged dysphagia due to listeria-rhombencephalitis with brainstem abscess and acute polyradiculoneuritis. J Infect. 2006;52(6):e165–7.
7. Lee SU, Kim HJ, Choi JY, Kim JK, Kim JS. Acute vestibular syndrome associated with anti-GQ1b antibody. Neurology. 2019;93(11):e1085–92.
8. Moreno-de-Jesús C, Prieto-Sánchez-de-Puerta L, Mármol-Szombathy I, Domínguez-Durán E. Paraneoplastic acute vestibular syndrome: what, when and how. Acta Otorrinolaringol Esp. 2021;
9. Sunara D, Krnić Martinić M, Lovrić Kojundžić S, Marčić L. Vestibular neuronitis in a vestibular schwannoma patient. Auris Nasus Larynx; 2021.
10. Vibert D, Häusler R. Acute peripheral vestibular deficits after whiplash injuries. Ann Otol Rhinol Laryngol. 2003;112(3):246–51.
11. Kattah JC, Dhanani SS, Pula JH, Mantokoudis G, Tehrani ASS, Toker DEN. Vestibular signs of thiamine deficiency during the early phase of suspected Wernicke encephalopathy. Neurol Clin Pract. 2013;3(6):460–8.
12. Kattah JC, Talkad AV, Wang DZ, Hsieh YH, Newman-Toker DE. HINTS to diagnose stroke in the acute vestibular syndrome: three-step bedside oculomotor examination more sensitive than early MRI diffusion-weighted imaging. Stroke. 2009;40(11):3504–10.
13. Kattah JC. Use of HINTS in the acute vestibular syndrome. An overview. Stroke Vasc Neurol. 2018;3(4):190–6.

14. Newman-Toker DE, Kattah JC, Alvernia JE, Wang DZ. Normal head impulse test differentiates acute cerebellar strokes from vestibular neuritis. Neurology. 2008;70(24 Pt 2):2378–85.
15. Jorns-Haderli M, Straumann D, Palla A. Accuracy of the bedside head impulse test in detecting vestibular hypofunction. J Neurol Neurosurg Psychiatry. 2007;78(10):1113–8.
16. Korda A, Carey JP, Zamaro E, Caversaccio MD, Mantokoudis G. How good are we in evaluating a bedside head impulse test? Ear Hear. 2020a;41(6):1747–51.
17. Korda A, Sauter TC, Caversaccio MD, Mantokoudis G. Quantifying a learning curve for video head impulse test: pitfalls and pearls. Front Neurol. 2020b;11:615651.
18. Newman-Toker DE, Saber Tehrani AS, Mantokoudis G, Pula JH, Guede CI, Kerber KA, Blitz A, Ying SH, Hsieh YH, Rothman RE, Hanley DF, Zee DS, Kattah JC. Quantitative video-oculography to help diagnose stroke in acute vertigo and dizziness: toward an ECG for the eyes. Stroke. 2013;44(4):1158–61.
19. Korda A, Wimmer W, Wyss T, Michailidou E, Zamaro E, Wagner F, Caversaccio MD, Mantokoudis G. Artificial intelligence for early stroke diagnosis in acute vestibular syndrome. Front Neurol. 2022a;13:919777.
20. Korda A, Wimmer W, Zamaro E, Wagner F, Sauter TC, Caversaccio MD, Mantokoudis G. Videooculography "HINTS" in acute vestibular syndrome: a prospective study. Front Neurol. 2022b;13:920357.
21. Mantokoudis G, Saber Tehrani AS, Wozniak A, Eibenberger K, Kattah JC, Guede CI, Zee DS, Newman-Toker DE. Impact of artifacts on VOR gain measures by video-oculography in the acute vestibular syndrome. J Vestib Res. 2016;26(4):375–85.
22. Ahmadi SA, Vivar G, Navab N, Möhwald K, Maier A, Hadzhikolev H, Brandt T, Grill E, Dieterich M, Jahn K, Zwergal A. Modern machine-learning can support diagnostic differentiation of central and peripheral acute vestibular disorders. J Neurol. 2020;267(Suppl 1):143–52.
23. Zamaro E, Saber Tehrani AS, Kattah JC, Eibenberger K, Guede CI, Armando L, Caversaccio MD, Newman-Toker DE, Mantokoudis G. VOR gain calculation methods in video head impulse recordings. J Vestib Res. 2020;30(4):225–34.
24. Mantokoudis G, Tehrani AS, Wozniak A, Eibenberger K, Kattah JC, Guede CI, Zee DS, Newman-Toker DE. VOR gain by head impulse video-oculography differentiates acute vestibular neuritis from stroke. Otol Neurotol. 2015;36(3):457–65.
25. Morrison M, Korda A, Zamaro E, Wagner F, Caversaccio MD, Sauter TC, Kalla R, Mantokoudis G. Paradigm shift in acute dizziness: is caloric testing obsolete? J Neurol. 2021;
26. Athanasia K, Wilhelm W, Ewa Z, Franca W, Thomas C Sauter, Marco D Caversaccio, Georgios MM (2022) Videooculography HINTS in Acute Vestibular Syndrome: A Prospective Study. Front Neurol 12:13:920357. eCollection PMID: 35903121 PMCID: PMC9314570. https://doi.org/10.3389/fneur.2022.920357
27. Siegrist S, Wyss T, Korda A, Mantokoudis G. Clinical assessment of the nystagmus fixation suppression test: an experimental study. J Int Adv Otol. 2024;20(3):241–6.
28. Mantokoudis G, Korda A, Zee DS, Zamaro E, Sauter TC, Wagner F, Caversaccio MD. Bruns' nystagmus revisited: a sign of stroke in patients with the acute vestibular syndrome. Eur J Neurol. 2021a;
29. Mantokoudis G, Wyss T, Zamaro E, Korda A, Wagner F, Sauter TC, Kerkeni H, Kalla R, Morrison M, Caversaccio MD. Stroke prediction based on the spontaneous nystagmus suppression test in dizzy patients: a diagnostic accuracy study. Neurology. 2021b;97(1):e42–51.
30. Korda A, Zamaro E, Wagner F, Morrison M, Caversaccio MD, Sauter TC, Schneider E, Mantokoudis G. Acute vestibular syndrome: is skew deviation a central sign? J Neurol. 2021a;
31. Korda A, Zee DS, Wyss T, Zamaro E, Caversaccio MD, Wagner F, Kalla R, Mantokoudis G. Impaired fixation suppression of horizontal vestibular nystagmus during smooth pursuit: pathophysiology and clinical implications. Eur J Neurol. 2021b;28(8):2614–21.
32. Eggers SDZ, Bisdorff A, von Brevern M, Zee DS, Kim JS, Perez-Fernandez N, Welgampola MS, Della Santina CC, Newman-Toker DE. Classification of vestibular signs and examination techniques: nystagmus and nystagmus-like movements. J Vestib Res. 2019;29(2–3):57–87.

33. Newman-Toker DE, Curthoys IS, Halmagyi GM. Diagnosing stroke in acute vertigo: the HINTS family of eye movement tests and the future of the “eye ECG”. Semin Neurol. 2015;35(5):506–21.
34. Leigh RJ, Zee DS. The neurology of eye movements. Oxford: Oxford University Press; 2015.
35. Morrison M, Kerkeni H, Korda A, Rass S, Caversaccio MD, Abegg M, Schneider E, Mantokoudis G. Automated alternate cover test for ‘HINTS’ assessment: a validation study. Eur Arch Otorrinolaringol. 2022;279(6):2873–9.
36. Joutsijoki H, Varpa K, Iltanen K, Juhola M. Machine learning approach to an otoneurological classification problem. Annu Int Conf IEEE Eng Med Biol Soc. 2013;2013:1294–7.
37. Juhola M. On machine learning classification of otoneurological data. Stud Health Technol Inform. 2008;136:211–6.
38. Juhola M, Laurikkala J, Viikki K, Kentala E, Pyykkö I. Classification of patients on the basis of otoneurological data by using Kohonen networks. Acta Otolaryngol Suppl. 2001;545:50–2.
39. Lim EC, Park JH, Jeon HJ, Kim HJ, Lee HJ, Song CG, Hong SK. Developing a diagnostic decision support system for benign paroxysmal positional vertigo using a deep-learning model. J Clin Med. 2019;8(5)
40. Varpa K, Iltanen K, Juhola M. Machine learning method for knowledge discovery experimented with otoneurological data. Comput Methods Prog Biomed. 2008;91(2):154–64.
41. Visscher RMS, Feddermann-Demont N, Romano F, Straumann D, Bertolini G. Artificial intelligence for understanding concussion: retrospective cluster analysis on the balance and vestibular diagnostic data of concussion patients. PLoS One. 2019;14(4):e0214525.
42. Shaikh AG, Bronstein A, Carmona S, Cha YH, Cho C, Ghasia FF, Gold D, Green KE, Helmchen C, Ibitoye RT, Kattah J, Kim JS, Kothari S, Manto M, Seemungal BM, Straumann D, Strupp M, Szmulewicz D, Tarnutzer A, Tehrani A, Tilikete C, Welgampola M, Zalazar G, Kheradmand A. Consensus on virtual management of vestibular disorders: urgent versus expedited care. Cerebellum. 2021;20(1):4–8.
43. Trottet C, Vogels T, Keitel K, Kulinkina AV, Tan R, Cobuccio L, et al. Modular clinical decision support networks (MoDN)—updatable, interpretable, and portable predictions for evolving clinical environments. PLOS Digit Health. 2023:2(7):e0000108. https://doi.org/10.1371/journal.pdig.000010

Chapter 6
Video Head Impulse Test Findings in Central Vestibular Syndrome

Francisco Jose Gualtieri

Keywords Video HIT · Vestibuloocular reflex · Peripheral vestibulopathy · HINTS · Skew deviation

6.1 Introduction

Head impulse tests were introduced in 1988 to evaluate semicircular canal (SCC) function, rapidly becoming one of the most useful clinical tools to diagnose vestibular disorders [1]. The use of scleral search coils accurately and reliably detects impaired semicircular canal function, but is mostly limited to clinical research as the technique is complex and invasive. In 2009, the video head impulse test (vHIT) was introduced, recording each canal separately, and was quickly adopted as the method of choice in most clinics around the world [2]. Positive vHIT can be observed in lesions involving the direct VOR pathway, a 3-neuron reflex arc that includes: primary vestibular afferents, fibers connecting the vestibular nuclei to the ocular motor nuclei, and ocular motor nerves. Because HITs are mostly normal in lesions that spare the direct VOR pathway, they can be used in patients with acute spontaneous vertigo to differentiate central from peripheral vestibulopathies. However, recent observations have suggested that central vestibular disorders may also show positive or abnormal vHITs [3, 4]. In some patients with central lesions, normal clinical findings, but abnormal quantitative HIT findings have been recorded, suggesting a discrepancy between clinical (bedside) and quantitative HIT results.

Therefore, it is important to determine which characteristics of the VOR are altered in central lesions and can be assessed using both clinical and quantitative HIT, to identify those patients who will need to undergo a brain MRI.

Physicians conducting neurological evaluations should be familiar with the quantitative features of the head impulse VOR, which is observed in disorders involving the central nervous system (CNS).

F. J. Gualtieri (✉)
FLENI, Buenos Aires, Argentina
e-mail: fgualtieri@fleni.org.ar

P. Ravat et al. (eds.), *Advances in Translational Neuroscience of Vestibular and Eye Movement Disorders*, Contemporary Clinical Neuroscience,
https://doi.org/10.1007/978-3-032-07539-0_6

We present vHIT findings in several cases of central lesions, discuss relevant implications, and review the scientific literature.

6.2 Acute Vestibular Syndrome

vHIT gains recorded in patients presenting an acute vestibular syndrome and cerebellar lesions, primarily related to an acute stroke, will fall within normal values, differentiating central lesions of this type from acute unilateral peripheral vestibulopathy, with high sensitivity and specificity [3–5].

6.2.1 Case 1

A 54-year-old female patient presented to the emergency room with acute vestibular syndrome. Spontaneous nystagmus to the left was observed, with saccadic intrusions, but no other disturbances were recorded during a clinical head impulse test. The video head impulse test revealed normal gains with frequent saccadic intrusions. Brain MRI showed ischemic involvement of the middle cerebellar peduncle, extending to the ipsilateral cerebellar hemisphere (Fig. 6.1).

Some patients with acute CNS involvement may present vHIT findings mimicking those of bilateral vestibulopathy, even when the lesion is unilateral [5, 6].

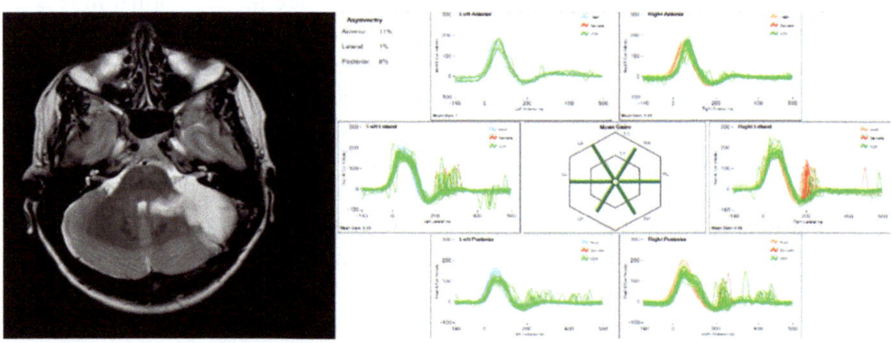

Fig. 6.1 A 54-year-old patient with acute vestibular syndrome; the video head impulse test shows normal gains with frequent saccadic intrusions. On brain MRI, an infarction involving the left middle cerebellar peduncle, extending to the ipsilateral cerebellar hemisphere, is observed

6.2.2 Case 2

An 18-year-old female patient presented with an acute vestibular syndrome showing spontaneous horizontal nystagmus to the right, nausea, and vomiting. The vHIT showed a bilateral decrease in gain with a peripheral pattern on the left and a central pattern on the right. MRI revealed a small lesion in the entry zone of the nerve and of the vestibular nucleus on the left side (Fig. 6.2), suggesting involvement of vestibulo-cerebellar pathways modulating VOR function.

Several reports have shown lesions in the cerebellum, or its connections to the vestibular system may or may not alter the VOR (vestibulo-ocular reflex), resulting in hypo- or hyperfunction [7, 8].

6.3 Combined Central and Peripheral Vestibulopathy

Lesions affecting the central nervous system as well as the labyrinth pose a particular diagnostic challenge, as central and peripheral signs will overlap. A high degree of suspicion is warranted in these cases, as well as careful patient workup, to identify underlying causes and provide adequate treatment. Vascular causes are more common in acute cases, especially anterior inferior cerebellar artery (AICA) infarction. Cerebellopontine angle tumors are seen more often in chronic cases [9]. Superposition of peripheral and central patterns can be observed in video HITs.

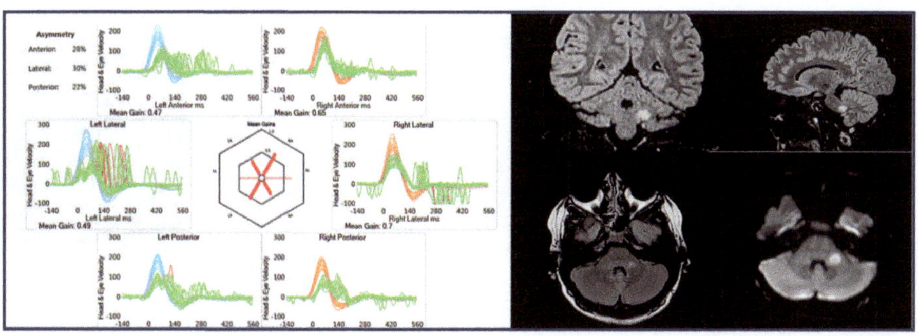

Fig. 6.2 An 18-year-old patient with acute vestibular syndrome. The vHIT shows peripheral-type involvement in the left horizontal semicircular canal, with great gain reduction and large amplitude compensatory saccades. The remaining semicircular canals show moderate gain reduction with low-amplitude compensatory saccades, attributable to vestibulo-cerebellar pathway involvement, modulating VOR. MRI shows a single lesion at the nerve root entry zone and the vestibular nucleus

6.3.1 Case 3

A patient with a lesion in the vestibular nucleus and at the nerve root entry with associated involvement of the cerebellar peduncles, exhibited a pattern suggesting peripheral involvement (Fig. 6.3a) on the same side of the lesion, and a central involvement pattern on the contralateral side, where no lesion was observed on brain MRI (Fig. 6.3b), presumably attributed to vestibulo-cerebellar pathway dysfunction affecting the VOR. Figure 6.3c shows the MRI with the lesion of the condition described.

6.3.2 Case 4

A 64-year-old man with progressive imbalance and chronic left-sided hearing loss shows on brain MRI a large vestibular schwannoma compressing the brainstem and left cerebellum. On the side involving the nerve (Fig. 6.4a), greatly reduced gains with compensatory saccades of large amplitude were recorded, as expected in peripheral pathology, while on the contralateral side, gains showed moderate reduction associated with compensatory saccades of low to moderate amplitude (Fig. 6.4b). Contralateral involvement of the VOR (vestibulo-ocular reflex) was attributed to impairment of modulating vestibulo-cerebellar pathways, altering the reflex. However, cerebellar lesions or damage to these pathways may, on occasion, not reveal any alteration.

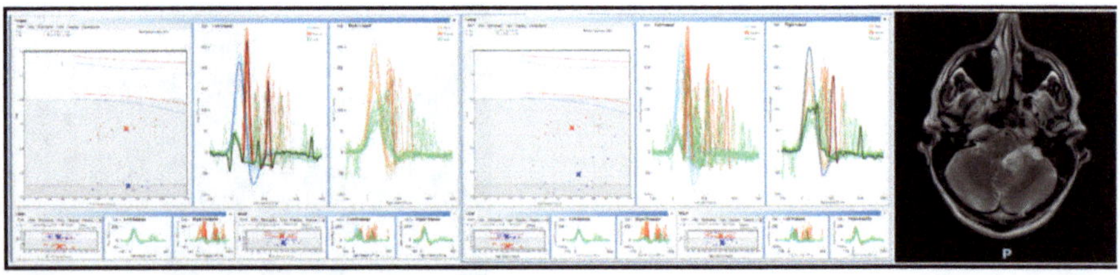

A B C

Fig. 6.3 Patient with vestibular nucleus and nerve root entry zone lesion and associated involvement of the cerebellar peduncles. A peripheral pattern (**a**) was recorded on the same side of the lesion, and a central involvement pattern on the contralateral side, where no lesion on brain MRI was observed (**b**), presumably attributed to vestibulo-cerebellar pathway dysfunction affecting the VOR. Brain MRI shows the corresponding lesion (**c**)

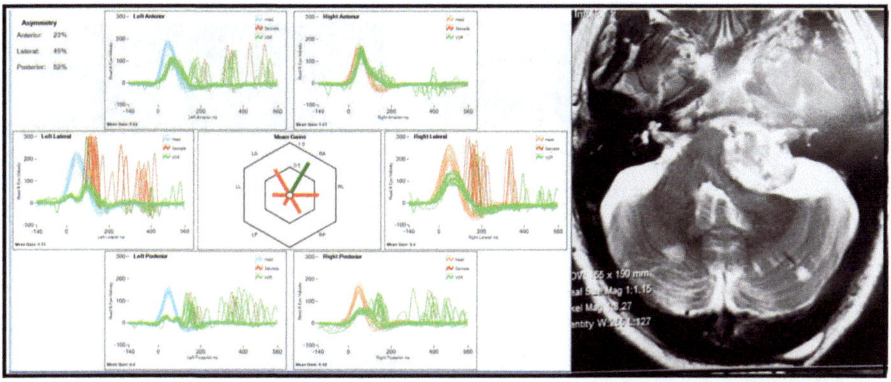

Fig. 6.4 Patient with progressive imbalance and left-sided hearing loss. Brain MRI shows a large vestibular schwannoma compressing the brainstem and cerebellum on the left. Due to peripheral injury, greatly reduced gains with compensatory saccades of large amplitude were observed on the left, as expected in peripheral pathology (**a**); on the contralateral side, gains showed moderate reduction, associated with low to moderate amplitude compensatory saccades (**b**)

6.4 Hyperactive VOR and Reversed Corrective Saccades

Patients with diffuse cerebellar dysfunction show hyperactive VORs with corrective backup saccades during low-frequency, low-acceleration, and velocity stimuli, such as sinusoidal rotations. Likewise, hyperactive VOR has been observed during HITs in patients with cerebellar degeneration [3]. The presence of hyperactive VOR and reversed corrective saccades has also been described in peripheral pathologies, including Meniere's syndrome, otosclerosis, and vestibular migraine [10].

We have seen this unusual pattern in patients with superficial hemosiderosis and hereditary ataxias (Fig. 6.5a, b).

6.5 Medial Longitudinal Fasciculus Lesions

Disruption of the medial longitudinal fasciculus (MLF) produces internuclear ophthalmoplegia (INO), a dysconjugate horizontal eye movement disorder, classically characterized by abducting eye nystagmus and adducting eye slowing during horizontal saccades. In patients with lesions of the medial longitudinal fasciculus, a decrease in VOR gain has been described, which is more significant for the posterior than the anterior semicircular canal [11]. In a patient diagnosed with multiple sclerosis who presented various brainstem lesions and bilateral internuclear ophthalmoplegia, generalized gain impairment was observed, mainly in both posterior semicircular canals (Fig. 6.6a–c).

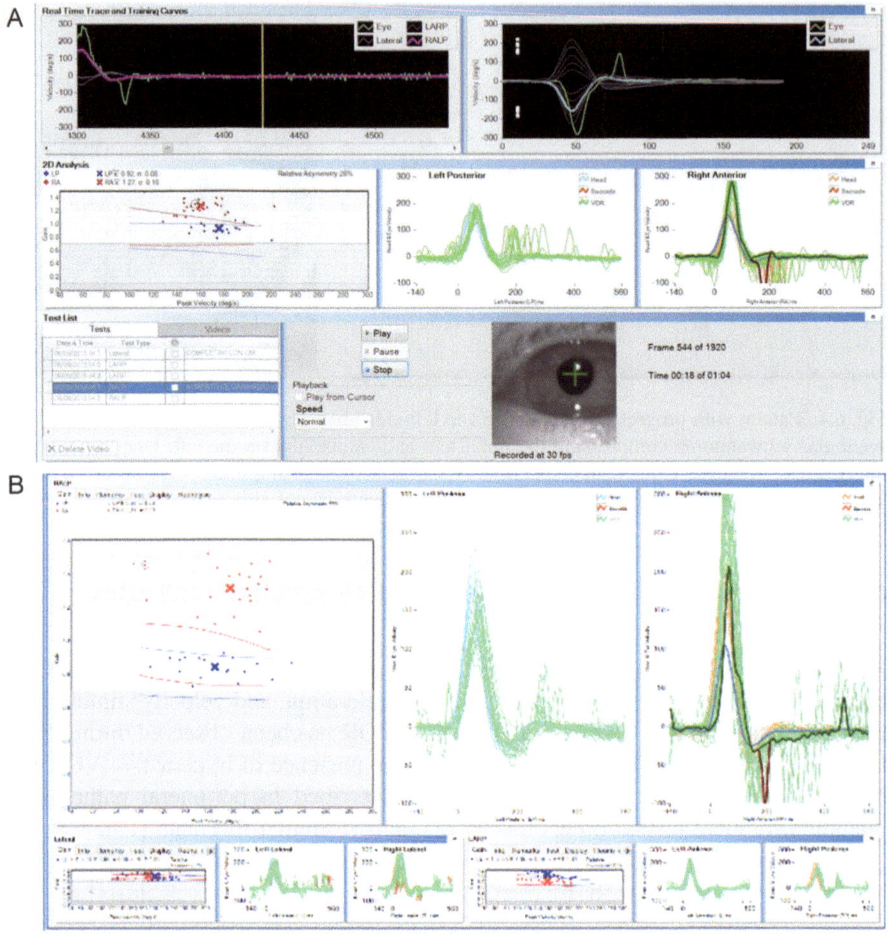

Fig. 6.5 A 65-year-old patient with superficial brain hemosiderosis presenting hyperactive VORs with corrective backup saccades on right anterior semicircular canal recording (**a** and **b**)

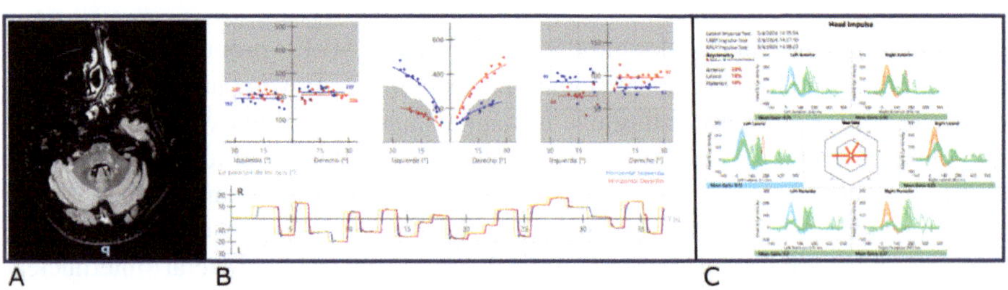

Fig. 6.6 Patient with multiple sclerosis presenting various brainstem lesions on MRI (**a**). VNG recordings show bilateral INO (**b**) and generalized gain impairment on vHIT, mainly in both posterior semicircular canals (**c**)

6.6 Discussion

Recent quantitative studies evaluating HITs using magnetic search coils or video-based techniques have demonstrated specific abnormal HIT patterns associated with central vestibular disorders [3–6]. Although normal clinical HITs in acute vestibular syndrome are typical of central lesions, discrepancies have been observed between clinical findings and quantitative HIT results in patients with pathology of the central nervous system. Horizontal head impulse VOR gains can be significantly reduced, unilaterally or bilaterally (positive HITs), in lesions involving the vestibular nucleus, the nucleus prepositus hypoglossi, or the flocculus. In diffuse cerebellar lesions, the VOR gain during horizontal head impulses may increase (hyperactive) with corrective saccades directed in the opposite direction. The presence of cross-coupled vertical corrective saccades during horizontal HITs is also suggestive of diffuse cerebellar lesions [3]. Lesions involving the vestibular nucleus, the medial longitudinal fasciculus, and the cerebellum may show decreased or increased gains of the VOR during vertical HITs [7, 8].

The presence of clearly positive HITs with large left-right asymmetry and large amplitude unidirectional corrective saccades is typical of peripheral vestibular hypofunction. On the other hand, different response patterns, including normal, positive, hyperactive, and cross-coupled head impulse signs, have been observed in central vestibulopathies [3, 4]. HITs recordings may even reveal alterations in temporal VOR characteristics, such as premature decelerations, with or without early acceleration. While negative HITs are a clear indication of a central lesion in acute vestibular syndromes, positive HIT findings do not always indicate a peripheral lesion. Therefore, other central signs, such as direction-changing nystagmus and skew deviation, should be carefully sought in patients with acute vestibular syndrome and positive HITs, especially in disorders involving peripheral and central vestibular structures, simultaneously. Likewise, in anterior inferior cerebellar artery (AICA) infarctions and cerebellopontine angle tumors, the HINTS (head impulse, nystagmus, and test of skew) may not be sufficient to detect a central lesion. Additional testing may be required (e.g., HINTS Plus) in patients presenting vestibulopathy together with unexplained hearing loss.

References

1. Halmagyi GM, Curthoys IS. A clinical sign of canal paresis. Arch Neurol. 1988;45:737–9.
2. Halmagyi GM, Chen L, MacDougall HG, Weber KP, McGarvie LA, Curthoys IS. The video head impulse test. Front Neurol. 2017;8:258.
3. Choi JY, Kim HJ, Kim JS. Recent advances in head impulse test findings in central vestibular disorders. Neurology. 2018;90(13):602–12.
4. Kim SH, Lee SU, Cho BH, Cho KH, Yu S, Kim BJ, Kim JS. Analyses of head impulse tests in patients with posterior circulation stroke and vestibular neuritis. Neurology. 2023;100(23): e2374–85.

5. Chen L, Halmagyi GM. Central lesions with selective semicircular canal involvement mimicking bilateral vestibulopathy. Front Neurol. 2018;9:264.
6. Zuma e Maia F, Luis L. Inferior peduncle lesion presenting with bilaterally impaired vestibular responses to horizontal and posterior head impulses. Laryngoscope. 2015;125(10):2386–7.
7. Kremmyda O, Kirchner H, Glasauer S, Brandt T, Jahn K, Strupp M. False-positive head-impulse test in cerebellar ataxia. Front Neurol. 2012;3:162.
8. Park HK, Kim JS, Strupp M, Zee DS. Isolated floccular infarction: impaired vestibular responses to horizontal head impulse. J Neurol. 2013;260(6):1576–82.
9. Choi SY, Kim HJ, Kim JS. Chasing dizzy chimera: diagnosis of combined peripheral and central vestibulopathy. J Neurol Sci. 2016;371:69–78.
10. Vargas-Alvarez A, Ninchritz-Becerra E, Goiburu M, Betances F, Rey-Martinez J, Altuna X. Clinical prevalence of enhanced vestibulo-ocular reflex responses on video head impulse test. Otol Neurotol. 2021;42(8):e1160–9.
11. Lee SH, Kim SH, Kim SS, Kang KW, Tarnutzer AA. Preferential impairment of the Contralesional posterior Semicircular Canal in Internuclear Ophthalmoplegia. Front Neurol. 2017;8:502.

Chapter 7
Exercise Prescription and Vestibular Physical Therapy for Persons with Vestibular Hypofunction

Susan L. Whitney

Keywords Exercise · Vestibular · Hypofunction · Bilateral · Technology

The objectives of this chapter are to provide an overview of the process that created the vestibular hypofunction clinical practice guideline, to describe some of the action statements from the updated vestibular rehabilitation clinical practice guideline, and to discuss the quantification and recording of exercise activity in persons receiving vestibular physiotherapy [1–3]. There is continuing emerging data to support the use of vestibular rehab as one of the optimal choices for intervention in persons with peripheral vestibular hypofunction [1].

A consensus was reached in several sites in Europe that data collected from multiple sites would be used to answer clinical questions related to the effectiveness of various interventions for persons living with vestibular disorders [4–6]. *Creating large data banks* that are collected systematically by clinicians and researchers worldwide is enhancing our knowledge of what are the negative prognostic factors for recovery, the optimal timing of rehab interventions, and which diagnostic groups are most amenable to improvements with rehab [4, 5]. *DIZZYNET was* created in Munich in 2014 as part of the European Network for vertigo and balance research to encourage multidisciplinary collaborations to answer questions about vestibular conditions, especially those that are less common [4–7].

S. L. Whitney (✉)
Departments of Physical Therapy and Otolaryngology, Clinical and Translational Sciences, University of Pittsburgh, Pittsburgh, PA, USA
e-mail: whitney@pitt.edu

P. Ravat et al. (eds.), *Advances in Translational Neuroscience of Vestibular and Eye Movement Disorders*, Contemporary Clinical Neuroscience,
https://doi.org/10.1007/978-3-032-07539-0_7

7.1 Vestibular Exercise Prescription

Klatt et al. [8] created a flow chart to try to better understand how to prescribe exercises for people with vestibular disorders. They included vestibulo-ocular reflex (VOR) exercises in different stance positions, whether people were moving or not moving [8]. Since 2015, convergence exercises, habituation exercises, and virtual reality exercise programs have been added to the list of potential treatment options for persons living with vestibular hypofunction [9–11]. Vibrotactile feedback, as well as galvanic stimulation, have been incorporated into exercise programs to facilitate recovery of stance and gait [12–15].

One of the main concerns about exercise prescription is that not all people with vestibular disorders have the same symptoms, physical impairments, and activity restrictions. Some people have an imbalance, some have difficulty with their gait, others with dizziness and vertigo, and others with visual blurring [16, 17]. Depending on the presenting signs and symptoms, the person should have their exercise prescription developed, and if needed, modified to best address their needs in a customized exercise program [18–20].

Changing foot/upper extremity/head position during the exercise or exercising with eyes open versus eyes closed are only a few of the options in advancing or making an exercise less difficult [8]. More recently, people have been studying dual tasking to try to determine how the dual task affects postural control in people with a vestibular disorder and or mild traumatic brain injury (mTBI)/concussion. [21–29] In persons with mTBI, asking them to do dual tasking is very sensitive at differentiating healthy athletes from persons with mTBI [28]. Balance typically recovers rapidly within the first week or two after mTBI [30], yet the ability to do dual tasking may continue to be impaired in persons with dizziness and mTBI longer than 2 weeks [31].

The vestibular hypofunction clinical practice guideline was updated and published in 2022 [1]. This guideline helps to provide information about what is the best practice and where we need additional data to try and answer questions to improve care for people living with vestibular hypofunction. The question of the clinical practice guideline (CPG) was, "Is exercise effective at enhancing recovery of function in people with peripheral vestibular hypofunction?" [1] Sixty-seven papers were critically appraised, and key statements from the clinical practice guideline include that current evidence supports that vestibular physiotherapy provides benefit to people with unilateral and bilateral vestibular loss. In addition, there were a variety of balance training methods described in the literature, but none of the balance training methods were superior to the others described [1]. Treatment options such as virtual reality, optokinetic stimulation, vibrotactile feedback, and many low technology treatments are available to rehabilitate people with vestibular hypofunction, either unilateral or bilateral.

Basic components of a vestibular rehabilitation program include gaze stabilization exercises based on trying to regain or improve the vestibulo-ocular reflex (VOR) gain. Other exercises include substitution, habituation exercises, balance and gait training, and walking for endurance [17]. There are many other iterations and combinations of the exercises above that are possible [8]. Cognitive behavioral therapy seems to be an excellent adjunct to vestibular rehabilitation in persons with chronic vestibular loss as it appears to augment the effectiveness of the exercise program, especially in persons with persistent dizziness and those with coexisting anxiety and/or depression [32–38].

Typically, the VOR × 1 exercise is prescribed in either the yaw or the pitch plane. The VOR × 1 exercise consists of focusing on a target at varying distances and moving the head up and down, right/left, or in the roll plan at different speeds while keeping the target in focus. In addition, people progress the VOR × 1 exercise to VOR × 2, where the target is moving in the opposite direction of the head, whereas the eyes try to maintain the target in focus. The VOR × 2 exercise can be done in the pitch and yaw plane.

Substitution exercises [39] are used for people who have low or no gain with their VOR. Substitution exercises are ideal for persons with bilateral vestibular loss [40] and can include the eye/head movement between targets and an exercise called the "imaginary target" or remembered exercise [39]. The eye/head movement between targets exercise was designed to train the brain to use a saccade before moving the head to decrease visual blurring in people with bilateral vestibular loss. The imaginary target exercise was designed to activate the cervical ocular reflex to signal to the brain where the head is in space as a mechanism to substitute for vestibular inputs [39]. Low VOR gain typically results in visual blurring with head movements. The guideline states that clinicians should not provide saccadic exercises or smooth pursuit exercises alone to try to increase the gain of the VOR. Smooth pursuit and saccadic exercises by themselves have not been shown to increase the VOR gain [1].

More recently, others [9, 41, 42] have developed the VOR × 1 with convergence exercise to increase the VOR gain in persons after a vestibular disorder (Fig. 7.1). During convergence with VOR × 1 exercises, one holds a target at arm's length and brings the target in toward the eye. As they are moving the head in either pitch or yaw plane, they are looking at the target as it comes in closer to their eyes. This convergence with the VOR × 1 exercise has been shown to increase the gain of the VOR within the session [9]. Additionally, there is preliminary evidence that incremental VOR gain adaptation can occur with either passive or active head movement training paradigms [43–47].

All exercises that are done with the eye and head movements can be done either in a sitting or standing position, and some can be accomplished during walking [8]. Methods to increase the difficulty of the exercise include having people walk forwards or backward, and doing dual tasking during gait [8].

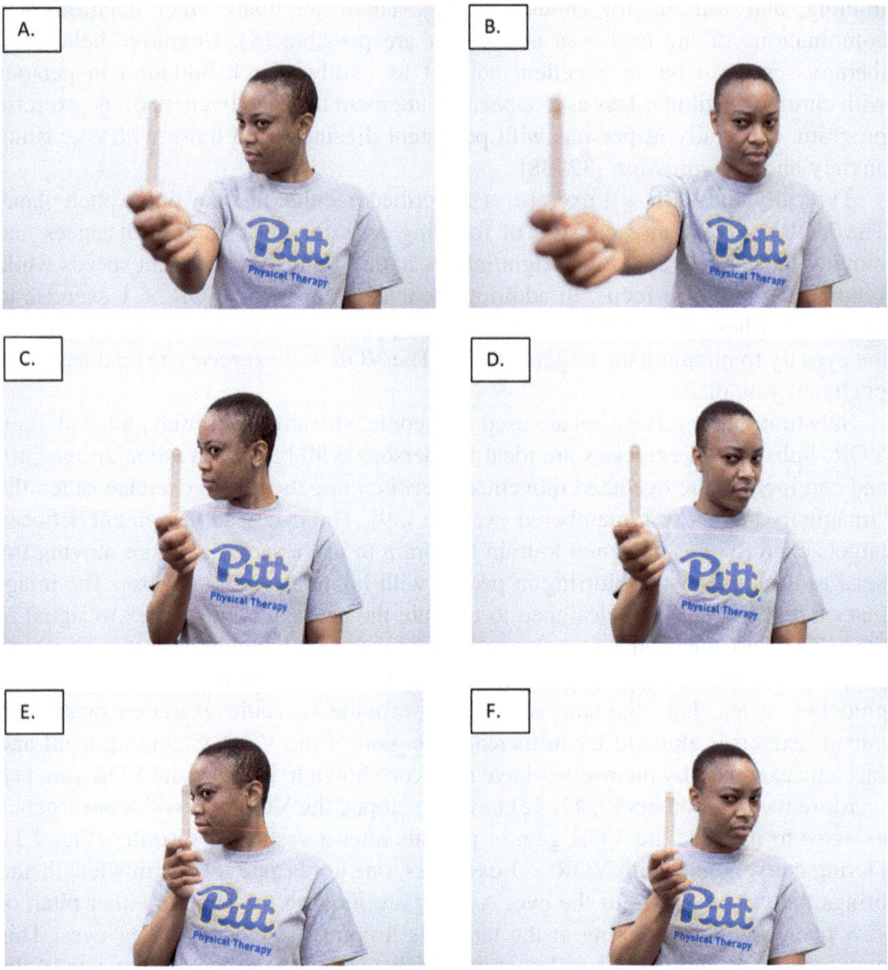

Fig. 7.1 The patient is focused on the stick with a letter as they move their outstretched arm to the right and left (**a**, **b**). In (**c**) and (**d**) and (**e**) and (**f**), the target is moved closer to their eyes as they keep their eyes focused throughout the exercise sequence and then is gradually moved away from their face

7.2 Acuity and Vestibular Physiotherapy

Evidence suggests that vestibular physiotherapy should be offered to people with acute and subacute vestibular hypofunction [2, 48–51]. Lacour et al. reported that early initiation of vestibular physiotherapy within the first 2 weeks after acute onset of vestibular hypofunction is optimal [48, 49]. There are significant changes in VOR gain based on whether people are seen within the first 2 weeks versus 4 weeks after the vestibular injury [48, 49].

7.3 Risks, Harms, and Cost of Vestibular Physiotherapy

Some of the risks, harms, costs, and benefits of that were identified within the guideline. Therapy can make people dizzy and sometimes sick with emesis, as well as potentially increasing their fall risk because of the exercises that they are prescribed to improve their postural control. The above risks are especially apparent during the acute stages after vestibular hypofunction, and fall risk is always a concern with people with significant bilateral vestibular loss [52–56].

Early postoperatively following vestibular surgery, it is important to assess for bleeding and/or a cerebral spinal fluid (CSF) leak when people are seen in the hospital before beginning vestibular exercises. Checking for the "halo" or "double ring" sign is a quick method to assess if there is a CSF leak. After using a sterile gauze pad to soak up some of the leaking fluid from the nose or ear, one may visualize a halo shape on the gauze pad if there is at least a mixture of 30% CSF to blood [57]. If a clear fluid is leaking from the ear or nose postsurgically, the physician needs to be contacted immediately as the "halo" or "double ring" sign is not 100% sensitive for identifying a CSF leak [57]. These CSF leaks are relatively rare after surgery but can occur [58]. Asking the patient to avoid blowing their nose until the surgeon has eliminated the possibility of a CSF leak is imperative. Current evidence suggests that beta-2 transferrin is the optimal test to identify a CSF leak, and a high-resolution CT is optimal to localize the site of the leak [59].

Additionally, there is the increased cost and time spent traveling to receive vestibular physiotherapy. There is evidence that telehealth related to vestibular physiotherapy can be advantageous [60–68]. There needs to be more data about telehealth, including how to determine what patient characteristics a person needs to possess that result in optimal recovery with the use of telehealth. Not all patients have access to the internet, especially in rural and underdeveloped areas of the world; some are fearful to try a telehealth visit, and others have limited experience with the technology necessary to perform their exercises [62, 69]. Key to the telehealth exam is taking a detailed history and performing key elements of the physical exam, often with the use of a cell phone so that the eye exam can be completed virtually [69–72]. There are apps and video recordings on cell phones that have been developed that the patient can create a video of their eye movements, which can be shared with the clinician, that are helpful in the diagnosis and management of vestibular conditions [71, 73].

It is recommended that clinicians explore the delivery of physiotherapy via telehealth [66]. In the future, we may be providing vestibular physiotherapy directly online with individuals, plus using teaching resources from the World Wide Web to render care [68].

7.4 Exercise Monitoring and Chronicity

The vestibular hypofunction guideline suggests that clinicians should offer supervised physiotherapy for people with unilateral and bilateral vestibular loss [1]. However, a systematic review did not show clear evidence that supervision of the exercise program was essential [74]. It remains unclear as to what type of exercise supervision is necessary. The level of necessary supervision may be dependent on the personal characteristics of the patient and their exercise environment, including support within the home. It appears that people older than 50 years of age may benefit more from supervision, and those who are fearful of falling may not do well if they are not supervised with their exercise program. Some type of monitoring may be necessary, even if they are remotely monitored for their home exercise program, especially if people have cognitive impairment [75, 76].

Physiotherapy should be offered to people with chronic, unilateral hypofunction, regardless of how long people have had symptoms [1]. There is good evidence that targeted exercise to address the person's limitations in activities and participation is better than just providing a list of home exercises [20, 77, 78]. Additionally, there are many new innovative strategies that people around the globe are using for physiotherapy, including vibrotactile systems to improve posture control [13, 14, 79–82], and using balance exercises through commercially available software programs that can enhance balance and postural control. Virtual reality systems are becoming more available, cheaper, and provide improved feedback to people about whether the person is successful with their exercise program [1, 10, 16, 66, 83, 84]. Motivating people is always critical to improve performance. Newer technologies can provide immediate and often ongoing feedback to persons with vestibular disorders which can aid in motivating the individual to continue their exercise program [39, 63].

7.5 Challenges to Rehab Success

There are certain factors that influence rehabilitation outcomes. Age and gender do not generally affect rehabilitation outcome, although there is some evidence that age affects outcomes with people living with benign paroxysmal positional vertigo [1, 85–87]. It also appears that vestibular physiotherapy is especially important for those over the age of 50 [51, 88].

Vestibular exercises can improve outcomes regardless of time from onset, but there is potentially harm in delaying intervention [48, 49]. Some comorbidities that have clearly been shown to affect outcome include anxiety, depression, peripheral neuropathy, migraine, abnormal binocular vision, and impaired cognition [16, 68, 89–93]. All have been shown to negatively impact patient outcomes. Anxiety and/or depression can develop with long-term vestibular problems, possibly resulting in less-than-optimal recovery [85, 94–101]. If anxiety and depression can be dealt with earlier in the rehab process, perhaps with cognitive behavioral therapy [37, 38], outcomes may be optimized.

Persons with cognitive dysfunction are more likely to have a delayed recovery as there is poor carryover of their learning and they may require longer to get better. Klatt et al. [76] have outlined ideas for treatment planning and strategies to improve exercise adherence in persons with mild cognitive impairments to improve outcomes. In addition, those people who are experiencing frequent episodes of vertigo, such as in active Meniere's disease, need to be medically managed before the start of rehabilitation. If an older adult has movement restrictions, especially in the cervical spine, it may preclude them from doing some of the eye/head movements that are often prescribed for persons with unilateral and bilateral vestibular hypofunction, which may limit their success with their rehab program.

Fear avoidance is another factor that negatively affects functional recovery. The vestibular activities avoidance instrument 9 (VAAI-9) was developed to determine if fear of doing exercises affects a person's activity and participation abilities after a vestibular disorder [101]. Dunlap at al [92, 94] reported that there were significant predictors of activity and participation limitations at 3 months, including whether the person was very dizzy when they started their initial plan of care, had depression, had fear-avoidant behaviors, was younger, and if they were taking multiple medications. Younger people were found to be more affected by the sequalae of a vestibular disorder than older adults [92], possibly because dizziness impacts their daily lives more as they may have additional responsibilities with caregiving, work, and leisure activities than older adults. Fear avoidance had the strongest association with disability at 3 months and should be carefully considered when constructing an exercise program [92].

Others have recently reported that worse scores on a dizziness fear avoidance questionnaire were related to persons living with chronic dizziness compared to better scores received by persons with acute or episodic vestibular conditions [102]. Fear avoidance is a construct that should be carefully considered when providing an exercise prescription [103]. If a person believes that exercise will make them worse, they are much less likely to do exercises independently in their homes.

Negotiating what is reasonable for the person to do for their exercise program and when it is optimal for them to do their exercises is one important step in improving exercise adherence. Kalderon et al. [104] reported on facilitators and barriers as perceived by both patients and physical therapists related to the performance of home exercise programs. The barriers included patient comorbidities, lack of time, the exercises increasing their symptoms, a lack of motivation, and a need for more feedback about their exercises when doing the exercises at home. Conversely, the facilitators for exercise performance included proper education instruction, a safe and comfortable place to exercise, motivation to get better, time to work the exercises into their daily schedule, and the use of relaxation techniques/mindfulness [104].

The short-term use of vestibular suppressant medications in acute vestibular hypofunction can assist in the management of vegetative symptoms including nausea and emesis [105]. Short-term use of low dose antihistamines in people with chronic vestibular dysfunction may help to control symptoms without having a negative impact on performance outcomes [105].

7.6 Emerging Technologies

Emerging technologies are being developed in virtual reality and with wearable-enabled devices [3, 10, 39, 63, 66, 83, 84, 106–109]. Technology has been developed to provide the patient feedback about their vestibular exercises that can be assessed by the therapist when they come back into the clinic [108]. Others have developed technology that provides feedback to the patient and physiotherapist via the cloud for virtual monitoring of exercise performance [39, 63, 93]. The use of eye and facial recognition software is now being used with apps and within virtual reality head-mounted devices to record eye movements and their accuracy. [110]

Habituation to moving scenes in a virtual environment in persons with chronic peripheral vestibular disorders has promise as a technology that can decrease perceived dizziness [111, 112]. A recent randomized trial compared the use of habituation, adaptation, and a combination of both in people with chronic, unilateral vestibular loss [74]. The three different groups did daily exercise at home three times a day. The group that did both the habituation and adaptation exercises got better faster than the adaptation or the habituation group alone [74]. There may be benefits in providing multiple kinds of exercises to patients to improve their functional recovery.

7.7 Vestibular Hypofunction, Falls, and Benign Paroxysmal Positional Vertigo (BPPV)

Unilateral vestibular loss is associated with BPPV; thus, performing the Dix-Hallpike is an essential element of the clinical exam [113]. There is now clear evidence from a systematic review and meta-analysis that BPPV increases the odds of falling and negatively affects the spatial-temporal parameters of gait [114]. The canalith reposition maneuver (CRM) can decrease fear of falling, improve level walking, and decrease falls. Pauwels et al. [114] suggested that performing a successful CRM can result in decreased fall reports, lessen fear of falling, and improve level walking. Others have reported that following treatment with the CRM for BPPV, there was a 64% reduction in the number of reported falls [115]. There is always a possibility of having vestibular loss plus BPPV, suggesting that best management dictates the performance of the Dix-Hallpike and the CRM if indicated by a positive Dix-Hallpike result.

7.8 Summary

Movement appears to be the key ingredient to vestibular exercise programs, most likely head and eye movement. The optimal exercise program for persons with unilateral or bilateral vestibular loss remains unclear. The VOR with convergence exercises increases the VOR gain and should be incorporated into vestibular exercise programs for people with unilateral and bilateral vestibular loss. Early referrals to begin a vestibular exercise program are essential, as outcomes are better with people who are referred within the first two weeks after onset. The use of technology may enhance motivation and can objectively record exercise progression over time. Finally, we need to develop precise exercise prescriptions unique to the individual to facilitate recovery and treat patients based on their presenting complaints and functional limitations to optimize functional outcomes in persons living with vestibular loss.

References

1. Hall CD, Herdman SJ, Whitney SL, Anson ER, Carender WJ, Hoppes CW, Cass SP, Christy JB, Cohen HS, Fife TD, Furman JM, Shepard NT, Clendaniel RA, Dishman JD, Goebel JA, Meldrum D, Ryan C, Wallace RL, Woodward NJ. Vestibular rehabilitation for peripheral vestibular hypofunction: an updated clinical practice guideline from the academy of neurologic physical therapy of the American Physical Therapy Association. J Neurol Phys Ther. 2022;46(2):118–77. https://doi.org/10.1097/npt.0000000000000382. PubMed PMID: 34864777; PMCID: PMC8920012.
2. McDonnell MN, Hillier SL. Vestibular rehabilitation for unilateral peripheral vestibular dysfunction. Cochrane Database Syst Rev. 2015;1(1):Cd005397. https://doi.org/10.1002/14651858.CD005397.pub4. Epub 20150113. PubMed PMID: 25581507; PMCID: PMC11259236.
3. Reynard P, Ortega-Solís J, Tronche S, Darrouzet V, Thai-Van H. Guidelines of the French Society of Otorhinolaryngology and Head and Neck Surgery (SFORL) for vestibular rehabilitation in children with vestibular dysfunction. A systematic review. Arch Pediatr. 2024;31(4):217–23. https://doi.org/10.1016/j.arcped.2024.02.006. Epub 20240501. PubMed PMID: 38697883.
4. Grill E, Akdal G, Becker-Bense S, Hübinger S, Huppert D, Kentala E, Strobl R, Zwergal A, Celebisoy N. Multicenter data banking in management of dizzy patients: first results from the DizzyNet registry project. J Neurol. 2018;265(Suppl 1):3–8. https://doi.org/10.1007/s00415-018-8864-1. Epub 20180416. PubMed PMID: 29663119.
5. Zwergal A, Brandt T, Magnusson M, Kennard C. DIZZYNET—a European network initiative for vertigo and balance research: visions and aims. J Neurol. 2016;263(Suppl 1):S2–9. https://doi.org/10.1007/s00415-015-7912-3. Epub 20160415. PubMed PMID: 27083879; PMCID: PMC4833796.
6. Zwergal A, van de Berg R, Dieterich M. DIZZYNET 2020: basic and clinical vestibular research united. J Neurol. 2020;267(Suppl 1):1–2. https://doi.org/10.1007/s00415-020-10317-4. PubMed PMID: 33205298; PMCID: PMC7718177.
7. Jahn K, Lopez C, Zwergal A, Zur O, Cakrt O, Kellerer S, Kerkeni H, Tjernström F, Meldrum D. Vestibular rehabilitation therapy in Europe: chances and challenges. J Neurol. 2019;266

(Suppl 1):9–10. https://doi.org/10.1007/s00415-019-09368-z. Epub 20190517. PubMed PMID: 31102020.
8. Klatt BN, Carender WJ, Lin CC, Alsubaie SF, Kinnaird CR, Sienko KH, Whitney SL. A conceptual framework for the progression of balance exercises in persons with balance and vestibular disorders. Phys Med Rehabil Int. 2015;2(4). Epub 20150428. PubMed PMID: 27489886; PMCID: PMC4968039.
9. Chang TP, Schubert MC. Convergence vestibulo-ocular reflex in unilateral vestibular hypofunction: behavioral evidence in support of a novel gaze stability exercise. J Neurol Phys Ther. 2021;45(1):3–11. https://doi.org/10.1097/npt.0000000000000335. PubMed PMID: 33065632.
10. Hasimova Z, Sahbaz T, Karacay BC, Karan A. Effectiveness of virtual reality therapy in chronic unilateral vestibular hypofunction: a randomized controlled study. Turk J Phys Med Rehabil. 2023;69(3):286–93. https://doi.org/10.5606/tftrd.2023.12360. Epub 20230707. PubMed PMID: 37674792; PMCID: PMC10478541.
11. Lilios A, Chimona T, Papadakis C, Chatziioanou I, Nikitas C, Skoulakis C. Different vestibular rehabilitation modalities in unilateral vestibular hypofunction: a prospective study. Otol Neurotol. 2023;44(4):e246–55. https://doi.org/10.1097/mao.0000000000003836. Epub 20230216. PubMed PMID: 36946364.
12. Appiah-Kubi KO, Galgon A, Tierney R, Lauer R, Wright WG. Concurrent vestibular activation and postural training recalibrate somatosensory, vestibular and gaze stabilization processes. PLoS One. 2024;19(7):e0292200. https://doi.org/10.1371/journal.pone.0292200. Epub 20240705. PubMed PMID: 38968181; PMCID: PMC11226066.
13. Bao T, Klatt BN, Carender WJ, Kinnaird C, Alsubaie S, Whitney SL, Sienko KH. Effects of long-term vestibular rehabilitation therapy with vibrotactile sensory augmentation for people with unilateral vestibular disorders – a randomized preliminary study. J Vestib Res. 2019;29 (6):323–34. https://doi.org/10.3233/ves-190683. PubMed PMID: 31609716; PMCID: PMC9249282.
14. Bao T, Noohi F, Kinnaird C, Carender WJ, Barone VJ, Peethambaran G, Whitney SL, Seidler RD, Sienko KH. Retention effects of long-term balance training with vibrotactile sensory augmentation in healthy older adults. Sensors (Basel) 2022;22(8). https://doi.org/10.3390/s22083014. Epub 20220414. PubMed PMID: 35459000; PMCID: PMC9027305.
15. Pires A, Silva TR, Torres MS, Diniz ML, Tavares MC, Gonçalves DU. Galvanic vestibular stimulation and its applications: a systematic review. Braz J Otorhinolaryngol. 2022;88(Suppl 3(Suppl 3)):S202–s11. https://doi.org/10.1016/j.bjorl.2022.05.010. Epub 20220705. PubMed PMID: 35915031; PMCID: PMC9760994.
16. Dunlap PM, Holmberg JM, Whitney SL. Vestibular rehabilitation: advances in peripheral and central vestibular disorders. Curr Opin Neurol. 2019;32(1):137–44. https://doi.org/10.1097/wco.0000000000000632. PubMed PMID: 30461465
17. Sulway S, Whitney SL. Advances in vestibular rehabilitation. Adv Otorhinolaryngol. 2019;82:164–9. https://doi.org/10.1159/000490285. Epub 20190115. PubMed PMID: 30947180.
18. Pavlou M. The use of optokinetic stimulation in vestibular rehabilitation. J Neurol Phys Ther. 2010;34(2):105–10. https://doi.org/10.1097/NPT.0b013e3181dde6bf. PubMed PMID: 20588097
19. Telian SA, Shepard NT, Smith-Wheelock M, Kemink JL. Habituation therapy for chronic vestibular dysfunction: preliminary results. Otolaryngol Head Neck Surg. 1990;103(1):89–95. https://doi.org/10.1177/019459989010300113. PubMed PMID: 2117736.
20. Vereeck L, Wuyts FL, Truijen S, De Valck C, Van de Heyning PH. The effect of early customized vestibular rehabilitation on balance after acoustic neuroma resection. Clin Rehabil. 2008;22(8):698–713. https://doi.org/10.1177/0269215508089066. PubMed PMID: 18678570.
21. Danneels M, Van Hecke R, Leyssens L, Cambier D, van de Berg R, Van de Velde L, Van Rompaey V, Maes L. 2BALANCE: test-retest reliability of a cognitive-motor dual-task

protocol. J Vestib Res. 2022;32(4):341–53. https://doi.org/10.3233/ves-210069. PubMed PMID: 34974447.

22. Danneels M, Van Hecke R, Leyssens L, Degeest S, Cambier D, van de Berg R, Van Rompaey V, Maes L. 2BALANCE: a cognitive-motor dual-task protocol for individuals with vestibular dysfunction. BMJ Open. 2020;10(7):e037138. https://doi.org/10.1136/bmjopen-2020-037138. Epub 20200714. PubMed PMID: 32665391; PMCID: PMC7365489.
23. Pavlou M, Costafreda SG, Galsworthy W, Korres G, Bamiou DE. The interplay between cognition, functional and dual-task gait in persons with a vestibular disorder versus healthy controls. Sci Rep. 2023;13(1):10130. https://doi.org/10.1038/s41598-023-35904-z. Epub 20230622. PubMed PMID: 37349351; PMCID: PMC10287746.
24. Roberts JC, Cohen HS, Sangi-Haghpeykar H. Vestibular disorders and dual task performance: impairment when walking a straight path. J Vestib Res. 2011;21(3):167–74. https://doi.org/10.3233/ves-2011-0415. PubMed PMID: 21558642; PMCID: PMC3543698.
25. Ingriselli JM, Register-Mihalik JK, Schmidt JD, Mihalik JP, Goerger BM, Guskiewicz KM. Outcomes, utility, and feasibility of single task and dual task intervention programs: preliminary implications for post-concussion rehabilitation. J Sci Med Sport. 2014;17(6):580–5. https://doi.org/10.1016/j.jsams.2013.11.006. Epub 20131128. PubMed PMID: 24373772; PMCID: PMC9266434.
26. Yang S, Johnson PK, Hansen CR, Wilde EA, Cortez MM, Dibble LE, Fino PC, Raffegeau TE. Talking while walking after concussion: acute effects of concussion on speech pauses and gait speed. medRxiv. 2024. https://doi.org/10.1101/2024.08.06.24311478. Epub 20240807. PubMed PMID: 39211856; PMCID: PMC11361251.
27. Howell DR, Osternig LR, Chou LS. Single-task and dual-task tandem gait test performance after concussion. J Sci Med Sport. 2017;20(7):622–6. https://doi.org/10.1016/j.jsams.2016.11.020. Epub 20170124. PubMed PMID: 28169147.
28. Oldham JR, Howell DR, Knight CA, Crenshaw JR, Buckley TA. Single-task and dual-task tandem gait performance across clinical concussion milestones in collegiate student-athletes. Clin J Sport Med. 2021;31(6):e392–e7. https://doi.org/10.1097/jsm.0000000000000836. PubMed PMID: 32852299; PMCID: PMC7887129.
29. Van Deventer KA, Seehusen CN, Walker GA, Wilson JC, Howell DR. The diagnostic and prognostic utility of the dual-task tandem gait test for pediatric concussion. J Sport Health Sci. 2021;10(2):131–7. https://doi.org/10.1016/j.jshs.2020.08.005. Epub 20200812. PubMed PMID: 32795624; PMCID: PMC7987555.
30. Furman GR, Lin CC, Bellanca JL, Marchetti GF, Collins MW, Whitney SL. Comparison of the balance accelerometer measure and balance error scoring system in adolescent concussions in sports. Am J Sports Med. 2013;41(6):1404–10. https://doi.org/10.1177/0363546513484446. Epub 20130412. PubMed PMID: 23585486; PMCID: PMC4899870.
31. Berkner J, Meehan WP 3rd, Master CL, Howell DR. Gait and quiet-stance performance among adolescents after concussion-symptom resolution. J Athl Train. 2017;52(12):1089–95. https://doi.org/10.4085/1062-6050-52.11.23. Epub 20171120. PubMed PMID: 29154694; PMCID: PMC5759692.
32. Axer H, Finn S, Wassermann A, Guntinas-Lichius O, Klingner CM, Witte OW. Multimodal treatment of persistent postural-perceptual dizziness. Brain Behav. 2020;10(12):e01864. https://doi.org/10.1002/brb3.1864. Epub 20200928. PubMed PMID: 32989916; PMCID: PMC7749543.
33. Conder R, Conder AA. Neuropsychological and psychological rehabilitation interventions in refractory sport-related post-concussive syndrome. Brain Inj. 2015;29(2):249–62. https://doi.org/10.3109/02699052.2014.965209. Epub 20141007. PubMed PMID: 25291459.
34. Johansson M, Akerlund D, Larsen HC, Andersson G. Randomized controlled trial of vestibular rehabilitation combined with cognitive-behavioral therapy for dizziness in older people. Otolaryngol Head Neck Surg. 2001;125(3):151–6. https://doi.org/10.1067/mhn.2001.118127. PubMed PMID: 11555746.

35. Naber CM, Water-Schmeder O, Bohrer PS, Matonak K, Bernstein AL, Merchant MA. Interdisciplinary treatment for vestibular dysfunction: the effectiveness of mindfulness, cognitive-behavioral techniques, and vestibular rehabilitation. Otolaryngol Head Neck Surg. 2011;145(1):117–24. https://doi.org/10.1177/0194599811399371. PubMed PMID: 21493331.
36. Zang J, Zheng M, Chu H, Yang X. Additional cognitive behavior therapy for persistent postural-perceptual dizziness: a meta-analysis. Braz J Otorhinolaryngol 2024;90(3):101393. https://doi.org/10.1016/j.bjorl.2024.101393. Epub 20240124. PubMed PMID: 38350404; PMCID: PMC10867767.
37. Herdman D, Norton S, Murdin L, Frost K, Pavlou M, Moss-Morris R. The INVEST trial: a randomised feasibility trial of psychologically informed vestibular rehabilitation versus current gold standard physiotherapy for people with Persistent Postural Perceptual Dizziness. J Neurol. 2022;269(9):4753–63. https://doi.org/10.1007/s00415-022-11107-w. Epub 20220410. PubMed PMID: 35397754; PMCID: PMC8994825.
38. Herdman D, Norton S, Pavlou M, Murdin L, Moss-Morris R. Protocol for a randomised controlled feasibility study of psychologically informed vestibular rehabilitation for people with persistent dizziness: INVEST trial. Pilot Feasibility Stud. 2021;7(1):156 https://doi.org/10.1186/s40814-021-00896-y. Epub 20210816. PubMed PMID: 34399847; PMCID: PMC8364941.
39. Hall CD, Flynn S, Clendaniel RA, Roberts DC, Stressman KD, Pu W, Mershon D, Schubert MC. Remote assessment and management of patients with dizziness: development, validation, and feasibility of a gamified vestibular rehabilitation therapy platform. Front Neurol. 2024;15: 1367582. https://doi.org/10.3389/fneur.2024.1367582. Epub 20240530. PubMed PMID: 38872821; PMCID: PMC11169667.
40. Whitney SL, Alghadir AH, Anwer S. Recent evidence about the effectiveness of vestibular rehabilitation. Curr Treat Options Neurol. 2016;18(3):13. https://doi.org/10.1007/s11940-016-0395-4. PubMed PMID: 26920418.
41. Lewis RF, Clendaniel RA, Zee DS. Vergence-dependent adaptation of the vestibulo-ocular reflex. Exp Brain Res. 2003;152(3):335–40. https://doi.org/10.1007/s00221-003-1563-9. Epub 20030723. PubMed PMID: 12879175.
42. Migliaccio AA, Cremer PD, Aw ST, Halmagyi GM. Vergence-mediated changes in Listing's plane do not occur in an eye with superior oblique palsy. Invest Ophthalmol Vis Sci. 2004;45 (9):3043–7. https://doi.org/10.1167/iovs.04-0014. PubMed PMID: 15326119.
43. Büki B, Tamás LT, Todd CJ, Schubert MC, Migliaccio AA. Absence of a vergence-mediated vestibulo-ocular reflex gain increase does not preclude adaptation. J Vestib Res. 2021;31(2): 109–17. https://doi.org/10.3233/ves-201560. PubMed PMID: 33427708.
44. Todd CJ, Schubert MC, Figtree WVC, Migliaccio AA. Incremental vestibulo-ocular reflex adaptation training dynamically tailored for each individual. J Neurol Phys Ther. 2019;43 (Suppl 2):S2–s7. https://doi.org/10.1097/npt.0000000000000269. PubMed PMID: 30883486.
45. Todd CJ, Schubert MC, Rinaudo CN, Migliaccio AA. Unidirectional vertical vestibuloocular reflex adaptation in humans using 1D and 2D scenes. Otol Neurotol. 2022;43(9):e1039–e44. https://doi.org/10.1097/mao.0000000000003684. Epub 20220909. PubMed PMID: 36075099.
46. Migliaccio AA, Schubert MC. Pilot study of a new rehabilitation tool: improved unilateral short-term adaptation of the human angular vestibulo-ocular reflex. Otol Neurotol. 2014;35 (10):e310–6. https://doi.org/10.1097/mao.0000000000000539. PubMed PMID: 25122595.
47. Rinaudo CN, Schubert MC, Cremer PD, Figtree WVC, Todd CJ, Migliaccio AA. Improved oculomotor physiology and behavior after unilateral incremental adaptation training in a person with chronic vestibular hypofunction: a case report. Phys Ther. 2019;99(10): 1326–33. https://doi.org/10.1093/ptj/pzz083. PubMed PMID: 31197314.
48. Lacour M, Tardivet L, Thiry A. Posture deficits and recovery after unilateral vestibular loss: early rehabilitation and degree of hypofunction matter. Front Hum Neurosci. 2021;15:776970.

https://doi.org/10.3389/fnhum.2021.776970. Epub 20220204. PubMed PMID: 35185493; PMCID: PMC8855301.

49. Lacour M, Thiry A, Tardivet L. Two conditions to fully recover dynamic canal function in unilateral peripheral vestibular hypofunction patients. J Vestib Res. 2021;31(5):407–21. https://doi.org/10.3233/ves-201557. PubMed PMID: 33749626.
50. Kamo T, Ogihara H, Azami M, Momosaki R, Fushiki H. Effects of early vestibular rehabilitation in patients with acute vestibular disorder: a systematic review and meta-analysis. Otol Neurotol. 2023;44(9):e641–e7. https://doi.org/10.1097/mao.0000000000004006. PubMed PMID: 00129492-202310000-00003.
51. Tokle G, Mørkved S, Bråthen G, Goplen FK, Salvesen Ø, Arnesen H, Holmeslet B, Nordahl SHG, Wilhelmsen KT. Efficacy of vestibular rehabilitation following acute vestibular neuritis: a randomized controlled trial. Otol Neurotol. 2020;41(1):78–85. https://doi.org/10.1097/mao.0000000000002443. PubMed PMID: 00129492-202001000-00020
52. Dobbels B, Lucieer F, Mertens G, Gilles A, Moyaert J, van de Heyning P, Guinand N, Pérez Fornos A, Herssens N, Hallemans A, Vereeck L, Vanderveken O, Van Rompaey V, van de Berg R. Prospective cohort study on the predictors of fall risk in 119 patients with bilateral vestibulopathy. PLoS One. 2020;15(3):e0228768. https://doi.org/10.1371/journal.pone.0228768. Epub 20200309. PubMed PMID: 32150553; PMCID: PMC7062241.
53. Herdman SJ, Schubert MC, Tusa RJ. Strategies for balance rehabilitation: fall risk and treatment. Ann N Y Acad Sci. 2001;942:394–412. PubMed PMID: 11710480.
54. Herssens N, How D, van de Berg R, McCrum C. Falls among people with bilateral vestibulopathy: a review of causes, incidence, injuries, and methods. JAMA Otolaryngol Head Neck Surg. 2022;148(2):187–92. https://doi.org/10.1001/jamaoto.2021.3673. PubMed PMID: 34989780.
55. Schniepp R, Schlick C, Schenkel F, Pradhan C, Jahn K, Brandt T, Wuehr M. Clinical and neurophysiological risk factors for falls in patients with bilateral vestibulopathy. J Neurol. 2017;264(2):277–83. https://doi.org/10.1007/s00415-016-8342-6. Epub 20161122. PubMed PMID: 27878442.
56. Wuehr M, Decker J, Schenkel F, Jahn K, Schniepp R. Impact on daily mobility and risk of falling in bilateral vestibulopathy. J Neurol. 2022;269(11):5746–54. https://doi.org/10.1007/s00415-022-11043-9. Epub 20220314. PubMed PMID: 35286481; PMCID: PMC9553788.
57. Ray AM. Halo sign is neither sensitive nor specific for cerebrospinal fluid leak. Ann Emerg Med. 2009;53(2):288. https://doi.org/10.1016/j.annemergmed.2008.06.474. PubMed PMID: 19167628.
58. Ben Ammar M, Merkus P, Di Lella F, Sanna M. Management of CSF leak after vestibular schwannoma surgery. Otol Neurotol. 2012;33(3):491–2. https://doi.org/10.1097/MAO.0b013e31824881c2. PubMed PMID: 22377648.
59. Oakley GM, Alt JA, Schlosser RJ, Harvey RJ, Orlandi RR. Diagnosis of cerebrospinal fluid rhinorrhea: an evidence-based review with recommendations. Int Forum Allergy Rhinol. 2016;6(1):8–16. https://doi.org/10.1002/alr.21637. Epub 20150915. PubMed PMID: 26370330.
60. Aldawsary N, Almarwani M. The combined effect of gaze stability and balance exercises using telerehabilitation in individuals with vestibular disorders during the COVID-19 pandemic: a pilot study. PLoS One. 2023;18(5):e0282189. https://doi.org/10.1371/journal.pone.0282189. Epub 20230505. PubMed PMID: 37146078; PMCID: PMC10162509.
61. Grillo D, Zitti M, Cieślik B, Vania S, Zangarini S, Bargellesi S, Kiper P. Effectiveness of telerehabilitation in dizziness: a systematic review with meta-analysis. Sensors (Basel) 2024;24(10). https://doi.org/10.3390/s24103028. Epub 20240510. PubMed PMID: 38793883; PMCID: PMC11125243.
62. Harrell RG, Schubert MC, Oxborough S, Whitney SL. Vestibular rehabilitation telehealth during the SAEA-CoV-2 (COVID-19) pandemic. Front Neurol. 2021;12:781482. https://doi.org/10.3389/fneur.2021.781482. Epub 20220120. PubMed PMID: 35126289; PMCID: PMC8811028.

63. Hovareshti P, Roeder S, Holt LS, Gao P, Xiao L, Zalkin C, Ou V, Tolani D, Klatt BN, Whitney SL. VestAid: a tablet-based technology for objective exercise monitoring in vestibular rehabilitation. Sensors (Basel) 2021;21(24). https://doi.org/10.3390/s21248388. Epub 20211215. PubMed PMID: 34960480; PMCID: PMC8706065.
64. Kriz M, Conover S, Cameron N, Conover M, Simon J, Prueter J, Samy R, Backous D. Comparing in-clinic versus telehealth for vestibular physical therapy. J Vestib Res. 2024;34(4):215–22. https://doi.org/10.3233/ves-220094. PubMed PMID: 38905068.
65. Noda M, Kuroda T, Nomura A, Ito M, Yoshizaki T, Fushiki H. Smartphone-assisted medical care for vestibular dysfunction as a telehealth strategy for digital therapy beyond COVID-19: scoping review. JMIR Mhealth Uhealth. 2023;11:e48638. https://doi.org/10.2196/48638. Epub 20230911. PubMed PMID: 37695671; PMCID: PMC10496931.
66. Perez-Heydrich CA, Creary-Miller I, Spann M, Agrawal Y. Remote delivery of vestibular rehabilitation for vestibular dysfunction: a systematic review. Otol Neurotol. 2024;45(6): 608–18. https://doi.org/10.1097/mao.0000000000004218. PubMed PMID: 38865717.
67. Rosenberg K. Web-guided self-treatment of benign paroxysmal positional vertigo. Am J Nurs. 2023;123(6):62. https://doi.org/10.1097/01.Naj.0000938744.32996.31. PubMed PMID: 37233142.
68. van Vugt VA, van der Wouden JC, Essery R, Yardley L, Twisk JWR, van der Horst HE, Maarsingh OR. Internet based vestibular rehabilitation with and without physiotherapy support for adults aged 50 and older with a chronic vestibular syndrome in general practice: three armed randomised controlled trial. BMJ. 2019;367:l5922. https://doi.org/10.1136/bmj.l5922. Epub 20191105. PubMed PMID: 31690561; PMCID: PMC6829201.
69. Kim HJ, Kim JS, Choi KD, Choi SY, Lee SH, Jung I, Park JH. Effect of self-treatment of recurrent benign paroxysmal positional vertigo: a randomized clinical trial. JAMA Neurol. 2023;80(3):244–50. https://doi.org/10.1001/jamaneurol.2022.4944. PubMed PMID: 36648931; PMCID: PMC10011937.
70. Shaikh AG, Bronstein A, Carmona S, Cha YH, Cho C, Ghasia FF, Gold D, Green KE, Helmchen C, Ibitoye RT, Kattah J, Kim JS, Kothari S, Manto M, Seemungal BM, Straumann D, Strupp M, Szmulewicz D, Tarnutzer A, Tehrani A, Tilikete C, Welgampola M, Zalazar G, Kheradmand A. Consensus on virtual management of vestibular disorders: urgent versus expedited care. Cerebellum. 2021;20(1):4–8. https://doi.org/10.1007/s12311-020-01178-8. PubMed PMID: 32794025; PMCID: PMC7426203.
71. Melliti A, van de Berg M, van de Berg R. Capturing nystagmus during vertigo attacks using a smartphone: adherence, characteristics, pearls and pitfalls. J Neurol. 2023;270(12):6044–56. https://doi.org/10.1007/s00415-023-11965-y. Epub 20230831. PubMed PMID: 37653139; PMCID: PMC10632223.
72. Barreto RG, Yacovino DA, Cherchi M, Nader SN, Teixeira LJ, Silva DAD, Verdecchia DH. The role of the smartphone in the diagnosis of vestibular hypofunction: a clinical strategy for teleconsultation during the COVID-19 pandemic and beyond. Int Arch Otorhinolaryngol. 2021;25(4):e602–e9. https://doi.org/10.1055/s-0041-1736340. Epub 20210804. PubMed PMID: 34777592; PMCID: PMC8580156.
73. Young AS, Lechner C, Bradshaw AP, MacDougall HG, Black DA, Halmagyi GM, Welgampola MS. Capturing acute vertigo: a vestibular event monitor. Neurology. 2019;92 (24):e2743–e53. https://doi.org/10.1212/wnl.0000000000007644. Epub 20190515. PubMed PMID: 31092626.
74. Lilios A, Chimona T, Nikitas C, Papadakis C, Chatziioannou I, Skoulakis C. The effect of supervision in vestibular rehabilitation in patients with acute or chronic unilateral vestibular dysfunction: a systematic review. Otol Neurotol. 2021;42(10):e1422–e31. https://doi.org/10.1097/mao.0000000000003354. PubMed PMID: 00129492-202112000-00004.
75. Varriano B, Sulway S, Wetmore C, Dillon W, Misquitta K, Multani N, Anor C, Martinez M, Cacchione E, Rutka J, Tartaglia MC. Vestibular exercises as a fall prevention strategy in patients with cognitive impairment. Can J Neurol Sci. 2020;47(1):126–30. https://doi.org/10.1017/cjn.2019.309. Epub 20191118. PubMed PMID: 31735189.

76. Klatt BN, Ries JD, Dunlap PM, Whitney SL, Agrawal Y. Vestibular physical therapy in individuals with cognitive impairment: a theoretical framework. J Neurol Phys Ther. 2019;43 (Suppl 2(Suppl 2 Spec INTERNATIONAL CONFERENCE ON VESTIBULAR REHABILITATION)):S14–s19. https://doi.org/10.1097/npt.0000000000000266. PubMed PMID: 30883488; PMCID: PMC6465065.
77. Smółka W, Smółka K, Markowski J, Pilch J, Piotrowska-Seweryn A, Zwierzchowska A. The efficacy of vestibular rehabilitation in patients with chronic unilateral vestibular dysfunction. Int J Occup Med Environ Health. 2020;33(3):273–82. https://doi.org/10.13075/ijomeh.1896.01330. Epub 20200326. PubMed PMID: 32235946.
78. Telian SA, Shepard NT. Update on vestibular rehabilitation therapy. Otolaryngol Clin N Am. 1996;29(2):359–71. PubMed PMID: 8860934.
79. Candreia C, Rust HM, Honegger F, Allum JHJ. The effects of vibro-tactile biofeedback balance training on balance control and dizziness in patients with persistent postural-perceptual dizziness (PPPD). Brain Sci 2023;13(5). https://doi.org/10.3390/brainsci13050782. Epub 20230511. PubMed PMID: 37239254; PMCID: PMC10216252.
80. Dozza M, Wall C 3rd, Peterka RJ, Chiari L, Horak FB. Effects of practicing tandem gait with and without vibrotactile biofeedback in subjects with unilateral vestibular loss. J Vestib Res. 2007;17(4):195–204. PubMed PMID: 18525145; PMCID: PMC2474458.
81. Janssen M, Pas R, Aarts J, Janssen-Potten Y, Vles H, Nabuurs C, van Lummel R, Stokroos R, Kingma H. Clinical observational gait analysis to evaluate improvement of balance during gait with vibrotactile biofeedback. Physiother Res Int. 2012;17(1):4–11. https://doi.org/10.1002/pri.504. Epub 20110105. PubMed PMID: 21207506.
82. Janssen M, Stokroos R, Aarts J, van Lummel R, Kingma H. Salient and placebo vibrotactile feedback are equally effective in reducing sway in bilateral vestibular loss patients. Gait Posture. 2010;31(2):213–7. https://doi.org/10.1016/j.gaitpost.2009.10.008. Epub 20091118. PubMed PMID: 19926286.
83. Heffernan A, Abdelmalek M, Nunez DA. Virtual and augmented reality in the vestibular rehabilitation of peripheral vestibular disorders: systematic review and meta-analysis. Sci Rep. 2021;11(1):17843. https://doi.org/10.1038/s41598-021-97370-9. Epub 20210908. PubMed PMID: 34497323; PMCID: PMC8426502.
84. Xie M, Zhou K, Patro N, Chan T, Levin M, Gupta MK, Archibald J. Virtual reality for vestibular rehabilitation: a systematic review. Otol Neurotol. 2021;42(7):967–77. https://doi.org/10.1097/mao.0000000000003155. PubMed PMID: 33782257.
85. Whitney SL, Sparto PJ, Furman JM. Vestibular rehabilitation and factors that can affect outcome. Semin Neurol. 2020;40(1):165–72. https://doi.org/10.1055/s-0039-3402062. Epub 20191230. PubMed PMID: 31887754.
86. Whitney SL, Wrisley DM, Marchetti GF, Furman JM. The effect of age on vestibular rehabilitation outcomes. Laryngoscope. 2002;112(10):1785–90. https://doi.org/10.1097/00005537-200210000-00015. PubMed PMID: 12368616.
87. Babac S, Djeric D, Petrovic-Lazic M, Arsovic N, Mikic A. Why do treatment failure and recurrences of benign paroxysmal positional vertigo occur? Otol Neurotol. 2014;35(6):1105–10. https://doi.org/10.1097/mao.0000000000000417. PubMed PMID: 24853239.
88. Ismail EI, Morgan AE, Abdel Rahman AM. Corticosteroids versus vestibular rehabilitation in long-term outcomes in vestibular neuritis. J Vestib Res. 2018;28(5–6):417–24. https://doi.org/10.3233/ves-180645. PubMed PMID: 30714984; PMCID: PMC9249297.
89. Pavlou M, Acheson J, Nicolaou D, Fraser CL, Bronstein AM, Davies RA. Effect of developmental binocular vision abnormalities on visual vertigo symptoms and treatment outcome. J Neurol Phys Ther. 2015;39(4):215–24. https://doi.org/10.1097/npt.0000000000000105. PubMed PMID: 26371531.
90. Vitkovic J, Winoto A, Rance G, Dowell R, Paine M. Vestibular rehabilitation outcomes in patients with and without vestibular migraine. J Neurol. 2013;260(12):3039–48. https://doi.org/10.1007/s00415-013-7116-7. Epub 20130924. PubMed PMID: 24061769.

91. Wrisley DM, Whitney SL, Furman JM. Vestibular rehabilitation outcomes in patients with a history of migraine. Otol Neurotol. 2002;23(4):483–7. https://doi.org/10.1097/00129492-200207000-00016. PubMed PMID: 12170150.
92. Dunlap PM, Sparto PJ, Marchetti GF, Furman JM, Staab JP, Delitto A, Klatt BN, Whitney SL. Fear avoidance beliefs are associated with perceived disability in persons with vestibular disorders. Phys Ther 2021;101(9). https://doi.org/10.1093/ptj/pzab147. PubMed PMID: 34120180.
93. Harrell RG, Cassidy AR, Klatt BN, Hovareshti P, Whitney SL. Vestibular rehabilitation in cerebellar ataxia with neuropathy and vestibular areflexia syndrome (CANVAS) – a case report. J Otol. 2023;18(4):199–207. https://doi.org/10.1016/j.joto.2023.06.004. Epub 20230720. PubMed PMID: 37877066; PMCID: PMC10593570.
94. Dunlap PM, Marchetti GF, Sparto PJ, Staab JP, Furman JM, Delitto A, Whitney SL. Exploratory factor analysis of the vestibular activities avoidance instrument. JAMA Otolaryngol Head Neck Surg. 2021;147(2):144–50. https://doi.org/10.1001/jamaoto.2020.4203. PubMed PMID: 33237289; PMCID: PMC7689567.
95. Goto F, Nomura K, Taka F, Arai M, Sugaya N. Analysis of factors affecting the outcomes of in-hospitalized vestibular rehabilitation in patients with intractable dizziness. Otol Neurotol. 2017;38(3):368–72. https://doi.org/10.1097/mao.0000000000001297. PubMed PMID: 27984359.
96. Kim TS, Lee WH, Heo Y. Prevalence and contributing factors of anxiety and depression in patients with vestibular migraine. Ear Nose Throat J. 2024;103(5):305–12. https://doi.org/10.1177/01455613231181219. Epub 20230617. PubMed PMID: 37329273.
97. MacDowell SG, Wellons R, Bissell A, Knecht L, Naquin C, Karpinski A. The impact of symptoms of anxiety and depression on subjective and objective outcome measures in individuals with vestibular disorders. J Vestib Res. 2018;27(5–6):295–303. https://doi.org/10.3233/ves-170627. PubMed PMID: 29125530.
98. Søberg HL, Andelic N, Langhammer B, Tamber AL, Bruusgaard KA, Kleffelgaard I. Effect of vestibular rehabilitation on change in health-related quality of life in patients with dizziness and balance problems after traumatic brain injury: a randomized controlled trial. J Rehabil Med. 2021;53(4):jrm00181. https://doi.org/10.2340/16501977-2823. Epub 20210421. PubMed PMID: 33842981; PMCID: PMC8814830.
99. Staab JP. Psychiatric considerations in the management of dizzy patients. Adv Otorhinolaryngol. 2019;82:170–9. https://doi.org/10.1159/000490286. Epub 20190115. PubMed PMID: 30947209.
100. Van Laer L, Dunlap PM, Vereeck L, Hendriks E, Sluydts M, Whitney SL. Fear avoidance beliefs, anxiety, and depression in healthy individuals and persons with vestibular disorders across cultures. Front Neurol. 2023;14:1296411. https://doi.org/10.3389/fneur.2023.1296411. Epub 20231201. PubMed PMID: 38107631; PMCID: PMC10722427.
101. Dunlap PM, Alradady FA, Costa CM, Delitto A, Terhorst L, Sparto PJ, Furman JM, Marchetti GF, Staab JP, Chueh J, Whitney SL. The psychometric properties of the 9-item vestibular activities avoidance instrument. Phys Ther 2023;103(12). https://doi.org/10.1093/ptj/pzad094. PubMed PMID: 37459243.
102. La Touche R, Castillejos-Carrasco-Muñoz R, Tapia-Toca MC, Pardo-Montero J, Lerma-Lara S, de la Rosa-Díaz I, Sorrel-Luján M, Paris-Alemany A. Development and validation of the dizziness fear-avoidance behaviours and beliefs inventory for patients with vestibular disorders. PeerJ. 2023;11:e15940. https://doi.org/10.7717/peerj.15940. Epub 20230829. PubMed PMID: 37663281; PMCID: PMC10473040.
103. Sever E, Kiliç G, Algun ZC. The effects of vestibular rehabilitation on kinesiophobia and balance with individuals who has vestibular hypofunction. Indian J Otolaryngol Head Neck Surg. 2022;74(Suppl 3):4319–24. https://doi.org/10.1007/s12070-021-02979-x. Epub 20211113. PubMed PMID: 34804903; PMCID: PMC8589630.
104. Kalderon L, Kaplan A, Wolfovitz A, Levy-Tzedek S, Gimmon Y. Barriers and facilitators of vestibular rehabilitation: patients and physiotherapists' perspectives. J Neurol Phys Ther.

2024;48(3):140–50. https://doi.org/10.1097/npt.0000000000000470. Epub 20240301. PubMed PMID: 38426842; PMCID: PMC11208053.
105. Basta D, Borsellino L, Ernst A. Antivertiginous drug therapy does not hinder the efficacy of individualized vibrotactile neurofeedback training for vestibular rehabilitation – a randomized trial. Int J Rehabil Res. 2017;40(4):333–8. https://doi.org/10.1097/mrr.0000000000000245. PubMed PMID: 28777127.
106. Meldrum D, Herdman S, Vance R, Murray D, Malone K, Duffy D, Glennon A, McConn-Walsh R. Effectiveness of conventional versus virtual reality-based balance exercises in vestibular rehabilitation for unilateral peripheral vestibular loss: results of a randomized controlled trial. Arch Phys Med Rehabil. 2015;96(7):1319–28.e1. https://doi.org/10.1016/j.apmr.2015.02.032. Epub 20150402. PubMed PMID: 25842051.
107. Meldrum D, Jahn K. Gaze stabilisation exercises in vestibular rehabilitation: review of the evidence and recent clinical advances. J Neurol. 2019;266(Suppl 1):11–8. https://doi.org/10.1007/s00415-019-09459-x. Epub 20190805. PubMed PMID: 31385017.
108. Meldrum D, Murray D, Vance R, Coleman S, McConnell S, Hardiman O, McConn WR. Toward a digital health intervention for vestibular rehabilitation: usability and subjective outcomes of a novel platform. Front Neurol. 2022;13:836796. https://doi.org/10.3389/fneur.2022.836796. Epub 20220329. PubMed PMID: 35422750; PMCID: PMC9001890.
109. D'Silva LJ, Phongsavath T, Partington K, Pickle NT, Marschner K, Zehnbauer TP, Rossi M, Skop K, Roos PE. A gaming app developed for vestibular rehabilitation improves the accuracy of performance and engagement with exercises. Front Med (Lausanne). 2023;10:1269874. https://doi.org/10.3389/fmed.2023.1269874. Epub 20231124. PubMed PMID: 38076248; PMCID: PMC10704144.
110. Schuetz I, Fiehler K. Eye tracking in virtual reality: vive pro eye spatial accuracy, precision, and calibration reliability. J Eye Mov Res 2022;15(3). https://doi.org/10.16910/jemr.15.3.3. Epub 20220907. PubMed PMID: 37125009; PMCID: PMC10136368.
111. Hondebrink MS, Mert A, van der Lint R, de Ru JA, van der Wurff P. Motion-based equilibrium reprocessing therapy a novel treatment method for chronic peripheral vestibulopathies: a pilot study. Medicine (Baltimore). 2017;96(24):e7128. https://doi.org/10.1097/md.0000000000007128. PubMed PMID: 28614234; PMCID: PMC5478319.
112. Goodwin N, Powell G, Loizides F, Derry-Sumner H, Rajenderkumar D, Sumner P. Feasibility of gamified visual desensitisation for visually-induced dizziness. Sci Rep. 2024;14(1):17864. https://doi.org/10.1038/s41598-024-67745-9. Epub 20240801. PubMed PMID: 39090130; PMCID: PMC11294564.
113. Pollak L, Davies RA, Luxon LL. Effectiveness of the particle repositioning maneuver in benign paroxysmal positional vertigo with and without additional vestibular pathology. Otol Neurotol. 2002;23(1):79–83. https://doi.org/10.1097/00129492-200201000-00018. PubMed PMID: 11773852.
114. Pauwels S, Casters L, Lemkens N, Lemmens W, Meijer K, Meyns P, van de Berg R, Spildooren J. Gait and falls in benign paroxysmal positional vertigo: a systematic review and meta-analysis. J Neurol Phys Ther. 2023;47(3):127–38. https://doi.org/10.1097/npt.0000000000000438. Epub 20230307. PubMed PMID: 36897200; PMCID: PMC10521788.
115. Jumani K, Powell J. Benign paroxysmal positional vertigo: management and its impact on falls. Ann Otol Rhinol Laryngol. 2017;126(8):602–5. https://doi.org/10.1177/0003489417718847. PubMed PMID: 28718303.

Chapter 8
Incremental VOR Adaptation: Clinical Trial Results

Michael C. Schubert and Americo Migliaccio

Keywords VOR gain adaptation · Gaze stability training · Vestibulo-ocular reflex · Rehabilitation · Vestibular rehabilitation · Compensatory saccade · Mild traumatic brain injury · Video head impulse

8.1 Introduction

Modern vestibular rehabilitation was borne from the exercises developed in the 1940's by British Otolaryngologist Terence Cawthorne, when he hypothesized that the root cause of dizziness and imbalance symptoms in patients with vestibular hypofunction (and mild traumatic brain injury) was impairment of the vestibular system [1]. Cawthorne asked F.S. Cooksey to develop an exercise regimen to treat the dizziness and imbalance symptoms, which was published in 1946 and became the gold standard of care for the rehabilitative treatment of vestibular disorders for decades [2]. These exercises were groundbreaking and verified as effective for

Submitted in honor of Sergio Carmona MD, as part of the La Fundación Iberoamericana de Neurotología (FINO) Conference, April 2023.

M. C. Schubert (✉)
Laboratory of Vestibular NeuroAdaptation, Department of Otolaryngology—Head and Neck Surgery, Johns Hopkins University, Baltimore, MD, USA

Department of Physical Medicine and Rehabilitation, School of Medicine, Johns Hopkins University, Baltimore, MD, USA
e-mail: mschube1@jhmi.edu

A. Migliaccio
Graduate School of Biomedical Engineering, University of New South Wales, Sydney, NSW, Australia

Department of Otolaryngology—Head and Neck Surgery, Johns Hopkins University, Baltimore, MD, USA
e-mail: americo.migliaccio@gmail.com

P. Ravat et al. (eds.), *Advances in Translational Neuroscience of Vestibular and Eye Movement Disorders*, Contemporary Clinical Neuroscience,
https://doi.org/10.1007/978-3-032-07539-0_8

improving many different behavioral outcome measures. The Cawthorne-Cooksey exercises, however, were not targeted to rehabilitate the lesioned vestibulo-ocular reflex (VOR). It was not until the mid-1990s that exercises targeting the VOR were prescribed and revealed to have benefits. Susan Herdman and Rick Clendaniel conducted one of the earliest randomized-controlled trials (RCT) investigating the role of prescribing explicit gaze stabilization exercises [3]. Compared against eye-movement-only exercises (no head motion—control group), Herdman et al. illustrated both improved balance and reduced perception of imbalance. Later, Herdman et al. illustrated that similar gaze stability exercise training (compared with a control group doing eye-movement-only exercises) improved the ability to see clearly during head rotation, known as dynamic visual acuity (DVA), in participants with unilateral and bilateral vestibular hypofunction [4, 5]. The mechanism for improving DVA when the VOR is lesioned is putatively related to both enhancement of the gain (gain = eye/head velocity, bringing it closer to unity) of the VOR, but also the presence of compensatory saccades that reduce the gaze position error created when the eye falls off target during an ipsilesional head rotation [6, 7].

Gaze stabilization exercises are a collective of four unique exercises: VORx1, VORx2, gaze shift, and imaginary target. The VORx1 exercise teaches patients to rotate their head while visually fixating on a still target, which requires a gain of one for visual stability. The VORx2 exercise teaches the patient to rotate their head exactly opposite in direction to a moving target while fixating the moving target, which requires a gain of two. The Gaze Shift exercise teaches patients to actively gaze shift between two stationary targets [8]. The imaginary target exercise has patients attempt to keep their eyes fixated on a still target, after moving their head with eyes closed.

Gaze stability training represents the uniquely 'vestibular' component for the rehabilitative management of patients with vestibular hypofunction. While balance exercises are also incorporated, it is the gaze stability component that distinguishes vestibular rehabilitation as valuable in the treatment of vestibular hypofunction [9]. We have developed a safe, noninvasive rehabilitation technique that, after a single 15-min session, increases the gain of the VOR (Fig. 8.1).

Known as incremental vestibulo-ocular reflex adaptation (IVA), users wear a head-mounted module that projects a laser target onto a wall that moves at an adjustable fraction of head velocity, Fig. 8.2.

We have illustrated that the lesioned gain of the VOR retains the ability to be increased [10, 11]. The IVA method is robust and can create a vestibular asymmetry by concomitantly increasing VOR gain for head rotations to one side, while reducing gain for head rotations to the opposite side [12]. Recently, the IVA method was used to unidirectionally increase the vertical VOR gain in healthy controls during vertical IVA training [13].

The IVA method holds promise for improving the function of individuals with vestibular hypofunction, yet clinical trials using the IVA method have only recently been completed, and the superiority of methods between traditional vestibular rehabilitation (VPT) vs IVA is unknown. The purpose of this paper is to review

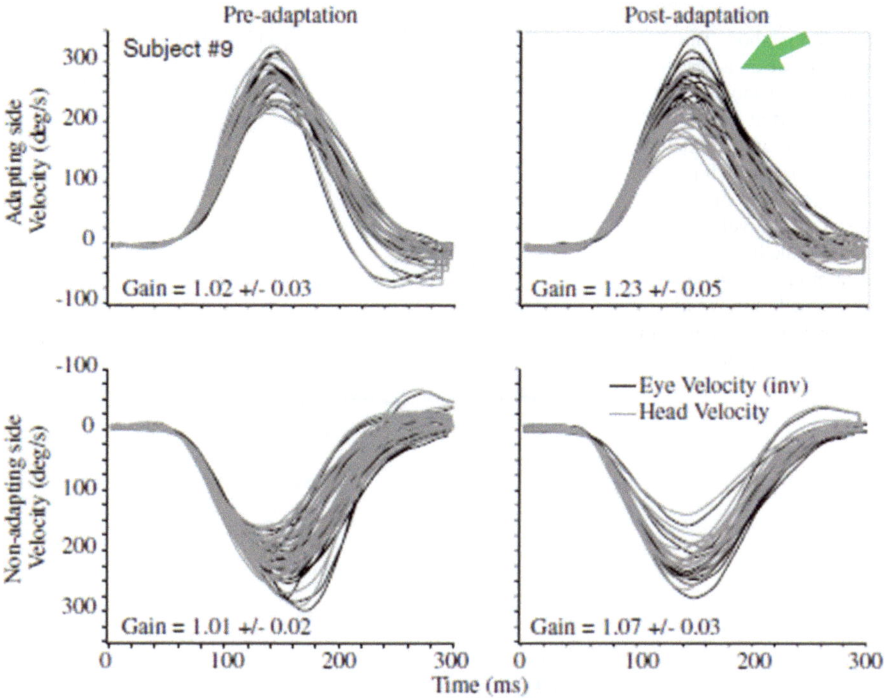

Fig. 8.1 Unilateral VOR adaptation. The upper right panel (adapting side) black eye velocity traces (green arrow) are higher than the head velocity traces (gray), as evidenced by the VOR gain being 1.23 (16% increase). A VOR gain of 1.0 is normal. We are able to drive the VOR gain up on one side only, a critical rehabilitation advance for treating unilateral vestibular symptoms

the existing clinical trial data using the IVA method, as presented at the 2023 La Fundación Iberoamericana de Neurotología (FINO) meeting in honor of Dr. Sergio Carmona.

8.2 Results

8.2.1 One Week Randomized Clinical Trial

Twenty-four subjects aged 18–85 years with chronic (>6 months) stable peripheral vestibular hypofunction with a minimum of two abnormal video head impulse tests as verified by a neurologist, consented for participation. The control and experimental groups performed VOR exercises in the laboratory for 15 min/day for 3 days in 1 week. The control group performed VORx1 exercises using StableEyes to project a space-fixed target, whereas the experimental group performed IVA exercises using StableEyes to control the laser target motion [14].

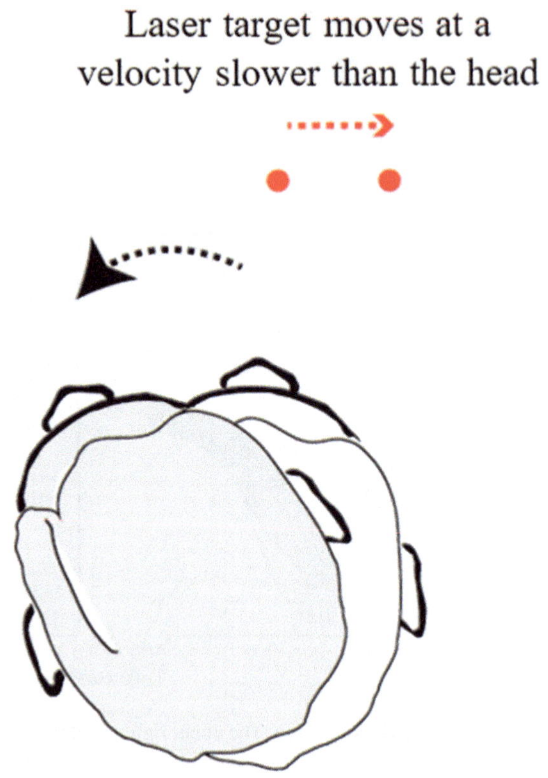

Fig. 8.2 Incremental VOR adaptation method. The participant wears a head-mounted laser module that contains inertial measurement units to detect head velocity, which is programmable to move the laser target at a velocity relative to the head velocity. The laser target can be programmed to move in a direction that can increase or decrease VOR gain

The primary outcome measure was VOR gain as measured using video-oculography (Interacoustics Middelfart, Denmark). Secondary outcome measures included: metrics of compensatory saccades (measured using scleral search coils), head impulse dynamic visual acuity, postural control (stand on firm/foam surface, eyes open/closed), fall risk (dynamic gait index), gait speed, gait kinematics per GaitRITE mat (Franklin NJ, USA) and subjective scales (dizziness handicap inventory).

8.2.1.1 Oculomotor Improvement

VOR gain increased significantly during both passive and active head rotation only for those participants in the IVA group by 12.1 ± 19.9% and 13.4 ± 16.3%, respectively. In contrast, the VOR gain of participants in the control group did not change (−2.8 ± 10.7% passive or −5.4 ± 9.6% active).

The mean latency of compensatory saccades (CS) during ipsilesional passive head rotations significantly reduced over the week for the control group from 271 ± 98 ms (Day 1) to 193 ± 117 (Day 5) ($P < 0.05$). In contrast, those in the

IVA group did not change significantly, though tended to increase, 152 ± 63 ms (Day 1) to 184 ± 82 ms (Day 5), ($P = 0.28$).

8.2.1.2 Functional Improvement

Dynamic visual acuity during active head impulse rotation improved similarly for both groups, and there was no difference in the rate of DVA improvement per day (0.04 Logarithm of the minimal angle resolvable (LogMAR)).

Subjective perception of how disabling participants judged their dizziness, as measured by the Dizziness Handicap Inventory (DHI), was significantly improved for the IVA group. Only those with IVA training showed a significant reduction from baseline for total score ($F1,13 = 12.5$, $P < 0.01$), functional score ($F1,13 = 12.1$, $P < 0.01$), and emotional score ($F1,13 = 5.4$, $P < 0.04$).

Similarly, gait and posture were significantly better in the IVA group:

- IVA increased gait speed from days 1 to 5 by 0.08 ± 0.24 m/s, whereas VORx1 reduced gait speed by 0.17 ± 0.32 m/s ($P < 0.02$). [15, 16]
- IVA increased step length by 5.6 ± 5.9 cm, which was more than double VORx1 (2.07 ± 6.65 cm) ($P < 0.005$).
- The ability for subjects to stand on a firm surface with eyes closed was significantly increased for the IVA group (from days 1 to 5 by 0.84 ± 3.02 s), whereas for the VOR x1 group, it decreased by 3.22 ± 5.92 s ($P < 0.05$).
- More subjects in the IVA group had improved head rotation task scores as per the dynamic gait index (DGI) at day 5 (walking with yaw head turn, walk with pitch head turn) ($P < 0.05$) than in the control group. However, the improvement in total DGI scores was similar between groups, though only the IVA group achieved a normal score (mean > 20) on Day 5.

8.2.2 Two-Year Randomized Clinical Trial

For the 2-year RCT, $n = 21$ subjects with chronic peripheral vestibular hypofunction were recruited following the same inclusion criteria as the 1-week RCT [17]. Unique to the long-term RCT was that each treatment arm was 6 months in duration with a 6-month washout. VOR gain was measured once monthly. One group performed the VORx1 (control) exercise using StableEyes to project the space-fixed laser target for 15 min once/day for 6 months, followed by 6 months washout, then crossed over into IVA training using StableEyes for 15 min once/day for 6 months, followed by a final 6-month washout. The other group followed the same protocol, but with the training order swapped, i.e., IVA training before VORx1 training. All treatments were completed at home (a StableEyes unit was provided to each participant). Similar outcome measures to the 1-week trial above were collected, and video-oculography was used for VOR gain and compensatory saccade metrics (i.e., not scleral search coils).

8.2.2.1 Oculomotor Improvement

As was the result in the 1-week trial, only IVA training showed improved VOR gain during both passive and active head impulse testing, 30.6% and 20.6% ($P < 0.01$), respectively. In contrast, VORx1 (control) training showed no change in VOR gain for passive (−0.7%) or active (−2.4%) head impulse testing ($P = 0.49$). On average, IVA training resulted in 27% retention of the VOR gain increase at the end of the 6-month washout. However, not all subjects lost their improved VOR gain; a few subjects continued to increase even though IVA training had ended. For example, one subject showed VOR gain of 0.35 ± 0.04 at the beginning of washout, which increased to 0.43 ± 0.07 after 6 months, representing at 22% improvement during the washout. The extent of VOR gain increase due to IVA training was dependent on group assignment order. Higher VOR gain increases were achieved when participants were randomized to IVA as their first treatment arm (i.e., before cross-over to VORx1) compared to those who had IVA as their second treatment arm (i.e., after cross-over from VORx1). Further evidence for this cross-over effect is that the DHI score was better when IVA occurred before VORx1, compared to after VORx1.

Like the results of compensatory saccades latency from the one-week RCT, VORx1 resulted in a greater reduction of CS latency (45 ms) compared to IVA training (20 ms) ($P < 0.03$).

8.2.2.2 Functional Improvement

The long-term trial suffered six dropouts (COVID-19, travel, personal issues), prompting reanalysis using multiple imputation. After this reanalysis, neither gait speed nor step length was different between IVA or VORx1 training. Before reanalysis, both were significantly better after IVA training.

The impulse DVA scores were similarly affected by the dropout rate, as the final analysis did not show any significant difference between IVA or VORx1 training, even though the mean scores after IVA training improved by 30% for passive and 33% for active head impulse testing. Impulse DVA scores were better during active compared to passive head rotations for both VORx1 and IVA training ($P < 0.05$).

The DHI total score and DHI physical subdomain were better in those subjects who were randomized first to IVA training followed by VORx1. Whereas, when participants were randomized first to VORx1 followed by IVA, DHI was no different.

8.2.3 INVENT

The INVENT clinical trial (incremental velocity error as a new treatment for vestibular rehabilitation) is a double-blinded randomized, cross-over design comparing traditional vestibular rehabilitation (VPT) with IVA rehabilitation [18]. The INVENT clinical trial is still recruiting participants, and thus, the descriptive data below represent mean data without knowing which treatment may be superior. Separating itself from the prior clinical trials, the traditional VPT exercises were not delivered using the StableEyes training device. The StableEyes VORx1 training exceeds current VORx1 clinical guidelines on head velocity because it employs a high-frequency head rotation stimulus that predominantly requires the VOR for vision stabilization, in addition to providing a high visual-contrast environment shown to be crucial for VOR adaptation [19]. Instead, the gaze stability training as part of the VPT was delivered via paper/pencil following the current recommended clinical practice guidelines [9]. Additionally, the INVENT trial is recruiting participants with chronic peripheral vestibular hypofunction or mild traumatic brain injury (mTBI) as long as either diagnosis includes a "vestibular" symptom (i.e., dizziness, imbalance, gaze stability, etc).

8.2.3.1 Oculomotor Improvement

The mean VOR gain for $n = 24$ participants with chronic vestibular hypofunction during passive impulses toward the affected side is 0.49 ± 0.32; contralesional passive impulses toward the unaffected side are 0.79 ± 0.26 (<0.8 is abnormal). For the subjects with mTBI ($N = 7$), VOR gain for passive impulses toward the right side is 0.8 ± 0.42 (<0.8 is abnormal) and 1.1 ± 0.21 to the left side.

To date, most subjects with peripheral vestibular hypofunction are showing increased VOR gain that is changing from between 0.18 and 0.54, which represents a 13–200% percent improvement.

8.2.3.2 Functional Improvement

Both patient cohorts appear to have similar gait and posture metrics as measured to date, Table 8.1. In general, it appears that reports of subjective confidence in balance and perception of dizziness disability are similar across the cohorts. The mean

Table 8.1 Posture and gait metrics from both cohorts of the INVENT trial

	mCTSIB s	Gait speed m/s	FGA	TUG Ipsi/right s	TUG contra/left s
Peripheral vestibular hypofunction	109 ± 13.9	1.1 ± 0.21	22.6 ± 4.6	9.6 ± 2.4	9.6 ± 2.1
mTBI	97 ± 20.7	1.07 ± 0.13	25 ± 7	6.3 ± 1.8	6.6 ± 0.93

mTBI mild traumatic brain injury, *CTSIB* Clinical Test of the Sensory Interaction on Balance, normal score is 120 s, *TUG* Timed Up and Go, normal scores >13.5 s, *FGA* Functional Gait Assessment Score, normal score is >22, *Ipsi* ipsilesional whole body turn, *Contra* contralesional whole body turn

Activity of Balance Confidence score is 85 ± 19 for participants with mTBI, and 76 ± 17 for participants with chronic peripheral vestibular hypofunction (abnormal is <80). Total scores on the DHI for participants with mTBI are 29 ± 30 (categorized as moderately disabling), while the peripheral hypofunction participants report a total score of 41.8 ± 20.6 (categorized as moderately disabling).

8.3 Discussion

Current standard of care for VPT attempts to change the VOR gain via gaze stability exercises [20], yet few studies have objectively demonstrated a change in the VOR. [6, 21, 22] Most studies do not objectively measure the VOR with eye-tracking systems, but rather consider balance, dynamic visual acuity, and subjective measures of impairment as evidence for the effectiveness of VPT. [22–26] Given that the VOR training response is both context dependent [27, 28] and VOR adaptation is frequency selective [19], efforts to modify the physiologically relevant high-frequency VOR must involve high-frequency head rotations, yet traditional VPT typically uses lower frequency gaze stability exercises.

To date, we have illustrated the IVA method for changing the gain of the VOR in 11 healthy control studies and six vestibular hypofunction studies, with VOR gain being the primary focus, not clinical outcomes. These early studies in humans demonstrated significant increases in VOR gain with just 15 min of training. To date, we have not had any adverse effects using the IVA method.

8.3.1 Short-Term IVA Training

Relative to the x1 gaze stability exercise, participants who completed 3 days of IVA training in the laboratory showed improved VOR gain and had a longer latency compensatory saccade, suggesting a decreased reliance on compensatory saccades due to this improvement in gain. In addition to these oculomotor advantages, our 3-day clinical trial revealed that participants in the IVA group improved their time to stand on a firm surface with eyes closed, improved their gait speed, doubled their step length, and improved their gait stability during head turns. Subjective perception of disability was also better after IVA training.

8.3.2 Longer-Term IVA Training

Relative to x1 gaze stability training, 6 months of IVA training on average 4 days/week revealed increased VOR gain during passive (30.6%) and active (20.6%) head impulse testing, whereas the VORx1 control exercise group had no change in VOR gain. Although a majority of the change in VOR due to IVA was reduced after 6 months, a mean 27% was retained. Additionally, the amount of retention did vary across subjects, suggesting that IVA is a training response that may need to be augmented with repeat 'booster' exposures. In addition, the largest increases in VOR gain occurred during the first 4–8 weeks of IVA training, which corresponds to the duration of typical current vestibular rehabilitation programs. Such a program could be repeated as needed and preferably performed concurrently with balance and gait exercises. The 2-year clinical trial suffered from dropout and a reanalysis that affected some of the early, significantly improved posture and gait outcome measures. Subjective perception of disability, however, was still significantly better after IVA training. Results from the INVENT trial are pending.

8.3.3 What's Next

Dosing of vestibular rehabilitation, including the IVA method, is still largely unknown. For example, would IVA be as effective if it were delivered every other day? Only recently has the literature suggested that early onset gaze stability training leads to improved oculomotor function and functional behavior [29]. Does an optimal window exist for IVA training in the setting of acute vestibular hypofunction?

We have not applied the IVA method to patients without peripheral vestibular deficits and mTBI, though soon it will be compared against traditional VPT in participants with multiple sclerosis.

In conclusion, the IVA holds promise for advancing VPT by providing a method to improve the VOR gain when deficient. Results from the first clinical trials suggest that changing VOR gain does, in fact, help improve posture, gait, and subjective perception of disability. Certainly, the IVA method is also a great method to measure plasticity of the brain, and it also holds promise as a useful research instrument.

Acknowledgments The author acknowledges Carlo N Rinaudo PhD, Chris J Todd BE, M Muntaseer Mahfuz PhD, Jennifer Millar MS phyiscal therapy (PT), Amrita Mukunda Bachelor of Science (BS), Isaac Thorman MS, and Colin R Grove PT PhD, for their help in data collection.

Funding Michael C Schubert (MCS) is funded by the Department of Defense under (Grant award #W81XWH-15-1-0442). Americo Aniello Migliaccio (AAM) was supported by The Garnett Passe and Rodney Williams Memorial Foundation Senior/Principal Research Fellowship and Project Grant (2013-15), and National Health and Medical Research Council (NHMRC) Development Grant APP105550. Carlo N Rinaudo (C.N.R.) was supported by an Australian Postgraduate Award, the Chiropractors Association of Australia, and a Neuroscience Research Australia scholarship.

Michael C Schubert (MCS) and Americo Aniello Migliaccio (AAM) hold US and Australian patents on the StableEyes device through Johns Hopkins University and Neuroscience Research Australia.

References

1. Cawthorne T. Vestibular injuries. Proc R Soc Med. 1946;39(5):270–3. PMID:19993268.
2. Cooksey FS. Rehabilitation in vestibular injuries. Proc R Soc Med. 1946;39(5):273–8.
3. Herdman SJ, Clendaniel RA, Mattox DE, Holliday MJ, Niparko JK. Vestibular adaptation exercises and recovery: acute stage after acoustic neuroma resection. Otolaryngol Head Neck Surg. 1995;113(1):77–87. https://doi.org/10.1016/s0194-5998(95)70148-6.
4. Herdman SJ, Schubert MC, Das VE, Tusa RJ. Recovery of dynamic visual acuity in unilateral vestibular hypofunction. Arch Otolaryngol Head Neck Surg. 2003;129(8):819–24.
5. Herdman SJ, Hall CD, Schubert MC, Das VE, Tusa RJ. Recovery of dynamic visual acuity in bilateral vestibular hypofunction. Arch Otolaryngol Head Neck Surg. 2007;133(4):383–9.
6. Schubert MC, Migliaccio AA, Clendaniel RA, Allak A, Carey JP. Mechanism of dynamic visual acuity recovery with vestibular rehabilitation. Arch Phys Med Rehabil. 2008;89(3): 500–7. PMID: 18295629
7. Schubert MC, Hall CD, Das V, Tusa RJ, Herdman SJ. Oculomotor strategies and their effect on reducing gaze position error. Otol Neurotol. 2010;31(2):228–31. https://doi.org/10.1097/MAO.0b013e3181c2dbae.
8. Scherer M, Schubert MC. High-velocity angular vestibulo-ocular reflex adaptation to position error signals. J Neurol Phys Ther. 2010;34(2):82–6. PMID: 20588093; PMCID: PMC2954113.
9. Hall CD, Herdman SJ, Whitney SL, Anson ER, Carender WJ, Hoppes CW, Cass SP, Christy JB, Cohen HS, Fife TD, Furman JM, Shepard NT, Clendaniel RA, Dishman JD, Goebel JA, Meldrum D, Ryan C, Wallace RL, Woodward NJ. Vestibular rehabilitation for peripheral vestibular hypofunction: an updated clinical practice guideline from the academy of neurologic physical therapy of the American Physical Therapy Association. J Neurol Phys Ther. 2022;46 (2):118–77. https://doi.org/10.1097/NPT.0000000000000382. PMID: 34864777; PMCID: PMC8920012.
10. Schubert MC, Della Santina CC, Shelhamer M. Incremental angular vestibulo-ocular reflex adaptation to active head rotation. Exp Brain Res. 2008;191(4):435–46. https://doi.org/10.1007/s00221-008-1537-z. Epub 2008 Aug 20. PMID: 18712370; PMCID: PMC2946619.
11. Migliaccio AA, Schubert MC. Unilateral adaptation of the human angular vestibulo-ocular reflex. J Assoc Res Otolaryngol. 2013;14(1):29–36.
12. Gimmon Y, Migliaccio AA, Todd CJ, Figtree WVC, Schubert MC. Simultaneous and opposing horizontal VOR adaptation in humans suggests functionally independent neural circuits. J Neurophysiol. 2018;120(4):1496–504. https://doi.org/10.1152/jn.00134.2018. Epub 2018 June 27
13. Todd CJ, Schubert MC, Rinaudo CN, Migliaccio AA. Unidirectional vertical vestibuloocular reflex adaptation in humans using 1D and 2D scenes. Otol Neurotol. 2022;43(9):e1039–44. https://doi.org/10.1097/MAO.0000000000003684. Epub 2022 Sept 9
14. Rinaudo CN, Schubert MC, Cremer PD, Figtree WVC, Todd CJ, Migliaccio AA. Once-daily incremental vestibular-ocular reflex adaptation training in patients with chronic peripheral vestibular hypofunction: a 1-week randomized controlled study. J Neurol Phys Ther. 2021;45 (2):87–100. https://doi.org/10.1097/NPT.0000000000000348.
15. Wellons RD, Duhe SE, MacDowell SG, Hodge A, Oxborough S, Levitzky EE. Estimating the minimal clinically important difference for balance and gait outcome measures in individuals with vestibular disorders. J Vestib Res. 2022;32(3):223–33. https://doi.org/10.3233/VES-201630.
16. Thorman IB, Loyd BJ, Clendaniel RA, Dibble LE, Schubert MC. The minimal clinically important difference for gait speed in significant unilateral vestibular hypofunction after vestibular rehabilitation. J Otol. 2023;18(1):15–20. https://doi.org/10.1016/j.joto.2022.11.001. Epub 2022 Nov 10. PMID: 36820158; PMCID: PMC9937836.
17. Rinaudo CN, Schubert MC, Cremer PD, Figtree WVC, Todd CJ, Migliaccio AA. Comparison of incremental vestibulo-ocular reflex adaptation training versus x1 training in patients with

chronic peripheral vestibular hypofunction: a two-year randomized controlled trial. J Neurol Phys Ther. 2021;45(4):246–58. https://doi.org/10.1097/NPT.0000000000000369.
18. Ervin AM, Schubert MC, Migliaccio AA, Perin J, Coulibaly H, Millar JL, Roberts D, Shelhamer M, Gold D, Beauregard S, Pinto R, Brungart D, Ward BK, INVENT VPT Research Group. Incremental velocity error as a new treatment in vestibular rehabilitation (INVENT VPT) trial: study protocol for a randomized controlled crossover trial. Trials. 2021;22(1):908. https://doi.org/10.1186/s13063-021-05876-4. PMID: 34895314; PMCID: PMC8666079.
19. Rinaudo CN, Schubert MC, Figtree WVC, Todd CJ, Migliaccio AA. Human vestibulo-ocular reflex adaptation is frequency selective. J Neurophysiol. 2019;122(3):984–93. https://doi.org/10.1152/jn.00162.2019. Epub 2019 Jul 24
20. Herdman SJ, Clendaniel R. Vestibular rehabilitation (contemporary perspectives in rehabilitation). 4th ed. Philadelphia: FA Davis.2014
21. Verdecchia DH, Mendoza M, Sanguineti F, Binetti AC. Outcomes after vestibular rehabilitation and Wii® therapy in patients with chronic unilateral vestibular hypofunction. Acta Otorrinolaringol Esp. 2014;65(6):339–45. English, Spanish. https://doi.org/10.1016/j.otorri.2014.02.012. Epub 2014 July 1.
22. Millar JL, Gimmon Y, Roberts D, Schubert MC. Improvement after vestibular rehabilitation not explained by improved passive VOR gain. Front Neurol. 2020;11:79. https://doi.org/10.3389/fneur.2020.00079. PMID: 32153490; PMCID: PMC7044341.
23. Hall CD, Herdman SJ. Reliability of clinical measures used to assess patients with peripheral vestibular disorders. J Neurol Phys Ther. 2006;30(2):74–81.
24. Hall CD, Heusel-Gillig L, Tusa RJ, Herdman SJ. Efficacy of gaze stability exercises in older adults with dizziness. J Neurol Phys Ther. 2010;34(2):64–9. PMID:20588090
25. Roller RA, Hall CD. A speed-based approach to vestibular rehabilitation for peripheral vestibular hypofunction: a retrospective chart review. J Vestib Res. 2018;28(3–4):349–57. https://doi.org/10.3233/VES-180633. PMID: 29689764; PMCID: PMC9249287.
26. Arnold SA, Stewart AM, Moor HM, Karl RC, Reneker JC. The effectiveness of vestibular rehabilitation interventions in treating unilateral peripheral vestibular disorders: a systematic review. Physiother Res Int 2017;22(3). https://doi.org/10.1002/pri.1635. Epub 2015 June 25.
27. Schubert MC, Migliaccio AA, Minor LB, Clendaniel RA. Retention of VOR gain following short-term VOR adaptation. Exp Brain Res. 2008;187(1):117–27. https://doi.org/10.1007/s00221-008-1289-9. Epub 2008 Jan 30. PMID: 18231780; PMCID: PMC2964937.
28. Yakushin SB, Raphan T, Cohen B. Context-specific adaptation of the vertical vestibuloocular reflex with regard to gravity. J Neurophysiol. 2000;84:3067–71.
29. Lacour M, Tardivet L, Thiry A. Rehabilitation of dynamic visual acuity in patients with unilateral vestibular hypofunction: earlier is better. Eur Arch Otorrinolaringol. 2020;277(1): 103–13. https://doi.org/10.1007/s00405-019-05690-4. Epub 2019 Oct 21.

Chapter 9
Minimum Stimulus Strategy: A Step-by-Step Diagnostic Approach to BPPV

Giacinto Asprella Libonati, Fernanda Asprella Libonati, and Rudi Pecci

Keywords Benign paroxysmal positional vertigo · BPPV · Minimum stimulus strategy · Upright BPPV protocol · Upright head pitch test · Upright head roll test · Upright LARP/RALP test · Geotropic nystagmus · Apogeotropic nystagmus

Benign paroxysmal positional vertigo (BPPV) is the most common vestibular disorder, mechanically affecting the posterior labyrinth, which we can refer to as "maculo-canalopathy," where otoconial debris dislodges from the utricular macula and enters one or more semicircular canals (SCs), the latter becoming sensitive to gravity, which inappropriately activates the involved ampullary receptors [1, 2].

Two main pathophysiological mechanisms are widely recognized as responsible for BPPV: canalolithiasis, with free-floating particles within the SC, and cupulolithiasis, with particles attached to the cupula of the affected SC [1–4]. Specifically, in canalolithiasis, following head position changes, otoconial particles freely move back and forth within the canal, pushing on the endolymphatic column like a piston, thus provoking endolymph displacement and cupula deflection, and generating either excitatory or inhibitory stimuli and short-lasting paroxysmal positional nystagmus and vertigo [5]. Conversely, in cupulolithiasis, otoconial particles adhere to the cupula, rendering it heavy and gravity sensitive (heavy cupula), so that both positional nystagmus and vertigo following head position changes are more long-lasting [1–5].

Another uncommon variant of BPPV is that of canalith jam, in which otoconia aggregate in a clot, plugging the canal lumen and blocking the endolymphatic flow

G. Asprella Libonati (✉)
Ospedale "Madonna delle Grazie", U.O.C. di Otorinolaringoiatria e Chirurgia Cervico-Facciale, Matera, Italy

F. Asprella Libonati
Independent Researcher, Rome, Italy

R. Pecci
Azienda Ospedaliero-Universitaria Careggi, SOD di Audiologia e Otorinolaringoiatria, Florence, Italy

P. Ravat et al. (eds.), *Advances in Translational Neuroscience of Vestibular and Eye Movement Disorders*, Contemporary Clinical Neuroscience,
https://doi.org/10.1007/978-3-032-07539-0_9

within the SC [6–9]. In this case, the hydrostatic pressure between the otoconial clot and the cupula of the affected canal results in a persistent cupular deflection that generates a direction-fixed nystagmus, regardless of the head position.

We can topographically classify BPPV according to the affected SC and the involved arm [10, 11], thus identifying the Posterior Semicircular Canal (PSC)-Benign Paroxysmal Positional Vertigo (BPPV), with the geotropic and apogeotropic variants, the Lateral Semicircular Canal (LSC)-BPPV, similarly with the geotropic and apogeotropic variants, the Anterior Semicircular Canal (ASC)-BPPV, and the Multicanal (MC)-Benign Paroxysmal Positional Vertigo (BPPV). More frequently, BPPV involves a single SC, although more SCs might be simultaneously affected, and on one or both sides. Furthermore, the most frequent variant is PSC-BPPV, accounting for 80–90% of cases; whereas, those involving the Horizontal Semicircular Canal (HSC) and ASC account for 10–20% and 3% of all BPPV, respectively [10].

As for the diagnosis, observation of the triggered positional nystagmus is the key feature to identify the affected ear and the involved canal and arm in BPPV. Indeed, according to the Vestibulo-Ocular Reflex (VOR) pathways, the deflection of each cupula, either exciting or inhibiting the corresponding ampullary afferents, induces the contraction of specific extraocular muscle couples, thus generating the pathognomonic nystagmus [12]. In order to trigger the positional nystagmus, we need to perform equally specific positioning tests, which we refer to as diagnostic maneuvers, moving the patient's head along the plane of the examined SC; these maneuvers induce cupular deflection in the affected canal, directly in the case of cupulolihiasis by exploiting gravity, or by moving free-floating otoconia using both gravity and inertial/centrifugal forces in the case of canalolithisis [3, 4, 10, 11, 13]. However, concerning this, we must keep in mind that a diagnostic maneuver specific to an SC could evoke a positional nystagmus specific to another SC; therefore, diagnosis should always be based on the triggered nystagmus rather than on the diagnostic maneuver [1, 2].

Another important aspect is that the positional nystagmus is more easily triggered and observed if the patient is examined after a short delay from the onset of positional vertigo, that is, within the "active period" (24/72 h). Nevertheless, in this period, the patient suffers from more intense vestibular, autonomic, and anxiety-related symptoms, so that it may be very hard to perform the diagnostic maneuver, and also to detect the affected ear and the involved canal and arm; this is especially true for Horizontal Semicircular Canal (HSC)-Benign Paroxysmal Positional Vertigo (BPPV), where repeated positionings are usually needed.

The solution to this limitation of the traditional bedside vestibular examination is the "Minimum Stimulus Strategy" (MSS), a nystagmus-based approach carried out with ongoing monitoring of the triggered nystagmus by video-Frenzel goggles according to the head position changes, and that allows to reduce the number of diagnostic and therapeutic maneuvers required for BPPV management and consequently the patient's discomfort [14–18].

9.1 Minimum Stimulus Strategy (MSS)

This is a strategy adopted when the case history suggests BPPV, and a rational diagnostic and therapeutic approach to BPPV, aiming at solving it within the first treatment session with a "step by step" procedure [14] (Fig. 9.1).

9.1.1 First Step

The examination begins by looking for any horizontal nystagmus with the patient in a sitting position and the head in neutral position, that is, pseudo-spontaneous nystagmus (PSN). If there is no nystagmus, PSN may be triggered by a mild horizontal head shaking. If, again, there is no nystagmus, it has to be seen if any nystagmus is trigged by changing the bending angle of the head with the patient still in sitting position, which we refer to as upright head pitch test (uHPT, Fig. 9.2). If an already present or a triggered horizontal nystagmus changes its beating direction with the uHPT, it can be considered a PSN, due to a LSC-BPPV; instead, if a vertical-torsional nystagmus is triggered with the uHPT, it suggests a PSC-BPPV, and the Dix-Hallpike maneuver is immediately performed (Flowchart 9.1).

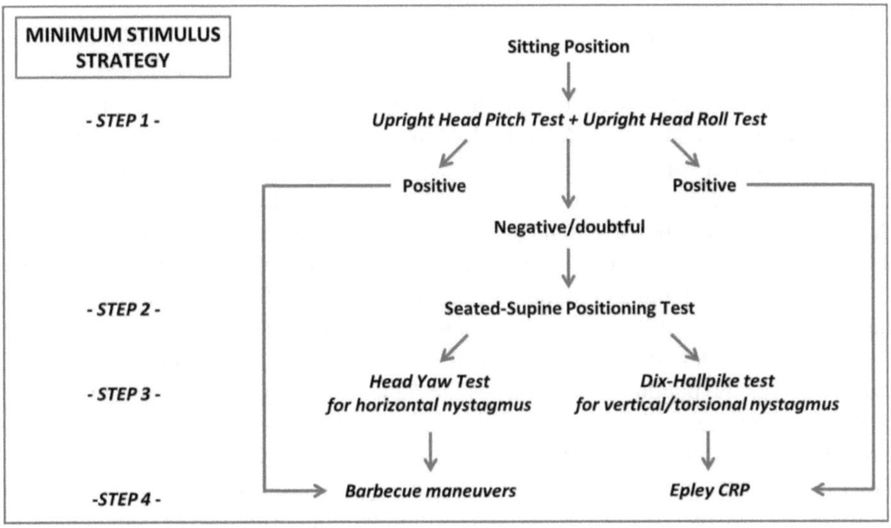

Fig. 9.1 Minimum Stimulus Strategy (MSS; modified from NEUROTOLOGY, by Sergio Carmona and Giacinto Asprella Libonati, 3rd edition. Buenos Aires: Librería Akadia Editorial, 2014)

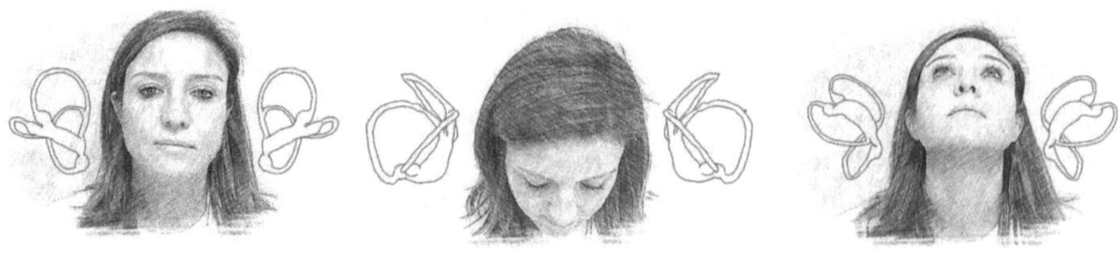

Fig. 9.2 Upright head pitch test. (uHPT; modified from [20])

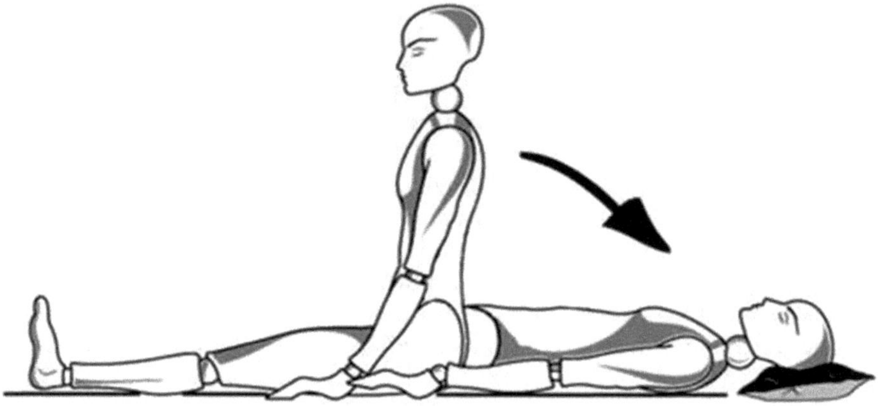

Flowchart 9.1 Strategy of Minimum Stimulus Strategy: first step. (Reproduced from NEUROTOLOGY, by Sergio Carmona and Giacinto Asprella Libonati, 3rd edition. Buenos Aires: Librería Akadia Editorial, 2014)

9.1.2 Second Step

In the second step, the patient is brought from the sitting to the supine position, with the head straight and bent 30° forward, looking for a positional nystagmus, which we refer to as seated supine positioning test (SSPT, Fig. 9.3). Such a maneuver may trigger the movement of free-floating otoconia in canalolithiasis, as well as the deflection of heavy-made cupula in cupulolithias, in both LSC and PSC-BPPV, and thus we also refer to as "Asprella single maneuver" (Fig. 9.4; Flowchart 9.2).

If a vertical-torsional nystagmus is triggered with the SSPT, the patient's head is turned 45° to the right if the torsional component of the nystagmus is directed to the right, and to the left if it is directed to the left; thereafter, the headrest is lowered rapidly so that the patient's head reaches the right or left Dix-Hallpike position. If the nystagmus persists or becomes more intense, the diagnosis of PSC-BPPV is confirmed, and the Epley's canalith repositioning procedure (CRP) is performed. This latter is the most appropriate and suitable technique to use, as it allows one to go on from the already reached Dix-Hallpike position, without bringing the patient back to the sitting position and avoiding other vertigo, according to the MSS. Furthermore, and equally paramount, during the CRP, the patient's eye movements may be

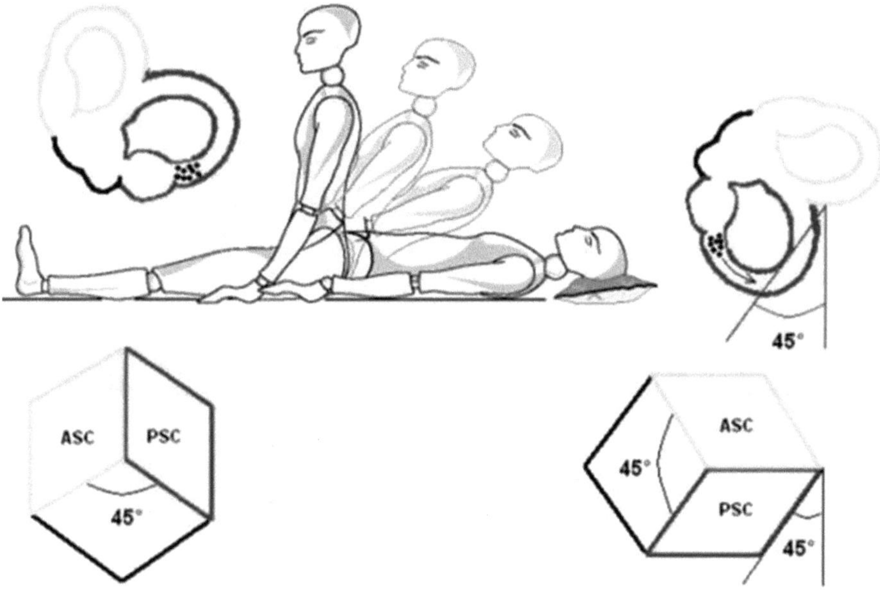

Fig. 9.3 Seated supine positioning test (SSPT; modified from [2])

followed under "step by step" video-Frenzel goggles monitoring, according to the nystagmus-based approach, looking for good prognostic signs: specifically, a nystagmus beating toward the same direction in both the 1° (ipsilateral, diagnostic, Dix-Hallpike) and 2° (contralateral Dix-Hallpike) position points to an ampullofugal movement of the debris, and therefore to the maneuver's effectiveness; likewise, a nystagmus beating downward in both the 3° (face down on the healthy side) and 4° (sitting) position points to a gravitating movement of the debris through the common crus, and therefore, again, to the maneuver's effectiveness (Flowchart 9.3).

Conversely, if a horizontal nystagmus is triggered with the SSPT, LSC-BPPV may be supposed: specifically, a right beating nystagmus may indicate either a left geotropic variant or a right apogeotropic one, and vice versa. Indeed, when the patient is brought in supine position with the head straight and bent 30° forward, the LSC is located on a vertical plane, and therefore the otoconial particles are pushed downward because of the brisk deceleration caused by the maneuver; in so doing, if they are in the posterior arm of the LSC, they move away from the ampulla, thus generating an ampullofugal, and for that inhibitory, endolymphatic current and a nystagmus beating toward the healthy side; instead, if they are in the anterior arm of the LSC, they move toward the ampulla, thus generating an ampullopetal, and for that excitatory, endolymphatic current and a nystagmus beating toward the affected side (Fig. 9.5).

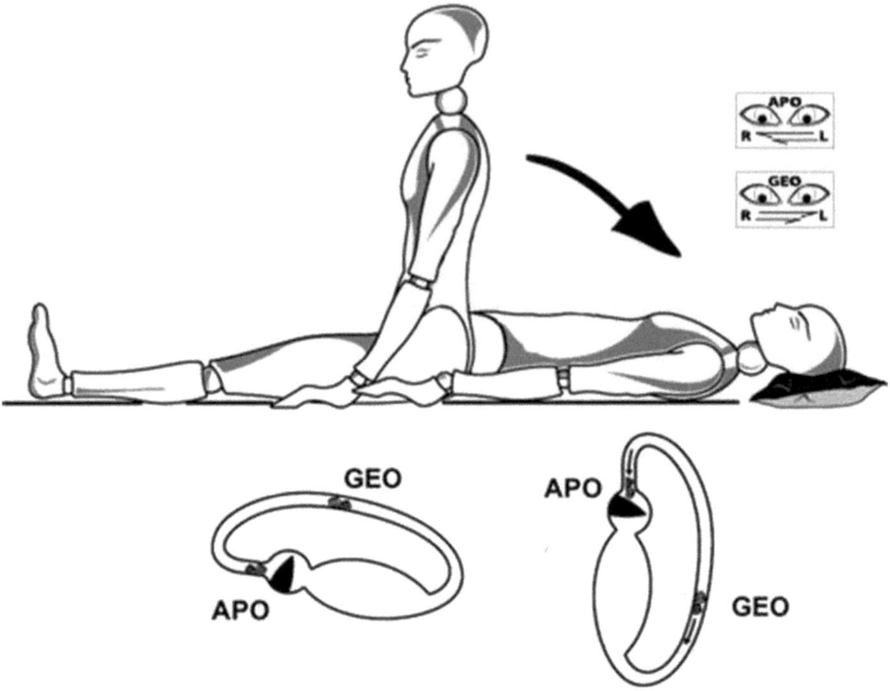

Fig. 9.4 Seated supine positioning maneuver or Asprella single maneuver (reproduced from NEUROTOLOGY, by Sergio Carmona and Giacinto Asprella Libonati, 3rd edition. Buenos Aires: Librería Akadia Editorial, 2014)

9.1.3 *Third Step*

In the third step, the patient's head is turned 180° from side to side in supine position, which we refer to as the supine head yaw test (sHYT). The latter allows for identifying both variant (geotropic or apogeotropic) and side (right or left) of the LSC-BPPV; specifically, the sHYT allows for looking for the third lateralization sign used to recognize the affected side, that is, the ear toward the stronger nystagmus beats. Furthermore, during the sHYT, it is more appropriate and suitable to turn the patient's head toward the same direction of the horizontal nystagmus triggered by the SSPT; indeed, if the nystagmus persists or increases in intensity, a geotropic variant, and of the opposite side, of LSC-BPPV may be diagnosed; instead, if the nystagmus decreases, stops and then reverses its direction, an apogeotropic variant, and of the same side, of LSC-BPPV may be diagnosed. Even more paramount, in both cases, in so doing, the patient's head is rotated toward the side on which the nystagmus is less intense, that is, the healthy side in the geotropic form, and the affected side in the apogeotropic one, according to the MSS once more.

Finally, if a geotropic vâriant of LSC-BPPV is diagnosed, the Vannucchi-Asprella maneuver is performed, going on toward the healthy side; if an

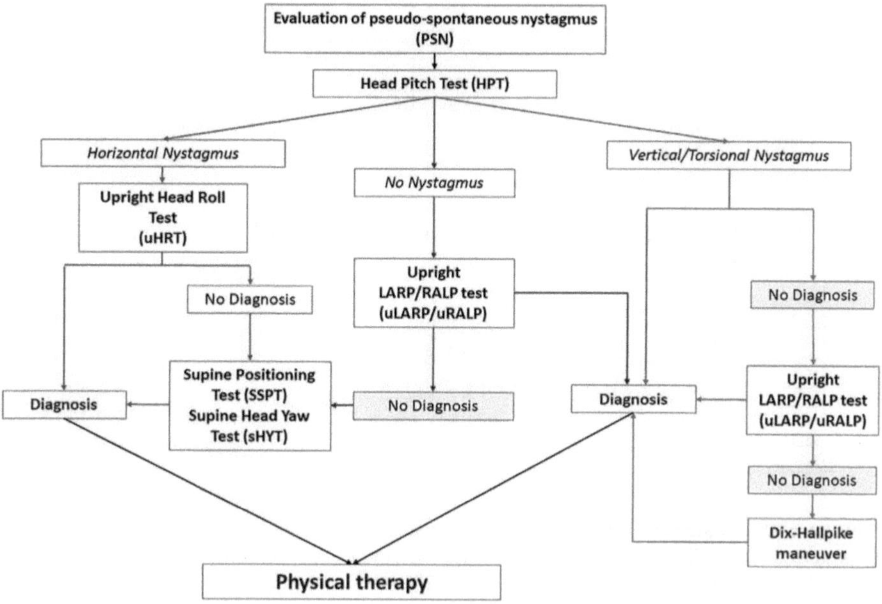

Flowchart 9.2 Strategy of Minimum Stimulus Strategy: second step. (Reproduced from NEUROTOLOGY, by Sergio Carmona and Giacinto Asprella Libonati, 3rd edition. Buenos Aires: Librería Akadia Editorial, 2014)

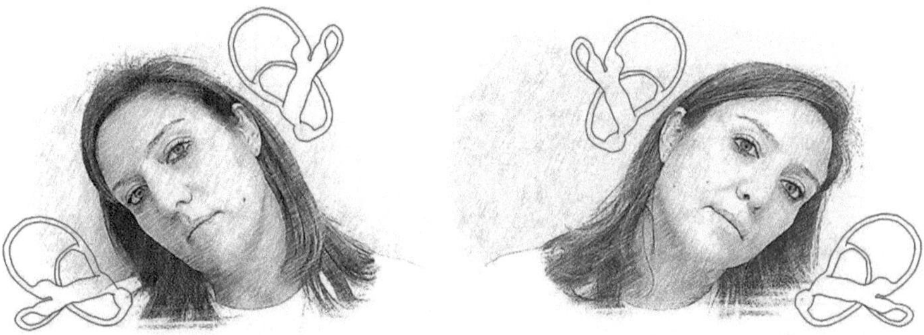

Flowchart 9.3 Strategy of Minimum Stimulus Strategy: second step and PSC therapy. (Reproduced from NEUROTOLOGY, by Sergio Carmona and Giacinto Asprella Libonati, 3rd edition. Buenos Aires: Librería Akadia Editorial, 2014)

apogeotropic variant of LSC-BPPV is diagnosed, first of all it is transformed in a geotropic one by a very fast rotation of the patient's head toward the healthy side (180° from side to side), and immediately after this, it is solved again with the Vannucchi-Asprella maneuver, which is performed again toward the healthy side. This maneuver is the most appropriate and suitable technique to use, exactly because it can immediately solve both geotropic and apogeotropic forms, as well as transform

Fig. 9.5 LSC-BPPV seated supine positioning nystagmus right side. (reproduced from [2])

Flowchart 9.4 Strategy of Minimum Stimulus Strategy: second and third steps and LSC therapy. (Reproduced from NEUROTOLOGY, by Sergio Carmona and Giacinto Asprella Libonati, 3rd edition. Buenos Aires: Librería Akadia Editorial, 2014)

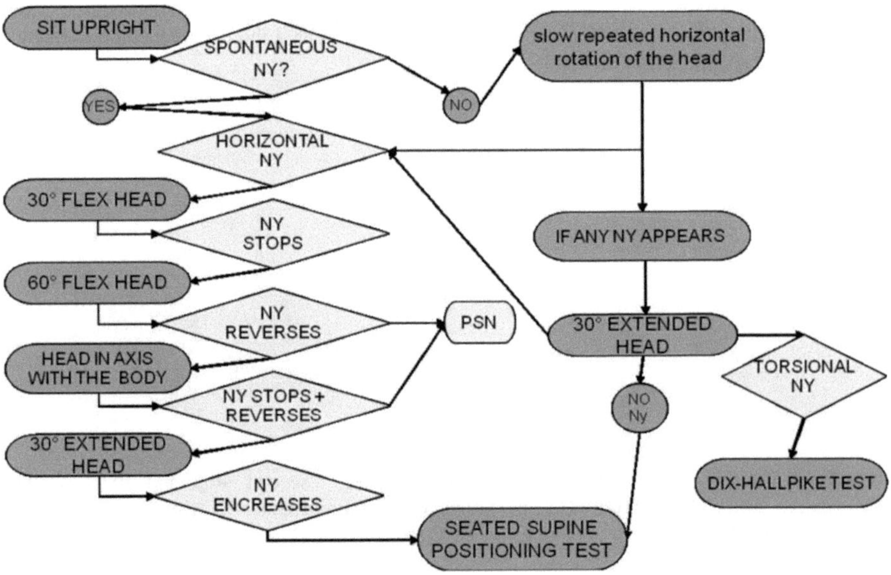

Flowchart 9.5 Minimum Stimulus Strategy: second and third steps and LSC therapy. (Reproduced from NEUROTOLOGY, by Sergio Carmona and Giacinto Asprella Libonati, 3rd edition. Buenos Aires: Librería Akadia Editorial, 2014)

apogeotropic forms into geotropic ones (Flowcharts 9.4 and 9.5), always according to the MSS. Furthermore, and equally paramount, also during this maneuver, the patient's eye movements may be followed under "step by step" video-Frenzel goggles monitoring, according to the nystagmus-based approach, looking for good prognostic signs: specifically, a horizontal nystagmus beating to the healthy side points to an ampullofugal movements of the debris, and when this nystagmus weakens or disappears it means that the debris has been pushed out of the canal and the therapy can be stopped.

Other techniques, such as the barbecue rotation or the Gufoni maneuver, can be used to solve the few cases that do not work well with the Vannucchi-Asprella maneuver.

9.2 Upright BPPV Protocol (UBP)

Recently, the approach in sitting position to the patient with BPPV has been enriched with other details that have been defined "Upright BPPV Protocol" (UBP) [19, 20] (Fig. 9.6), a key point of MSS consisting of diagnostic tests performed by keeping the patient in sitting position and bending his head slowly along the spatial axes, in order to move the debris by gravity along the involved SC.

The rotational movements of the head are described based on the three orthogonal planes of space, pitch, roll, and yaw, each identified by an axis around which the rotation of the head occurs, Y, X, and Z, respectively. Two additional planes exist, aligned with each couple of vertical semicircular canals (SCCs), that is, the left anterior and right posterior, LARP plane, and the right anterior and left posterior, RALP plane, respectively. Such a protocol consists of slowly bending the patient's head along all the SCCs' planes, pitch plane, roll plane, LARP and RALP planes, thus inducing canaliths gravitation, always monitoring positional nystagmus by video-Frenzel goggles.

The UBP [20] starts looking for PSN, with the eyes in primary gaze position, looking straight ahead while sitting: the first UBP step.

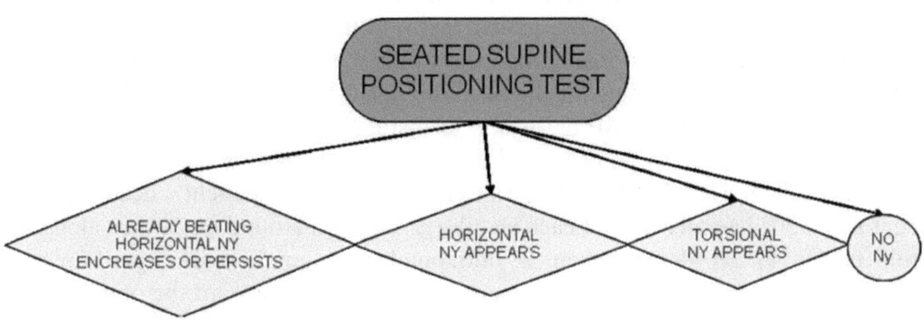

Fig. 9.6 Diagnostic algorithm for BPPV according to the Minimum Stimulus Strategy (MSS; reproduced from [20])

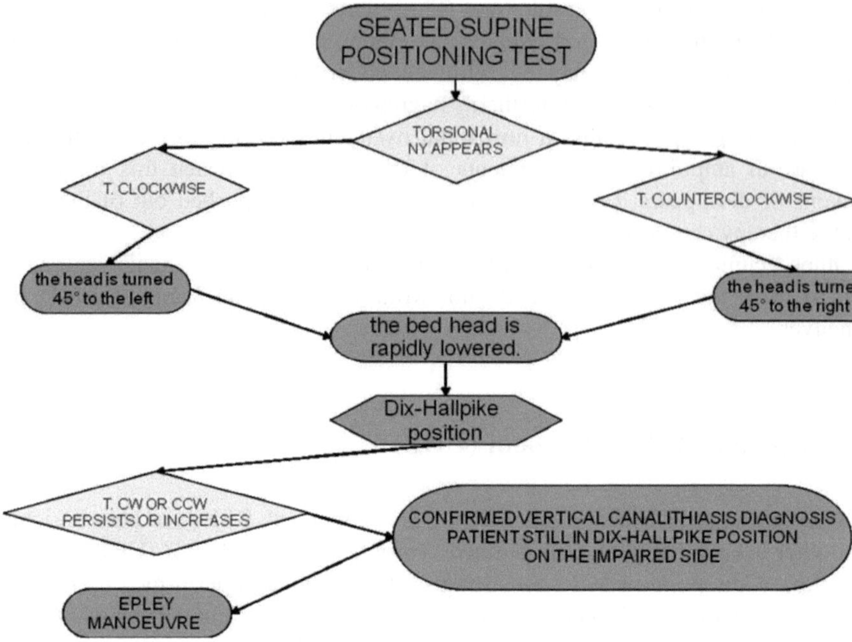

Fig. 9.7 Upright head roll test (uHRT; modified from [20])

If a horizontal persistent nystagmus is observed in the first step, it should be checked performing the uHPT, the second UBP step (Fig. 9.2). In this test, the patient's head is slowly bent 60° forward and then 30° backward slowly rotating around the Y-axis, along the pitch plane; the head should be held in each position until nystagmus modifies its direction, usually within 30 seconds. Sometimes, a horizontal nystagmus in the primary position, head erect, is not yet visible, but such a nystagmus can be elicited by performing the uHPT. The latter nystagmus must equally be considered a PSN if it changes its beating direction with the movements of the head along the pitch plane. The uHPT can sometimes evoke a vertical-torsional nystagmus, typical for a BPPV of the posterior canal. Such a diagnosis can be confirmed by bringing the patient down to the Dix-Hallpike position on the side toward which the torsional component of the nystagmus beats. When in such a position, a typical nystagmus for the posterior canal BPPV is evoked once again.

The third step of the UBP allows us to distinguish, in the presence of a PSN, whether this is due to a geotropic or apogeotropic variant of the lateral canal. It consists of the upright head roll test (uHRT) (Fig. 9.7). The patient's head is slowly bent about 30° laterally toward each shoulder, turning it around the X-axis along the roll plane; in so doing when a horizontal nystagmus (either geotropic or apogeotropic) is elicited, the same maneuver is performed toward the opposite side to check if the nystagmus reverses its direction. Even in this case, the head should be held in each position until any nystagmus appears, usually within 30 seconds. When UBP allows a reliable diagnosis of the side and variant of LC-BPPV, the appropriate therapeutic maneuvers may follow, and specifically, it should be recommended to proceed with a therapeutic procedure starting from the sitting position, such as the Gufoni maneuver. Otherwise, nystagmus evaluation must be continued by bringing the patient down to the supine position (SSPT), looking for seated supine positioning nystagmus (SSPN), and performing the sHYT. This last test allows us to confirm the diagnosis already made in the sitting position of BPPV of the lateral canal in the geotropic or apogeotropic variant, or to clarify the diagnosis if, in the third step of the UBP, this had not been concluded or was not entirely clear.

Conversely, vertical-torsional direction-changing nystagmus, elicited by the uHPT, just allows a reliable diagnosis of BPPV affecting vertical SCCs. Specifically, in geotropic PC-BPPV, the uHPT usually evokes a down-beating nystagmus with torsional component toward the healthy side—not always present—when the patient's head is bent forward, and an up-beating nystagmus with torsional component toward the affected side in backward head bending. Instead, in both apogeotropic variants of PC-BPPV and AC-BPPV, bending the patient's head along the pitch plane (forward and/or backward) usually evokes a down-beating nystagmus with no or poor torsional component. This torsional component, if present with the head tilted back, beats toward the healthy side in the apogeotropic variant of PC-BPPV and toward the affected side in the AC-BPPV. Moreover, the uHPT becomes even more sensitive for vertical canals involvement by moving the patient's head along the planes of the vertical SCCs couples, that is, right anterior—left posterior (RALP) plane and left anterior–right posterior (LARP) plane, thus performing the upright RALP/LARP test (uRALP/uLARP test) (Figs. 9.8 and 9.9).

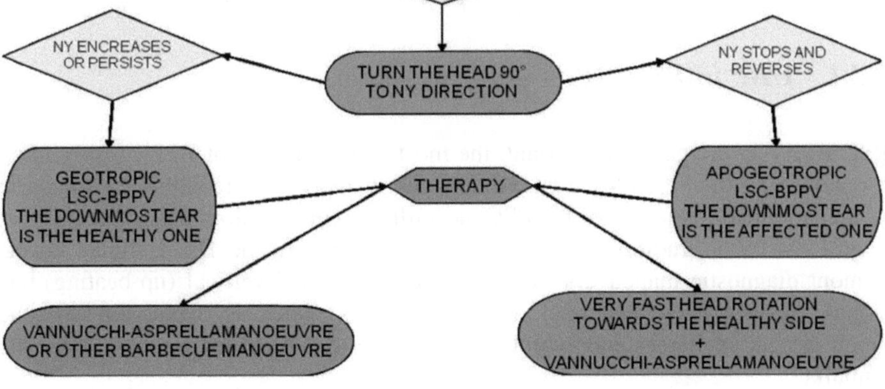

Fig. 9.8 Upright right anterior-left posterior test (uRALP test; modified from [20])

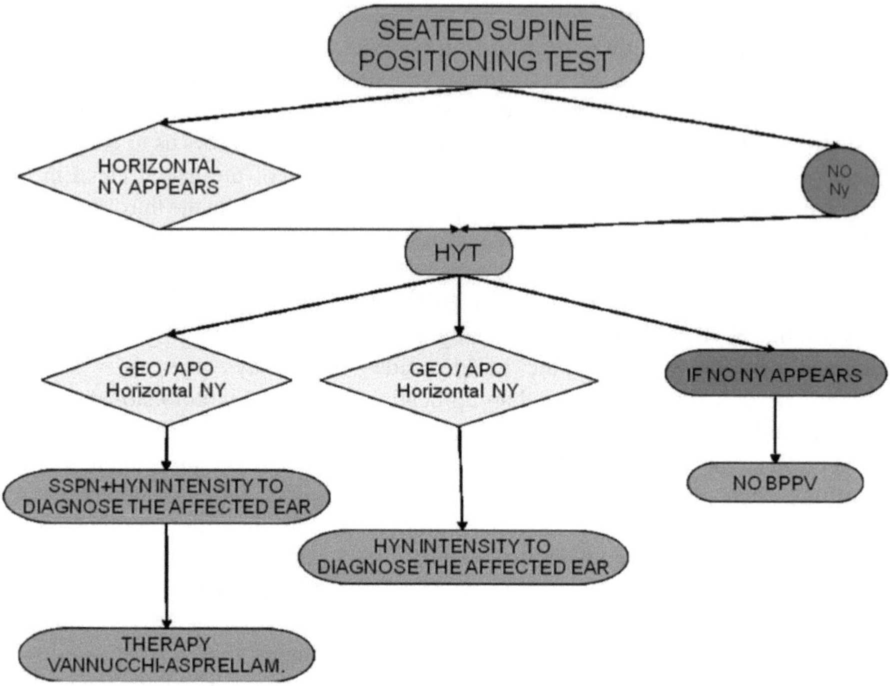

Fig. 9.9 Upright left anterior-right posterior test (uLARP test; modified from [20])

In so doing, right PC-BPPV can be diagnosed by slowly bending the patient's head backward and forward along the LARP plane, uLARP test, whereas the uRALP test may easily detect the involvement of the left PC. Once the affected SCC has been identified, it is possible to proceed with the suitable therapeutic maneuver, and the most used therapeutic techniques for vertical canals are the Epley CRP and the Semont liberatory maneuver.

9.2.1 UBP in PSC-BPPV

PSC-BPPV in the geotropic variant, the most common form of BPPV, accounting for about 80% of all BPPV cases, is due to debris freely floating within the long arm of the ampullary emicanal of the PSC, according to the canalolithiasis theory.

PSC-BPPV is traditionally diagnosed by performing the Dix-Hallpike or the Semont diagnostic maneuvers, which trigger the typical vertical (up-beating)-torsional (ipsilesional) positional nystagmus; this latter is generated by the ampullofugal excitatory endolymphatic flow, due to debris moving away from the ampulla.

The correct diagnosis of PSC-BPPV can be easily achieved by the UBP in most cases, particularly in those with a recent onset.

Indeed, during the uHPT, when the patient's head is bent backward, the typical excitatory nystagmus (vertical up-beating with the torsional component beating toward the affected ear) is commonly observed; whereas, when the patient's head is bent forward, the opposite inhibitory nystagmus (vertical down-beating with the torsional component beating toward the healthy ear) can be observed.

Moreover, if the uHPT is ineffective to trigger the vertical-torsional nystagmus, or its torsional component is not clear, the test effectiveness significantly increases if the patient's head is moved along the RALP and LARP planes, that is, performing the uRALP and uLARP tests.

Specifically, in the uRALP test, since the RALP plane approximately aligns with the left PSC, the patient's head is rotated 45° to the left side from the neutral position, and then bent 60° forward and 30° backward; in so doing, the patient's head in moved along the plane of the affected canal, thus making the test more effective in inducing either inhibitory (ampullopetal) or excitatory (ampullofugal) endolymphatic flows, and for that more effective in generating either inhibitory (down-beating/right-torsional) or excitatory (up-beating/left-torsional) nystagmus, compared to the uHPT (Fig. 9.8). In case of PSC-BPPV of the right ear, the uRALP test should not trigger any nystagmus, because in this case the head movements are performed along, and the gravity vector is aligned with, a plane orthogonal to the right PSC plane, thus preventing any debris movements.

Similarly, in the uLARP test, head bending and gravity vector are aligned with the right PSC plane, performing it with the patient's head rotated 45° to the right side from the neutral position; in this case, if the affected canal is the right one, forward head bending results in inhibitory down-beating/left-torsional nystagmus, whereas backward head bending results in excitatory up-beating/right-torsional nystagmus (Fig. 9.9). Accordingly, in case of PSC-BPPV of the left ear, the uLARP test should not be able to trigger any nystagmus, both head movements and gravity axis being aligned with a plane orthogonal to left PSC plane.

In PSC-BPPV (Benign Paroxysmal Positional Vertigo), the UBP (Uprigth BPPV Protocol) allows for achieving the correct diagnosis in 75.23% of cases with the sole uHPT, and in 87.16% of cases just by adding uRALP/uLARP tests. Moreover, the correct diagnosis is achieved in 100% of patients with symptoms onset from less than 7 days (64/64), and in 68.89% of patients suffering from BPPV for more than 7 days (31/45) ($p < 0.001$); this means, as expected, that the diagnostic accuracy of UBP is closely related to the time elapsed since the onset of symptoms, that is, the more recent the onset of BPPV, the higher the sensitivity of the Uprigth BPPV Protocol (UBP [21].

When the affected side has been identified with the UBP, it should be verified by one of the traditional diagnostic tests, that is, the Dix-Hallpike maneuver or the Semont diagnostic maneuver. With the first maneuver, the patient's head is turned 45° toward the affected side to align the ipsilateral PSC with the sagittal plane, and then the patient is rapidly brought into the supine position with the head hyperextended approximately 20° off the bed. With the second maneuver, the patient's head is rotated 45° toward the healthy side, this time to align the affected PSC with the frontal plane, and then the patient is rapidly brought toward the affected side with the nose pointing upward. In both cases, the typical paroxysmal positional nystagmus due to PSC excitation is triggered, so that the PSC-BPPV diagnosis is confirmed, and its treatment can immediately be carried out, by the Epley CRP in the first case, and the Semont liberatory maneuver in the second case.

9.2.2 UBP in LSC-BPPV

The second most common type of BPPV is that involving the LSC, accounting for about 15–25% of all BPPV cases.

The diagnosis is traditionally based on identifying the typical and purely horizontal direction-changing positional nystagmus while performing the sHYT. Depending on the direction of the horizontal nystagmus, which in turn depends on the debris position within the LSC, two variants of LSC-BPPV can be identified [2–4, 10, 11, 13]:

1. The geotropic one, accounting for about 75% of all LSC-BPPV, with debris freely floating within the nonampullary emicanal of the LSC, and which we refer to as "geotropic canalolithiasis"; indeed, when the patient's head is rotated toward the affected side in supine position, the debris moves toward the ampulla, so generating an ampullopetal (excitatory) endolymphatic flow, and for that triggering a nystagmus beating toward the undermost ear, which is the affected one; conversely, when the patient's head is rotated toward the healthy side in supine position, the debris moves away from the ampulla, so generating an ampullofugal (inhibitory) endolymphatic flow, and for that once again triggering a nystagmus beating toward the undermost ear, which this time is the healthy one; in both positions, the triggered horizontal nystagmus is directed toward the ground, so that, it is called "geotropic."
2. The apogeotropic one, accounting for about 25% of all LSC-BPPV, with debris either freely floating within the long arm of the ampullary emicanal of the LSC, and which we refer to as "apogeotropic canalolithiasis," or attached to the cupula of the LSC, and which we refer to as "cupulolithiasis" or "heavy cupula"; this time, when the patient's head is rotated toward the affected side in supine position, the debris moves away from the ampulla in case of canalolithisis, and the cupula deflects away from the ampulla in case of cupulolithiasis, so both generating an ampullofugal (inhibitory) endolymphatic flow, and for that triggering a nystagmus beating toward the uppermost ear, which is the healthy one; conversely, when the patient's head is rotated toward the healthy side in supine position, the debris moves toward the ampulla in case of canalolithisis and the cupula deflects toward the ampulla in case of cupulolithiasis, so both generating an ampullopetal (excitatory) endolymphatic flow, and for that once again triggering a nystagmus beating toward the uppermost ear, which this time is the

affected one; in both positions, the triggered horizontal nystagmus is directed away from the ground, so that, it is called "apogeotropic."

Based on the second Ewald's law, postulating that the response to an excitatory stimulus, which in the LSC is due to an ampullopetal cupula deflection, is always more intense than an inhibitory stimulus, which in the LSC is due to an ampullofugal cupula deflection, the affected side can be identified by comparing the nystagmus intensity on each side lateral position during the sHYT: indeed, the more intense triggered nystagmus always beats toward the affected ear, so that the affected side is the one where the nystagmus is more intense in geotropic variant, and the one where the nystagmus is less intense in apogeotropic variant.

Nevertheless, unfortunately, the difference in nystagmus intensity during the sHYT is not always easy to detect [14–19, 22, 23]. Moreover, if this is the case, repeating the test more times, in order to identify the affected side, usually results in severe autonomic symptoms and consequent significant discomfort in many patients, especially in those with recent BPPV onset, or in those more anxious or even phobic, thus preventing the correct diagnosis and treatment.

These limitations of the traditional approach can be avoided with the UBP [19, 22].

The first step of this protocol is the evaluation of PSN, that can detected in 50–75% of patients with BPPV involving the LSC; indeed, this latter lies on a backward sloping plane, drawing a 30° front-open-angle with the horizontal plane, so that the free-floating debris gravitates within the canal in case of canalolithiasis, and the heavy-made cupula is persistently displaced in case of cupulolithiasis. Therefore, in geotropic variant the debris moves away from the ampulla, and PSN is usually directed toward the healthy side; conversely, in apogeotropic variant the debris moves toward the ampulla in case of apogeotropic canalolithias, and the cupula is deflected toward the ampulla in case of cupulolithiasis, so that in both conditions PSN is usually directed toward the affected side [14–19, 22–26].

In the second step of the UBP, the uHPT is performed: in this test, bending the patient's head 30° backward increases the angle between the LSC and the horizontal plane, either enhancing the intensity of an already present PSN, or triggering a PSN, both beating toward the healthy side in geotropic variants and toward the affected side in apogeotropic ones; instead, bending the patient's head 30° forward, that is achieving the neutral position, gets the LSC aligning with the horizontal plane, nullifying the angle between the two, so that the PSN disappears; finally, bending the patient's head 60° forward reverses the angle between the LSC and the horizontal plane, either reversing the direction of an already present PSN, or triggering a PSN, in this case both beating toward the affected side in geotropic variants and toward the healthy side in apogeotropic ones [14–19, 22, 27–30]. This is the key point of the second step, because the examiner can restrict the diagnostic hypotheses to only two options by means of the uHPT, just performing it in only one direction along the pitch plane, forward or backward, and detecting only one horizontal nystagmus, right beating or left beating: the geotropic variant of one side or the apogeotropic variant of the opposite side.

Then, the examiner can distinguish between these two options with the third step of the UBP, that is, the uHRT [19, 22]: indeed, tilting the patient's head toward each shoulder allows the debris to gravitate within the affected LSC, triggering a positional horizontal nystagmus beating, on both sides, either toward the undermost ear in the geotropic variants or toward the uppermost ear in the apogeotropic ones. This is the key point of the third step, because the examiner can restrict the diagnostic hypotheses to only one option by means of the uHRT, just performing it in only one direction along the roll plane, rightward or leftward, and detecting only one horizontal nystagmus, geotropic or apogeotropic.

In so doing, combining the data provided by both the uHPT and the uHRT, the examiner can achieve the correct diagnosis by means of one nystagmus per plane: rightward or leftward in the pitch plane, and geotropic or apogeotropic in the roll plane. For example, in the case of a right geotropic LSC-BPPV, due to non-ampullary arm canalolithiasis, the PSN, if present, should beat leftward, the uHPT should result in right beating nystagmus in forward head bending (or left beating nystagmus in backward head bending), and the uHRT should result in geotropic right beating nystagmus in rightward head tilt (or geotropic left beating nystagmus in leftward head tilt). Conversely, in the case of right apogeotropic LSC-BPPV, due to either ampullary emicanal long arm canalolithiasis or cupulolithiasis, the PSN, if present, should beat rightward, the uHPT should trigger a left beating nystagmus in forward head bending (or right beating nystagmus in backward head bending), and the uHRT should trigger an apogeotropic left beating nystagmus in rightward head tilt (or apogeotropic right beating nystagmus in leftward head tilt) (Figs. 9.5, 9.6, 9.7, and 9.8).

If the UBP does not provide effective or sufficient signs to achieve a correct diagnosis, its efficacy may be enhanced by applying some slight inertial accelerations to the patient's head, thus promoting the movement of otoconia, which start to gravitate along LSC, when performing both the uHPT and uHRT. Moreover, other signs of lateralization may be looked for, such as the SSPN, which can be observed by bringing the patient from the sitting to the supine position (along the pitch plane), with the head bent 30° forward [15]. In this position, the LSC aligns with the gravity vector, allowing the debris to gravitate within the arm of the affected canal: specifically, in geotropic variants, the debris moves away from the ampulla along the nonampullary arm, generating an inhibitory endolymphatic flow, which results in contralesional nystagmus; in apogeotropic variants, either the free-floating debris moves toward the ampulla along the ampullary arm, in case of canalolithiasis, or the heavy-made cupula bends toward the ampulla, in case of cupulolithiasis, both generating an excitatory endolymphatic flow, which results in ipsilesional nystagmus. Once the SSPN has been triggered by the SSPT, the sHYT may be performed, turning the patient's head 180° to either side, and evaluating the direction and intensity of the positional nystagmus.

The UPB allows to achieve a correct diagnosis of LSC-BPPV in 95.5% of cases, regardless of the delay from symptom onset to diagnosis, against the 85.1% of the sHYT [19, 22].

When the affected side of the LSC-BPPV has been identified, several therapeutic maneuvers can be used, aiming at promoting the ampullofugal progression of the debris within the canal. These maneuvers include techniques working by angular acceleration with negative inertia (Baloh and Lempert barbecue rotation maneuvers, Fife technique and Vannucchi-Asprella maneuver [14, 31–35]), techniques working by linear acceleration with positive inertia (Gufoni maneuver [36] and its variants [37, 38]), techniques working by gravitational settling (Vannucchi forced prolonged position and shortened forced position [39, 40]), and mixed techniques (Asprella maneuver [2, 41, 42] and Zuma maneuver [2, 43–45]). However, when a confident diagnosis has been established with the UBP, it should be recommended to proceed with a technique starting from the sitting position, such as the Gufoni maneuver, according to the MSS.

9.2.3 UBP in BPPV Presenting with Positional Torsional Down-Beating Nystagmus (pTDBN)

Two forms of BPPV are mainly associated with pTDBN: the ASC-BPPV, the most uncommon subtype of BPPV, and the apogeotropic PSC-BPPV (aPSC-BPPV), with debris located in the distal portion of the nonampullary arm of the PSC. This latter has been named "apogeotropic variant" [46, 47], because the positional nystagmus triggered in the provoking positions beats away from the ground with its vertical component, being a down-beating nystagmus, that is, in the opposite direction to that of the classical PSC-BPPV, whose positional nystagmus triggered in the provoking positions beats toward the ground with its vertical component, being an up-beating nystagmus, and for that being a geotropic nystagmus.

The pTDBN can be triggered in both forms of BPPV, bringing the patient into one or both Dix-Hallpike positionings, or into the straight head hanging (SHH) position.

According to the MSS, the same nystagmus can be observed using the UBP: indeed, the vertical down-beating component of the positional nystagmus can be triggered during the uHPT by bending the patient's head forward or backward, while the uRALP/uLARP test can facilitate the detection of its torsional component. Unfortunately, the pTDBN poses a diagnostic challenge for two main reasons: first, the torsional component of the positional nystagmus may be missing, making it hard to identify the affected SCC; second, also in the presence of the torsional component, the positional nystagmus is generated by the same extraocular muscles both in ASC-BPPV and aPSC-BPPV of the opposite side, making it hard to distinguish between the two forms of BPPV (for example, a patient presenting with a positional Down-Beating Nystagmus (pDBN) with a right beating torsional component may be affected by a right ASC-BPPV or a left aPSC-BPPV).

Regarding the therapy of these forms of BPPV, the most used maneuver for ASC-BPPV is that proposed by Yacovino [48], while the demi-Semont maneuver, the 45°-Forced Prolonged Position, and the Quick Liberatory Rotation are the most used techniques for aPSC-BPPV [46, 47].

In conclusion, the UBP allows for identifying the affected canal in BPPV already in the upright position, so that the clinician may avoid unpleasant maneuvers to the patients, immediately proceeding with the most proper physical therapy; this makes the diagnostic and therapeutic procedure even more comfortable and minimally invasive, according to the principle of the Minimum Stimulus Srategy (MMS) (Figs. 9.1 and 9.6).

The major limitation of this protocol is differentiating BPPV subtypes with paroxysmal nystagmus due to canalolithiasis from those with persistent nystagmus due to cupulolithiasis; in fact, nystagmus triggered by the uHPT and uHRT is generally weak, being generated by a low acceleration, such as gravity. This limitation has to be kept in mind also when dealing with direction-changing positional nystagmus, which can also be due to central disorders mimicking BPPV. For these reasons, in patients with a clinical history not properly typical of BPPV, or in those not responding to therapeutic maneuvers, it is strongly recommended to also perform the traditional diagnostic tests.

References

1. Epley JM. Human experience with canalith repositioning maneuvers. Ann N Y Acad Sci. 2001;942:179–91.
2. Asprella Libonati G. Benign paroxysmal positional vertigo and positional vertigo variants. AIJOC. 2012;4:25–40.
3. Kim JS, Zee DS. Clinical practice. Benign paroxysmal positional vertigo: N. Engl J Med. 2014;370:1138–47.
4. Nuti D, Masini M, Mandalà M. Benign paroxysmal positional vertigo and its variants. Handb Clin Neurol. 2016;137(137):241–56.
5. Epley JM. Positional vertigo related to semicircular canalithiasis. Otolaryngol Head Neck Surg. 1995;112:154–61.
6. Martellucci S, Castellucci A, Malara P, et al. Spontaneous jamming of horizontal semicircular canal combined with canalolithiasis of contralateral posterior semicircular canal. J Audiol Otol. 2021;26(1):55–60.
7. Ko KM, Song MH, Kim JH, et al. Persistent spontaneous nystagmus following a canalith repositioning procedure in horizontal semicircular canal benign paroxysmal positional vertigo. JAMA Otolaryngol Head Neck Surg. 2014;140:250–2.
8. Chang YS, Choi J, Chung WH. Persistent direction-fixed nystagmus following canalih repositioning maneuver for horizontal canal BPPV: a case of canalith jam. Clin Exp Otorhinolaryngol. 2014;2:138–41.
9. Castellucci A, Malara P, Brandolini C, et al. Isolated horizontal canal hypofunction differentiating a canalith jam from an acute peripheral vestibular loss. Am J Otolaryngol. 2019;40:319–22.
10. Bhattacharyya N, Gubbels SP, Schwartz SR, et al. Clinical practice guideline: benign paroxysmal positional vertigo (update). Otolaryngol Head Neck Surg. 2017;156(3 suppl):S1–47.
11. von Brevern M, Bertholon P, Brandt T, et al. Benign paroxysmal positional vertigo: diagnostic criteria. J Vestib Res. 2015;25:105–17.

12. Bronstein AM, Patel M, Arshad Q, et al. A brief review of the clinical anatomy of the vestibular-ocular connections-how much do we know? Eye (Lond). 2015;29:163–70.
13. von Brevern M, Radtke A, Lezius F, et al. Epidemiology of benign paroxysmal positional vertigo: a population based study. J Neurol Neurosurg Psychiatry. 2007;78:710–5.
14. Asprella Libonati G, Gagliardi G, Cifarelli D, et al. "Step by step" treatment of lateral semicircular canal canalolithiasis under videonystagmoscopic examination. Acta Otorhinolaryngol Ital. 2003;23:10–5.
15. Asprella Libonati G. Diagnostic and treatment strategy of lateral semicircular canal canalolithiasis. Acta Otorhinolaryngol Ital. 2005;25:277–83.
16. Asprella Libonati G. Pseudo-spontaneous nystagmus: a new sign to diagnose the affected side in lateral semicircular canal benign paroxysmal positional vertigo. Acta Otorhinolaryngol Ital. 2008;28:73–8.
17. Asprella Libonati G. Lateral semicircular canal benign paroxysmal positional vertigo diagnostic signs. Acta Otorhinolaryngol Ital. 2010;30:222.
18. Asprella Libonati G. Lateral canal BPPV with pseudo-spontaneous nystagmus masquerading as vestibular neuritis in acute vertigo: a series of 273 cases. J Vestib Res. 2014;24:343–9.
19. Martellucci S, Malara P, Castellucci A, et al. Upright BPPV protocol: feasibility of a new diagnostic paradigm for lateral semicircular canal benign paroxysmal positional vertigo compared to standard diagnostic maneuvers. Front Neurol. 2020;11:1–12.
20. Asprella Libonati G, Martellucci S, Castellucci A, et al. Minimum stimulus strategy: a step by step diagnostic approach to BPPV. J Neurol Sci. 2022;434:1–10.
21. Martellucci S, Castellucci A, Malara P, et al. Is it possible to diagnose posterior Semicircular Canal BPPV from the sitting position? The role of the head pitch test and the upright tests along the RALP and LARP planes. Am J Otolayngol. 2022;43(4):103474.
22. Malara P, Castellucci A, Martellucci S. Upright head roll test: a new contribution for the diagnosis of lateral semicircular canal benign paroxysmal positional vertigo. Audiol Res. 2020;20:236.
23. Califano L, Melillo MG, Mazzone S, Vassallo A. "Secondary signs of lateralization" in apogeotropic lateral canalolithiasis. Acta Otorhinolaryngol Ital. 2010;30:78–86.
24. Lee SU, Kim HJ, Kim JS. Pseudo-spontaneous and head-shaking nystagmus in horizontal canal benign paroxysmal positional vertigo. Otol Neurotol. 2014;35:495–500.
25. Lee HJ, Kim YH, Hong SK, et al. Pseudo-spontaneous nystagmus in lateral semicircular canal benign paroxysmal positional vertigo. Clin Exp Otorhinolaryngol. 2012;5:201–6.
26. Im DH, Yang YS, Choi H, et al. Pseudo-spontaneous nystagmus in horizontal semicircular canal canalolithiasis. Medicine. 2017;96:e7849.
27. Choung YH, Shin YR, Kahng H, et al. "Bow and lean test" to determine the affected ear of horizontal canal benign paroxysmal positional vertigo. Laryngoscope. 2006;116:1776–81.
28. Lee SH, Choi KD, Jeong SH, et al. Nystagmus during neck flexion in the pitch plane in benign paroxysmal positional vertigo involving the horizontal canal. J Neurol Sci. 2007;256:75–80.
29. Lee JB, Han DH, Choi SJ, et al. Efficacy of the "bow and lean test" for the management of horizontal canal benign paroxysmal positional vertigo. Laryngoscope. 2010;120:2339–46.
30. Marcelli V. Nystagmus intensity and direction in bow and lean test: an aid to diagnosis of lateral semicircular canal benign paroxysmal positional vertigo. Acta Otorhinolaryngol Ital. 2016;36:520–6.
31. Baloh RW, Jacobson K, Honrubia V. Horizontal semicircular canal variant of benign positional vertigo. Neurology. 1993;43:2542–9.
32. Lempert T. Horizontal benign positional vertigo. (Letter). Neurology. 1994;44:2213–4.
33. Baloh RW. Reply to the letter by Lempert: horizontal benign positional vertigo. Neurology. 1994;44:2214.
34. Fife TD. Horizontal canal benign positional vertigo. Advancing the science of vestibular diagnostics. ENG report, ed. ICS Medical; 1995. p. 1–4.
35. Vannucchi P, Giannoni B. Terapia della vertigine parossistica posizionale del canale semicircolare laterale. Tecniche a confronto. In: Nuti D, Passali D, editors. VII Giornata di

vestibologia pratica "La terapia fisica delle vertigini periferiche". Milan: CSS-Formenti; 1998. p. 61–73.
36. Gufoni M, Mastrosimone L. Trattamento con manovra di riposizionamento per la canalolitiasi orizzontale. Acta Otorhinolaringol Ital. 1998;18(6):363–7.
37. Gufoni M, Mastrosimone L, Di Nasso F. Repositioning maneuver in benign paroxysmal vertigo of horizontal semicircular canal. Acta Otorhinolaryngol Ital. 1998;18:363–7.
38. Mandalà M, Pepponi E, Santoro GP, et al. Double-blind randomized trial on the efficacy of the Gufoni maneuver for treatment of lateral canal BPPV. Laryngoscope. 2013;123:1782–6.
39. Vannucchi P, Giannoni B, Giuffreda P et al. The therapy of benign paroxysmal positional vertigo of the horizontal Semicircular Canal. International workshop on eye movements. Ed CM Pavia; 1994p. 321–4.
40. Giannoni B, Pecci R, Pollastri F, et al. Treating benign paroxysmal positional vertigo of the lateral semicircular canal with a shortened forced position. Front Neurol. 2023;6:14.
41. Asprella Libonati G. Horizontal benign paroxysmal positional vertigo. In: Dispenza F, De Stefano A, edirors. Understanding benign paroxysmal positional vertigo. Jaypee Brothers Medical Publishers Pvt Limited. 1st ed.; 2017. pages 320.
42. Asprella Libonati G, Gufoni M. In: Nuti D, Pagnini P, Vicini C, editors. Revisione critica di venti anni di vertigine parossistica posizionale benigna. Milano: Comitato Simposi Scientifici Formenti; 1999. p. 321–36.
43. Vannucchi P, Asprella Libonati G, Gufoni M. The physical treatment of lateral semicircular canal canalolithiasis. Audiol Med. 2005;3:52–6.
44. Zuma e Maia F. New treatment strategy for apogeotropic horizontal canal benign paroxysmal positional vertigo. Audiol Res. 2016;6:163.
45. Ramos BF, Cal R, Brock CM, et al. Apogeotropic variant of horizontal semicircular canal benign paroxysmal positional vertigo: where are the particles? Audiol Res. 2019;9:228.
46. Vannucchi P, Pecci R, Giannoni B, et al. Apogeotropic posterior semicircular canal benign paroxysmal positional vertigo: some clinical and therapeutic considerations. Audiol Res. 2015;31:130.
47. Vannucchi P, Pecci R, Giannoni B. Posterior semicircular canal benign paroxysmal positional vertigo presenting with torsional down-beating nystagmus: an apogeotropic variant. Int J Otolaryngol. 2012;20:413603.
48. Yacovino DA, Hain TC, Gualtieri F. New therapeutic maneuver for anterior canal benign paroxysmal positional vertigo. J Neurol. 2009;256:1851–5.

Chapter 10
Complications After Benign Paroxysmal Positional Vertigo Repositioning Maneuvers

Rosa Maria Castillo Ramos and Jose Carlos Rosmaninho Seabra

Keywords benign paroxysmal positional vertigo (BPPV) · Repositioning maneuvers · Recurrences · Reentry · Canal switch · Canal jam

10.1 Introduction

Benign paroxysmal positional vertigo (BPPV) is a common disorder that causes short episodes of vertigo (a false sensation of movement or spinning) triggered by changes in head position, namely when the patient lies down or gets up from bed, rolls over in bed, tilts his head back, or bends forward, among others. It is by far the most common cause of recurrent rotational vertigo.

The fundamental pathophysiological process in BPPV involves the displacement of the otoconia of the utricular macula, which then enter the semicircular canals. Thence, when there is a change in the static position of the head with respect to gravity, the otolith debris moves to a new position within the semicircular canals, leading to the excitation or inhibition of the cresta ampullary of the canal evolved causing the false sensation of rotation, that is, the vertigo crisis.

This process affects, for gravity-dependent reasons, much more commonly the posterior semicircular canal, representing between 60–90% of all cases according to different authors. However, the other canals may also have evolved. The semicircular horizontal canal may be involved in about 10% of cases, but its incidence may be underestimated, since spontaneous remission of symptoms is more likely to occur when this is the canal involved, compared to cases with posterior canal involvement. The anterior semicircular canal is rarely involved due to its higher position in the labyrinth so the otolith debris rarely enter to this canal by gravity reasons, making the debris less likely to be trapped in this canal [1].

BPPV is usually of unknown origin; it is said to be idiopathic, although in some cases it can be associated with head trauma or a prolonged reclining position.

R. M. C. Ramos (✉) · J. C. R. Seabra
ENT Department: Vertigo Unit, Hospital Cuf Porto/Hospital Luz Setubal, Lisboa, Porto, Portugal

P. Ravat et al. (eds.), *Advances in Translational Neuroscience of Vestibular and Eye Movement Disorders*, Contemporary Clinical Neuroscience,
https://doi.org/10.1007/978-3-032-07539-0_10

Clinically, BPPV is characterized by a brief but very strong sensation of spinning vertigo, usually lasting less than 1 min, triggered by a change in the position of the head in relation to gravity, like lying down or getting up from bed, rolling over one side when lying in bed, tilting the head up or down, among others. Sometimes patients with BPPV occasionally report persistent dizziness and imbalance, but careful anamnesis almost always reveals that their symptoms are worse with changes in head position.

Diagnosis of this condition is based on clinical findings and especially the provocative tests, namely and foremost the Dix-Hallpike test, but also the McClure test. These tests in patients suffering from BPPV provoke, after a short latency, a vertigo crisis lasting a few seconds with the typical nystagmus response, which is canal dependent—torsional (clock or anticlockwise) and upbeat in case of posterior canal, the same but with downbeat component in the anterior (superior) canal, and horizontal in the horizontal canal involvement.

Sometimes the condition resolves spontaneously, but most patients need repositioning maneuvers, which are specific for each canal, making so the correct diagnosis of the canal involved is the key of the treatment with the right maneuver. Complications rarely arise when performing the maneuvers, but we must recognize the problem as early as possible and resolve the complication effectively.

The recurrence of BPPV may be linked to some systemic diseases, but true recurrence should be differentiated from persistence of canalithiasis, which is often a complication of the maneuver due to a reflux of otoliths into the canal [2]. Difference between recurrence and "reentry" of the debris is that in recurrence there is an interval free of symptoms (the otholits are in the correct place for a variable period of time causing no symptoms) where in the "reentry" there is not this time free of symptoms, and the otholits are close to the entrance of the canal so a little movement or even spontaneously can reenter again in the canal immediately after maneuver or in the next few days.

Drug therapy is useless unless for symptomatic purposes (to relieve severe nausea or vomiting).

10.2 Repositioning Maneuvers

BPPV patients may have a variable otolith mass, ranging from fine particles unable to elicit clinically evident nystagmus to high-mass particles visible with the operating microscope. The mass of otoliths may have a role in determining not only the difficulty of obtaining particle repositioning but also the type and likelihood of complications. Of utmost importance is the quality of the maneuver performed—the angles achieved, the stiffness and speed, among others. An improperly performed maneuver leads to complications such as inadequate resolution of symptoms and worsening of the psychological impact of the situation in the patient with a significantly poor quality of life.

According to these factors, one must choose between the different possibilities of maneuvers indicated for that particular situation, the one with which one feels most comfortable and trained to perform.

If maneuvers are done properly, you can assume that, when several or more than one maneuver is required to “clean” the canal (negative Dix-Hallpike test), that is probably due to the fact that there are a large number of low-mass otoliths dispersed into the canal or attached to its walls.

10.3 Repositioning Maneuver Complications

Although repositioning maneuvers are free of major complications, we will discuss mainly three types of complications associated with the repositioning maneuvers: reentry BPPV, canal switch, and canal jam.

10.3.1 Reentry

Reentry BPPV is when the debris after being out of the canal reenters the same canal. This is theoretically easy to admit, but in daily practice, it is difficult to distinguish from recurrence of the disease, being the main difference the time lapse free from symptoms between the maneuver and the new occurrence of symptoms.

Canal switch is a “reentry” of the debris in another canal, for instance, from posterior to horizontal or from posterior to anterior/superior. This is naturally easier to recognize. Between 2020 and 2022, we made a prospective analysis of the outcome of our repositioning maneuvers in 965 patients. We found that 870 (90%) had involvement of the posterior semicircular canal (PSC). In these cases, we performed mostly the Epley maneuver and found that in almost 5% of the patients, we had a canal switch as a complication (43 of the 965 patients considered).

In our practice, we noted two forms of canal switch. The first is immediate, occurring some minutes after the repositioning session. The second is delayed, occurring 1 or 2 days after the maneuvers were done. Undoubtedly, we assumed that the same can occur with debris reentering the same canal, although, as mentioned before, this could be more difficult to distinguish with BPPV recurrence.

As a general rule, we do not do more than one maneuver per session. However, in patients with an immediate reentry episode, particles after the treatment procedure were probably located near the openings of the common crus and horizontal semicircular canal (HSC), so that reentry is simpler in case of high-mass particles rather than dispersed otoliths.

Patients with delayed reentry/canal switch had more than two maneuvers in the same session and a negative final Dix-Hallpike/Roll test. The canal reentry or switch was evident after an average of 2–3 days; it is possible that this time lapse was necessary to assemble particles into the utricle before casual reentry. Undoubtedly, to distinguish delayed canal reentry from recurrence can be, as already mentioned, very difficult.

In our series, there was evidence that performing a new Dix-Hallpike test after the repositioning maneuver (to check if the problem had been solved) enhanced the probability of a reentry. Indeed, the cause of reentry BPPV or canal switch could be related to the repetition of the Dix-Hallpike test to assess cleaning of the canal.

We also believe that the time elapsed between the repositioning maneuver and the verification Dix-Hallpike test should be greater than 10 min to reduce the risk of immediate reentry/canal switch. Delayed canal reentry is not preventable by increasing the time before performing the verification test, but it likely depends on a casual position/movement done by the patients.

There are several studies showing that the Dix-Hallpike test performed after the repositioning maneuver can either increase or decrease relapses. This was the reason why we began to analyze this problem.

In our study, we initially thought that the Dix-Hallpike test performed after the repositioning maneuver to check if the problem had been solved, followed by a second Epley maneuver performed more quickly (all in the same session), could help to prevent some recurrences. However, further in our study with more cases, we found that, on the contrary, there were fewer recurrences in patients who did not undergo the verification Dix-Hallpike test (DH) in the same session, and in those who did the Dix-Hallpike, a quick second Epley maneuver does not seem to prevent recurrences. Therefore, although the study is preliminary and we still need to continue with a larger number of patients, our results seem to recommend not performing the Dix-Hallpike test immediately after the repositioning maneuver in order to minimize recurrences.

10.3.2 Canal Switch

During repositioning maneuvers, suspended fragments can migrate from one semicircular canal to another. The most common situation is migration from the posterior to the lateral canal. In this case, the nystagmus, which was initially horizonto-rotatory (clockwise or anticlockwise) and vertical upbeating, becomes horizontal and geotropic, characteristic of lateral canalithiasis. This situation should be reversed with the maneuver for treating lithiasis of the horizontal semicircular canal (Lempert or Gufoni maneuver).

Much less common is the migration of otoliths from the posterior to the anterior (superior) canal. In this case, the initially upper vertical vector turns into a lower vertical one (downbeating), and one possible treatment is to perform the Yacovino maneuver.

It is very likely that a new episode of BPPV in another canal of the same side after a few days following treatment affecting a different canal is due to reentry of otoliths.

10.3.3 Canal Jam

During the repositioning maneuver, incomplete migration of fragments of otoliths, which end up hitting the common crura, occurs in the case of vertical canals or within the canal itself. In the case of the Epley maneuver, the complication occurs when the patient turns his head toward the ground. This mass of impacted fragments, called jam, maintains a constant thrust force on the canal dome as a result of negative pressure. The pressure exerted by the jam causes a persistent deflection of the dome of the semicircular canal, keeping it depolarized (vertical canals) or hyperpolarized (lateral canals). This stimulus triggers a sensation of intense dizziness and nystagmus regardless of the position of the head. Treatment consists of the reverse maneuver in an attempt to eliminate the blockage and return the crest to its normal resting position. Vibration applied to the mastoid and prolonged lateral decubitus in the position in which the patient feels most comfortable can also be useful.

References

1. Kim JS, et al. Benign paroxysmal positional vertigo. New Engl J Med. 2014;370:148.
2. De Stefano A, Dispenza F, Suarez H, et al. A multicenter observational study on the role of comorbidities in the recurrent episodes of benign paroxysmal positional vertigo. Auris Nasus Larynx. 2014;41:31–6.
3. Scotto di Santillo L, et al. Canal switch: a possible complication of physical therapeutic manoeuvers for posterior canal benign paroxysmal positional vertigo. Acta Otorhinolaryngol Ital. 2023; PMID: 36860150
4. Foster CA, Zaccaro K, Strong D. Canal conversion and reentry: a risk of Dix-Hallpike during canalith repositioning procedures. Otol Neurotol. 2012;33:199–203.
5. Neri G, Filograna Pignatelli GR, Pacella A, et al. Recurring paroxysmal positional vertigo: evaluation of the vascular factor. Acta Otorhinolaryngol Ital. 2021;41:77–83.

Chapter 11
Persistent Postural Perceptual Dizziness

David Herdman and Diego Kaski

Keywords Persistent postural perceptual dizziness · Psychologically informed rehabilitation · Cognitive physiotherapy · Predictive coding · Vestibular

11.1 Introduction

'Dizziness' is amongst the most common complaints across all medical settings, and lifetime prevalence of moderate to severe dizziness may be as high as 30% [17, 18]. Around a quarter of those with acute dizziness following a 'peripheral vestibular disorder' go on to experience persistent symptoms at 1 year [28]. When dizziness persists beyond the expected timeline, recovery is often difficult. After a mean follow-up time of 8.5 years, Huppert et al. [12] reported that only 27% of patients reported a complete remission.

The most common cause of chronic dizziness is 'persistent postural perceptual dizziness' (PPPD). The term 'PPPD' was created by the Barany Society in 2017 to amalgamate the pre-existing, but overlapping diagnoses used around the world up to that point, such as 'phobic postural vertigo' in Germany and 'chronic subjective dizziness' in America [24]. It is a chronic neuro-vestibular 'functional' disorder, meaning that the disorder is characterised by a change in functioning rather than macro-structural deficits [9]. That is not to say that people with PPPD cannot have structural deficits, but rather that the symptoms represent a maladaptation in terms of how the brain processes sensory and postural information rather than any physical damage itself. Whilst functional neurological disorders sit within the DSM-IV

D. Herdman
Sense Research Unit, Department of Clinical and Movement Neurosciences, University College London, London, UK

Audiovestibular Department, St George's University Hospital Foundation Trust, London, UK

D. Kaski (✉)
Sense Research Unit, Department of Clinical and Movement Neurosciences, University College London, London, UK
e-mail: d.kaski@ucl.ac.uk

P. Ravat et al. (eds.), *Advances in Translational Neuroscience of Vestibular and Eye Movement Disorders*, Contemporary Clinical Neuroscience,
https://doi.org/10.1007/978-3-032-07539-0_11

(psychiatry), there is increasing recognition that the underlying mechanisms are in fact neurological in origin, with neuro-psychological factors that predispose or perpetuate symptoms.

The main features of the diagnostic criteria for PPPD include (1) Dizziness, unsteadiness, or swaying/rocking sensations lasting 3 months or more; (2) symptoms that are present most days for most of the day (hours-long periods); (3) symptoms are exacerbated by upright posture, self-motion, and visual motion; and that (4) it is triggered by any condition that causes vestibular symptoms or disrupts balance. It is now recognised as the most common condition seen in specialist outpatient neurotology clinics [25] and accounts for around a fifth of all patients with vestibular complaints [32], with women outnumbering men by 2:1 [1].

There is ongoing debate whether PPPD may present a group of clinically distinct syndromes, since not all patients may experience the strict myriad of exacerbating factors (visual, self-motion, and upright posture) necessary to confirm a definite diagnosis [20]. Further, it can often be indistinguishable from other similar neurological disorders, such as the 'spontaneous' form of Mal de Debarquement, or the chronic variant of Vestibular Migraine [22, 26]. Nevertheless, it appears to be robust and is generally acceptable to patients [11].

At present, little is known about the optimal way to treat this condition. Current treatment approaches draw from the understanding that PPPD encompasses vestibular, postural, cognitive, emotional, and motor control factors that influence the development and maintenance of symptoms. As such, patients are most often treated with vestibular rehabilitation or cognitive-behavioural therapy (CBT). Antidepressant medications, most notably serotonin norepinephrine reuptake inhibitors (SNRI) and selective serotonin reuptake inhibitors (SSRI), may play a role as adjuncts to CBT [33], particularly in individuals with strong psychological symptoms (anxiety, depression) and sleep disturbance, but there is insufficient evidence base for these in PPPD in isolation. In this chapter, we offer an argument for why psychological approaches should be adopted as core features of any treatment approach.

11.2 Psychological Impact

Persistent dizziness can have profound effects on quality of life, including employment, loss of independence, social isolation, disability, and medical consultations [3, 8, 19]. The process of seeking medical care can add to this burden, often resulting in costly and/or unnecessary investigations [7]. Patients can wait an average of 4 years from the onset of dizziness symptoms to receive a specialist diagnosis [14]. Further, Klarendic et al. [14] reported an alarming 90% non-specialist misdiagnosis rate. Patients with comorbid anxiety are at particularly high risk of reporting greater subjective impairment and healthcare utilisation [30], thus creating its own vicious cycle in the context of diagnostic delay and consequences of incorrect diagnoses.

Dizziness can be a distressing feeling that may induce fear and anxiety, irrespective of inherent trait anxiety levels [21]. There is a tendency for rapid conditioning, fear generalisation, and avoidance behaviour to develop. Almost half of patients with vestibular symptoms suffer from psychiatric comorbidity, and these patients show more severe psychosocial impairment [15]. There is a huge variation in the subjective impact of vestibular conditions of the same objective severity. Two patients may have the same degree of physical disease or damage yet have markedly different physical and psychological responses.

The variation in the ways in which patients make sense of and respond to illness has been examined according to Leventhal's Common Sense Model (CSM) of self-regulation, which describes the dimensions along which lay people make sense of illness, and how these dimensions influence the adoption of behaviours to reduce health threats [16]. It has been suggested that the illness beliefs held by the patient may be a key element driving the development of PPPD [27]. Patients who view their illness as more serious and with more negative consequences experience higher levels of dizziness-related disability [10, 31]. It is important to note that in some instances, patients' symptom interpretations may indeed be accurate, but the overriding tendency to view symptoms of dizziness negatively seems to be unhelpful. Therefore, it is important to ascertain how patients think about or understand their condition and have some idea of whether that is an accurate or balanced view or not. In practice, this can be as easy as asking a patient: "What do *you* think is going on?"

It is not only the overall representation of the illness that is important but also the day-to-day interpretation of symptoms, which could be even more strongly associated with self-reported handicap than beliefs about the illness as a whole. Specific illness-related cognitions such as fear avoidance are highly prevalent in patients with vestibular symptoms [5] and are strong predictors of poor long-term outcomes after vestibular neuritis [29]. In addition to fear avoidance, Herdman et al. [10] also found that dizziness-related disability was highly correlated with symptom focusing, catastrophising, beliefs that dizziness was a sign of physical damage, and embarrassment.

These cognitive interpretations can contribute to the maintenance of PPPD by fostering maladaptive patterns of sensory processing and behaviours that sustain symptoms over time. Likewise, 'cognitive-behavioural' responses seem to be critical in both the development and perpetuation of PPPD, specifically through mechanisms like balance vigilance and visual dependency. 'Balance vigilance' denotes an increased, often conscious, monitoring of postural stability. In individuals with PPPD, balance vigilance may arise as an adaptive reaction to an acute vestibular event or dizzy episode. Over time, this hypervigilant state may become maladaptive, since increased awareness of even small shifts in balance or movement can exacerbate symptoms of unsteadiness or vertigo. This continual emphasis on physical signals might impair automatic postural control systems, resulting in prolonged symptoms as balance becomes increasingly effortful and reliant on conscious attention rather than reflexive responses [6].

Visual Dependency is a core feature of PPPD, although it is not specific to this condition. It describes the tendency to rely excessively on visual inputs to maintain balance and spatial orientation. This over-reliance on vision can become problematic, as it heightens sensitivity to visual motion (e.g., crowded environments or scrolling screens) and has been found to be highly predictive of chronic dizziness following a vestibular event [4, 29]. The dependence on visual information creates a maladaptive feedback loop: as visual dependency increases, so does the likelihood of experiencing dizziness in visually complex or dynamic environments, further reinforcing the perception of imbalance.

Together, balance vigilance and visual dependency exemplify how cognitive-behavioural responses can perpetuate PPPD. Rather than assisting in recovery, these adaptations reinforce maladaptive perceptual patterns and maintain a state of chronic dizziness. Understanding the role of these responses is essential in guiding where therapy should be targeted.

11.3 Differential Diagnosis

One of the challenges in diagnosing PPPD is that it can be triggered by an array of precipitating events. Although not necessary for the diagnosis, it is helpful to take a detailed history of the onset to begin to formulate how PPPD may have taken hold, and to set expectations and/or allay potential fears of recurrence. For example, if the condition was precipitated by a panic attack, then consideration of psychological intervention should be prioritised.

PPPD is not a diagnosis of exclusion. Rather, the clinician should try their best to formulate a complete picture and a practical approach to any comorbidity, such as migraine, should be taken. Nevertheless, it should be made clear the difference between 'precipitating' and 'perpetuating' factors. That is to say what may have triggered it is now different to what is keeping the symptoms going. This distinction is vital, particularly for patients seeking further diagnostic work-up, which, particularly as you get further from the time of symptom onset, is either unlikely to yield any meaningful abnormalities or discover incidental findings that may increase health anxiety and/or lead to iatrogenic harm.

In their retrospective study, Watts et al. [30] found that diagnostic uncertainty actually increased after vestibular testing as neuro-otologists tended to overdiagnose otological causes and underestimate functional and/or psychological ones. To counter this, vestibular function testing should be used more judiciously when attempting to answer a clear clinical question, and clinicians must avoid the urge to overinterpret borderline findings.

Similarly, many people with PPPD will obtain imaging at some point in their journey. If the bedside examination is normal, this is probably not needed, but should be considered in cases of progressive symptoms and/or when there is no clearly identifiable triggering event. Caution should be used when using imaging for reassurance purposes only, since anxiety often increases when people are waiting for investigations. Whilst it can be reassuring for the doctor, it rarely actually reassures people with health anxiety and can lead to a greater chance of picking up unrelated signs. This has to be balanced with the need to adequately reassure a patient that there is no other explanation for the symptoms, and it is often helpful to have such open discussions with patients early on.

Another important distinction is that not all causes of chronic dizziness are PPPD. It is important, therefore, to base the diagnosis on the presence of the 'positive' features outlined in the diagnostic criteria rather than use PPPD as a 'waste-basket syndrome'. Although PPPD is associated with psychological comorbidity and can be precipitated by psychological events like panic attacks, it is not itself a psychiatric disorder. However, 'psychogenic' forms of dizziness do occur, and the clinician should try to recognise this, since they will not respond favourably to physiotherapy. For example, a patient with PPPD may be 'avoiding' visiting busy restaurants due to the visual stimulation exacerbating dizziness and a fear of losing control or embarrassment. However, a patient with agoraphobia, for example, could also be avoiding busy restaurants due to dizziness, but rather their fear is linked to being trapped and unable to escape.

Other forms of chronic dizziness can share some of the proposed mediating factors, such as conscious attention towards walking and balance, as also seen in people with 'fear of falling' but should still be considered distinct clinical profiles that warrant their own targeted therapy. The benefits of sticking with a strict criterion means that it will help refine therapeutic targets, whilst not diluting treatment effects through a more diverse or heterogeneous patient group but also allow us to discover and classify new and emerging forms of 'unexplained' dizziness.

11.4 Cognitive Physiotherapy (CPT)

It is logical to assert that there is no vestibular disorder that cannot potentially be assisted by a cognitive-behavioural approach. Cognitive-behavioural therapy (CBT) can be applied to the assessment and treatment of almost every chronic medical problem, even though it originated as a treatment for depression and anxiety. This is because, at its core, CBT is concerned with trying to understand how people think (cognition), feel (emotions), and what they do (behaviour) in the context of the human body (physiology) and environment. A change in any one of these domains may bring about changes in another.

A very large part of vestibular rehabilitation therapy (VRT) is behaviour change. In other words, trying to understand how we help people engage in behaviours that will help them feel better. The similarities between VRT and CBT have already been documented by authors like Staab [23] and Beidel and Horak [2]. As psychological factors appear to make a significant contribution to PPPD, the benefits of exercise therapy might be enhanced by augmenting these implicit psychological elements. A similar approach to integrating CBT and physiotherapy has recently been shown to produce large improvements for people living with chronic back pain [13].

However, CBT, and VRT for that matter, is not just one thing. To be truly effective, treatments must be empirically derived and precisely targeted. For example, gaze stabilization exercises, designed to improve the vestibulo-ocular reflex (VOR), may not be effective unless there is clear, uncompensated VOR dysfunction. When the VOR is intact, but dizziness persists due to perceptual or cognitive factors, these exercises do little to address the core issue, which is not the vestibular deficit itself but the misinterpretation of sensory information and maladaptive predictions about threat and balance.

Vestibular rehabilitation therapy may inadvertently increase balance vigilance, heightening the patient's attention to their symptoms and reinforcing maladaptive predictions. When individuals are repeatedly asked to focus on balance-related exercises, their brain prioritizes predictions that balance is fragile or prone to disruption. This heightened attentional bias toward balance-related sensory information may reinforce the perception that balance is constantly at risk. As a result, rather than resolving the symptoms, such attention may lead to further hypervigilance and perpetuate the cycle of dizziness and instability.

'Cognitive physiotherapy' (CPT) offers a more effective alternative for PPPD by directly targeting the cognitive and perceptual distortions that maintain the symptoms. Instead of focusing primarily on the vestibular system, therapy needs to acknowledge that the brain's maladaptive predictions about balance and threat are central to the disorder. A key part of cognitive approaches needs to be normalising balance vigilance through re-directing attention, and direct exposure to threatening situations so that the patient can overcome avoidance.

By exposing patients to situations that they have learned to avoid due to perceived threat (e.g. crowded places, visually complex environments), therapy allows the brain to update its predictions in light of new, non-threatening sensory information. Over time, this exposure reduces the reliance on hypervigilant balance monitoring and correcting the perception of constant threat or imbalance. By integrating psychological strategies alongside physical activities, CPT can better manage both the cognitive and physical dimensions of PPPD, which are deeply intertwined through maladaptive prediction models.

11.5 Summary

Psychosocial response to illness rarely correlates highly with vestibular function. Psychological variables are particularly important in understanding the way in which patients respond to challenging symptoms associated with PPPD. The literature in this area highlights the importance of beliefs, and it is this which makes CBT and VRT ideal approaches to helping patients to overcome the disorder. The evidence base for this approach is currently lacking, and physiotherapists are not always clear on how they might tailor their assessments and treatments to take account of the idiosyncratic elements of working with someone with PPPD. This requires a careful approach to training and development, in addition to future trials and clearly documented protocols.

References

1. Adamec I, Juren Meaški S, Krbot Skorić M, Jažić K, Crnošija L, Milivojević I, Habek M. Persistent postural-perceptual dizziness: clinical and neurophysiological study. J Clin Neurosci. 2020;72:26–30. https://doi.org/10.1016/j.jocn.2020.01.043.
2. Beidel DC, Horak FB. Behavior therapy for vestibular rehabilitation. J Anxiety Disord. 2001;15 (1):121–30. https://doi.org/10.1016/S0887-6185(00)00046-3.
3. Benecke H, Agus S, Kuessner D, Goodall G, Strupp M. The burden and impact of vertigo: findings from the REVERT patient registry. Front Neurol. 2013;4:136. https://doi.org/10.3389/fneur.2013.00136.
4. Cousins S, Kaski D, Cutfield N, Arshad Q, Ahmad H, Gresty MA, Seemungal BM, Golding J, Bronstein AM. Predictors of clinical recovery from vestibular neuritis: a prospective study. Ann Clin Transl Neurol. 2017, May;4(5):340–6. https://doi.org/10.1002/acn3.386.
5. Dunlap PM, Sparto PJ, Marchetti GF, Furman JM, Staab JP, Delitto A, Klatt BN, Whitney SL. Fear avoidance beliefs are associated with perceived disability in persons with vestibular disorders. Phys Ther. 2021;101(9). https://doi.org/10.1093/ptj/pzab147.
6. Ellmers TJ, Kal EC. Exploring the role of attention towards balance in chronic dizziness: development of the balance vigilance questionnaire. Eur J Neurol. 2024;31(3):e16148. https://doi.org/10.1111/ene.16148.
7. Gandolfi MM, Reilly EK, Galatioto J, Judson RB, Kim AH. Cost-effective analysis of unilateral vestibular weakness investigation. Otol Neurotol. 2015;36(2):277–81. https://doi.org/10.1097/mao.0000000000000649.
8. Gopinath B, McMahon CM, Rochtchina E, Mitchell P. Dizziness and vertigo in an older population: the Blue Mountains prospective cross-sectional study. Clin Otolaryngol. 2009;34 (6):552–6. https://doi.org/10.1111/j.1749-4486.2009.02025.x.
9. Hallett M, Aybek S, Dworetzky BA, McWhirter L, Staab JP, Stone J. Functional neurological disorder: new subtypes and shared mechanisms. Lancet Neurol. 2022;21(6):537–50. https://doi.org/10.1016/s1474-4422(21)00422-1.
10. Herdman D, Norton S, Pavlou M, Murdin L, Moss-Morris R. Vestibular deficits and psychological factors correlating to dizziness handicap and symptom severity. J Psychosom Res. 2020;132:109969. https://doi.org/10.1016/j.jpsychores.2020.109969.
11. Herdman D, Evetovits A, Everton HD, Murdin L. Is 'persistent postural perceptual dizziness' a helpful diagnostic label? A qualitative exploratory study. J Vestib Res. 2021;31(1):11–21. https://doi.org/10.3233/ves-201518.
12. Huppert D, Strupp M, Rettinger N, Hecht J, Brandt T. Phobic postural vertigo. J Neurol. 2005;252(5):564–9. https://doi.org/10.1007/s00415-005-0699-x.
13. Kent P, Haines T, O'Sullivan P, Smith A, Campbell A, Schutze R, Attwell S, Caneiro JP, Laird R, O'Sullivan K, McGregor A, Hartvigsen J, Lee D-CA, Vickery A, Hancock M. Cognitive functional therapy with or without movement sensor biofeedback versus usual care for chronic, disabling low back pain (RESTORE): a randomised, controlled, three-arm, parallel group, phase 3, clinical trial. Lancet. 2023;401(10391):1866–77. https://doi.org/10.1016/S0140-6736(23)00441-5.
14. Klarendic M, Joffily L, Rodrigues FA, Gomes TA, Edlow J, Koohi N, Kaski D. The dizzy patient: duration from symptom onset to specialist review. J Neurol. 2024. https://doi.org/10.1007/s00415-024-12652-2.
15. Lahmann C, Henningsen P, Brandt T, Strupp M, Jahn K, Dieterich M, Eckhardt-Henn A, Feuerecker R, Dinkel A, Schmid G. Psychiatric comorbidity and psychosocial impairment among patients with vertigo and dizziness. J Neurol Neurosurg Psychiatry. 2015;86(3): 302–8. https://doi.org/10.1136/jnnp-2014-307601.
16. Leventhal HM, Nerenz D. The common sense representation of illness danger. In: Rachman S, editor. Medical psychology, vol. 2. Oxford: Pergamon Press; 1980.
17. Murdin L, Schilder AG. Epidemiology of balance symptoms and disorders in the community: a systematic review. Otol Neurotol. 2015;36(3):387–92. https://doi.org/10.1097/mao.0000000000000691.

18. Neuhauser HK. The epidemiology of dizziness and vertigo. Handb Clin Neurol. 2016;137:67–82. https://doi.org/10.1016/B978-0-444-63437-5.00005-4.
19. Neuhauser HK, Radtke A, von Brevern M, Lezius F, Feldmann M, Lempert T. Burden of dizziness and vertigo in the community. Arch Intern Med. 2008;168(19):2118–24. https://doi.org/10.1001/archinte.168.19.2118.
20. Park JH, Nguyen TT, Kim SH, Park JY, Na S, Jeon EJ, Seo JW, Cho CG, Oh SJ, Choi SW, Choi KD, Choi SY, Choi JE, Hong SK, Chung WH, Cho YS, Lee HH, An YH, Han KH, Lee H, Kim HA, Lee HY, Lee JD, Lee SA, Oh SY. Clinical characteristics of persistent postural-perceptual dizziness and its visual subtype in Korean patients: a multicenter cross-sectional study. Brain Behav. 2024;14(2):e3389. https://doi.org/10.1002/brb3.3389.
21. Pollak L, Klein C, Rafael S, Vera K, Rabey JM. Anxiety in the first attack of vertigo. Otolaryngol Head Neck Surg. 2003, Jun;128(6):829–34. https://doi.org/10.1016/s0194-59980300454-6.
22. Pubudu A, David M-A, Dolores V-MM, Peter JG. Migraine and PPPD: how can we differentiate between these entities? J Neurol Neurosurg Psychiatry. 2024;95(Suppl 2):A71. https://doi.org/10.1136/jnnp-2024-ABN.235.
23. Staab JP. Behavioral aspects of vestibular rehabilitation. NeuroRehabilitation. 2011;29(2):179–83. https://doi.org/10.3233/NRE-2011-0693.
24. Staab JP, Eckhardt-Henn A, Horii A, Jacob R, Strupp M, Brandt T, Bronstein A. Diagnostic criteria for persistent postural-perceptual dizziness (PPPD): consensus document of the committee for the classification of vestibular disorders of the Bárány Society. J Vestib Res. 2017;27(4):191–208. https://doi.org/10.3233/ves-170622.
25. Strupp M, Dlugaiczyk J, Ertl-Wagner BB, Rujescu D, Westhofen M, Dieterich M. Vestibular disorders. Dtsch Arztebl Int. 2020;117(17):300–10. https://doi.org/10.3238/arztebl.2020.0300.
26. Tarnutzer AA, Kaski D. What's in a name? Chronic vestibular migraine or persistent postural perceptual dizziness? Brain Sci. 2023;13(12). https://doi.org/10.3390/brainsci13121692.
27. Trinidade A, Harman P, Stone J, Staab JP, Goebel JA. Assessment of potential risk factors for the development of persistent postural-perceptual dizziness: a case-control pilot study [original research]. Front Neurol. 2021;11. https://doi.org/10.3389/fneur.2020.601883.
28. Trinidade A, Cabreira V, Goebel JA, Staab JP, Kaski D, Stone J. Predictors of persistent postural-perceptual dizziness (PPPD) and similar forms of chronic dizziness precipitated by peripheral vestibular disorders: a systematic review. J Neurol Neurosurg Psychiatry. 2023;94(11):904–15. https://doi.org/10.1136/jnnp-2022-330196.
29. Van Laer L, Hallemans A, De Somer C, Janssens de Varebeke S, Fransen E, Schubert M, Van Rompaey V, Vereeck L. Predictors of chronic dizziness in acute unilateral vestibulopathy: a longitudinal prospective cohort study. Otolaryngol Head Neck Surg. 2024. https://doi.org/10.1002/ohn.964.
30. Watts E, Lindley K, Irving R, Dalton L. Balance testing: does it make a difference? J Laryngol Otol. 2023;137(5):506–514. https://doi.org/10.1017/S0022215122001967.
31. Wiltink J, Tschan R, Michal M, Subic-Wrana C, Eckhardt-Henn A, Dieterich M, Beutel ME. Dizziness: anxiety, health care utilization and health behavior—results from a representative German community survey. J Psychosom Res. 2009;66(5):417–24. https://doi.org/10.1016/j.jpsychores.2008.09.012.
32. Wolf J, Limburg K, Sattel H, Lahmann C. Perceived illness consequences predict the long-term course of handicap in patients with vertigo and dizziness beyond vestibular abnormality. J Psychosom Res. 2023;172:111401. https://doi.org/10.1016/j.jpsychores.2023.111401.
33. Xue H, Chong Y, Jiang ZD, Liu ZL, Ding L, Yang SL, Wang L, Xiang WP. Etiological analysis on patients with vertigo or dizziness. Zhonghua Yi Xue Za Zhi. 2018;98(16):1227–30. https://doi.org/10.3760/cma.j.issn.0376-2491.2018.16.008.
34. Yu YC, Xue H, Zhang YX, Zhou J. Cognitive behavior therapy as augmentation for sertraline in treating patients with persistent postural-perceptual dizziness. Biomed Res Int. 2018;2018:8518631. https://doi.org/10.1155/2018/8518631.

Chapter 12
Pharmacotherapy in Disorders of Eye Movements and the Vestibular System

Alexander A. Tarnutzer

Keywords Shared decision-making · Therapeutic outcomes · N-acetyl-L-leucine · 4-Aminopyridine · Betahistine

12.1 Introduction

Vertigo and dizziness are among the most frequent leading symptoms in patients presenting to the emergency department (ED) or the physician's office, making up to 2.1–7.1% of all ED-consultations [1–4]. Based on national surveys, the 1-year prevalence for dizziness ranged between 14.8% and 35.6% [5–8], underlining the impact of vertigo, dizziness, and gait imbalance on the general population. While the differential diagnosis in the dizzy/ataxic patient is very broad and reaches across almost all specialties, a substantial fraction is related to either otologic/vestibular disorders (32.9%) or neurologic disorders, including cerebrovascular disease (11.2%) [2]. With a high burden of disease [5] and a substantial impact on the patient's quality of life [9], treating the dizzy patient is the key to reducing, e.g., the risk of falls and fall-related injuries, impairment of vision due to oscillopsia, and secondary complications such as anxiety disorders and panic attacks.

Often, a combination of both drug treatments and non-pharmacological interventions, including vestibular rehabilitation (physical therapy, repositioning maneuvers in patients with benign paroxysmal positional vertigo (BPPV)), cognitive behavioral therapy/psychotherapy, and readjustments of the work-life balance, is combined. This chapter focuses on the pharmacotherapy of neuro-otologic disorders. Both established and emerging drug treatments will be reviewed, and current evidence will be summarized. This also includes the critical discussions of current gaps in knowledge and failed treatment approaches. Treatment recommendations for both

---

A. A. Tarnutzer (✉)
Cantonal Hospital of Baden, Baden, Switzerland

Faculty of Medicine, University of Zurich, Zurich, Switzerland
e-mail: alexander.tarnutzer@ksb.ch

P. Ravat et al. (eds.), *Advances in Translational Neuroscience of Vestibular and Eye Movement Disorders*, Contemporary Clinical Neuroscience,
https://doi.org/10.1007/978-3-032-07539-0_12

peripheral-vestibular disorders (including Menière‘s disease (MD), acute unilateral vestibulopathy (AUV), vestibular paroxysmia (VP), vestibular schwannoma (VS), and benign paroxysmal positional vertigo (BPPV) and central vestibular disorders (including cerebellar ataxias, functional dizziness, and vestibular migraine) will be discussed. Furthermore, I will also summarize current evidence for pharmacotherapy of distinct nystagmus patterns as seen in various ocular motor disorders, including downbeat nystagmus (DBN), upbeat nystagmus (UBN), acquired pendular nystagmus (APN), periodic alternating nystagmus (PAN), and saccadic intrusions (SI).

12.2 Definitions

Before addressing pharmacotherapy in specific neuro-otologic disorders, several eye movement findings need to be defined. This includes nystagmus and saccadic intrusions.

Nystagmus is defined as a spontaneous, involuntary eye movement [10]. Often, a fast phase and a slow phase can be distinguished. Thereby, a slow drift of the eyes away from the desired position, termed “slow phase“, can be observed, followed by a corrective eye movement that can be of a jerk or pendular type. In case of a jerk nystagmus, the slow phase is followed by a fast corrective (centrally triggered) eye movement, termed “fast phase“. The direction of the fast phase defines the beating direction of a jerk nystagmus. Noteworthy, in the case of a pendular nystagmus, no fast phase and slow phase can be distinguished.

Saccadic intrusions (SI) are defined as initially fast (saccadic) eye movements driving the eyes away from central fixation or disrupting ongoing eye movements such as pursuit eye movements or eccentric gaze holding. Two main categories of saccadic intrusions can be distinguished based on the presence/absence of an intersaccadic interval. Those saccadic intrusions presenting with an intersaccadic interval are referred to as square-wave jerks (SWJ) or macrosaccadic oscillations. Noteworthy, based on the amplitude of the SWJs, micro (<3°) and macro (>3°) SWJs are distinguished. If no intersaccadic interval is seen, a further distinction is made based on the beating direction of the SI. Whereas an ocular flutter (OF) is restricted to a single (usually horizontal) plane, an opsoclonus is defined as a SI in multiple planes.

12.3 Pharmacotherapy in Neuro-Otology: Treatment Principles

Before initiating drug treatment in patients with peripheral or central vestibular disorders, several key issues should be considered. This includes identifying the underlying disorder and assessing comorbidities (such as epilepsy, migraine, psychiatric disorders, liver disease, or impaired kidney function) and contraindications (e.g., cardiac arrhythmia, pregnancy) that may substantially impact/limit drug choice. For neuro-otologic disorders presenting with nystagmus, determining the pattern of the nystagmus observed is important. For acquired nystagmus patterns, the first choice is usually drug treatment; thus, the most suitable substance(s) need to be identified, which is disease-dependent. The use of drugs in vestibular and oculomotor disorders is often based on open-label, noncontrolled studies and expert opinion. Many of the drug treatments discussed in this chapter are off-label. Thus, the use of off-label medication should always be disclosed to the patient, and a shared decision should be reached. At an individual level, a risk-benefit assessment should be made, and country-specific drug approval and coverage by health insurance companies need to be taken into account. As a general treatment principle, any drug treatment should start at a low dose and should be gradually increased until one of three endpoints is reached [10]. Either (a) effective treatment is confirmed, (b) drug intake has to be stopped due to side effects, or (c) the maximal dose of the specific medication has been reached without effective relief as reported by the patient and/or as observed on clinical examination (e.g., persistent down-beating nystagmus).

12.4 Pharmacotherapy in Peripheral-Vestibular Disorders

Drug treatment is a cornerstone of various peripheral acute vestibular syndromes and episodic vestibular syndromes. In this section, various drug treatment approaches to selected peripheral vestibular disorders are discussed (see also [11]).

12.4.1 Acute Unilateral Vestibulopathy (AUV, Vestibular Neuritis)

Acute unilateral loss of peripheral-vestibular function due to either inflammation or ischemia of the vestibular nerve is one of the most important differential diagnoses in the patient presenting with an acute vestibular syndrome (AVS). According to the Bárány consensus criteria, it is defined as an acute or subacute onset of sustained spinning or non-spinning vertigo lasting for at least 24 hours with a spontaneous peripheral vestibular nystagmus and unambiguous evidence of reduced VOR (vestibulo-ocular reflex) function on the side opposite the direction of the fast phase of the spontaneous nystagmus. At the same time, no evidence for acute central neurologic, otologic, or audiologic symptoms or signs is present [12]. By use of video-head-impulse testing, superior-branch and inferior-branch involvement of the vestibular nerve can be distinguished.

In confirmed cases, drug treatment has three principal goals. First, prescription of antiemetics should be considered to treat any emerging or persistent nausea and vomiting. Second, an anti-inflammatory therapy should be evaluated, and third, central vestibular plasticity and adaptation could be facilitated by drug treatment. For treating nausea and vomiting, various anti-vertiginous drugs, including dimenhydrinate, metoclopramide, and ondansetron, are available. Importantly, contraindications need to be considered as, e.g., avoiding the prescription of metoclopramide in patients suffering from Parkinson's disease. As anti-vertiginous drugs may inhibit central compensation, intake should be restricted to the first 2 to 3 days after symptom onset [13]. A prescription of anti-inflammatory drugs should be considered in patients presenting with an AUV within 72 hours after symptom onset and moderate or severe symptom intensity. Various treatment protocols have been evaluated in clinical trials, and several societies have published consensus statements. Whereas in a large prospective, randomized, double-blind, two-by-two factorial trial the intake of methylprednisolone 100 mg/d with tapering by 20 mg every fourth day (total duration of intake 22 days) was associated with significantly better long-term outcome of peripheral-vestibular function (as assessed by caloric irrigation) 12 months after symptom onset compared to placebo, treatment with the antiviral agent valacyclovir in the same study did not improve the outcome [14]. An important limitation of this seminal study, however, is the lack of reported patient-related outcome measures; thus, the impact of steroid treatment on quality of life after AUV remains unclear. A recent consensus statement issued by the German ENT society recommended treatment of AUV with high-dose prednisolone (250 mg, intravenously or orally) for initial treatment [15]. In contrast, in a Cochrane analysis published in 2011 [16] and systematic reviews published in 2021 [17] and in 2024 [18], no treatment recommendation for steroids was issued due to insufficient evidence. Thus, in clinical practice, an individual risk-benefit analysis should be made. Early presentation, high symptom intensity, and lack of preexisting conditions that may worsen when steroids are prescribed (e.g., diabetes, psychiatric disorders, dementia) may favor the use of steroids in AUV. Different treatment schemes have been proposed, ranging from 250 mg methylprednisolone intravenously to 50 mg prednisolone orally with distinct tapering strategies (see [17] for details). While epidemiologic data indicated an increased risk for AUV in patients with cardiovascular risk factors, it currently remains unclear whether vascular processes present a causal case of AUV [18]. Thus, prescribing anti-thrombotic drugs such as aspirin or clopidogrel in patients with isolated AUV is not established [18]. Experimental trials in rodents have indicated that central plasticity and adaptation can be facilitated and thus functional recovery can be accelerated by the intake of betahistine [19] and Ginkgo biloba extract EGb 761 [20, 21]. Furthermore, there is also promising data on N-acetyl-L-leucine in animal studies demonstrating accelerated vestibular compensation [22].

12.5 Menière's Disease (MD)

MD is defined as an episodic vestibular syndrome requiring at least two spontaneous episodes of vertigo (duration 20 min to 12 hours) associated with low- to medium-frequency sensorineural hearing loss and fluctuating aural symptoms (hearing loss, tinnitus, and/or fullness) in the affected ear according to the Bárány consensus diagnostic criteria [23]. In the treatment of patients with MD, prevention of further attacks (or at least a significant reduction in their frequency of occurrence) has priority in order to improve quality of life and to avoid emergence and/or progression of (persistent) vestibulo-cochlear deficits. Currently, there are over 3500 treatment studies on MD published, with various drug treatment approaches investigated, including dietary modifications (salt restriction), oral pharmacotherapy (diuretics, betahistine), intratympanic pharmacotherapy (dexamethasone or methylprednisolone, gentamicin), and surgical therapy (endolymphatic sac surgery) [24, 25]. However, a recent Cochrane review on systemic treatments in MD concluded "evidence for systemic pharmacological interventions for Menière's disease is very uncertain. There are few RCTs that compare these interventions to placebo or no treatment, and the evidence that is currently available from these studies is of low or very low certainty" [26]. Likewise, these authors concluded that "the evidence for intratympanic corticosteroids in the treatment of Ménière's disease is uncertain" [27] and that "the evidence for the use of intratympanic gentamicin in the treatment of Ménière's disease is very uncertain" [28].

Traditionally, betahistine, an H1 agonist and H3 antagonist, has been considered a first choice in the secondary prevention of MD. Recently, the role of betahistine in the treatment of MD has been investigated in a multicenter, double-blind, randomized, placebo-controlled, three-arm, parallel group, phase III, dose-defining superiority trial (BEMED) [29]. Comparing long-term prophylactic treatment with betahistine 3× 24 mg or 3× 48 mg with placebo over 9 months, no change in the time course of vertigo episodes related to MD relative to placebo was observed. Thus, prescription of betahistine in MD is not supported by this negative study. This negative finding may well be explained by the fact that betahistine is rapidly metabolized by intestinal and hepatic MAO-B enzymes. Thus, possible future strategies include the use of even higher doses of betahistine (3× 96 mg or more) [30] or a combined intake of betahistine and a MAO-B inhibitor (such as selegiline) to reduce the betahistine dosage needed [31].

A recent survey assessed the impact of the BEMED trial on prescription strategies in otologists and neuro-otologist subspecialists in the United Kingdom before and after publication of the BEMED trial [32]. Noteworthy, no changes in prescription strategies could be identified despite the negative results from the BEMED trial.

Alternatively, intratympanic pharmacotherapy can be considered. Intratympanic application of steroids demonstrated a benefit in only two out of three randomized controlled trials (RCTs), and gentamicin showed a benefit in three out of three RCTs. Whereas steroids have a transient effect only and application may need to be repeated, hearing loss is a significant side-effect of gentamicin application. In a recent consensus statement, it was concluded that more studies with a high level of evidence are needed to evaluate the drug type, efficacy, and therapeutic endpoint of intratympanic therapy for MD [33].

Overall, there is currently limited evidence to support the use of betahistine in MD treatment. Instead, patients with active MD may be considered for intratympanic pharmacotherapy (preferentially with steroids). This, however, requires patient referral for treatment to an ENT-specialist. The combination of betahistine with an MAO-B inhibitor at this time may be considered, but it is an off-label treatment and side effects of MAO-B inhibitors, such as insomnia, drowsiness, dry mouth, and headache, must be taken into account, as well as significant interactions with antidepressants (SSRIs, SNRIs) and an increased risk for serotonin syndrome.

12.5.1 Vestibular Paroxysmia

In patients diagnosed with vestibular paroxysmia according to the Bárány consensus diagnostic criteria (i.e., presenting with at least ten attacks of spontaneous (non)-spinning vertigo with a duration of <1 min and stereotyped phenomenology and with response to treatment with carbamazepine or oxcarbazepine) [34], drug treatment focuses on reducing the ephaptic stimulation of the vestibular nerve due to the nerve-vessel conflict (most often by the anterior inferior cerebellar artery). Thus, membrane-stabilizing substances have been preferred. While there is established evidence for the treatment with carbamazepine (400–800 mg per day) [35] and oxcarbazepine (600–900 mg per day) [36], adherence to these treatments is often limited due to significant side effects such as sedation, dizziness, confusion, ataxia, nausea, and fatigue. In a single trial with oxcarbazepine, the dropout rate reached 60% [36]. Alternatively, lacosamide (200–400 mg per day) may be considered, with fewer contraindications and side effects reported [37]. At the same time, lacosamide proved effective in patients who did not tolerate Carbamazepine/Oxcarbazepine. Importantly, both carbamazepine, oxcarbazepine, and lacosamide are off-label treatments in patients with vestibular paroxysmia, and before treatment initiation, only a diagnosis of probable vestibular paroxysmia can be made, as response to treatment is a mandatory requirement for a diagnosis of vestibular paroxysmia according to the diagnostic criteria by the Classification Committee of the Bárány Society.

12.5.2 Vestibular Schwannoma

Whereas surgical resection and radiotherapy represent the preferred treatment approaches in patients suffering from vestibular schwannoma, pharmacotherapy may be considered as well in conjunction. While slowly progressive impairment of hearing or peripheral-vestibular function or both is characteristic of vestibular schwannoma, often considerable residual semicircular canal and otolith function may be observed. Thus, by surgical resection, a sudden drop of peripheral-vestibular function is caused, resulting in an iatrogenic acute unilateral vestibulopathy with intense vertigo, nausea/vomiting, and gait imbalance postoperatively. This may prolong hospitalization duration and compromise participation in rehabilitation. Therefore, the concept of vestibular pre-habilitation has been introduced. Specifically, presurgical reduction of ipsilateral peripheral-vestibular function was promoted to ameliorate symptoms after surgery and to facilitate recovery. By application of intratympanic gentamicin, a reduction of peripheral-vestibular function is achieved. While this procedure has been implemented at some ENT departments, no RCTs have been published [38]. In a retrospective case series from a single academic hospital, application of intratympanic gentamicin resulted ina moderate reduction of peripheral-vestibular function, confirming the feasibility of this approach [39]. In a recent systematic review of the literature, a positive effect on postoperative recovery was identified in four out of eight studies on presurgical intratympanic gentamicin application [38]. However, the few studies that assessed whether pre-habilitation resulted in an improved quality of life were all negative [40–42].

12.5.3 Benign Paroxysmal Positional Vertigo (BPPV)

Canalith repositioning maneuvers have been proven to be highly efficient in the treatment of BPPV [43]. Pharmacotherapy thus has only a supportive and additive role in this treatment procedure, aiming at reducing treatment-related side effects such as nausea, vomiting, and anxiety. However, recent data suggests that Vitamin D3-supplementation may be effective in reducing the risk of recurrence of BPPV, as indicated in several RCTs and two systematic reviews [44, 45]. Specifically, by bringing serum vitamin D3 to levels >10 ng/ml, the recurrence rate (within 18 months) could be reduced from 43% to 14% in one study [46]. In another study, the combined prescription of Vitamin D3 (800 IE/d) and calcium (1000 mg/d) reduced the risk of recurrence of BPPV within 12 months from 47% to 38% with a number-needed to treat of 3.7 [47]. Thus, vitamin D3 supplementation is recommended in recurrent BPPV and low serum levels of vitamin D3 [44].

12.6 Pharmacotherapy for Central Vestibular Disorders

Pharmacotherapy plays an important role in the acute treatment and prevention of a broad range of acute, episodic, and chronic central vestibular disorders, ranging from vertebrobasilar stroke and malnutrition to vestibular migraine and cerebellar ataxia to functional dizziness. Here, we review current treatment approaches for the most frequent central vestibular diseases.

12.6.1 Vertebrobasilar Stroke

About 25% of all acute vestibular syndromes (AVS) are due to a central cause, with the majority (about 80%) being related to an acute ischemic stroke in the vertebrobasilar territory [48]. Pharmacotherapy plays an important role both in the acute treatment and in the secondary prevention of ischemic strokes and should be in line with current guidelines (see e.g., AHA/ASA guidelines [49]). For AVS patients, however, Lee and Kim proposed that "isolated vascular vertigo is not indicated for thrombolysis due to its low disability score" [50]. Thus, the authors concluded that the use of alteplase (recurrent tissue plasminogen activator, rtPA) for intravenous thrombolysis (IVT) should be considered in the acute phase only if the NIHSS is >4 or if a clinically relevant deficit is observed (e.g., severe gait imbalance) that is not evaluated by the NIHSS [50]. Importantly, there are no prospective controlled trials on the long-term benefits of IVT in central AVS patients; thus, treatment decisions are made on an individual case basis. In the secondary prevention after vertebrobasilar stroke, antiplatelet therapy and/or anticoagulation and the modification of established risk factors (including dyslipidemia, hypertension, and diabetes) by pharmacotherapy are essential and should be in adherence to current guidelines [51].

12.6.2 Wernicke Encephalopathy

In acute thiamine deficiency, oculomotor deficits are often observed in the early course of disease before other—more obvious—deficits such as ataxia or encephalopathy are noticeable. The combination of a bilaterally abnormal head-impulse test and gaze-evoked nystagmus and/or vertical spontaneous nystagmus in the patient with AVS is characteristic of acute thiamine deficiency. High-dose intravenous thiamine supplementation (600–1500 mg per day for at least 3–5 days) should be started immediately in cases of suspected Wernicke encephalopathy due to the risk of persistent neurological deficits if treatment is delayed or missed [52, 53]. Tolerability and safety of thiamine supplementation are excellent.

12.6.3 Vestibular Migraine

While mostly neglected for a long time, research on the diagnosis and treatment of vestibular migraine (VM) has grown substantially in recent years. Providing consensus diagnostic criteria by both the IHS and the Bárány society has facilitated its recognition considerably, allowing faster treatment initiation [54]. Whereas there is little data on specific treatment strategies in patients with VM in the acute phase, several studies have addressed prophylactic treatment in VM patients. Thus, for treating the acute attacks ("abortive treatments"), the same selection of drugs (including paracetamol, nonsteroidal anti-inflammatory drugs (NSAIDs), anti-emetics, vestibular sedatives, and triptans) is considered for migraine treatment [55]. Analgesics should be used as the first choice, and triptans may be added second. Importantly, intake should be limited to a maximum of 10–12 days per month. In cases with isolated vestibular symptoms, i.e., no headaches, the use of triptans is discouraged [56]. Vestibular sedatives may be used in patients with intense vertigo or dizziness and antiemetics (e.g., prochlorperazine, cyclizine, cinnarizine, or domperidone) to treat nausea. The role of gepants (e.g., ubrogepant, zavegepant, rimagepant), which are small molecules that antagonize the calcitonin gene-related peptide (CGRP) receptors, in the acute treatment of VM is currently unclear.

For prevention, several classes of substances have been investigated in the past. However, many recently published trials in the field have significant shortcomings or have been terminated prematurely, as the PROVEMIG trial (investigated drug: metoprolol) [57]. Emerging treatment options include botulinumtoxin A (PRE-EMPT Scheme, total dose of 155 units [58]) and monoclonal antibodies against CGRP (as erenumab, galcanezumab, fremanezumab) [59]. However, data is preliminary, and use of these substances for VM is currently off-label. Thus, reimbursement may be denied by health insurance companies. While, based on a systematic review and a meta-analysis published in 2021, no recommendations regarding preferred treatment strategies could be made because of low quality of evidence and heterogeneity of study design and outcome reporting [60], a recent comprehensive practical review addresses this gap [56]. Propranolol, tricyclics, and flunarizine were indicated as preferred preventive treatments (see Table 12.1 for details). Vestibular rehabilitation may offer benefit in patients refractory or intolerant of pharmacological treatment and in patients desiring a non-pharmacological approach [61]. In general, lifestyle modifications should be discussed in all patients, optimizing sleep habits, adherence to regular exercises, reduction of stress, and avoidance of fasting and potential dietary triggers such as alcohol, cheese, and caffeine [62].

12.6.4 Cerebellar Ataxia

For treating oculomotor, vestibular, and balance complaints in cerebellar ataxia, various drugs can be considered. This includes 4-aminopyridine (4-AP), sustained release 4-AP (SR-4AP), N-acetyl-DL-leucine (Tanganil), acetazolamide (Diamox), and chlorzoxazone. In most circumstances, evidence for specific treatment protocols is very limited and originates from case series or single case studies. Use of these substances is off-label in most cases, and thus, the benefits and risks of starting a specific treatment have to be balanced carefully. In the following, I will discuss the evidence for selected drugs in the treatment of both sporadic, acquired, and hereditary cerebellar ataxias.

Table 12.1 Preventive treatment of vestibular migraine

Substance	Starting dose/ titration/target dose	Synergistic effects for comorbidities	Side effects/ contraindications	Comments
First choice (best efficacy evidence, lower rate of serious unwanted side effects)				
Betablocker (propranolol)	*Initial dose:* 20 mg twice daily *Titration*: Increase by 20–40 mg twice daily every 1–2 weeks *Target dose range:* 20–80 mg twice daily	Tachycardia, hypertension, tremor,	Asthma, bradycardia, hypotension, depression, cautious use in type 1 diabetes mellitus	Generally well-tolerated. No superiority of metoprolol over placebo was found in a single study [57]
Calcium channel blocker (flunarizine)	*Initial dose:* 10 mg at night in younger patients/ start at 5 mg at night in those aged >65 years and increase by 5 mg after 1 week *Target dose:* 10 mg	Comorbid insomnia	Somnolence, weight gain, worsening of parkinsonian symptoms, depression	Low discontinuation rates despite side effects in many patients
Tricyclics (amitriptyline, nortriptlyline	*Initial dose:* 10 mg at night *Titration*: Increase by 10 mg per day every 1–2 weeks/ *Target dose range:* 10–150 mg at night	Comorbid pain, insomnia	Somnolence, dry mouth, orthostatic hypotension, tachycardia	Patients with anxiety/depression may benefit from higher doses (>75 mg at night)
Second choice (evidence of efficacy, but risk of more serious unwanted effects)				
Antiepileptics (topiramate)	*Initial dose:* 10 mg at night *Titration*: Increase by 25 mg every 2 weeks *Target dose range:* 25–100 mg twice daily	Comorbid obesity	Contraindicated in women of childbearing potential and in those with uncontrolled low mood. Avoid in underweight patients	
Antiepileptics (sodium valproate)	*Initial dose:* 200 mg twice a day *Titration*: Increase by 200–400 mg every 1–2 weeks *Target dose range:* 200–1000 mg twice daily	Comorbid bipolar disorders	Contraindicated in women of childbearing potential, worsening of tremor	

(continued)

Table 12.1 (continued)

Substance	Starting dose/ titration/target dose	Synergistic effects for comorbidities	Side effects/ contraindications	Comments
SNRIs (venlafaxine)	*Initial dose:* 37.5 mg daily *Titration*: Increase by 37.5–75 mg every 2–4 weeks *Target dose range:* 37.5 mg–225 mg per day	Comorbid pain, insomnia, low mood	Risk of withdrawal syndrome, can raise blood pressure	Avoid sudden cessation due to risk of withdrawal syndrome
Third choice (limited efficacy evidence in VM—could be considered if failure of multiple other options)				
Antiepileptics (lamotrigine)	*Initial dose:* 25 mg daily for 2 weeks *Titration*: Increase by 25 mg every 2 weeks (see also national dosage instructions) *Target dose range:* 50–100 mg twice daily	Comorbid bipolar disorder, epilepsy	Risk of rash/Stevens-Johnson syndrome during titration	Generally well-tolerated, but very slow titration is mandatory. Take interaction with sodium valproate into account
AT2-antagonist (candesartan)	*Initial dose:* 2 mg daily *Titration*: Increase by 2–4 mg every 4 weeks *Target dose range:* 2–16 mg daily	Comorbid hypertension		Safe and well-tolerated
Botulinum toxin A	Use PREEMPT scheme (31 injection points, 5 units per point)	Prominent headache		Likely more effective for headache than for vestibular symptoms
Nutraceuticals (riboflavin/coenzyme Q10/magnesium)	Riboflavin 400 mg daily; coenzyme Q-10150 mg daily; magnesium 400–600 mg daily (start with 100 mg, titrate up by 50 mg every 3–7 days)	Comorbid muscle cramps		Safe and well-tolerated

Modified after [56]

12.6.4.1 Overview of Selected Drugs

4-aminopyridine (4-AP) is a potassium-channel blocking drug that increases the resting discharge rate and excitability of cerebellar Purkinje cells (PCs) [63]. This results in an improvement in the synchronicity of the firing rate of cerebellar PCs. By such increased activity of PCs, the GABA-ergic inhibitory influence on the deep cerebellar nuclei mediated by the vestibulo-cerebellum is enhanced. As a result, a reduction of downbeat nystagmus and gait ataxia can be observed [64]. 4-AP is usually started at 5 mg per day and titrated up to a dose of 5 mg three times a day. Contraindications include a history of epilepsy or cardiac arrhythmia. The first dose should be given under surveillance, including an ECG. Alternatively, a **sustained-release (SR) formulation of 4-AP** can be used (fampridine), which was shown to have fewer side effects with no mandatory ECG after first application.

N-Acetyl-DL-leucine (Tanganil) and N-Acetyl-L-leucine (the L-enantiomer of N-Acetyl-DL-leucine) are acetylated derivatives of the amino acid leucine, acting as prodrugs. Both substances can cross the blood-brain barrier and have multimodal modes of acting, including increased ATP (adenosine triphosphate) production and reducing mitochondrial reactive oxygen species (ROS), and reducing lipid storage in neuronal and nonneuronal tissues (see [65] for more details). No significant side effects have been reported for N-Acetyl-DL-leucine, whereas no data has been published yet on the L-enantiomer.

Chlorzoxazone (CHZ) is a small conductance calcium-activated potassium channel activator shown to reduce irregular PCs firing in CACNA1A mutant mice [66]. Potential side effects included abdominal discomfort.

Acetazolamide, a carbonic anhydrase inhibitor, has been used to treat cerebellar disorders based on the hypothesis of a gain-of-function effect on the calcium channel CaV2.1 [65]. Doses of 250–1000 mg daily have been considered, and substantial side effects, including tingling paresthesia, taste disturbance, and gastrointestinal complaints, have been reported.

12.6.4.2 Cerebellar Disorders: Treating Ataxia and Oculomotor Complaints

An improvement in cerebellar ataxic gait under 4-AP has been demonstrated in various cerebellar disorders, including multiple systems atrophy (MSA), sporadic adult-onset ataxia (SAOA), and cerebellar stroke [67, 68]. Similarly, for sustained release (SR) 4-AP (fampridine), moderate short-term gait improvements were seen [69]. Alternatively, intake of N-acetyl-DL-leucine (Tanganil) at a dose of 5 g per day resulted in an improved SARA score, i.e., reduced cerebellar symptoms including gait, speech, alternating movements in various cerebellar ataxias [70], and improved gait and reaching in Niemann-Pick Type C [71]. Noteworthy, in the ALCAT cross-over trial, including 108 patients with (non)hereditary cerebellar ataxia (excluding Friedreich ataxia, MSA cerebellar type (MSAc), rapidly progressive ataxia), no treatment response was seen for N-acetyl-DL-leucine over a treatment period of 6 weeks [72].

In episodic cerebellar ataxias, the primary treatment goal is to reduce the frequency, intensity, and duration of attacks. For episodic ataxia type 2 (EA2), which is characterized by a CACNA1A mutation and recurrent attacks of ataxia for hours and interictal mild cerebellar symptoms (including gaze-evoked nystagmus and downbeat nystagmus) use of 4-AP has been shown to be effective in two trials. In a randomized placebo-controlled trial (cross-over design) use of 4-AP (5 mg three times daily) in patients with genetically confirmed EA2 or other familial episodic ataxias resulted in a significant reduction of the monthly median attack frequency (from 6.5 to 1.7) and to an improved quality of life [73], leading to a treatment recommendation by the AAN (1 class I study) [74]. More recently, sustained-release 4-AP (2× 10 mg per day) has been shown to have fewer side effects than acetazolamide 750 mg/d and comparable efficacy in a randomized, placebo-controlled cross-over trial (EAT2TREAT trial) [75]. Likewise, in spinocerebellar ataxia 27B (SCA27), which has recently been linked to an autosomal dominant intronic GAA repeat expansion in the FGF14 gene and presents as a pancerebellar syndrome, a treatment response to 4-AP with relevance for everyday living was reported by 86% of treated patients [76].

12.6.5 Functional Dizziness

12.6.5.1 Persistent Postural-Perceptual Dizziness (PPPD)

For PPPD, which is defined as a chronic vestibular syndrome with exacerbation of symptoms in upright posture, during active or passive motion, or when exposed to moving visual stimuli (see [77] for detailed definition), treatment strategies focus on multimodality treatment plans, albeit no large-scale randomized controlled trials of any treatments are available [78, 79]. Combining specially adapted vestibular rehabilitation, serotonergic medications such as SSRIs (sertraline, paroxetine, fluoxetine, citalopram) or SNRIs (venlafaxine, duloxetine), and cognitive behavior therapy has been most promising (see Table 12.2).

12.6.5.2 Mal de débarquement Syndrome (MdDS)

In MdDS, which is considered a disorder of postmotion-induced persistent oscillating vertigo with a duration exceeding 48 hours (see [80] for diagnostic criteria), accompanying symptoms include chronic fatigue, visually induced dizziness, headaches, tinnitus, and anxiety. Thus, treatment strategies should ideally address the entire spectrum of reported complaints. Besides experimental neuromodulatory protocols and lifestyle modifications, SSRIs and benzodiazepines have been proposed for treatment (see Table 12.2). Noteworthy, drug treatment in MdDS reduces symptom intensity, but typically does not result in remission. Starting at very low doses and slow titration is recommended for drug treatment in MdDS [81]. Since the

Table 12.2 Treatment recommendations for selected drugs in functional dizziness[a]

Persistent postural-perceptual dizziness (PPPD)		
	Initial dose	**Therapeutic dose range**
SSRIs (Selective Serotonin Reuptake Inhibitors)		
Sertraline	25 mg daily	50–150 mg daily
Escitalopram	5 mg daily	10–20 mg daily
Fluoxetine	10 mg daily	20–60 mg daily
SNRIs (Serotonin and Norepinephrine Reuptake Inhibitors)		
Venlafaxine (extended release)	37.5 mg daily	75 mg–225 mg daily
Duloxetine	20–30 mg daily	40–60 mg daily
Mirtazapine	7.5 mg nightly	15–30 mg nightly
Mal de débarquement syndrome (MdDS)b		
	Recommended drugs	**Limitations**
Antidepressants	SSRIs (citalopram, escitalopram, sertraline, paroxetine, fluoxetine) were much more effective than tricyclic antidepressants (TCAs) and SNRIs (100)	
Benzodiazepines	Preferred use of clonazepam due to long half-life (start with 0.25 mg per day, increase up to 0.5 mg twice daily) Small doses (e.g. diazepam 1–2 mg) may be used for limiting postmotion exacerbation	Use of benzodiazepines should be restricted to avoid dependence

Adapted from [78]

[a]For treating coexisting psychiatric illnesses such as anxiety or depressive disorders, higher doses may be needed

[b]No specific dose recommendations identified in the literature

prevalence of vestibular migraine in MdDS is high (reaching up to 62% [82]), a decision should be made which symptoms dominate and thus should be treated in priority. Furthermore, treating MdDS as a vestibular migraine disorder (using verapamil, nortriptyline, or topiramate) has resulted in improved quality of life, also in patients not meeting diagnostic criteria for vestibular migraine [83].

12.7 Pharmacotherapy of Oculomotor Disorders

12.7.1 Downbeat Nystagmus

Presence of downbeat nystagmus (DBN) may be noted as part of a cerebellar syndrome (see above) or may be found in isolation (with the underlying pathophysiology often remaining unknown). DBN may result in substantial impairment of the patient's quality of life due to oscillopsia and gait imbalance, and thus, pharmacotherapy should be evaluated in affected patients to improve symptom control. For

Table 12.3 Pharmacotherapy of downbeat nystagmus (DBN)

Drug	Mechanism of action	Dosage	Comments, limitations, side effects
Clonazepam [84]	GABA-A agonist	0.5–2 mg per day, titrate up to maximum of 20 mg per day	No controlled studies available *Side effects* include drowsiness, ataxia, behavior problems
4-AP [85–87]	Potassium-channel blocker	5 mg per day, titrate up to three times 5 mg per day. Maximal daily dose 30 mg	4-AP crosses the brain-blood barrier more readily than 3,4-DAP *Side effects* include transient paresthesia, nausea, headache *Use should be avoided* in patients with cardiac arrhythmia (ECG needed before and 45 min after application of first dose) or epilepsy
Sustained release 4-AP (fampridine) [88]	Potassium-channel blocker	Two times 10 mg per day	No cardiac monitoring necessary, but much more expensive than 4-AP *Use shoud be avoided* in patients with epilepsy
Chlorzoxazone [66]	Activator of calcium-activated potassium channels	500 mg single dose on day 1, three times 500 mg on day 2 and following	Data from a single pilot study (class IV evidence) only *Side effects* include drowsiness, lightheadedness, dizziness, gastrointestinal bleeding
N-Acetyl-DL-Leucine [89]	Multiple modes of action (increased ATP production, Decreased mitochondrial reactive oxygen species, reduced lipid storage) [65]	1–3 g (up to 5 g) per day	Data from a single pilot study, non-blinded

treating DBN, various drugs have been shown to have some effect on the nystagmus intensity and frequency, targeting different receptors and channels. This includes GABA-A agonists (e.g., clonazepam) [84], potassium-channel blocking drugs (4-AP, 3–4-diamonopyridine (3–4-DAP), SR-4AP), and activators of calcium-activated potassium channels (chlorzoxazone) and amino acids (See Table 12.3 for details). Use of 4-AP reduced the intensity of DBN [85, 86], gaze-evoked nystagmus

[87], and postural instability [85], but had no effect on patient satisfaction [85]. Use of SR-4AP resulted in a reduced slow-phase velocity of DBN as well [88]. In a small case series of ten patients with DBN (idiopathic DBN in seven cases), a treatment trial with chlorzoxazone over 1 or 2 weeks significantly reduced the DBN slow-phase velocity and increased the visual acuity. Furthermore, a trend toward less postural sway was observed [66]. In patients with ataxia telangiectasia, intake of N-acetyl-DL-Leucine over 4 weeks resulted in a significant reduction of DBN slow-phase velocity and ataxia [89]. For other substances, only anecdotal evidence for treatment response is available, including baclofen, gabapentin, and memantine [10].

12.7.2 Upbeat Nystagmus (UBN)

An upbeat spontaneous nystagmus can be found in various brainstem disorders, including medullary or midbrain ischemic lesions, demyelinating brainstem lesions, and Wernicke Encephalopathy due to acute thiamine deficiency. UBN is usually transient in nature, but may be persistent and disabling due to oscillopsia and resulting gait imbalance. Treatment strategies depend on the underlying pathology [90, 91]. For acute thiamine deficiency, immediate high-dose supplementation of thiamine is key (see chapter Wernicke Encephalopathy for treatment recommendations). Importantly, application of only 100 mg–250 mg thiamine per day may not be sufficient to lower morbidity and mortality. Alternatively, baclofen (5–10 mg three times daily) [92], immediate-release 4-AP (5–10 mg three times daily [92] or SR-4AP (two times 10 mg daily) may be considered.

12.7.3 Acquired Pendular Nystagmus (APN)

In APN, characteristic to-and-fro slow ocular oscillations can be observed without the dichotomy of a slow phase and fast phase as seen in jerk nystagmus. The beating direction is typically horizontal, but a vertical and/or torsional component can be observed as well [10].

Both ischemic, hemorrhagic, and demyelinating brainstem lesions may result in APN, which is sometimes accompanied by palatal myoclonus, referred to as oculopalatal tremor (OPT) in these cases. Treatment is symptomatic in most cases, with gabapentin being considered a first-line drug (class B evidence, dose range: 900–2400 mg per day). Gabapentin is a nonselective GABA-ergic drug with likely other, non-GABA-ergic mechanisms of action, including inhibition of the NMDA-receptor. At a dose of 900 mg per day, it was effective in improving visual acuity and APN in a small case series ($n = 15$) [93, 94]. Side effects, including dizziness, somnolence, and fatigue, however, may limit its use. Alternatively, memantine, a noncompetitive NMDA-receptor antagonist, may be considered as a second-line treatment. At a dose of 20–40 mg per day, memantine effectively stopped the pendular nystagmus, resulting in improved visual acuity in half of the patients in one small case series ($n = 11$) [95]. Noteworthy, potential side effects, including dizziness, muscle weakness/pain, headache, confusion, and fatigue, need to be taken into account. As third-line treatment options, benzodiazepines (clonazepam), sodium valproate, scopolamine patches, trihexyphenidyl, D9-THC, and topiramate may be considered [10]. Overall, better treatment responses in APN due to demyelinating lesions than in the context of oculopalatal tremor have been reported, with often little treatment response in the latter case [90, 96].

Very rarely, a vertical pendular nystagmus with a seesaw pattern (elevation and intorsion of one eye and (at the same time) depression and extorsion of the other eye in one half-cycle and reversal during the other half of the cycle) has been observed in patients with chiasmal lesions and accompanying bitemporal hemianopia or with cerebellar lesions [10]. Response to treatment with alcohol, baclofen, clonazepam, carbamazepine, and gabapentin has been reported [91]. Furthermore, on rare occasions, a convergent-divergent form of APN can be seen in Whipple's disease in the setting of oculomasticatory myorhythmia, requiring antibiotic treatment for Tropheryma whipplei.

12.7.4 Periodic Alternating Nystagmus (PAN)

PAN is defined as a horizontal jerk nystagmus with spontaneous reversal of the nystagmus beating direction every 90–120 s (often with a few beats of vertical spontaneous nystagmus during the transition). PAN has been associated with lesions of the cerebellar nodulus or the cerebellar uvula, which are important for managing the time course of physiologic nystagmus and velocity storage. Patients suffering from PAN often report fluctuating oscillopsia. Drug treatment in PAN focuses on the inhibitory GABA-ergic pathways in order to help control velocity storage. Using the GABA-B agonist baclofen in an initial dose of three times 5 mg per day (titrate up to three times 10 mg per day and a maximal total daily dose of 80 mg), a positive effect has been reported [91]. Noteworthy, side effects of baclofen-use include drowsiness, weakness, fatigue, headache, insomnia, and nausea. Benefit from amantadine was reported in a single case [97].

12.7.5 Superior Oblique Myokymia

In superior oblique myokymia, paroxysmal monocular high-frequency oscillations can be observed. The beating direction is primarily torsional; an additional vertical component can be observed during adduction of the eye. Vascular compression of the fourth nerve may be responsible. Treatment is usually symptomatic. A treatment response to carbamazepine (200–400 mg three to four times per day), phenytoin (250–400 mg per day), and gabapentin (200–300 mg two times per day) has been reported [91].

12.7.6 *Saccadic Intrusions*

Saccadic intrusions are initiated by a fast saccadic eye movement and are often nonrhythmic and intermittent. Saccadic intrusions refer to a range of saccade types that occur and interfere with fixation. A broad distinction is made between SI with an intersaccadic interval (square-wave jerks) and SI without an intersaccadic interval (ocular flutter, opsoclonus) [10] (see also further above for more details). Oscillopsia and reduced visual acuity are characteristic complaints noted by affected patients. Saccadic intrusions can be observed in various diseases, including hereditary ataxias such as spinocerebellar ataxias (e.g., SCA 2, 3, 6, 17), ataxia telangiectasia and Friedreich ataxia [98], multiple sclerosis, tumors in the posterior fossa, including neuroblastoma in children, paraneoplastic syndromes, and parainfectious brainstem encephalitis. Overall, there is limited data on therapeutic options, with often small case series or single cases only [11]. In case of ocular flutter or opsoclonus, treatment with gabapentin (total daily dose 300–900 mg), carbamazepine (total daily dose 400–800 mg), or clonazepam (total daily dose 1–2 mg) can be considered. For (macro)-square-wave jerks and square-wave oscillations, a treatment effect of memantine (total daily dose 15–60 mg) has been demonstrated.

12.8 Conclusions

A broad range of substances and treatment strategies have been evaluated in various peripheral and central neuro-otologic disorders presenting with oculomotor and vestibular complaints, as reviewed here. Importantly, the treatment decision and drug selection should be made individually, taking into account the strength of evidence for a given treatment, treatment costs, co-morbidities, and potential side effects. Whenever possible, quantitative measurements (e.g., video-oculography) should be used to document treatment response [99]. While aminopyridines have been most broadly used to treat cerebellar disease and oculomotor disorders in the past, the search for new treatment strategies and substances is far from over. More insights are expected, for example, from derivatives of N-acetyl-DL-Leucine [65]. Importantly, pharmacotherapy in neuro-otology is often based on open-label, noncontrolled studies or case series and on expert opinion. Thus, there is a dire need for high-quality randomized, placebo-controlled, and double-blinded trials. At the same time, this is complicated by the fact that many of these disorders are rare, substances available are not under exclusive marketing protection anymore, and thus, funding for such research is difficult to achieve.

Furthermore, sometimes, drug treatment is not the solution to the patient's complaints but their underlying cause. Thus, medication history, current medication, and related side effects should always be evaluated.

References

1. Ljunggren M, Persson J, Salzer J. Dizziness and the acute vestibular syndrome at the emergency department: a population-based descriptive study. Eur Neurol. 2018;79(1–2):5–12.
2. Newman-Toker DE, Hsieh YH, Camargo CA Jr, Pelletier AJ, Butchy GT, Edlow JA. Spectrum of dizziness visits to US emergency departments: cross-sectional analysis from a nationally representative sample. Mayo Clin Proc. 2008;83(7):765–75.
3. Kerber KA, Meurer WJ, West BT, Fendrick AM. Dizziness presentations in U.S. emergency departments, 1995-2004. Acad Emerg Med. 2008;15(8):744–50.
4. Goeldlin M, Gaschen J, Kammer C, Comolli L, Bernasconi CA, Spiegel R, et al. Frequency, aetiology, and impact of vestibular symptoms in the emergency department: a neglected red flag. J Neurol. 2019;266(12):3076–86.
5. Neuhauser HK, Radtke A, von Brevern M, Lezius F, Feldmann M, Lempert T. Burden of dizziness and vertigo in the community. Arch Intern Med. 2008;168(19):2118–24.
6. Bisdorff A, Bosser G, Gueguen R, Perrin P. The epidemiology of vertigo, dizziness, and unsteadiness and its links to co-morbidities. Front Neurol. 2013;4:29.
7. Kerber KA, Callaghan BC, Telian SA, Meurer WJ, Skolarus LE, Carender W, et al. Dizziness symptom type prevalence and overlap: a US nationally representative survey. Am J Med. 2017;130(12):1465.
8. Chang J, Hwang SY, Park SK, Kim JH, Kim HJ, Chae SW, et al. Prevalence of dizziness and associated factors in South Korea: a cross-sectional survey from 2010 to 2012. J Epidemiol. 2018;28(4):176–84.
9. Mueller M, Schuster E, Strobl R, Grill E. Identification of aspects of functioning, disability and health relevant to patients experiencing vertigo: a qualitative study using the international classification of functioning, disability and health. Health Qual Life Outcomes. 2012;10:75.
10. Rucker JC. Nystagmus and saccadic intrusions. Continuum (Minneap Minn). 2019;25(5): 1376–400.
11. Zwergal A, Strupp M, Brandt T. Advances in pharmacotherapy of vestibular and ocular motor disorders. Expert Opin Pharmacother. 2019;20(10):1267–76.
12. Strupp M, Bisdorff A, Furman J, Hornibrook J, Jahn K, Maire R, et al. Acute unilateral vestibulopathy/vestibular neuritis: diagnostic criteria. J Vestib Res. 2022;32(5):389–406.
13. Beck R, Gunther L, Xiong G, Potschka H, Boning G, Bartenstein P, et al. The mixed blessing of treating symptoms in acute vestibular failure – evidence from a 4-aminopyridine experiment. Exp Neurol. 2014;261:638–45.
14. Strupp M, Zingler VC, Arbusow V, Niklas D, Maag KP, Dieterich M, et al. Methylprednisolone, valacyclovir, or the combination for vestibular neuritis. N Engl J Med. 2004;351(4): 354–61.
15. S2k-Leitlinie – Vestibuläre Funktionsstörungen. Available from: https://register.awmf.org/assets/guidelines/017-078l_S2k_Vestibulaere-Funktionsstoerungen_2021-05.pdf
16. Fishman JM, Burgess C, Waddell A. Corticosteroids for the treatment of idiopathic acute vestibular dysfunction (vestibular neuritis). Cochrane Database Syst Rev. 2011;(5): CD008607.
17. Leong KJ, Lau T, Stewart V, Canetti EFD. Systematic review and meta-analysis: effectiveness of corticosteroids in treating adults with acute vestibular neuritis. Otolaryngol Head Neck Surg. 2021;165(2):255–66.
18. Sousa FA, Alves CS, Pinto AN, Meireles L, Rego AR. Pharmacological treatment of acute unilateral vestibulopathy: a review. J Audiol Otol. 2024;28(1):18–28.
19. Antons M, Lindner M, Eilles E, Gunther L, Delker A, Branner C, et al. Dose- and application route-dependent effects of betahistine on behavioral recovery and neuroplasticity after acute unilateral labyrinthectomy in rats. Front Neurol. 2023;14:1175481.
20. Orendorz-Fraczkowska K, Pospiech L, Gawron W. Results of combined treatment for vestibular receptor impairment with physical therapy and Ginkgo biloba extract (Egb 761). Otolaryngol Pol. 2002;56(1):83–8.

21. Lindner M, Gosewisch A, Eilles E, Branner C, Kramer A, Oos R, et al. Ginkgo biloba extract EGb 761 improves vestibular compensation and modulates cerebral vestibular networks in the rat. Front Neurol. 2019;10:147.
22. Gunther L, Beck R, Xiong G, Potschka H, Jahn K, Bartenstein P, et al. N-acetyl-L-leucine accelerates vestibular compensation after unilateral labyrinthectomy by action in the cerebellum and thalamus. PLoS One. 2015;10(3):e0120891.
23. Lopez-Escamez JA, Carey J, Chung WH, Goebel JA, Magnusson M, Mandala M, et al. Diagnostic criteria for Meniere's disease. J Vestib Res. 2015;25(1):1–7.
24. Halmagyi GM, Akdal G, Welgampola MS, Wang C. Neurological update: neuro-otology 2023. J Neurol. 2023;270(12):6170–92.
25. Sharon JD, Trevino C, Schubert MC, Carey JP. Treatment of Meniere's disease. Curr Treat Options Neurol. 2015;17(4):341.
26. Webster KE, Galbraith K, Harrington-Benton NA, Judd O, Kaski D, Maarsingh OR, et al. Systemic pharmacological interventions for Meniere's disease. Cochrane Database Syst Rev. 2023;2(2):CD015171.
27. Webster KE, Lee A, Galbraith K, Harrington-Benton NA, Judd O, Kaski D, et al. Intratympanic corticosteroids for Meniere's disease. Cochrane Database Syst Rev. 2023;2 (2):CD015245.
28. Webster KE, Galbraith K, Lee A, Harrington-Benton NA, Judd O, Kaski D, et al. Intratympanic gentamicin for Meniere's disease. Cochrane Database Syst Rev. 2023;2(2): CD015246.
29. Adrion C, Fischer CS, Wagner J, Gurkov R, Mansmann U, Strupp M, et al. Efficacy and safety of betahistine treatment in patients with Meniere's disease: primary results of a long term, multicentre, double blind, randomised, placebo controlled, dose defining trial (BEMED trial). BMJ. 2016;352:h6816.
30. Lezius F, Adrion C, Mansmann U, Jahn K, Strupp M. High-dosage betahistine dihydrochloride between 288 and 480 mg/day in patients with severe Meniere's disease: a case series. Eur Arch Otorrinolaringol. 2011;268(8):1237–40.
31. Strupp M, Churchill GC, Naumann I, Mansmann U, Al Tawil A, Golentsova A, et al. Examination of betahistine bioavailability in combination with the monoamine oxidase B inhibitor, selegiline, in humans-a non-randomized, single-sequence, two-period titration, open label single-center phase 1 study (PK-BeST). Front Neurol. 2023;14:1271640.
32. Sutton L, Ghedia R, Harcourt J. Betahistine prescribing practices in England: an analysis of prescribing and National Spending pre- and post-BEMED trial. Otol Neurotol. 2023;44(6): e406–e11.
33. Li S, Pyykko I, Zhang Q, Yang J, Duan M. Consensus on intratympanic drug delivery for Meniere's disease. Eur Arch Otorrinolaringol. 2022;279(8):3795–9.
34. Strupp M, Lopez-Escamez JA, Kim JS, Straumann D, Jen JC, Carey J, et al. Vestibular paroxysmia: diagnostic criteria. J Vestib Res. 2016;26(5–6):409–15.
35. Hufner K, Barresi D, Glaser M, Linn J, Adrion C, Mansmann U, et al. Vestibular paroxysmia: diagnostic features and medical treatment. Neurology. 2008;71(13):1006–14.
36. Bayer O, Bremova T, Strupp M, Hufner K. A randomized double-blind, placebo-controlled, cross-over trial (Vestparoxy) of the treatment of vestibular paroxysmia with oxcarbazepine. J Neurol. 2018;265(2):291–8.
37. Strupp M, Elger C, Goldschagg N. Treatment of vestibular paroxysmia with lacosamide. Neurol Clin Pract. 2019;9(6):539–41.
38. Potdar T, Kontorinis G. Pre-habilitation with intratympanic gentamicin in vestibular schwannomas: a systematic review. J Laryngol Otol. 2023;137(9):985–91.
39. Tarnutzer AA, Bockisch CJ, Buffone E, Huber AM, Wettstein VG, Weber KP. Pre-habilitation before vestibular Schwannoma surgery-impact of Intratympanal gentamicin application on the Vestibulo-ocular reflex. Front Neurol. 2021;12:633356.
40. Balatkova Z, Cada Z, Hruba S, Komarc M, Cerny R. Assessment of visual sensation, psychiatric profile and quality of life following vestibular schwannoma surgery in patients prehabituated by chemical vestibular ablation. Biomed Pap Med Fac Univ Palacky Olomouc Czech Repub. 2020;164(4):444–53.

41. Cada Z, Balatkova Z, Chovanec M, Cakrt O, Hruba S, Jerabek J, et al. Vertigo perception and quality of life in patients after surgical treatment of vestibular Schwannoma with Pretreatment Prehabituation by chemical vestibular ablation. Biomed Res Int. 2016;2016:6767216.
42. Fellmann J, Bachinger D, Dalbert A, Roosli C, Huber A, Wettstein VG. Postural stability and handicap of dizziness after preoperative vestibular ablation and vestibular prehabilitation in patients undergoing vestibular schwannoma resection. J Vestib Res. 2022;32(1):49–56.
43. Bhattacharyya N, Gubbels SP, Schwartz SR, Edlow JA, El-Kashlan H, Fife T, et al. Clinical practice guideline: benign paroxysmal positional vertigo (update). Otolaryngol Head Neck Surg. 2017;156(3_suppl):S1–S47.
44. Jeong SH, Lee SU, Kim JS. Prevention of recurrent benign paroxysmal positional vertigo with vitamin D supplementation: a meta-analysis. J Neurol. 2022;269(2):619–26.
45. Hong X, Christ-Franco M, Moher D, Tse D, Lelli DA, Schramm D, et al. Vitamin D supplementation for benign paroxysmal positional vertigo: a systematic review. Otol Neurotol. 2022;43(7):e704.
46. Talaat HS, Kabel AM, Khaliel LH, Abuhadied G, El-Naga HA, Talaat AS. Reduction of recurrence rate of benign paroxysmal positional vertigo by treatment of severe vitamin D deficiency. Auris Nasus Larynx. 2016;43(3):237–41.
47. Jeong SH, Kim JS, Kim HJ, Choi JY, Koo JW, Choi KD, et al. Prevention of benign paroxysmal positional vertigo with vitamin D supplementation: a randomized trial. Neurology. 2020;95(9):e1117–e25.
48. Tarnutzer AA, Berkowitz AL, Robinson KA, Hsieh YH, Newman-Toker DE. Does my dizzy patient have a stroke? A systematic review of bedside diagnosis in acute vestibular syndrome. CMAJ. 2011;183(9):E571–92.
49. Powers WJ, Rabinstein AA, Ackerson T, Adeoye OM, Bambakidis NC, Becker K, et al. Guidelines for the early Management of Patients with Acute Ischemic Stroke: 2019 update to the 2018 guidelines for the early Management of Acute Ischemic Stroke: a guideline for healthcare professionals from the American Heart Association/American Stroke Association. Stroke. 2019;50(12):e344–418.
50. Lee SH, Kim JS. Acute diagnosis and Management of Stroke Presenting Dizziness or vertigo. Neurol Clin. 2015;33(3):687–98. xi
51. Kleindorfer DO, Towfighi A, Chaturvedi S, Cockroft KM, Gutierrez J, Lombardi-Hill D, et al. 2021 guideline for the prevention of stroke in patients with stroke and transient ischemic attack: a guideline from the American Heart Association/American Stroke Association. Stroke. 2021;52(7):e364–467.
52. Sechi G, Serra A. Wernicke's encephalopathy: new clinical settings and recent advances in diagnosis and management. Lancet Neurol. 2007;6(5):442–55.
53. Galvin R, Brathen G, Ivashynka A, Hillbom M, Tanasescu R, Leone MA, et al. EFNS guidelines for diagnosis, therapy and prevention of Wernicke encephalopathy. Eur J Neurol. 2010;17(12):1408–18.
54. Lempert T, Olesen J, Furman J, Waterston J, Seemungal B, Carey J, et al. Vestibular migraine: diagnostic criteria1. J Vestib Res. 2022;32(1):1–6.
55. Marmura MJ, Silberstein SD, Schwedt TJ. The acute treatment of migraine in adults: the american headache society evidence assessment of migraine pharmacotherapies. Headache. 2015;55(1):3–20.
56. Smyth D, Britton Z, Murdin L, Arshad Q, Kaski D. Vestibular migraine treatment: a comprehensive practical review. Brain. 2022;145(11):3741–54.
57. Bayer O, Adrion C, Al Tawil A, Mansmann U, Strupp M, investigators P. Results and lessons learnt from a randomized controlled trial: prophylactic treatment of vestibular migraine with metoprolol (PROVEMIG). Trials. 2019;20(1):813.
58. Oh SY, Kang JJ, Kim S, Lee JM, Kim JS, Dieterich M. A preliminary trial of botulinum toxin type a in patients with vestibular migraine: a longitudinal fMRI study. Front Neurol. 2022;13:955158.

59. Hoskin JL, Fife TD. New anti-CGRP medications in the treatment of vestibular migraine. Front Neurol. 2021;12:799002.
60. Byun YJ, Levy DA, Nguyen SA, Brennan E, Rizk HG. Treatment of vestibular migraine: a systematic review and meta-analysis. Laryngoscope. 2021;131(1):186–94.
61. Aydin İ, Gökçay F, Karapolat H, Eraslan S, Bilgen C, Kirazli T, et al. Effects of vestibular rehabilitation and pharmacological therapy in patients with vestibular migraine. Neurol Sci Neurophysiol. 2020;37(3):110–7.
62. Roberts RA, Watford KE, Picou EM, Hatton K, Trone TH, Brignola EY. Effects of lifestyle modification on vestibular migraine. Otol Neurotol. 2021;42(10):e1537–e43.
63. Kalla R, Strupp M. Aminopyridines and acetyl-DL-leucine: new therapies in cerebellar disorders. Curr Neuropharmacol. 2019;17(1):7–13.
64. Strupp M, Teufel J, Zwergal A, Schniepp R, Khodakhah K, Feil K. Aminopyridines for the treatment of neurologic disorders. Neurol Clin Pract. 2017;7(1):65–76.
65. Bremova-Ertl T, Hofmann J, Stucki J, Vossenkaul A, Gautschi M. Inborn errors of metabolism with ataxia: current and future treatment options. Cells. 2023;12(18)
66. Feil K, Claassen J, Bardins S, Teufel J, Krafczyk S, Schneider E, et al. Effect of chlorzoxazone in patients with downbeat nystagmus: a pilot trial. Neurology. 2013;81(13):1152–8.
67. Schniepp R, Wuehr M, Ackl N, Danek A, Brandt T, Strupp M, et al. 4-Aminopyridine improves gait variability in cerebellar ataxia due to CACNA 1A mutation. J Neurol. 2011;258(9):1708–11.
68. Schniepp R, Wuehr M, Neuhaeusser M, Benecke AK, Adrion C, Brandt T, et al. 4-aminopyridine and cerebellar gait: a retrospective case series. J Neurol. 2012;259(11):2491–3.
69. Giordano I, Bogdanow M, Jacobi H, Jahn K, Minnerop M, Schoels L, et al. Experience in a short-term trial with 4-aminopyridine in cerebellar ataxia. J Neurol. 2013;260(8):2175–6.
70. Strupp M, Teufel J, Habs M, Feuerecker R, Muth C, van de Warrenburg BP, et al. Effects of acetyl-DL-leucine in patients with cerebellar ataxia: a case series. J Neurol. 2013;260(10):2556–61.
71. Bremova-Ertl T, Claassen J, Foltan T, Gascon-Bayarri J, Gissen P, Hahn A, et al. Efficacy and safety of N-acetyl-l-leucine in Niemann–pick disease type C. J Neurol. 2022;269(3):1651–62.
72. Feil K, Adrion C, Boesch S, Doss S, Giordano I, Hengel H, et al. Safety and efficacy of acetyl-DL-leucine in certain types of cerebellar ataxia: the ALCAT randomized clinical crossover trial. JAMA Netw Open. 2021;4(12):e2135841.
73. Strupp M, Kalla R, Claassen J, Adrion C, Mansmann U, Klopstock T, et al. A randomized trial of 4-aminopyridine in EA2 and related familial episodic ataxias. Neurology. 2011;77(3):269–75.
74. Zesiewicz TA, Wilmot G, Kuo SH, Perlman S, Greenstein PE, Ying SH, et al. Comprehensive systematic review summary: treatment of cerebellar motor dysfunction and ataxia: report of the guideline development, dissemination, and implementation Subcommittee of the American Academy of neurology. Neurology. 2018;90(10):464–71.
75. Muth C, Teufel J, Schols L, Synofzik M, Franke C, Timmann D, et al. Fampridine and acetazolamide in EA2 and related familial EA: a prospective randomized placebo-controlled trial. Neurol Clin Pract. 2021;11(4):e438–e46.
76. Wilke C, Pellerin D, Mengel D, Traschutz A, Danzi MC, Dicaire MJ, et al. GAA-FGF14 ataxia (SCA27B): phenotypic profile, natural history progression and 4-aminopyridine treatment response. Brain. 2023;146(10):4144–57.
77. Staab JP, Eckhardt-Henn A, Horii A, Jacob R, Strupp M, Brandt T, et al. Diagnostic criteria for persistent postural-perceptual dizziness (PPPD): consensus document of the committee for the classification of vestibular disorders of the Barany society. J Vestib Res. 2017;27(4):191–208.
78. Staab JP. Persistent postural-perceptual dizziness. Semin Neurol. 2020;40(1):130–7.
79. Webster KE, Harrington-Benton NA, Judd O, Kaski D, Maarsingh OR, MacKeith S, et al. Pharmacological interventions for persistent postural-perceptual dizziness (PPPD). Cochrane Database Syst Rev. 2023;3(3):CD015188.

80. Cha YH, Baloh RW, Cho C, Magnusson M, Song JJ, Strupp M, et al. Mal de debarquement syndrome diagnostic criteria: consensus document of the classification Committee of the Barany Society. J Vestib Res. 2020;30(5):285–93.
81. Cha YH. Chronic dizziness. Continuum (Minneap Minn). 2021;27(2):420–46.
82. Beh SC, Chiang HS, Sanderson C. The interconnections of mal de Debarquement syndrome and vestibular migraine. Laryngoscope. 2021;131(5):E1653–E61.
83. Ghavami Y, Haidar YM, Ziai KN, Moshtaghi O, Bhatt J, Lin HW, et al. Management of mal de debarquement syndrome as vestibular migraines. Laryngoscope. 2017;127(7):1670–5.
84. Young YH, Huang TW. Role of clonazepam in the treatment of idiopathic downbeat nystagmus. Laryngoscope. 2001;111(8):1490–3.
85. Claassen J, Spiegel R, Kalla R, Faldon M, Kennard C, Danchaivijitr C, et al. A randomised double-blind, cross-over trial of 4-aminopyridine for downbeat nystagmus – effects on slowphase eye velocity, postural stability, locomotion and symptoms. J Neurol Neurosurg Psychiatry. 2013;84(12):1392–9.
86. Kalla R, Spiegel R, Claassen J, Bardins S, Hahn A, Schneider E, et al. Comparison of 10-mg doses of 4-aminopyridine and 3,4-diaminopyridine for the treatment of downbeat nystagmus. J Neuroophthalmol. 2011;31(4):320–5.
87. Kalla R, Glasauer S, Buttner U, Brandt T, Strupp M. 4-aminopyridine restores vertical and horizontal neural integrator function in downbeat nystagmus. Brain. 2007;130(Pt 9):2441–51.
88. Claassen J, Feil K, Bardins S, Teufel J, Spiegel R, Kalla R, et al. Dalfampridine in patients with downbeat nystagmus – an observational study. J Neurol. 2013;260(8):1992–6.
89. Brueggemann A, Bicvic A, Goeldlin M, Kalla R, Kerkeni H, Mantokoudis G, et al. Effects of acetyl-DL-leucine on ataxia and downbeat-nystagmus in six patients with ataxia telangiectasia. J Child Neurol. 2022;37(1):20–7.
90. Mehta AR, Kennard C. The pharmacological treatment of acquired nystagmus. Pract Neurol. 2012;12(3):147–53.
91. Straube A, Leigh RJ, Bronstein A, Heide W, Riordan-Eva P, Tijssen CC, et al. EFNS task force – therapy of nystagmus and oscillopsia. Eur J Neurol. 2004;11(2):83–9.
92. Dieterich M, Straube A, Brandt T, Paulus W, Buttner U. The effects of baclofen and cholinergic drugs on upbeat and downbeat nystagmus. J Neurol Neurosurg Psychiatry. 1991;54(7):627–32.
93. Averbuch-Heller L, Tusa RJ, Fuhry L, Rottach KG, Ganser GL, Heide W, et al. A double-blind controlled study of gabapentin and baclofen as treatment for acquired nystagmus. Ann Neurol. 1997;41(6):818–25.
94. Bandini F, Castello E, Mazzella L, Mancardi GL, Solaro C. Gabapentin but not vigabatrin is effective in the treatment of acquired nystagmus in multiple sclerosis: how valid is the GABAergic hypothesis? J Neurol Neurosurg Psychiatry. 2001;71(1):107–10.
95. Starck M, Albrecht H, Pollmann W, Straube A, Dieterich M. Drug therapy for acquired pendular nystagmus in multiple sclerosis. J Neurol. 1997;244(1):9–16.
96. Murphy OC, Calabresi PA, Gold DR. Teaching video NeuroImage: dramatic response to Topiramate in acquired Pendular nystagmus from multiple sclerosis. Neurology. 2023;100(1):47–8.
97. Lee SH, Lee SY, Choi SM, Kim BC, Kim MK, Kim JS. Resolution of periodic alternating nystagmus with amantadine. J Neurol Sci. 2016;364:65–7.
98. Garces P, Antoniades CA, Sobanska A, Kovacs N, Ying SH, Gupta AS, et al. Quantitative oculomotor assessment in hereditary ataxia: discriminatory power, correlation with severity measures, and recommended parameters for specific genotypes. Cerebellum. 2023;
99. Garces P, Antoniades CA, Sobanska A, Kovacs N, Ying SH, Gupta AS, et al. Quantitative oculomotor assessment in hereditary ataxia: systematic review and consensus by the ataxia global initiative working group on digital-motor biomarkers. Cerebellum. 2023;
100. Cha YH, Cui YY, Baloh RW. Comprehensive clinical profile of mal De Debarquement syndrome. Front Neurol. 2018;9:261.

Chapter 13
The Relevance of Vestibular Deficit in Machado-Joseph Disease and CANVAS: From a Simple Clinical Test to 20 Years of Studies

Carlos R. Gordon

Keywords Machado-Joseph disease · Spinocerebellar ataxia type 3 · Cerebellar · Vestibular · Head impulse test · Biomarker · CANVAS · Mentalizing

13.1 Introduction

Spinocerebellar ataxia type 3, also known as Machado-Joseph disease (SCA3/MJD), is a neurodegenerative autosomal dominant ataxia first described in the 1970s in families of Portuguese-Azorean extraction in the USA [1, 2]. Since then, SCA3/MJD has been reported in many ethnic backgrounds and is known to be the most common SCA worldwide [3, 4]. The clinical picture comprises slowly progressive cerebellar signs, including poor coordination of gait, dysarthria, and abnormal eye movements such as saccadic pursuit, gaze-evoked nystagmus, and saccadic dysfunction. The diagnosis is confirmed by genetic testing, which detects the disease-specific expansion of CAG trinucleotide repeats in the ATXN3 gene located on chromosome 14q32.12. While normal alleles have between 12 and 44 CAG repeats, most SCA3/MJD patients have one expanded allele with 51–87 CAG repeats. The mean age at onset is around 40 years, and the mean survival is 21 years after the onset of symptoms [4]. Nowadays, no disease-modifying approved treatment exists for SCA3/MJD, but new therapeutic strategies are being investigated, and several studies have stressed the importance of starting treatment in the preataxic stage of the disease [5–7].

The first cases of SCA3/MJD in Israel in a Yemenite Jewish family were reported in 1994, before the advent of genetic testing and based on the clinical features

C. R. Gordon (✉)
Faculty of Medicine, Department of Neurology, Tel Aviv University, Tel Aviv, Israel

Meir Medical Center, Kfar Saba, Israel
e-mail: cgordon@tauex.tau.ac.il

P. Ravat et al. (eds.), *Advances in Translational Neuroscience of Vestibular and Eye Movement Disorders*, Contemporary Clinical Neuroscience,
https://doi.org/10.1007/978-3-032-07539-0_13

compatible with the Azorean Spinocerebellar degeneration [8]. This clinical diagnosis was genetically confirmed two years later [9].

13.2 Vestibular Deficit Impairment in SCA3/MJD: A Part of the Disease Phenotype

Following the identification of the first genetically confirmed SCA3/MJD Yemenite family in Israel, I began to examine new patients and found that in addition to the typical cerebellar findings, they surprisingly also exhibited a vestibular impairment that was easily detected by the bedside Head Impulse Test (HIT), initially named the Rapid Doll's eye test or Head Thrust test. This test was first described by G. Michael Halmagyi and Ian Curthoys as a sign of canal paresis [10], meaning a peripheral vestibular loss, and was never described at that time in patients with central nervous system (CNS) lesions. This was, in fact, the beginning of 20 years of studies on vestibular deficit in SCA3/MJD and its clinical relevance.

In 2000, I attended the XVI Ibero-American Congress of Otoneurology in Rosario, Argentina, where I met G. Michael Halmagyi. I told him that I had some SCA3/MJD patients with a clear pathological HIT. Given that there were no prior reports of impaired HIT in cerebellar or other CNS diseases, he expressed his interest but advised me to consider the possibility that these patients might actually have peripheral vestibular loss. Upon returning to Israel, I continued to examine more patients, and in a pioneering study, we reported vestibulo-ocular areflexia that was easily detected by the bedside HIT and confirmed by the absence of vestibular response to both a standard caloric test and ice water ear irrigation in a group of seven SCA3/MJD patients from five different families, but not observed in other types of cerebellar patients [11].

In further studies using the bedside HIT and accurately recording eye and head movements with the magnetic search coil technique [12, 13], we confirmed that SCA3/MJD patients exhibited severely impaired horizontal Vestibulo-Ocular Reflex (VOR) gain (average 0.55; range 0.35–0.76). In addition, the bedside detection of bilateral pathological HIT aligned with the VOR gain measured using the magnetic search coil system. All eight patients with pathological bedside HIT had recorded VOR gain lower than 0.65, which was the lower limit of normal gain in our laboratory; whereas the two patients with normal bedside HIT had a recorded VOR gain of 0.76. The four patients with severe pathological bedside HIT exhibited VOR gain that was significantly lower than that of SCA3/MJD patients with slight pathological HIT [13]. Therefore, these results support the use of the HIT as an accurate bedside test for diagnosing VOR deficit in SCA3/MJD patients.

A later study using the video HIT (vHIT) confirmed the significantly low horizontal VOR gain of around 0.43 in SCA3/MJD but not in SCA1 and SCA2 [14].

Based on the aforementioned studies of impaired horizontal VOR, we raised the possibility that this deficit, as detected on bedside examination, could be part of the

SCA3/MJD phenotype. Consequently, in a broad clinical study of 53 examined patients from 17 families, we confirmed that the horizontal VOR deficit identified by the bedside HIT was present in 90% of these patients, suggesting that it is part of our SCA3/MJD population phenotype [15].

In all the cited studies, only the horizontal but not the vertical VOR function was tested. Furthermore, studies evaluating the otholitic function in SCA3/MJD by recording cervical Vestibular Evoked Potentials (cVEMP) are scarce and provide either contradictory or inconclusive results [16–19]. The advancement of computerized video HIT (vHIT) technology enables a precise and noninvasive evaluation of all semicircular canal functions [20]. In addition, a further developed Suppression Head Impulse test (SHIMP) provides an additional indicator of semicircular canal function. This test elicits a complementary saccadic pattern in healthy controls, where the presence of anti-compensatory saccades is a sign of remaining vestibular function [21]. Thus, to provide a more comprehensive evaluation of the vestibular function in SCA3/MJD, we designed a study measuring VOR gain of all six semicircular canals by vHIT, remaining horizontal VOR function by SHIMP, and saccular function by cVEMP in SCA3/MJD patients and healthy controls [22]. Figure 13.1 shows the results for vHIT, SHIMP, and cVEMP tests in a representative MJD patient and an age-matched healthy control.

All SCA3/MJD had significantly low VOR gain in all six semicircular canals, with a mean $\pm$ SEM of horizontal gain of 0.52 ± 0.04 and vertical gain of 0.57 ± 0.03 versus the Controls' gains of 0.95 ± 0.01 and 0.81 ± 0.02, respectively (Fig. 13.2).

SCA3/MJD also showed a significantly low VOR gain on the SHIMP test with left gain of 0.51 ± 0.04 and right gain of 0.46 ± 0.03 versus Controls' gain of 0.79 ± 0.01 and 0.83 ± 0.03, respectively. In contrast, SCA3/MJD had normal saccular function reflected by the presence of a cVEMP response in 18/20 patients and in 12/17 of Controls, with a nonsignificant difference between SCA3/MJD and healthy controls for P13 and N23 peaks latency. Receiver operating characteristic (ROC) analysis of horizontal VOR gain resulted in an area under the curve of 0.993, indicating that the average horizontal canals' VOR gain is an excellent classifier for distinguishing SCA3/MJD from healthy controls.

Considering all cited studies, we concluded that horizontal and vertical VOR impairment with preserved sacculo-collic function seems to be a distinctive feature of SCA3/MJD. Moreover, impaired horizontal VOR, easily detected by the bedside HIT, is part of the SCA3/MJD phenotype. It is emphasized once again that a horizontal VOR gain ≤ 0.65, as measured by the magnetic search coil system, can be detected by the bedside HIT [13]. Additionally, extensive clinical and laboratory experience with the vHIT has demonstrated that a horizontal VOR gain lower than 0.7 can be accurately detected in peripheral and central vestibular disorders by the HIT (unpublished personal data). Therefore, these findings support the use of the HIT as a reliable bedside test for the diagnosis of VOR deficits.

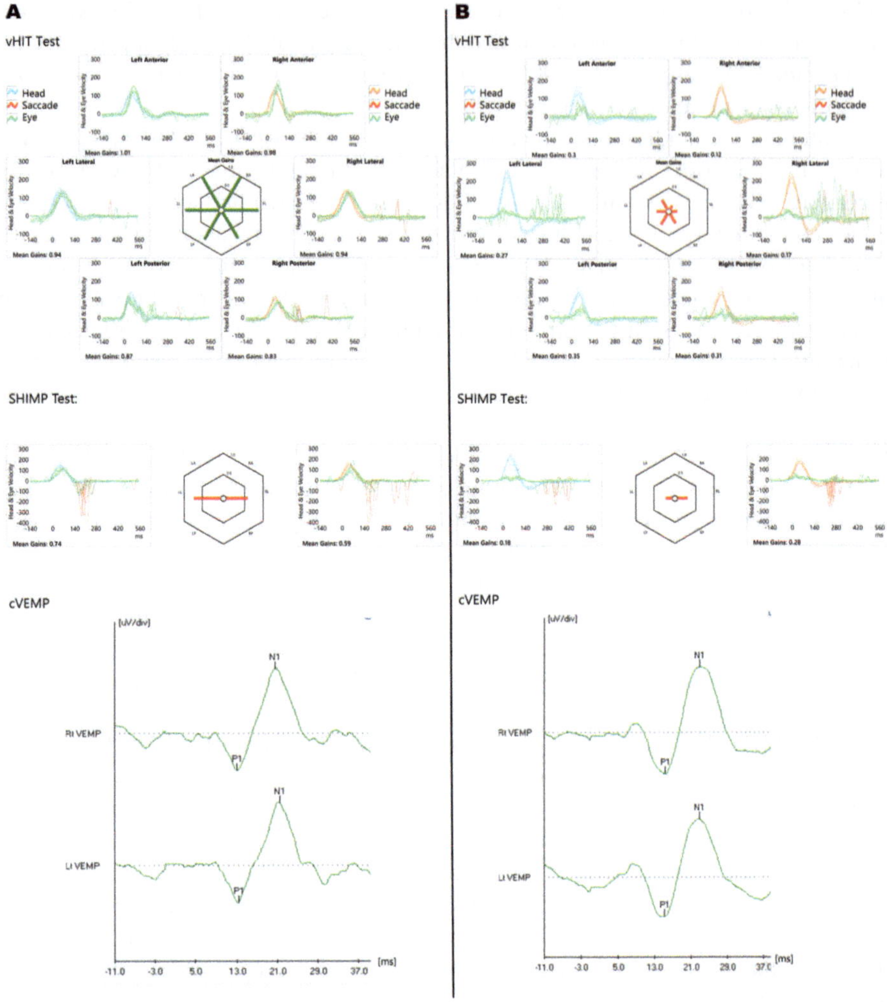

Fig. 13.1 Exemplary records of a representative. (**A**) Healthy control aged 70 years and (**B**) an MJD patient aged 72 years. vHIT and SHIMP tests: Velocity in degrees/sec of the head (light blue for the left side and orange for the right side), the eye (green), and the saccade when present (red) are shown for each semicircular canal together with the mean gain. The central hexagon is a graphic representation of the VOR gain relative to the age normative data, showing the normal range in green and lower values in red. cVEMP test: Average signal for the left (Lt) and right (Rt) sides with the peak marked by the operator (P1 = P13, N1 = N23). Amplitudes are not shown. (From D. Geisinger et al. [22])

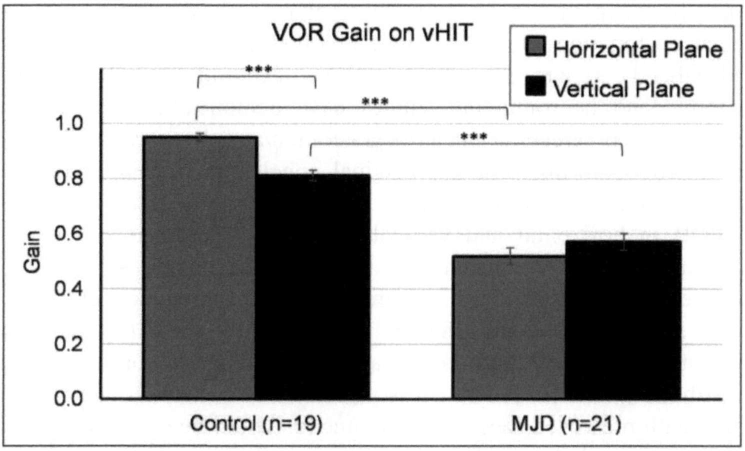

Fig. 13.2 Group comparison of VOR gain (mean ± SEM) in horizontal and vertical planes between MJD patients and healthy controls. *** $p < 0.001$. (From D. Geisinger et al. [22])

13.3 Where Is the Lesion in Terms of Neurological Localization?

How can we explain these findings in terms of neurological localization? Because our patients had no symptoms or signs of auditory or other cranial nerve impairment and based on their eye movement abnormalities, since our first report, we assumed that the VOR deficit of our SCA3/MJD patients is of central and not of peripheral vestibular origin [11]. We postulated that degeneration of the medial vestibular nucleus (MVN) and nucleus prepositus hypoglossi (NPH), the site of the so-called neural integrator and velocity storage mechanism, can account for the central vestibular loss and gaze-evoked nystagmus described in our patients. Although we had not corroborated neuropathological data, the involvement of the vestibular nucleus has been previously reported in some cases with clinical diagnosis of SCA3/MJD [23, 24]. In a review of the neuropathological findings of 18 necropsies designated as clinical SCA3/MJD, collected from the world literature until 1990, all eight cases in which vestibular nuclei were mentioned were considered "affected" [25]. Moreover, these investigators considered the involvement of the vestibular nuclei to be one of the neuropathological criteria for SCA3/MJD. In a detailed neuropathological study of four Dutch cases with clinical and genetically confirmed SCA3/MJD, degeneration of all five vestibular nuclei and their associated fibre tracts was found in all cases [26].

However, considering that the VOR is impaired in both horizontal and vertical planes, there may be a basis for an assumption that the deficit is of peripheral origin. Nevertheless, the cVEMP was present, meaning normal function of the saccular pathway, which runs in the inferior branch of the vestibular nerve with the posterior canal fibers, implying that at least part of the peripheral pathway is preserved. Since there is no direct explanation for how a cerebellar ataxia would present a selective semicircular dysfunction without otolithic involvement, these findings support our earliest idea that the impairment in SCA3/MJD is of central vestibular origin. Several pathoanatomic and neuroimaging studies have demonstrated central vestibular, brainstem, and deep cerebellar nuclei degeneration/atrophy in SCA3/MJD patients [27–30], which conforms with our original conclusion of central origin of the functional horizontal VOR deficit [11, 13]. Therefore, the finding of impaired angular VOR in horizontal and vertical planes but preserved saccular function could also suggest selective damage of the vestibular nuclei. Afferents from the peripheral vestibular organs show a degree of spatial segregation in the vestibular nuclei. The semicircular canal afferents predominantly reach the medial and the superior vestibular nuclei. In contrast, saccular afferents predominantly reach the lateral vestibular nuclei [31]. Thus, it is possible that horizontal and vertical VOR impairment with preserved sacculo-collic function could be explained by selective, predominantly medial and superior vestibular nuclei degeneration [22]. Pathoanatomic autopsy studies specifically investigating the central vestibular system in SCA3/MJD are relatively scarce (reviewed in [27]). Modern MRI studies have demonstrated atrophy of the brainstem and cerebellar peduncles; however, they do not specifically reference the vestibular nuclei, which remain challenging to visualize even with advanced MRI techniques [32–34]. Future high-resolution neuroimaging and neuropathological studies are needed to confirm selective degeneration within the vestibular nuclei in SCA3/MJD.

13.4 Horizontal VOR Gain Deficit as a Neurophysiological Biomarker of SCA3/MJD

In some of the studies mentioned above [13, 14], we had the impression that there was a negative correlation between VOR gain and the severity of the disease; that is, the lower the VOR gain, the higher the disease severity. Consequently, we raised the possibility that the horizontal VOR gain measured by the vHIT could be a potential neurophysiological biomarker of SCA3/MJD. This test has the advantage of being an easy, simple, and quick test requiring little cooperation from the patient.

Nowadays, no disease-modifying treatment is available for SCA3/MJD, but new therapeutic strategies are under investigation [5–7]. Genetic testing can identify individuals at risk of developing the disease, thus offering an opportunity to study the presymptomatic phase of SCA3/MJD, including early signs that herald the onset of the symptomatic phase. The search for a neurophysiological biomarker capable of identifying presymptomatic genetic carriers at the threshold of clinical diagnosis would be highly valuable for early diagnosis, monitoring disease progression, and evaluating the efficacy of new treatments [6, 35]. An ideal biomarker should be easy to measure, quantify, reproducible, stable in the general population, and unaffected by comorbid factors. For use in therapeutic trials, it must demonstrate a strong correlation with established clinical parameters of disease severity, exhibiting a concurrent change with disease progression [35].

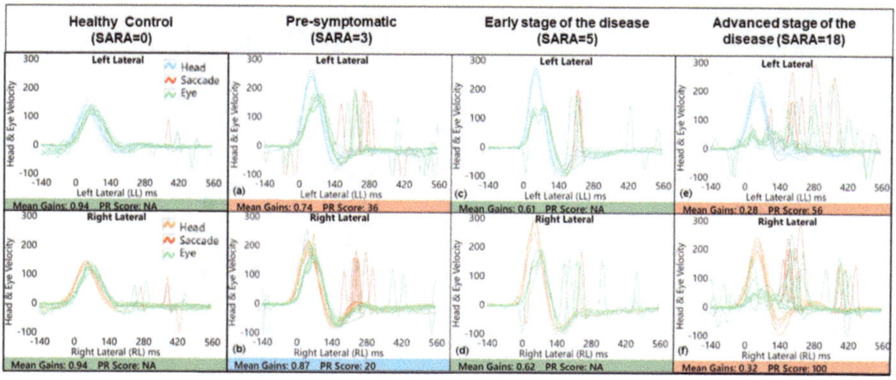

Fig. 13.3 Horizontal eye and head velocity and VOR gain on the video head impulse test (vHIT). Records presented for a healthy control subject, 70 y/o; a presymptomatic MJD carrier, 65 y/o; an MJD patient at an early stage of the disease, 70 y/o; and an MJD patient at an advanced stage of the disease, 72 y/o. Velocity (°/sec) of the head (blue and orange in the upper and lower figures, respectively), the eye (green), and the corrective saccades when present (red) are shown for ten trials along the calculated mean VOR gain. (From Z. Elyoseph et al. [36])

So, we designed a study to evaluate whether the horizontal VOR gain could serve as a reliable neurophysiological biomarker for the clinical onset, severity, and progression of SCA3/MJD [36]. Thirty-five SCA3/MJD patients, 11 presymptomatic genetically confirmed SCA3/MJD subjects, and 20 healthy controls underwent a detailed epidemiological and clinical neurological examination, including the Scale for the Assessment and Rating of Ataxia (SARA), which is used in almost all clinical trials of cerebellar disorders to measure ataxia severity [37]. VOR gain was measured using the vHIT system. Twenty of the SCA3/MJD patients were retested after a period of 1–3 years.

Figure 13.3 shows the horizontal VOR of a healthy control, a presymptomatic subject, and two patients at early and advanced stages of the disease.

The horizontal VOR gain was abnormal in 92% of SCA3/MJD cases, 54% of presymptomatic participants, and 0% of healthy controls. The SARA score revealed initial abnormalities (score between 0 and 3) in only three (27%) of the 11 presymptomatic participants. Thus, the rate of presymptomatic participants with reduced horizontal VOR gain was twofold that of those with positive SARA scores, even when including two asymptomatic participants with SARA scores of 3, which is generally considered the beginning of the ataxic stage. It should be noted that SARA is a valid method to measure the severity of ataxia, but not the disease onset [37], which is routinely set at the age of the beginning of symptoms of gait imbalance reported by patients or their relatives. Figure 13.4 illustrates the correlation between horizontal VOR gain and SARA score among all study participants.

Horizontal VOR gain in the SCA3/MJD group was significantly negatively correlated with the SARA score in the first ($r = 0.66$, $p < 0.001$) and second ($r = 0.61$, $p < 0.001$) examinations (Fig. 13.5a). Additionally, there was a significant negative correlation between the percentage change in horizontal VOR gain and the

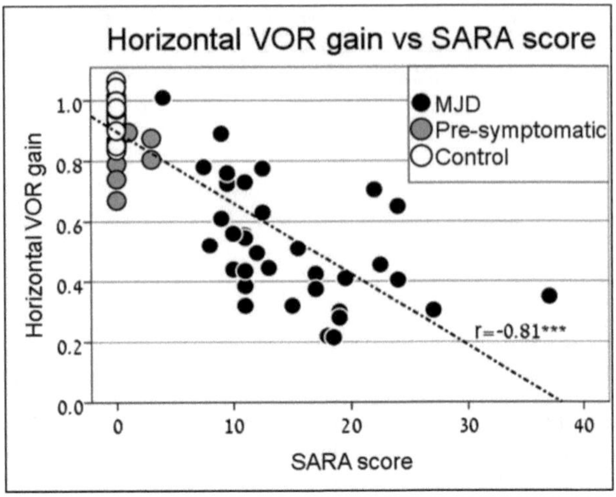

Fig. 13.4 Pearson correlation between horizontal VOR gain and SARA score in all study participants. The line represents the regression slope. $^{***}p < 0.001$. (From Z. Elyoseph et al. [36])

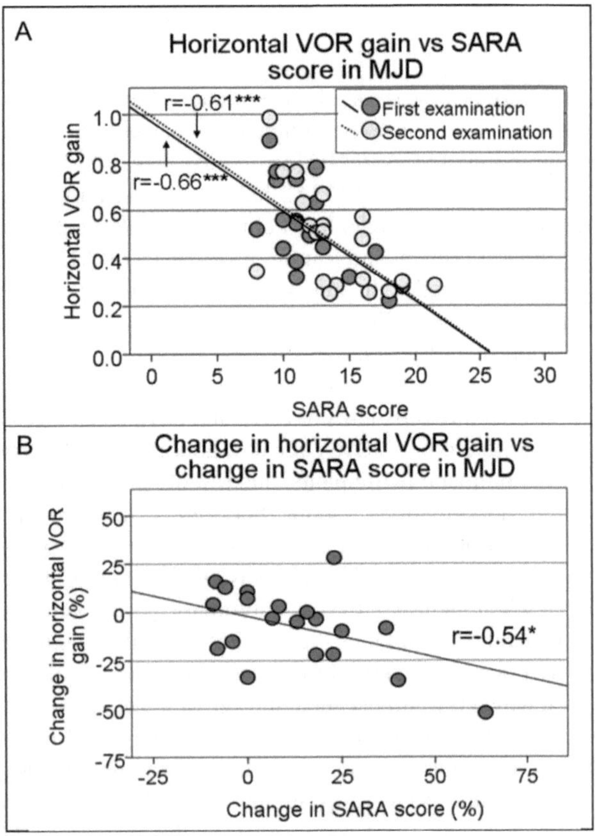

Fig. 13.5 Pearson correlations between horizontal VOR gain and SARA score in MJD patients ($N = 20$). (**A**) Correlations between horizontal VOR gain and SARA score from the first and second examinations conducted 9 to 24 months apart. (**B**) Correlation between the percentage change in horizontal VOR gain and the percentage change in SARA score over the two examinations carried out 9–24 months apart. $^{*}p < 0.05$, $^{***}p < 0.001$. (From Z. Elyoseph et al. [36])

percentage change in SARA score across the first and second examinations conducted 9–24 months apart ($r = -0.54$, $p < 0.05$) (Fig. 13.5b).

Interestingly, our SCA3/MJD population followed the well-known significant positive correlation between disease duration and disease severity assessed by the SARA score [37], but not between disease duration and the horizontal VOR gain deficit. Accordingly, we suggest that the contribution of horizontal VOR gain to the prediction of the SARA score is distinct from the contribution of disease duration. This assumption was confirmed by multiple regression analysis, which showed that both horizontal VOR gain and disease duration are significant independent predictors of SARA score. The combination of the horizontal VOR gain and disease duration is a better predictor of the SARA score than either one of them by itself. These findings are in line with those of another longitudinal study that measured different clinical ataxia scales, VOR, and eye movements in presymptomatic and symptomatic SCA3/MJD participants [38, 39]. VOR deficit progressed in the preataxic group when looking at natural history, along with the time to ataxia onset. Among different recorded eye movements, VOR was the best candidate to measure the natural history of the preataxic stage in SCA3/MJD [39].

All the above studies support the assumption that the horizontal VOR gain appears to be a reliable neurophysiological biomarker for the clinical onset, severity, and progression of SCA3/MJD and could be used in upcoming clinical studies. Most likely, a combination with other clinical, imaging, biochemical, and genomic profiling will be more beneficial in future therapeutic trials to identify early diagnosis and disease progression.

13.5 The Hidden Vestibular Symptomatology of SCA3/MJD

According to the Consensus document of the International Barany Society of Neuro-Otology [40], chronic Bilateral Vestibulopathy (BVP) is characterized by bilateral impairment of the VOR (such as horizontal canal gain less than 0.6 as measured by the vHIT) and symptoms of instability in walking, dizziness, and blurred vision. Given that the prominent symptoms of SCA3/MJD are signs of cerebellar ataxia, previous studies that examined the vestibular function in SCA3/MJD patients focused on the VOR measurement without any reference to the possibility that there may also be vestibular symptomatology, raising the possibility of incorporating the principles of vestibular rehabilitation in SCA3/MJD and classifying SCA3/MJD as a neurodegenerative/genetic disease with central BVP. Thus, we aimed to evaluate whether, beyond cerebellar ataxia, SCA3/MJD patients have typical vestibular symptomatology, which corresponds to the accepted diagnostic criteria of BVP according to the definition of the International Barany Society of Neuro-Otology [40]. Twenty-one SCA3/MJD, 12 clinically stable chronic Unilateral Vestibulopathy (UVP), 15 clinically stable chronic BVP, and 22 healthy controls underwent the vHIT evaluating VOR gain and filled out the following questionnaires related to

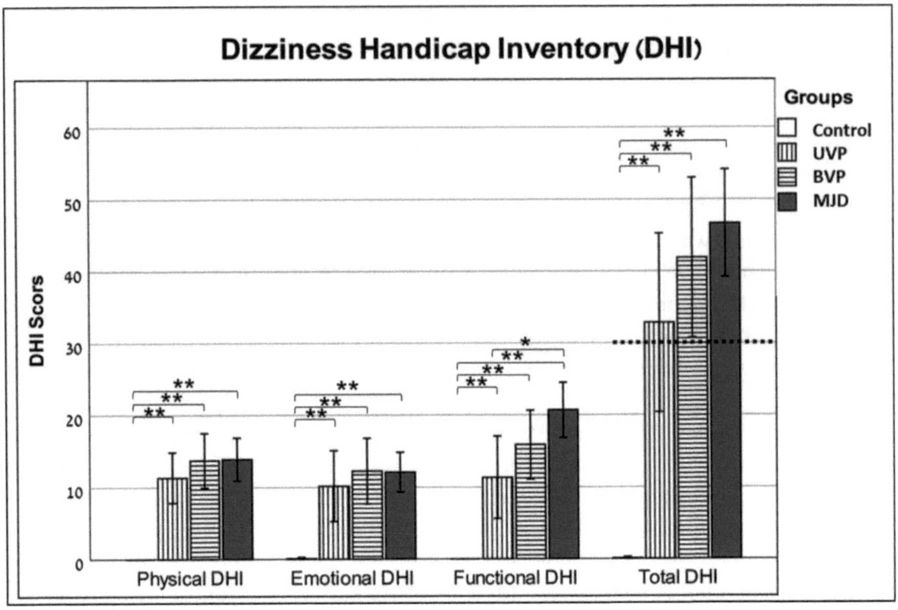

Fig. 13.6 The Dizziness Handicap Inventory (DHI). A score of the degree of disability caused by dizziness (mean ± SEM) in the MJD, BVP, UVP, and control groups. The dotted line indicates the clinical cut-off level. $p <$ **0.001, $p <$ *0.01 (From Z. Elyoseph et al. [41])

vestibular symptomatology: The Dizziness Handicap Inventory (DHI), the Activities-specific Balance Confidence Scale (ABC), the Vertigo Visual Scale (VVS), and the Beck Anxiety Inventory (BAI) [41]. The SCA3/MJD group demonstrated significant bilateral vestibular impairment with horizontal gain less than 0.6 in 71% of patients (0.54 ± 0.17). Similar to UVP and BVP, SCA3/MJD patients reported a significantly higher level of symptoms than Controls in the DHI that evaluates perceived handicapping effects imposed by dizziness and imbalance symptoms and is widely accepted in the assessment of vestibular symptoms (Fig. 13.6) as well in the, ABC, VVS, and BAI questionnaires that are also used in the evaluation of vestibular symptomatology.

We concluded that SCA3/MJD patients have, in addition to the cerebellar presentation, a significant impairment in VOR and typical symptoms that meet the diagnostic criteria of BVP of the International Barany Society of Neuro-Otology. We suggest that in a future version of the International Classification of Vestibular Disorders (ICVD), SCA3/MJD should be categorized under a separate section of central vestibulopathy with the heading of bilateral vestibulopathy. These findings are relevant to the clinical diagnosis process and possible treatment based on vestibular rehabilitation. As stated in the previous section, we and others [22, 36, 39, 42] suggested that the VOR impairment could precede the appearance of the accepted cerebellar clinical signs of SCA3/MJD and, therefore, can be used as a biomarker for SCA3/MJD disease onset. In addition, these findings support the idea that any dizziness or imbalance complaint in a patient with a family history of SCA3/

MJD but without cerebellar signs should be carefully examined for a vestibular deficit heralding the appearance of clear cerebellar signs. Since in SCA3/MJD, it is easier to detect cerebellar features in the clinical examination compared to vestibular deficits, clinicians tend not to check the possible vestibular signs (specifically not the bedside HIT), and in fact, signs of cerebellar ataxia may well mask the vestibular signs and symptoms. This may explain the lack of reports of vestibular symptoms in SCA3/MJD patients. It is well established that vestibular rehabilitation therapy, through exercises that improve VOR and balance with the help of visual fixation during various head and body movements, is effective in the syndrome of BVP [43]. Given that SCA3/MJD patients present with both vestibular impairment and vestibular symptoms, future studies applying vestibular rehabilitation techniques that help balance, improve VOR, and reduce dizziness and anxiety are needed to test the effectiveness of this treatment in SCA3/MJD.

13.6 Mentalizing Function Impairment in SCA3/MJD

Many studies have been conducted to characterize the role of the cerebellum in cognitive and affective functions. Schmahmann and Sherman demonstrated that cerebellar patients in general showed a Cerebellar Cognitive Affective Syndrome (CCAS) manifested in deficits in linguistic abilities, executive functions, and visuospatial skills, and includes affective symptoms, such as anxiety and depression [44, 45]. Several studies in SCA3/MJD patients have shown impairments in visual memory, verbal memory, visuospatial construction, verbal flow, executive functions, processing speed, procedural memory, and signs of anxiety and depression [46–48]. Despite the growing interest in the cognitive and affective aspects of SCA3/MJD, very little is known about mentalizing function.

During our clinical management of SCA3/MJD patients over the years, I had the impression that, in addition to cognitive and affective dysfunction, some of our patients exhibit a unique and unusual interpersonal behaviour that differs significantly from that seen in patients with other chronic neurodegenerative or cerebellar disorders. For example, some patients did not come to the scheduled clinic appointments without requesting to cancel or reschedule them, and only later demanded an urgent consultation that was not justified. Additionally, some patients have sent messages unrelated to their disease to the clinician's personal cell phone, assuming that their contact information is known to the physician. These behaviours align with characteristics described in the Theory of Mind [49]. Consequently, we decided to assess mentalizing functions in our SCA3/MJD patients. Mentalizing refers to the ability to understand and interpret the mental states of oneself and others in terms of thoughts, feelings, and intentions. It encompasses related concepts such as Theory of Mind, Social Cognition, Perspective Taking, and Empathy [50–52].

A total of 18 Jews of Yemenite origin with clinically and genetically confirmed SCA3/MJD (aged 54 ± 13 years), five presymptomatic SCA3/MJD with a positive genetic test (aged 35 ± 4 years), and 17 Jews of Yemenite origin healthy controls (aged 47 ± 17 years), completed the following battery of tests: (1) Reading the Mind in the Eyes (RME) which is a performance-based measure designed to assess the ability to accurately identify others' mental states, using 36 photos of the eye region of a human face. Participants are asked to choose the emotion or thought that corresponds to each picture out of four options. Impaired RME performance has been correlated with autism spectrum behaviours and deficient social skills [53]. (2) The Toronto Alexithymia Scale (TAS-20) is a self-report inventory designed to assess alexithymia, that is, one's difficulty in identifying and expressing feelings, alongside one's tendency to focus attention on concrete, external events (as opposed to inner experiences, such as one's emotions). Alexithymia is highly prevalent in autism spectrum disorder [54]. (3) False Belief test (FBt), which is a classic test to evaluate the ability to think about a situation from another person's point of view [55].

SCA3/MJD patients demonstrated significantly impaired performance in all mentalizing tests compared to healthy controls, while the presymptomatic SCA3/MJD carriers presented normal performance on all outcome variables. The observed impairment is reflected in the executive visual task (RME), the executive verbal story-based task (FBt), and in self-report alexithymia (TAS-20), and it reflects impairments in both identifying emotions of the other and in understanding the thought/perspective of the other (Figs. 13.7, 13.8, and 13.9). One-half of our SCA3/MJD patients showed abnormal clinical scores like those reported in the autism spectrum [56].

This study expands the neuropsychological presentation of SCA3/MJD patients, helps to understand their intrapersonal and interpersonal mental experience, and points to a significant widespread clinical impairment in mentalizing ability. This can help the patients themselves, family members, and medical staff to attribute

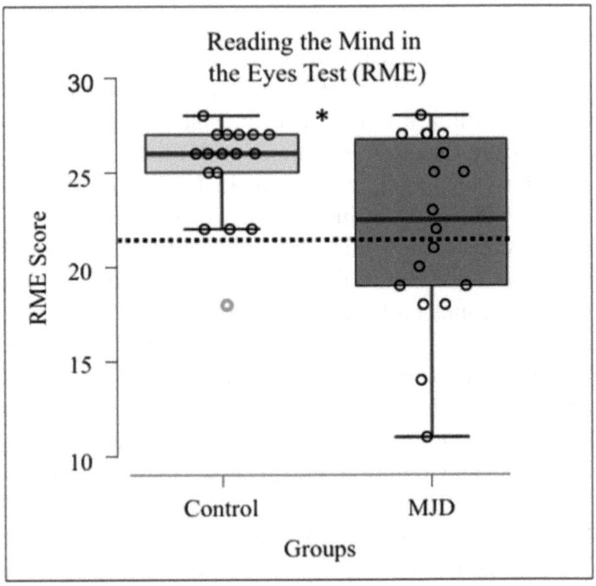

Fig. 13.7 Reading the Mind in the Eyes (RME) test scores for the MJD and control groups ($n = 18$ and 17, respectively). The horizontal line in the middle of the box represents the median value of the scores, while the lower and upper boundaries indicate the 25th and 75th percentiles, respectively. Maximum and minimum values are plotted, with outliers excluded and shown in grey. The dashed line indicates the clinical cut-off for an abnormally low RME score. *$p < 0.05$ (From Z. Elyoseph et al. [56])

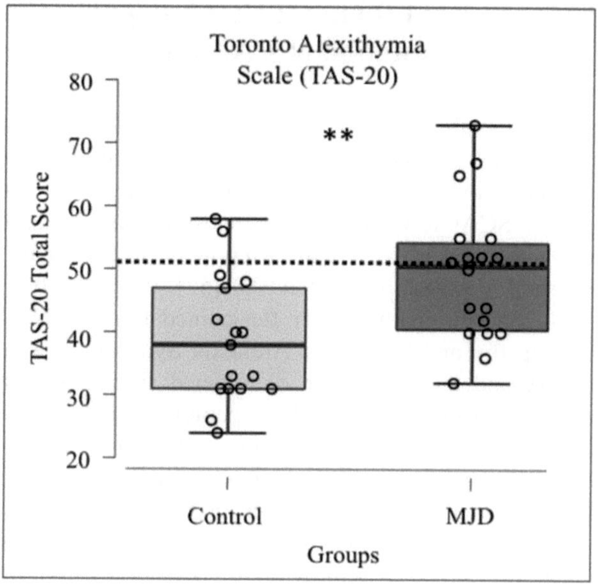

Fig. 13.8 The Toronto Alexithymia Scale (TAS-20) scores in the MJD and control groups ($n = 18$ and 17, respectively). The horizontal line in the middle of the box represents the median value of the scores, while the lower and upper boundaries indicate the 25th and 75th percentiles, respectively. Maximum and minimum values are plotted, and the dashed line marks the clinical cut-off for an abnormally high TAS-20 score. $^{**}p < 0.01$ (From Z. Elyoseph et al. [56])

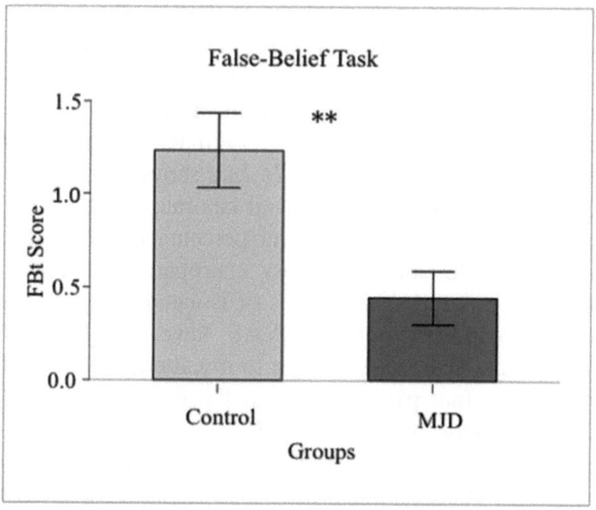

Fig. 13.9 The False Belief task (FBt) score (Mean ± SEM) in MJD and control groups ($n = 18$, 17, respectively. $^{**}p < 0.001$. (From Z. Elyoseph et al. [56])

interpersonal difficulties as an aspect of the disease and accordingly show more understanding and less judgment. In addition, one should consider using psychotherapeutic techniques that encourage the use of mentalization processes and the allocation of attentional resources to the intentions and feelings of the self and others, such as Mentalization-Based Therapy (MBT).

Although we have not clinically noted mentalizing dysfunction in our non-SCA3/MJD cerebellar patients, it is not clear whether the above findings could be part of the CCAS or are due to a combination of cerebellar and vestibular deficits. Further studies are needed to clarify this issue.

13.7 Isolated Bilateral Vestibulopathy Could Be the First Manifestation of CANVAS

Several months after our first publication on vestibular areflexia in SCA3/MJD [11], G. Michael Halmagyi emailed me to say that he had examined some cerebellar patients with bilateral vestibulopathy (BVP) who were negative for SCA3 and other genetic SCAs tests. The comprehensive clinical characterisation of these patients was published in 2004 with a report on four patients with Cerebellar Ataxia and Bilateral Vestibulopathy (CABV) [57], while the full description of the disorder, including neuropathy under the coined term CANVAS (Cerebellar Ataxia with Neuropathy and Vestibular Areflexia Syndrome), was published in 2011. Family history and the occurrence of affected sibling pairs indicated a likely late-onset recessive disorder [58, 59]. Histopathological studies demonstrated severe degeneration of Scarpa's ganglion and the vestibular nerve, which explains the impairment of vestibular function [60]. Autonomic manifestations and chronic cough have also been reported as major features of the disorder [61, 62]. Onset of neurological manifestations usually occurs between the ages of 50 and 60 years; however, chronic cough may precede them by up to 30 years [61–64]. In 2019, the genetic basis was identified as an abnormal biallelic intronic AAGGG repeat expansion in the Replication Factor Complex subunit 1 (RFC1). The normal alleles in the RFC1 locus usually have an average of about 11 repeats, while pathological expansions between 400 to over 2000 repeats are presented in CANVAS [63]. The onset and progression of the different clinical components of the CANVAS triad, along with the additional autonomic deficits and cough, vary, making diagnosis challenging when only one of the triad's signs is present [62]. In a series of 100 genetically confirmed CANVAS cases with available clinical and laboratory data, the complete triad was present in only 69% of cases, with lower percentages showing two components of the syndrome. While isolated sensory neuropathy was reported in 15% of cases, it was noted in this report that cases of isolated vestibular areflexia later developing the other components of CANVAS have not been documented [64]. However, G. Michael Halmagyi brought to my attention that in the original report of CABV, one of the four patients presented with BVP and only developed cerebellar signs approximately four years later [57].

Over the years, we examined four patients with chronic imbalance and BVP of unknown etiology who, after several years, developed cerebellar and neuropathic deficits. In all cases, the clinical diagnosis of isolated BVP was initially done by a detailed neuro-otological examination when noticing a clear impaired bilateral bedside HIT. The late appearance of mild cerebellar signs concomitant with sensory complaints raised the suspicion of CANVAS, and subsequent genetic tests revealed an abnormal biallelic expansion in RFC1, which confirmed the diagnosis [65]. These cases support the concept that CANVAS should be considered in every patient with BVP of unknown etiology, even without additional clinical cerebellar or neuropathic findings. This is of clinical importance given that about 50% of the cases in many BVP series are diagnosed as idiopathic cases, some of which may be undiagnosed

RFC1-related diseases that later develop the other components of CANVAS. Therefore, testing for the RFC1 expansion should be considered in cases of BVP of unknown origin.

13.8 Concluding Remarks

In this chapter, I review and discuss data from a series of studies that began when an impaired bedside HIT—a test described for peripheral vestibular loss—was unexpectedly detected in SCA3/MJD patients but not in other cerebellar diseases. Over the past 20 years, various studies have demonstrated that the horizontal VOR deficit, easily and accurately detected by the bedside HIT, is a characteristic feature of the SCA3/MJD phenotype. Subsequent laboratory studies using the magnetic search coil technique and vHIT have shown severely impaired VOR gain in both the horizontal and vertical planes, with preserved saccular function. Based on these and different histopathological studies, it is suggested that the VOR deficit in SCA3/MJD could be explained by selective degeneration, mainly of the medial and superior vestibular nuclei. Observations of a negative correlation between VOR gain and disease severity, along with the presence of impaired VOR in presymptomatic SCA3/MJD subjects, prompted a study demonstrating that the horizontal VOR gain measured by the vHIT could serve as a potential neurophysiological biomarker for SCA3/MJD, with the advantage of being an easy, simple, and quick test that requires little cooperation from the patient. Furthermore, it has been shown that SCA3/MJD patients exhibit typical vestibular symptomatology masked by cerebellar signs, as well as impaired mentalizing functions affecting both intrapersonal and interpersonal mental experiences. It is concluded that testing for VOR deficits is highly relevant for the diagnosis, management, and symptomatic treatment of SCA3/MJD patients and could be useful in forthcoming clinical studies, including therapeutic trials. Lastly, it is noted that an isolated BVP could be the initial presentation of CANVAS.

I would like to emphasize once again that these studies began twenty years ago when, at that time, I executed a new clinical test for canal paresis in SCA3/MJD patients. Today, the bedside HIT is mandatory in the neurological examination of cranial nerve function, and in my experience, it accurately detects VOR gain less than 0.7 in peripheral and central vestibular disorders, making it extremely useful in all cases of dizziness and imbalance.

References

1. Nakano KK, Dawson DM, Spence A. Machado disease: a hereditary ataxia in Portuguese emigrants to Massachusetts. Neurology. 1972;22:49–55.
2. Rosenberg RN, Nyhan WL, Coutinho P, Bay C. Joseph's disease: an autosomal dominant neurological disease in the Portuguese of the United States and the Azores Islands. Adv Neurol. 1978;21:33–57.

3. Schöls L, Bauer P, Schmidt T, Schulte T, Riess O. Autosomal dominant cerebellar ataxias: clinical features, genetics, and pathogenesis. Lancet Neurol. 2004;3:291–304.
4. Bettencourt C, Lima M. Machado-Joseph disease: from first descriptions to new perspectives. Orphanet J Rare Dis. 2011;6(1):35.
5. do Carmo Costa M. Recent therapeutic prospects for Machado Joseph disease. Curr Opin Neurol. 2020;33(4):519–26. https://doi.org/10.1097/WCO.0000000000000832.
6. Oliveira JBL, Martinez ARM, França MC Jr. Pharmacotherapy for the management of the symptoms of Machado-Joseph disease. Expert Opin Pharmacother. 2022;23(15):1687–94. https://doi.org/10.1080/14656566.2022.2135432.
7. Ghanekara SD, Kuo SH, Staffettia JS, Zesiewicza TA. Current and emerging treatment modalities for spinocerebellar ataxias. Expert Rev Neurother. 2022;22(2):101–14. https://doi.org/10.1080/14737175.2022.2029703.
8. Goldberg-Stern H, D'jaldetti R, Melamed E, Gadoth N. Machado-Joseph (Azorean) disease in a Yemenite Jewish family in Israel. Neurology. 1994;44:1298–301.
9. Lerer I, Merims D, Abeliovich D, Zlotogora J, Gadoth N. Machado-Joseph disease: correlation between the clinical features, the CAG repeat length and homozygosity for the mutation. Eur J Hum Genet. 1996;4:3–7.
10. Halmagyi GM, Curthoys IS. A clinical sign of canal paresis. Arch Neurol. 1988;45:737–9.
11. Gordon CR, Joffe J, Vainstein G, Gadoth N. Vestibulo-ocular arreflexia in families with spinocerebellar ataxia type 3 (Machado-Joseph disease). J Neurol Neurosurg Psychiatry. 2003;74:1403–6.
12. Gordon CR, Caspi A, Levite R, et al. Mechanisms of vestibulo-ocular reflex (VOR) cancellation in spinocerebellar ataxia type 3 (SCA-3) and episodic ataxia type 2 (EA-2). Prog Brain Res. 2008;171:519–25.
13. Gordon CR, Zivotofsky AZ, Caspi A. Impaired Vestibulo-Ocular Reflex (VOR) in Spinocerebellar Ataxia type 3 (SCA3): bedside and search coil evaluation. J Vestib Res. 2014;24:351–5.
14. Luis L, Costa J, Muñoz E, et al. Vestibulo-ocular reflex dynamics with head-impulses discriminates spinocerebellar ataxias types 1,2 and 3 and Friedreich ataxia. J Vestib Res. 2016;26(3):327–34. https://doi.org/10.3233/VES-160579.
15. Zaltzman R, Sharony R, Klein C, Gordon CR. Spinocerebellar ataxia type 3 in Israel: phenotype and genotype of a Jew Yemenite subpopulation. J Neurol. 2016;263:2207–14.
16. Takegoshi H, Murofushi T. Vestibular evoked myogenic potentials in patients with spinocereblIar degeneration. Acta Otolaryngol. 2000;120:821–4.
17. Murofushi T, Kaga K. Vestibular evoked myogenic potential. Its basics and clinical applications. Tokyo: Springer; 2009. p. 94.
18. Pedroso JL, Bor-Seng-Shu E, Braga-Neto P, et al. Neurophysiological studies and non-motor symptoms prior to ataxia in a patient with Machado–Joseph disease: trying to understand the natural history of brain degeneration. Cerebellum. 2014;13:447–51.
19. Ribeiro RS, Pereira MM, Pedroso JL, et al. Cervical and ocular vestibular evoked potentials in Machado–Joseph disease: functional involvement of otolith pathways. J Neurol Sci. 2015;358(1-2):294–8.
20. MacDougall HG, Weber KP, McGarvie LA, et al. The video head impulse test: diagnostic accuracy in peripheral vestibulopathy. Neurology. 2009;73(14):1134–41.
21. MacDougall HG, McGarvie LA, Halmagyi GM, et al. A new saccadic indicator of peripheral vestibular function based on the video head impulse test. Neurology. 2016;87(4):410–8.
22. Geisinger D, Elyoseph Z, Zaltzman R, Mintz M, Gordon CR. Angular vestibulo ocular reflex loss with preserved saccular function in Machado-Joseph disease. J Neurol Sci. 2021;424 https://doi.org/10.1016/j.jns.2021.117393.
23. Yuasa T, Ohama E, Harayama H, et al. Joseph's disease: clinical and pathological studies in a Japanese family. Ann Neurol. 1986;19:152–7.
24. Murofushi T, Mizuno M, Hatashida T, et al. Neuro otological and neuropathological findings in two cases with Machado-Joseph disease: the vestibular presentation. J Otolaryngol (Stockh). 1995;520(suppl):136–9.

25. Sequeiros J, Countinho P. Epidemiology and clinical as pects of Machado-Joseph disease. Adv Neurol. 1993;61:139 153.
26. Rüb U, Brunt ER, de Vos RA, et al. Degeneration of the central vestibular system in spinocerebellar ataxia type 3 (SCA3) patients and its possible clinical significance. Neuropathol Appl Neurobiol. 2004;30:402–14.
27. Rüb U, Brunt ER, Deller T. New insights into the pathoanatomy of spinocerebellar ataxia type 3 (Machado-Joseph disease). Curr Opin Neurol. 2008;21(2):111–6.
28. Scherzed W, Brunt ER, Heinsen H, et al. Pathoanatomy of cerebellar degeneration in Spinocerebellar Ataxia type 2 (SCA2) and type 3 (SCA3). Cerebellum. 2012;11:749–60.
29. Rezende TJR, de Paiva JLR, Martinez ARM, et al. Structural signature of SCA3: from presymptomatic to late disease stages. Ann Neurol. 2018;84:40–408.
30. Meira AT, Arruda WO, Ono SE, et al. Neuroradiological findings in the spinocerebellar ataxias. Tremor Other Hyperkinet Mov (NY). 2019;9 https://doi.org/10.7916/tohm.v0.682.
31. Barmack NH. Vestibular nuclei and their cerebellar connections. In: Gruol D, Koibuchi N, Manto M, Molinari M, Schmahmann J, Shen Y, editors. Essentials of cerebellum and cerebellar disorders. Springer; 2016. p. 69–78.
32. Yap KH, Abdul Manan H, Yahya N, Azmin S, Mohamed Mukari SA, Mohamed Ibrahim N. Magnetic resonance imaging and its clinical correlation in spinocerebellar ataxia type 3: a systematic review. Front Neurosci. 2022;16:859651. https://doi.org/10.3389/fnins.2022.859651.
33. Qiu H, Wu C, Liang J, et al. Structural alterations of spinocerebellar ataxias type 3: from pre-symptomatic to symptomatic stage. Eur Radiol. 2023;33:2881–94. https://doi.org/10.1007/s00330-022-09214-3.
34. Sha R, Su S, Hu M, et al. Global and regional brain grey and white matter morphometry alterations in type 1, 2, and 3 Spinocerebellar Ataxias (SCAs) patients. Cerebellum. 2025;24:4. https://doi.org/10.1007/s12311-024-01760-4.
35. Furtado GV, Oliveira CM, Bolzan G, Saute JA, Saraiva-Pereira ML, Jardim LB. State biomarkers for Machado Joseph disease: validation, feasibility and responsiveness to change. Genet Mol Biol. 2019;42(1):238–51. https://doi.org/10.1590/1678-4685-gmb-2018-0103.
36. Elyoseph Z, Geisinger D, Zaltzman R, Mintz M, Gordon CR. Horizontal vestibulo-ocular reflex deficit as a biomarker for clinical disease onset, severity, and progression of Machado-Joseph disease. Cerebellum. 2023;23(4):1273–9. https://doi.org/10.1007/s12311-023-01552-2.
37. Schmitz-Hübsch T, Du Montcel ST, Baliko L, Berciano J, Boesch S, Depondt C, Giunti P, Globas C, Infante J, Kang JS, Kremer B. Scale for the assessment and rating of ataxia. Neurology. 2006;66(11):1717–20. https://doi.org/10.1212/01.WNL.0000219042.60538.92.
38. de Oliveira CM, Leotti VB, Bolzan G, et al. Pre-ataxic changes of clinical scales and eye movement in Machado-Joseph disease: BIGPRO study. Mov Disord. 2021;36(4):985–94. https://doi.org/10.1002/mds.28466.
39. de Oliveira CM, Leotti VB, Cappelli AH, et al. Progression of clinical and eye movement markers in pre-ataxic carriers of Machado Joseph disease. Mov Disord. 2023;38(1):26–34. https://doi.org/10.1002/mds.29226.
40. Strupp M, Kim JS, Murofushi T, Straumann D, Jen JC, Rosengren SM, Della Santina CC, Kingma H. Bilateral vestibulopathy: diagnostic criteria consensus document of the Classification Committee of the Barany Society. J Vestib Res. 2017;27:177–89.
41. Elyoseph Z, Geisinger D, Zaltzman R, Mintz M, Gordon CR. The vestibular symptomatology of Machado-Joseph Disease. J Vestib Res. 2024;34(2–3):159–67.
42. Zivotofsky A, Caspi A, Gordon C. Impaired Vestibular Ocular Reflex (VOR) in presymptomatic Spinocerebellar Ataxia Type 3 (SCA-3). Invest Ophthalmol Vis Sci. 2013;54:174.
43. Porciuncula F, Johnson CC, Glickman LB. The effect of vestibular rehabilitation on adults with bilateral vestibular hypofunction: a systematic review. J Vestib Res. 2012;22:283–98.
44. Schmahmann JD, Sherman JC. The cerebellar cognitive affective syndrome. Brain. 1998;121:561–79.

45. Ahmadian N, van Baarsen K, van Zandvoort M, Robe PA. The cerebellar cognitive affective syndrome – a meta-analysis. Cerebellum. 2019;18:941–50.
46. Braga-Neto P, Pedroso JL, Alessi H, Dutra LA, Felício AC, Minett T, et al. Cerebellar cognitive affective syndrome in Machado Joseph disease: core clinical features. Cerebellum. 2012;11: 549–56.
47. Elyoseph Z, Mintz M, Vakil E, Zaltzman R, Gordon CR. Selective procedural memory impairment but preserved declarative memory in Spinocerebellar Ataxia Type 3. Cerebellum. 2020;19:226–34.
48. Lopes TM, D'Abreu A, Franca MC Jr, Yasuda CL, Betting LE, Samara AB, et al. Widespread neuronal damage and cognitive dysfunction in spinocerebellar ataxia type 3. J Neurol. 2013;260:2370–9.
49. Wellman HM, Cross D, Watson J. Meta-analysis of theory-of mind development: the truth about false belief. Child Dev. 2001;72:655–84.
50. Luyten P, Campbell C, Allison E, Fonagy P. The mentalizing approach to psychopathology: state of the art and future directions. Annu Rev Clin Psychol. 2020;16:297–325.
51. Choi-Kain LW, Gunderson JG. Mentalization: ontogeny, assessment, and application in the treatment of borderline personality disorder. Am J Psychiatry. 2008;165:1127–35.
52. Sokolov AA. The cerebellum in social cognition. Front Cell Neurosci. 2018;12:145. https://doi.org/10.3389/fncel.2018.00145.
53. Baron-Cohen S, Wheelwright S, Hill J, Raste Y, Plumb I. The "reading the mind in the eyes" test revised version: a study with normal adults, and adults with Asperger syndrome or high functioning autism. J Child Psychol Psychiatry. 2001;42:241–51.
54. Kinnaird E, Stewart C, Tchanturia K. Investigating alexithymia in autism: a systematic review and meta-analysis. Eur Psychiatry. 2019;55:80–9.
55. Birch SAJ, Bloom P. The curse of knowledge in reasoning about false beliefs. Psychol Sci. 2007;18:382–6. https://doi.org/10.1111/j1467-9280200701909.x.
56. Elyoseph Z, Geisinger D, Nave-Aival E, Zaltzman R, Gordon CR. "I do not know how you feel and how i feel about that": mentalizing impairments in Machado-Joseph disease. Cerebellum. 2023; https://doi.org/10.1007/s12311-023-01536-2.
57. Migliaccio AA, Halmagyi GM, McGarvie LA, et al. Cerebellar ataxia with bilateral vestibulopathy: description of a syndrome and its characteristic clinical sign. Brain. 2004;127:280–93.
58. Szmulewicz DJ, Waterston JA, Halmagyi GM, Mossman S, Chancellor AM, McLean CA, Storey E. Sensory neuropathy as part of the cerebellar ataxia neuropathy vestibular areflexia syndrome. Neurology. 2011;76:1903–10.
59. Szmulewicz DJ, Waterston JA, MacDougall HG, et al. Cerebellar ataxia, neuropathy, vestibular areflexia syndrome (canvas): a review of the clinical features and videooculographic diagnosis. Ann N Y Acad Sci. 2011;1233:139–47.
60. Ishai R, Seyyedi M, Chancellor AM, McLean CA, et al. The Pathology of the vestibular system in CANVAS. Otol Neurotol. 2021;42(3):e332–40. https://doi.org/10.1097/MAO.0000000000002985.
61. Wu TY, Taylor JM, Kilfoyle DH, et al. Autonomic dysfunction is a major feature of cerebellar ataxia, neuropathy, vestibular areflexia "CANVAS" syndrome. Brain. 2014;137:2649–56.
62. Szmulewicz DJ, McLean CA, MacDougall HG, Roberts L, Storey E, Halmagyi GM. CANVAS an update: clinical presentation, investigation and management. J Vestib Res. 2014;24:465–74.
63. Cortese A, Simone R, Sullivan R, et al. Biallelic expansion of an intronic repeat in RFC1 is a common cause of late-onset ataxia. Nat Genet. 2019;51:649–58.
64. Cortese A, Tozza S, Yau WY, Rossi S, Beecroft SJ, Jaunmuktane Z, et al. Cerebellar ataxia, neuropathy, vestibular areflexia syndrome due to RFC1 repeat expansion. Brain. 2020;143: 480–90.
65. Gordon CR, Zaltzman R, Geisinger D, Elyoseph Z, Gimmon Y. Bilateral vestibulopathy as the initial presentation of CANVAS. J Neurol Sci. 2024;460:122990.

Chapter 14
Posterior Circulation Transient Ischemic Attacks Manifested by Transient Vertigo or Ataxia. A 2025 Overview

Jorge C. Kattah

Keywords Posterior circulation transient ischemia · Acute vestibular syndrome · Acute ataxia syndrome · RAPID · Perfusion/diffusion in posterior circulation · Thrombolysis in posterior circulation stroke

14.1 Introduction

This chapter is written in honor of Sergio Carmona, M.D., who dedicated his career to Neurotology and Vestibular Neurology and shares with our Community an interest in vascular vertigo and ataxia.

The prognosis of posterior circulation strokes is highly variable; however, the potential for severe disability and mortality is high in the short and long term [1]. The initial symptoms frequently go underrecognized by patients and physicians alike. The diagnosis of transient ischemic attacks (TIA) involving the posterior fossa remains particularly difficult for two reasons: the first relates to the nature of the symptoms associated with episodic cranial nerve dysfunction, such as dizziness, vertigo, and truncal ataxia [2], which are common in episodic peripheral vestibulopathy; diplopia points to peripheral ocular motor abnormalities. Similarly, sudden hearing loss points more frequently to nonvascular cochlear etiologies. Other symptoms suggest, for example, blood pressure fluctuations with lightheadedness and presyncopal feeling. Sensory symptoms may overlap with those related to cervical radiculopathy. Neck pain is ubiquitous and often due to cervical spine degenerative disorders. Finally, symptoms secondary to lower cranial nerve symptoms overlap with local causes of dysarthria and dysphagia, etc. The second factor, a direct consequence of the reasons outlined above, is a delay in patients seeking medical attention; in general, most people overlook these symptoms and fail to consider them clear indicators of imminent stroke risk, particularly in comparison to the alarming carotid TIAs with transient weakness, aphasia, and visual loss [3]. As a

J. C. Kattah (✉)
Peoria, IL, USA
e-mail: kattah@uicomp.uic.edu

P. Ravat et al. (eds.), *Advances in Translational Neuroscience of Vestibular and Eye Movement Disorders*, Contemporary Clinical Neuroscience,
https://doi.org/10.1007/978-3-032-07539-0_14

result, clinicians often face a significant challenge confronting vertebrobasilar TIAs. In contrast, posterior fossa TIAs that begin with transient mono, hemi, triparesis, or quadriparesis are more easily recognized stroke warning symptoms [3].

The purpose of this chapter is to review recent progress in the evaluation of posterior fossa TIAs presenting with dizziness, vertigo, and ataxia. Even though cerebrovascular disorders involving the posterior circulation cause only 20–25% of the total number of strokes [4], it is important to establish a precise diagnostic/ management algorithm in the near future to improve outcomes. The possibility of mortality or severe disability, including quadriplegia and a locked-in syndrome, in these patients demands urgent recognition. In the New England Posterior Circulation Stroke Registry, the most common initial stroke symptom in the posterior circulation was dizziness, vertigo, and ataxia [4]. Specific intervention parameters for the management of neurovascular syndromes involving the anterior circulation are not available at the present time for posterior circulation stroke.

The prognosis of posterior circulation stroke may be modified by the timing of intervention; the worst scenario in posterior fossa ischemic stroke involves the basilar artery. Rapid progression of symptoms related to basilar artery occlusion (BAO) is amenable to treatment, and even in late cases, a potential poor prognosis in general justifies thrombolysis, regardless of the timing after presentation. There are four clinical treatment trials for *BAO: BEST, BASICS, ATTENTION*, and Basilar Artery Occlusion Endovascular Trial (BAOCHE). The latter two studies took place in China. In the first two studies, there was no greater benefit between intravenous thrombolysis (IT) versus endovascular thrombectomy (ET). The latter studies found greater benefit from ET, with a larger percentage of patients with an mRS of 3. An encouraging outcome result with intravenous thrombolysis (ivT) among 247 BAO patients with an average NIH stroke scale (NIHSS) of 18 yielded recanalization in 67% of them, and their three-month modified Rankin scale (mRS) was 3. Patients who underwent endovascular recanalization (ET) in this trial had poorer mRS scores and greater mortality [5]. In contrast, to my knowledge, outside a few case reports, only three retrospective ivT series involving posterior circulation lacunar strokes are available [6–8]. In one retrospective cohort, the clinical outcome of the ivT versus the untreated patients was similar. In addition, the PRISMS study targeting patients with non-disabling symptoms with an average NIH stroke scale score (NIHSS) of ≤ than 5 did not benefit from thrombolysis, and had a larger risk of side effects; a conclusion was not possible because the study terminated early [9]. In the second retrospective trial among 29 patients with acute vestibular syndrome (AVS), 15 underwent IvT and 14 did not. Most patients had a stroke, but vestibular neuritis, migraine, and TIA were a minority; there was no difference in outcome between treated and untreated patients (the mean NIHSS was 1.9 in treated and 0.8 in untreated patients). There were no complications from IvT [6]. The most recent study included 89 consecutive patients (NIHSS <5) who presented within 4.5 hours from symptom onset; 47 were treated and 42 were not. Often, there was associated weakness or hemiplegia in those treated. No difference in outcome was observed upon discharge, and brain hemorrhage occurred in 4/47 IvT patients [8]. A recently published meta-analysis review of the literature summarized the existing information on posterior circulation thrombolysis among acute vestibular syndrome (AVS) and acute imbalance syndrome (AIS) [10].

Examination of available literature on posterior fossa TIAs is germane to designing a prospective study to approach these cases. For clarity, I will divide this chapter into several sections.

1. *Definition of Transient Vascular Vertigo*

The term TIA was coined by Miller Fisher [11] in 1958, and he defined it initially as a "neurological deficit that lasted less than 24 hours". It is now clear that TIAs last often less than one hour. Importantly, the current critical requirement for diagnosis of a TIAs is tissue-based; thus, there must be an absence of acute imaging abnormalities to suggest infarction [12]. In general, if the deficit lasts less than 1 hour, the diagnosis of TIA versus stroke is often accurate. The Bárany Society recently defined the diagnosis of a TIA of vascular origin, to arrive at the diagnosis of vascular-mediated transient dizziness, vertigo, ataxia, not only the temporal profile should be < than 24-hour duration, but it must fulfill specific criteria [13]. There are a few studies analyzing isolated vertigo or vertigo/ataxia as TIAs. In one epidemiological study, the overall incidence of strokes after isolated vertigo was low (there was no specific estimate of TIAs of vascular origin). It involved a large retrospective study from the province of Ontario, Canada; however, their data suggested that stroke was frequently missed [14]. Choi et al., in 2017, in a small prospective study, found that clinical and vascular imaging of the posterior fossa were insufficient in the evaluation of patients with transient vascular vertigo, and noted in his small cohort that CT perfusion (CTP) imaging provided significant value in selecting patients at risk of cerebellar stroke [15]. An important recommendation from the Bárany Society in the evaluation of transient vascular vertigo was the need for reexamination of the patient within 24 hours after the first evaluation, as other episodic peripheral vestibular disorders last a similar period of time (vestibular migraine and Meniere's) [13]. In some instances, vertigo and ataxia patients due to stroke in evolution, when examined early, show new neurological deficits, often emerging during this period of time, thus providing clear central localization. In fact, frequent neuro-monitoring for additional neurological signs within this period suggests the need for detailed vascular imaging and possibly perfusion studies, hoping to identify those patients at risk, and still within a thrombolysis window [13, 15]. Most TIAs occur frequently within 48 hours preceding a stroke [16].

2. *Diagnostic Criteria for Probable Vascular Vertigo (from the Barany Society Consensus)*

Acute spontaneous vertigo, dizziness, or unsteadiness lasting less than 24 hours

1. *Focal central neurological symptoms or severe postural instability during the attack.*
2. *New onset of moderate to severe cranio-cervical pain during the attack (suggestive of vertebral artery dissection).*

3. *Increased risk for vascular events,* e.g., *ABCD² score of 4 or more, or atrial fibrillation.*
4. *Significant (>50%) narrowing of an artery of the vertebrobasilar system.*
5. *Not better accounted for by another disease or disorder.*

3. *Introduction to RAPID:* ***Ra****pid* ***P****rocessing of* ***P****erfusion and* ***D****iffusion as a Preliminary Step for Intervention, and Introduction of Posterior Circulation (PC ASPECTS score)*

Following an injection of intravenous contrast, cerebral perfusion parameters are studied, There are three basic maps in CTP: *Cerebral blood flow* (CBF) normal: 60 ml/100 g of gray matter and 25 ml/100 g of white matter, *Cerebral blood volume* (CBV), normal 4 ml/100 g for gray matter and 2 ml/100 g for white matter and *Mean Transient Time* (MTT), normal 4 sec in gray matter and 4.5 sec in white matter.

The *core infarct* is defined as cerebral tissue with a CBV of < than 40% of normal, with a CBF < than 30% of normal, and prolonged MTT.

The *ischemic penumbra* is defined as tissue with normal CBV, decreased CBF, and markedly prolonged MTT > 6 s.

The value in identifying patients with large ischemic penumbra and initiating thrombectomy has been well demonstrated, and found to be independent to some extent from former time constraints. While experience with RAPID in the anterior circulation is extensive [17], the same level of information is not available for the posterior circulation.

There are only a few studies utilizing CTP in posterior fossa strokes. In 2010, a study of TIA and nonlacunar excluded posterior circulation strokes had a low detection rate of 32%; however, it demonstrated the possibility of producing axial perfusion scans that avoided the beam hardening artifact [18]. A subsequent prospective noncontrast CT (NCCT), CT angiography (CTA), and CTP study involving $n = 88$ patients with posterior fossa TIA/stroke was identified by abnormal NCCT in 13 patients, abnormal CTA in 14 patients, and abnormal CTP with impaired maps in 31 patients (74%).

To evaluate the potential successful intervention in BAO, a CT hypoattenuation score was recently proposed to use along with the clinical findings: posterior circulation (pc) ASPECTS. This scale identifies and scores low-density lesions due to ischemia affecting the cerebellum, thalamus, and posterior cerebral artery; each will have a score of 1 if unilateral or 2 if bilateral. The number of lesions is then subtracted from 10, which indicates no NCCT lesions; hypoattenuation of the midbrain and pons scores 2. A cumulative score of 10 indicates a good prognosis, and a score < 6 is associated with a bad prognosis even if BA recanalization is achieved [19, 20]. Even though NCCT has low sensitivity to diagnose TIA or stroke in the posterior circulation, it may provide substantial information in the selection of IvT versus ET in patients with stroke in evolution.

4. *Potential Posterior Fossa Stroke Scale, and Posterior Fossa Strokes Outside BAO That Are Often Associated with a High Risk of Morbidity and Mortality*

With a few exceptions, the NIHSS is not particularly suited to identify posterior fossa stroke. A high NIHSS score, however, is associated with poorer outcomes in patients with basilar artery occlusion [20]. A recent posterior fossa stroke scale was proposed [21]. This scale did not include any of the Head Impulse N: Nystagmus and Test of skew (HINTS) plus findings [22–24], nor did it include sensitive ocular motor abnormalities that can be easily added (for example, ocular laterodeviation (OLD) with brief eyelid closure) [25]. HINTS and OLD should be scored with a high score in the scale, because of high specificity. In addition, it did not consider incubitus ataxia (grade 3) [26].

Implementation of a modified Adam's scale and National Institite of Health Stroke Scale (NIHSS), along with pc Alberta Stroke Progression Early CT scan Score (ASPECTS) from NCCT, CTA, and CT perfusion, could represent sensitive elements to stratify patients for proper intervention.

What posterior circulation strokes are particularly associated with poor prognosis besides BAO? Large cerebellar strokes in the acute phase frequently develop severe ischemic edema and cause mass effect on posterior fossa structures. Intensive care monitoring is necessary to monitor signs of deterioration. Currently, it is unknown if timely thrombolysis prevents this complication. Cerebellar stroke represents only 2% of all strokes in the USA [27], and yet they are associated with a disproportionately high risk of high stroke morbidity or mortality [28]. In a four-year prospective study involving the diagnosis of cerebellar infarct, published in 2016, 39 patients were misdiagnosed, and five of whom were discharged home without imaging. Only 12 of these patients (~ 10%) underwent IvT or ET. Fourteen patients developed ischemic edema and lethargy, five had hydrocephalus, seven hemorrhagic transformations, and four recurrent strokes. The initial presentation with nausea, vomiting, and posture abnormalities may be the main cause of the diagnostic difficulty. Dysarthria, curiously, when present, led to neuroimaging and correct diagnosis. The CT scan was abnormal in only 8 out of 27 when obtained within 24 hours of symptom onset. In contrast, MRI was performed at different stages within the first 24, 48 hours, and beyond, and it was abnormal in all patients. Of note, there were no MRIs obtained in less than 24 hours [29].

Penetrator small midline lacunar stroke or TIA generally have a good prognosis; however, it may precede significant atherosclerotic basilar occlusive artery disease, and early recognition may halt further progression. In a study, among 24 patients, TIAs preceded BAO, and vertigo was a common complaint in the preceding hours, and surprisingly, even in some cases for a few weeks before overt BAO [30]. Finally, large pontine infarcts are associated with a risk of locked-in syndrome, and early IvT and ET may prevent it [31].

A point to consider in the posterior fossa circulation is the high degree of vertebral artery asymmetry (about 30% of individuals on neuropathological studies) [32], and the high frequency of arterial dissection as a cause of stroke in individuals between the ages 18–45 [33]. Secondary stroke prevention is of significant utility, and a recent paper addressed the importance of proper management once a posterior circulation TIA diagnosis was established [34].

14.2 Conclusion

This chapter elucidates the development to date of clinical diagnostic tools and imaging protocols to precisely guide diagnosis and management of TIA and stroke in evolution involving the posterior circulation. Future work includes the design and validation of a supplemental posterior fossa stroke scale, increased availability to perform CTP in posterior fossa stroke, and detailed vascular studies. This information will then be used to manage patients with the best medical management, IvT, and ET depending on the findings present. ET will possibly be superior as a function of improved technology. A combined team of Vestibular Medicine and Neurovascular specialists ideally should join to design a common protocol.

References

1. Karvelas N, Palaiodimos L, Karamanis D, et al. Long-term outcomes after first-ever posterior circulation stroke and the prognostic significance of the new England Medical Center posterior circulation registry stroke classification: a prospective study from the Athens stroke registry. Eur Stroke J. 2024:23969873241302657.
2. Karatas M. Vascular vertigo: epidemiology and clinical syndromes. Neurologist. 2011;17:1–10.
3. Tao WD, Liu M, Fisher M, et al. Posterior versus anterior circulation infarction: how different are the neurological deficits? Stroke. 2012;43:2060–5.
4. Caplan LR. Verterbobasilar ischemia and Hemorrhage. United Kingdom by Bell and Bain Ltd: Cambridge University Press; 2015.
5. Raty S, Virtanen P, Ritvonen J, et al. IV thrombolysis in basilar artery occlusion: outcomes and comparison with endovascular Thrombectomy. Neurology. 2024;102:e209249.
6. Machner B, Choi JH, Neumann A, Trillenberg P, Helmchen C. What guides decision-making on intravenous thrombolysis in acute vestibular syndrome and suspected ischemic stroke in the posterior circulation? J Neurol. 2021;268:249–64.
7. Choi JH, Oh EH, Park MG, et al. Early MRI-negative posterior circulation stroke presenting as acute dizziness. J Neurol. 2018;265:2993–3000.
8. Shi T, Zhang Z, Jin B, et al. Choice of intravenous thrombolysis therapy in patients with mild stroke complaining of acute dizziness. Am J Emerg Med. 2022;52:20–4.
9. Khatri P, Kleindorfer DO, Devlin T, et al. Effect of Alteplase vs aspirin on functional outcome for patients with acute ischemic stroke and minor nondisabling neurologic deficits: the PRISMS randomized clinical trial. JAMA. 2018;320:156–66.
10. Edlow JA, Tarnutzer AA. Intravenous thrombolysis in patients with acute dizziness or imbalance and suspected ischemic stroke-systematic review. J Neurol. 2025;272:91.
11. Fisher C. Intermittent cerebral ischemia. In: Cerebral vascular disease. New York: Grune & Stratton; 1958. p. 81–97.
12. Al-Khaled M, Matthis C, Munte TF, Eggers J, Qug SSS. The incidence and clinical predictors of acute infarction in patients with transient ischemic attack using MRI including DWI. Neuroradiology. 2013;55:157–63.

13. Kim JS, Newman-Toker DE, Kerber KA, et al. Vascular vertigo and dizziness: diagnostic criteria. J Vestib Res. 2022;32:205–22.
14. Atzema CL, Grewal K, Lu H, Kapral MK, Kulkarni G, Austin PC. Outcomes among patients discharged from the emergency department with a diagnosis of peripheral vertigo. Ann Neurol. 2016;79:32–41.
15. Choi JH, Park MG, Choi SY, et al. Acute transient vestibular syndrome: prevalence of stroke and efficacy of bedside evaluation. Stroke. 2017;48:556–62.
16. Edlow JA, Kim S, Pelletier AJ, Camargo CA Jr. National study on emergency department visits for transient ischemic attack, 1992-2001. Acad Emerg Med. 2006;13:666–72.
17. Albers GW, Marks MP, Kemp S, et al. Thrombectomy for stroke at 6 to 16 hours with selection by perfusion imaging. N Engl J Med. 2018;378:708–18.
18. Lee IH, You JH, Lee JY, et al. Accuracy of the detection of infratentorial stroke lesions using perfusion CT: an experimenter-blinded study. Neuroradiology. 2010;52:1095–100.
19. Ouyang K, Kang Z, Liu Z, et al. Posterior circulation ASPECTS on CT angiography predicts futile recanalization of endovascular Thrombectomy for acute basilar artery occlusion. Front Neurol. 2022;13:831386.
20. Lin SF, Chen CI, Hu HH, Bai CH. Predicting functional outcomes of posterior circulation acute ischemic stroke in first 36 h of stroke onset. J Neurol. 2018;265:926–32.
21. Wisniewski A, Filipska K, Piec K, Jaskolski F, Slusarz R. Introducing Adam's scale of posterior stroke (ASPOS): a novel validated tool to assess and predict posterior circulation strokes. Brain Sci. 2021:11.
22. Kattah JC, Talkad AV, Wang DZ, Hsieh YH, Newman-Toker DE. HINTS to diagnose stroke in the acute vestibular syndrome: three-step bedside oculomotor examination more sensitive than early MRI diffusion-weighted imaging. Stroke. 2009;40:3504–10.
23. Newman-Toker DE, Kerber KA, Hsieh YH, et al. HINTS outperforms ABCD2 to screen for stroke in acute continuous vertigo and dizziness. Acad Emerg Med. 2013;20:986–96.
24. Tarnutzer AA, Gold D, Wang Z, et al. Impact of clinician training background and stroke location on bedside diagnostic accuracy in the acute vestibular syndrome – a meta-analysis. Ann Neurol. 2023;
25. Kattah JC. Concordant GRADE-3 truncal ataxia and ocular Laterodeviation in acute medullary stroke. Audiol Res. 2023;13:767–78.
26. Kattah JC, Martinez C, Zalazar G, Batuecas A, Lemos J, Carmona S. Role of incubitus truncal ataxia, and equivalent standing grade 3 ataxia in the diagnosis of central acute vestibular syndrome. J Neurol Sci. 2022;441:120374.
27. Datar S, Rabinstein AA. Cerebellar infarction. Neurol Clin. 2014;32:979–91.
28. Edlow JA, Newman-Toker DE, Savitz SI. Diagnosis and initial management of cerebellar infarction. Lancet Neurol. 2008;7:951–64.
29. Calic Z, Cappelen-Smith C, Anderson CS, Xuan W, Cordato DJ. Cerebellar infarction and factors associated with delayed presentation and misdiagnosis. Cerebrovasc Dis (Basel, Switzerland). 2016;42:476–84.
30. von Campe G, Regli F, Bogousslavsky J. Heralding manifestations of basilar artery occlusion with lethal or severe stroke. J Neurol Neurosurg Psychiatry. 2003;74:1621–6.
31. Janjua N, Wartenberg KE, Meyers PM, Mayer SA. Reversal of locked-in syndrome with anticoagulation, induced hypertension, and intravenous t-PA. Neurocrit Care. 2005;2:296–9.
32. Peeters JB, Idriceanu T, El Hage G, et al. A comprehensive review of the vertebral artery anatomy. Neurochirurgie. 2024;70:101518.
33. Gottesman RF, Sharma P, Robinson KA, et al. Clinical characteristics of symptomatic vertebral artery dissection: a systematic review. Neurologist. 2012;18:245–54.
34. Neto ACL, Kim JS, Bernardo WM, Bittar RSM. Vertigo and dizziness due to vertebrobasilar TIA: a prospective study. Front Stroke. 2024;3:1429068.

Chapter 15
Neuro-Otologic Manifestations of Multiple Sclerosis

Alfredo Laffue

Keywords Multiple sclerosis · Vertigo · Vestibular testing · Nystagmus · Oculomotor abnormalities

15.1 Introduction

Multiple sclerosis (MS) is one of the most prevalent demyelinating disorders of the central nervous system (CNS), affecting approximately 40 per 100,000 individuals in Latin America and 100 per 100,000 in North America and Northern Europe [1]. It is a complex autoimmune condition characterized by inflammation, demyelination, axonal degeneration, and progressive neuronal and structural atrophy. These pathological processes occur at all stages of the disease, affecting both white and gray matter, including the cerebral cortex and deep gray matter nuclei. The primary pathological targets in MS are myelin and oligodendrocytes. The pathogenesis of MS is thought to result from a combination of genetic susceptibility and environmental triggers, potentially of microbial origin [2].

MS can initially present as a clinically isolated syndrome (CIS). In most cases, the disease follows a relapsing-remitting course (RRMS). Over time, a subset of patients transitions to secondary progressive MS (SPMS), characterized by gradual neurological decline following an initial relapsing course, with or without superimposed relapses. A distinct form, primary progressive MS (PPMS), is marked by continuous disease progression from onset without preceding relapses [3].

Vestibular symptoms are a common manifestation of MS and may present early in the disease course. These symptoms are reported in 4–20% of patients during CIS and in 28–67% of those with RRMS. Involvement of vestibular and cerebellar pathways can lead to abnormal eye movements, most frequently nystagmus, and oculomotor abnormalities, such as saccadic oscillations or conjugate gaze disturbances [4].

A. Laffue (✉)
Fleni, Buenos Aires, Argentina

P. Ravat et al. (eds.), *Advances in Translational Neuroscience of Vestibular and Eye Movement Disorders*, Contemporary Clinical Neuroscience,
https://doi.org/10.1007/978-3-032-07539-0_15

Demyelination affecting the vestibulocochlear nerve and its connections can produce vertigo, which is often misdiagnosed as a peripheral otological disorder. Understanding these manifestations is crucial for differentiating between central and peripheral vestibular disorders. It is important to recognize that brainstem involvement serves as a strong prognostic factor for long-term disability in MS and, during the CIS phase, is associated with an increased risk of conversion to clinically definite MS [4–7].

This chapter explores the neurotological manifestations of MS, emphasizing the role of vestibular testing in diagnostic accuracy.

15.2 Vestibular Abnormalities in Multiple Sclerosis

15.2.1 Vertigo

Vertigo is a common symptom in MS, typically reflecting dysfunction in central vestibular pathways, including cerebellar involvement. The evaluation of vestibular symptoms in MS requires a careful approach due to the overlapping etiologies of central and peripheral vestibular disorders.

Three main clinical presentations of vertigo have been described in MS patients [8]:

- Acute Vestibular Syndrome (AVS): Characterized by the sudden onset of vertigo lasting days to weeks, often accompanied by nausea, vomiting, nystagmus, and gait instability.
- Positional Vertigo: Triggered by changes in head position, sometimes resembling benign paroxysmal positional vertigo (BPPV) but with central characteristics.
- Chronic or Subclinical Vestibular Dysfunction: Persistent vestibular abnormalities that may only be detected through detailed neurological and vestibular examination or objectively identified using videonystagmography.

When vertigo in MS arises from inflammatory demyelination, the medial vestibular nucleus and the root entry zone of the vestibulocochlear nerve are the most frequently affected neuroanatomical structures. However, MS patients may also develop vertigo due to other conditions, including peripheral vestibular disorders, vestibular migraine, or cerebrovascular events. Given the distinct therapeutic implications of these etiologies, careful differential diagnosis is essential to prevent misdiagnosis and ensure appropriate management [9].

15.2.2 Acute Vestibular Syndrome in MS

MS is an uncommon cause of acute vertigo, which is more often attributed to peripheral vestibular disorders or stroke. However, acute vestibular syndrome

(AVS) in MS can present as an initial symptom of the disease or, more commonly, occur during its course, particularly in RRMS. In MS, AVS may manifest as isolated vertigo but is frequently associated with internuclear ophthalmoplegia (INO) and, in some cases, spontaneous upbeat nystagmus, gait ataxia, or other brainstem signs, reinforcing its central origin [10]. It is critical to recognize that demyelinating lesions affecting the vestibular nuclei or their brainstem connections can mimic peripheral vestibular disorders, potentially leading to misdiagnosis. The recommended diagnostic approach involves the HINTS (Head Impulse, Nystagmus, Test of Skew), supplemented by auditory evaluation and gait assessment (HINTS Plus Ataxia) [8, 11]. A central etiology is suggested by the presence of at least one of the following findings:

- Normal head impulse test.
- Vertical or direction-changing nystagmus.
- Ocular tilt reaction (OTR) or skew deviation.
- Severe gait ataxia.
- Oculomotor abnormalities, such as INO.

The HINTS protocol appears to be a valid tool in MS for differentiating central from peripheral lesions, with the recognition that oculomotor disturbances are particularly prominent in this patient population.

15.2.3 Positional Vertigo

The most common cause of vertigo without other associated neurological signs is benign paroxysmal positional vertigo (BPPV), as diagnosed according to the Bárány Society criteria. However, demyelination in MS can also produce similar positional symptoms, mimicking BPPV [12].

In cases of MS, active lesions have been found near the fourth ventricle. The interruption of otolithic projections in the superior cerebellar peduncle, which connects the fastigial nucleus to the vestibular nuclei, has been postulated as the underlying mechanism. In MS, vertigo can be triggered by changes in head position, sometimes resembling BPPV but demonstrating central vestibular characteristics, such as [13, 14]:

- Positional downbeat nystagmus.
- Sustained positional upbeat nystagmus lasting more than 1 min.
- Apogeotropic nystagmus during the supine roll test.
- Association with other central ocular motor abnormalities, truncal imbalance, or both.

MS-related positional vertigo often does not respond to canalith repositioning maneuvers, further distinguishing it from BPPV and indicating central vestibular involvement [13–15].

15.2.4 Chronic or Subclinical Vestibular Dysfunction

Balance control is a complex process requiring the integration of visual, somatosensory, and vestibular inputs, along with the cerebellum and motor system. The vestibular system is one of the key components of postural stability. In patients with MS, balance disorders are often multifactorial, involving a combination of accumulated disability, muscle weakness, spasticity, gait impairment, cognitive dysfunction, and chronic or unrecognized vestibular pathology. Given these contributing factors, assessing vestibular dysfunction in MS patients is important, as it has a specific treatment with proven benefits in functional recovery, balance improvement, and fall prevention [4, 16].

Vestibular testing can be evaluated using:

- Videonystagmography (VNG).
- Video head impulse test (vHIT).
- Vestibular evoked myogenic potentials (VEMPs).
- Posturography.

Identifying and addressing vestibular impairment in MS patients are crucial for optimizing rehabilitation strategies and improving the overall quality of life [17].

15.2.5 Vestibular Testing in Multiple Sclerosis

The evaluation of vestibular symptoms in MS requires a differential approach due to the overlapping etiologies of central and peripheral vestibular disorders [4]. A review of the literature highlights frequent vestibular system abnormalities in MS; however, interpretation of these findings is often constrained by methodological limitations. Techniques such as vestibular evoked myogenic potentials (VEMPs), electronystagmography (ENG), and posturography often yield variable or nonreproducible results due to inherent technical and procedural challenges. Although historically used, these methods lack the precision and reliability necessary for accurate vestibular assessment [5, 18].

The use of VNG, particularly with a comprehensive protocol (see Appendix for our VNG and vHIT protocol), provides critical insights into both central and peripheral vestibular dysfunction, enabling more precise differentiation between MS-related manifestations and other vestibular disorders.

VNG allows for a detailed assessment of central vestibular dysfunction, which is prevalent in MS, revealing abnormalities such as vertical nystagmus, impaired vestibulo-ocular reflex (VOR) suppression, saccadic abnormalities, and INO. These findings are particularly valuable for detecting subtle central lesions that may have been overlooked in routine imaging, providing pivotal information for the diagnosis and appropriate management of neurotological symptoms in MS patients.

The vHIT, on the other hand, assesses high-frequency semicircular canal activity and is particularly useful for determining peripheral vestibular involvement (refer to Appendix for our VNG and vHIT techniques). Recent studies have demonstrated that vHIT can also identify certain central causes, distinguishing them from peripheral counterparts. The normal vHIT results observed in many MS cases with abnormal VNG findings underscore the central origin of symptoms. While peripheral vestibular hypofunction is characterized by clearly positive results in head impulse test results with marked asymmetry and large-amplitude unidirectional corrective saccades, central vestibulopathies exhibit a variety of response patterns [19, 20].

Although a normal clinical head impulse test in acute vestibular syndrome typically suggests a central lesion, vHIT alterations have been observed in patients with central nervous system disorders. Lesions affecting the vestibular nucleus, the nucleus prepositus hypoglossi, or the flocculus can result in a significant reduction in horizontal head impulse VOR gains, either unilaterally or bilaterally. VOR gains during vertical impulses on vHIT may be reduced or enhanced in lesions affecting the vestibular nucleus, medial longitudinal fasciculus, and cerebellum [20].

Despite the fact that positive head impulse test results may not always indicate a peripheral lesion, negative head impulse test results are a definitive sign of a central lesion in acute vestibular disorders. Consequently, patients with acute vestibular illness and positive head impulse test results should be closely monitored for additional central signs, such as INO, gaze-evoked nystagmus (GEN), vertical nystagmus, and impaired VOR suppression, particularly in conditions affecting both the peripheral and central vestibular systems simultaneously [20–23].

The combined use of VNG and vHIT represents a robust diagnostic strategy, enabling clinicians to effectively differentiate central from peripheral vestibular dysfunction. This integrative approach not only aids in the early identification of MS-related vestibular symptoms but also prevents the overdiagnosis of MS when other treatable vestibular conditions are present [20, 24].

15.3 Nystagmus and Oculomotor Abnormalities in Multiple Sclerosis

Oculomotor abnormalities are common in MS and can manifest as nystagmus, saccadic movements, or gaze conjugation issues. Visual oscillations are most frequently secondary to gaze-evoked nystagmus (GEN), vertical nystagmus, saccadic hypermetria, and pendular nystagmus. Except for upbeat nystagmus or internuclear ophthalmoplegia, which occur during acute attacks, other types of oculomotor abnormalities are typically observed in the chronic phase of the disease [25, 26].

15.4 Internuclear Ophthalmoplegia (INO)

INO is one of the most frequent oculomotor abnormalities in MS and is considered pathognomonic for the disease, particularly in cases of bilateral INO. However, other causes should be considered, such as vascular events, myasthenia gravis, post-radiation effects, AIDS, carcinomatous infiltration, paraneoplastic syndromes, medications, and head trauma. INO is characterized by paralysis of ipsilateral adduction, slow adducting saccade, and horizontal nystagmus of the contralateral eye during abduction, and sparing of ipsilateral adduction during convergence. However, this last dissociation may not always be present, as seen in bilateral INO cases, such as Wall-Eyed Bilateral Internuclear Ophthalmoplegia (WEBINO). INO results from a lesion in the medial longitudinal fasciculus (MLF), which transmits adduction signals from the contralateral abducens nucleus to the ipsilateral oculomotor nucleus during horizontal eye movements. In MS, INO is often bilateral, and slowed adducting saccades may be its only manifestation. In some cases, the range of adduction may appear intact without abduction nystagmus. Videonystagmography (VNG) facilitates the detection of subtle or clinically inconspicuous INO [25–28].

15.5 Pendular Nystagmus (PN)

PN can be considered a hallmark sign of MS and may result from damage to the cerebellum dentate efferents or olivary afferents. Clinically, it presents as sinusoidal oscillations of the eyes. Key features include frequencies ranging from 1 to 8 Hz, with an average of 3.5 Hz, oscillations in horizontal, vertical, or torsional planes, and possible disconjugate movements. PN in patients with MS is characterized by small amplitude, high frequency, and great regularity. In cases of central tegmental tract demyelination within the Guillain-Mollaret triangle, PN may be associated with palatal tremor, with MRI findings sometimes revealing inferior olivary hypertrophy [25, 26, 29].

15.6 Gaze-Evoked Nystagmus

GEN has been described in 16% of patients with MS and up to 44% of MS patients with visual dysfunction. GEN can occur with horizontal or upward vertical gaze. It results from a dysfunction of the neural integrator responsible for gaze holding (velocity-to-position integrator). In horizontal gaze, this neural network involves the prepositus hypoglossi nucleus and the medial vestibular nuclei. Vertical gaze parameters are integrated by the superior vestibular nuclei, the MLF, and the interstitial nucleus of Cajal in the midbrain. The cerebellar flocculus and paraflocculus are connected to the brainstem network, with a feedback mechanism controlling their function. Therefore, GEN can be caused by lesions in the brainstem and/or cerebellum [25–27].

15.6.1 Saccadic Abnormalities

Some oculomotor disorders in MS can be asymptomatic. Smooth pursuit is chronically impaired in 30% of MS patients, and impaired vestibulo-ocular reflex (VOR) suppression is reported in 70–75%. While smooth pursuit impairment can result from cerebellar, brainstem, and hemispheric lesions, the lack of VOR suppression is more specifically associated with chronic vestibulo-cerebellar involvement. A common chronic ocular movement abnormality in MS is saccadic dysmetria, present in approximately 30–40% of patients. Some may exhibit saccadic oscillations, such as square-wave jerks, saccadic intrusions, and ocular flutter [25, 27].

15.6.2 Auditory Symptoms in MS

Several review articles have examined auditory involvement in MS, highlighting a potential impact on hearing function in affected individuals. However, significant variability in data presentation and differences in study parameters make it difficult to draw definitive conclusions. Given these inconsistencies, further well-designed and standardized studies are needed to establish the prevalence and characteristics of auditory dysfunction in MS with greater reliability [5, 30].

15.7 Clinical Observations and Differential Diagnosis

Among 85 MS patients referred for neuro-otological evaluation at our center, vertigo was the most frequently reported vestibular symptom, affecting 65 patients (74%), followed by postural instability. Participants ranged in age from 21 to 75 years, with a mean age of 45 years, and 55 (64.7%) were female. The mean Expanded Disability Status Scale (EDSS) score at the time of vestibular symptom onset was 1.3 (range: 0–6.5). At one-year follow-up, 15 patients demonstrated a 1-point EDSS increase, with a mean EDSS of 1.45 (range: 0–8). Infratentorial brain lesions were observed on MRI in 68 cases (80%).

15.7.1 Etiology of Vestibular Symptoms

Among the MS patients referred for neuro-otological evaluation, the etiology of vestibular symptoms was attributed to MS in 47 cases (55.3%). The remaining 38 patients (44.7%) were diagnosed with other vestibular disorders. In all SPMS cases, vestibular symptoms were attributed to MS. Notably, vestibular symptoms marked the onset of MS in ten patients (11.7%), with six of these cases also exhibiting internuclear ophthalmoplegia (INO). For the remaining cases, the mean time from MS diagnosis to the appearance of vestibular symptoms was 6 years (range: 15 days to 40 years).

15.7.2 Alternative Vestibular Disorders

In 38 patients (44.7%), alternative vestibular disorders were identified, including:

Vestibular migraine: 24.8%.
Benign paroxysmal positional vertigo (BPPV): 10.7%.
Ménière's disease: 5.8%.
Acoustic neuroma: 2.3%.
Ramsay Hunt syndrome: 1.1%.

15.7.3 Vestibular Function Tests

Vestibular function tests, including videonystagmography (VNG) and video head impulse test (vHIT), were performed on this MS cohort (see Fig. 15.1).

15.7.4 Videonystagmography

Videonystagmography (VNG) was performed in 52 patients, revealing abnormal findings in 40 cases (77%). The most consistent abnormalities were categorized into nystagmus and oculomotor abnormalities and saccadic abnormalities (see Figs. 15.2 and 15.3).

A detailed summary of the abnormal findings detected through VNG is provided in Table 15.1:

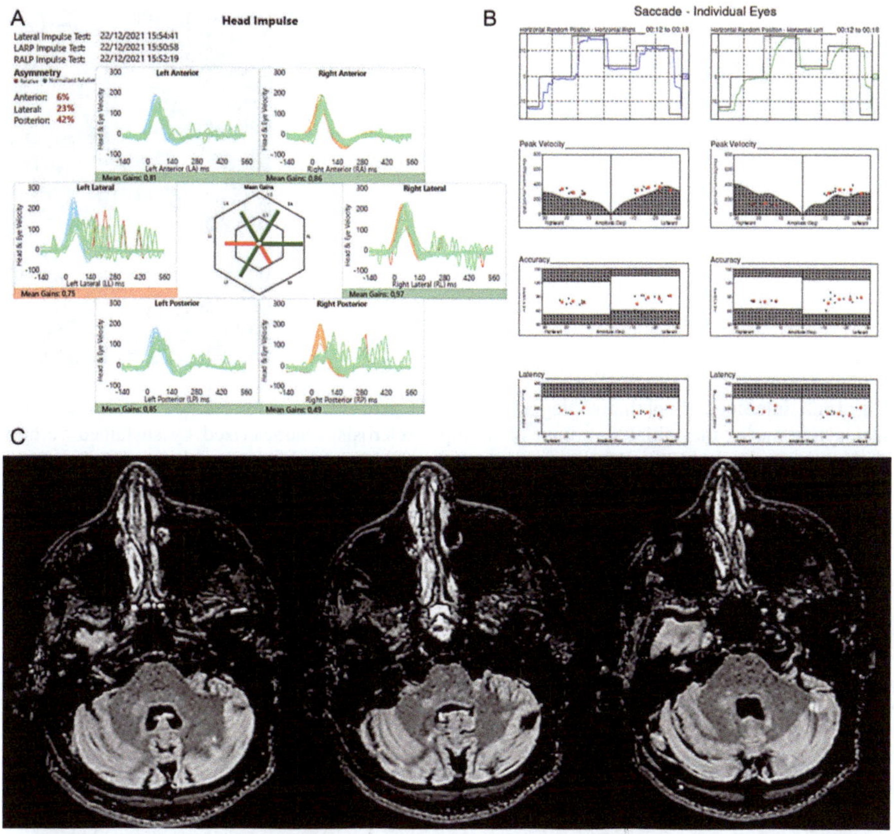

Fig. 15.1 Combined use of video head impulse test and videonystagmography. A 31-year-old male patient, diagnosed with multiple sclerosis following an acute vestibular syndrome 10 years ago, currently with an EDSS score of 2.0, was evaluated by neurotology due to dizziness with head movements, causing difficulties in sports activities. Vestibular studies revealed the following: vHIT showed a reduction in the gains of the left horizontal and right posterior semicircular canals with the presence of overt and covert saccades (**a**). VNG revealed a reduction in horizontal saccadic velocities in the left eye during adduction, compatible with left internuclear ophthalmoplegia (**b**). Brain MRI reveals multiple focal demyelinating lesions in the bilateral cerebellar hemispheres and the right cerebellar peduncle (**c**)

15.8 Video Head Impulse Test

The vHIT was performed in 40 cases, with normal results in 24 patients. Among those with normal vHIT results, six of eight patients whose vestibular symptoms were attributed to MS exhibited pathological findings on VNG, underscoring the importance of comprehensive vestibular testing. Within 16 patients with pathological vHIT findings, MS was identified as the underlying cause of vestibular

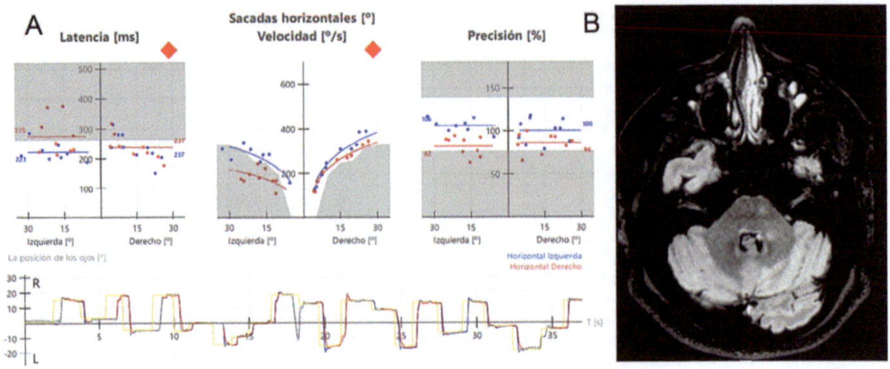

Fig. 15.2 Internuclear ophthalmoplegia (INO). A 51-year-old patient presented with acute vestibular syndrome as the initial symptom of multiple sclerosis, characterized by sustained vertigo, diplopia, and gait instability. Videonystagmography (VNG) revealed internuclear ophthalmoplegia (**a**), and brain MRI findings showed bilateral infratentorial lesions (**b**)

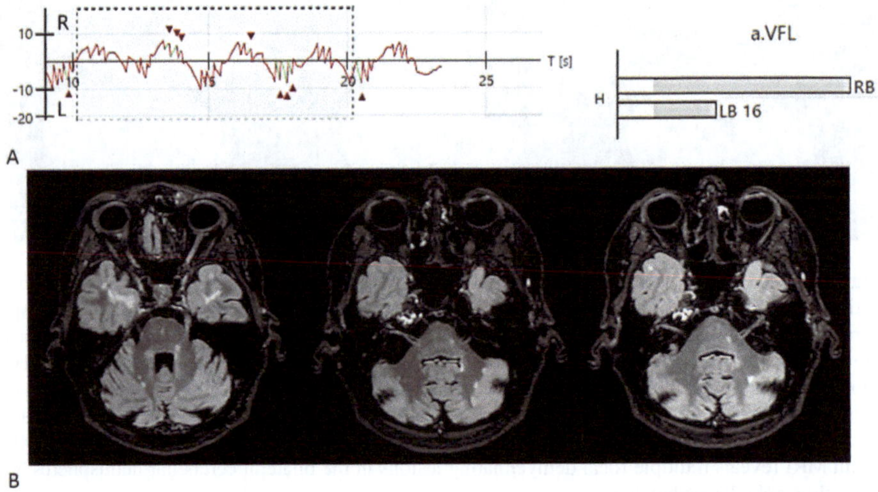

Fig. 15.3 Impaired vestibulo-ocular reflex (VOR) suppression. A 54-year-old female patient with secondary progressive multiple sclerosis (SPMS) and an Expanded Disability Status Scale (EDSS) score of 6.5 was evaluated for vertigo with positional changes and gait instability. Neurological examination revealed gaze-evoked nystagmus, abnormal vestibulo-ocular reflex (VOR) suppression (**a**), moderate crural paresis, and ambulation with a cane. The Dix-Hallpike maneuver on both sides elicited downbeat vertical nystagmus. Presented numerous brainstem lesions on brain MRI (**b**)

Table 15.1 VNG abnormal findings in 40 cases

Dysfunction type	Findings
Nystagmus and oculomotor abnormalities	Positional downbeat nystagmus: 7 Postural upbeat nystagmus: 5 PN: 4 cases Spontaneous upbeat nystagmus: 1 case Spontaneous downbeat nystagmus: 1 case INO: 6 cases Bilateral INO: 2 cases GEN: 4 cases
Saccadic abnormalities	Impaired vestibulo-ocular reflex (VOR) suppression: 15 cases Hypometric saccades: 5 cases Saccadic intrusions (square-wave jerks): 9 cases Ocular flutter: 1 case

In our cohort, the most frequent abnormalities detected on VNG included vertical nystagmus and impaired VOR suppression, followed by INO, GEN, and PN. Additionally, saccadic intrusions were commonly observed. For a more detailed perspective, for a more detailed perspective, please refer to the exemplary cases illustrated in Figs. 15.2 and 15.3. Abbreviations: *INO* internuclear ophthalmoplegia, *GEN* gaze-evoked nystagmus, *PN* pendular nystagmus

symptoms in ten cases (62.5%). In the subset of six cases with AVS, vHIT demonstrated abnormal VOR gains with overt saccades, while VNG confirmed accompanying INO. These findings highlight the importance of multimodal vestibular assessment in MS to distinguish central from peripheral etiologies and guide appropriate management (see Fig. 15.4). An analysis of the 16 abnormal vHIT cases in MS patients revealed a spectrum of vestibular impairments. For clarity, these cases were categorized based on unilateral or bilateral involvement, emphasizing the presence of saccades and specifying the affected semicircular canal (see Table 15.2):

15.9 Conclusions

Not all vertigo in MS patients should be attributed to the disease itself. Given the overlapping etiologies of central and peripheral vestibular disorders, a nuanced diagnostic approach is essential. The combined use of VNG and vHIT offers a robust diagnostic strategy, enabling clinicians to effectively differentiate central from peripheral vestibular dysfunction. Notably, nearly half of the patients presenting with vertigo or dizziness are ultimately diagnosed with alternative vestibular disorders, emphasizing the importance of a comprehensive evaluation. Additionally, VNG and vHIT are particularly valuable for detecting subtle central lesions that may be overlooked in routine neuroimaging, ultimately enhancing clinical decision-making and optimizing patient management.

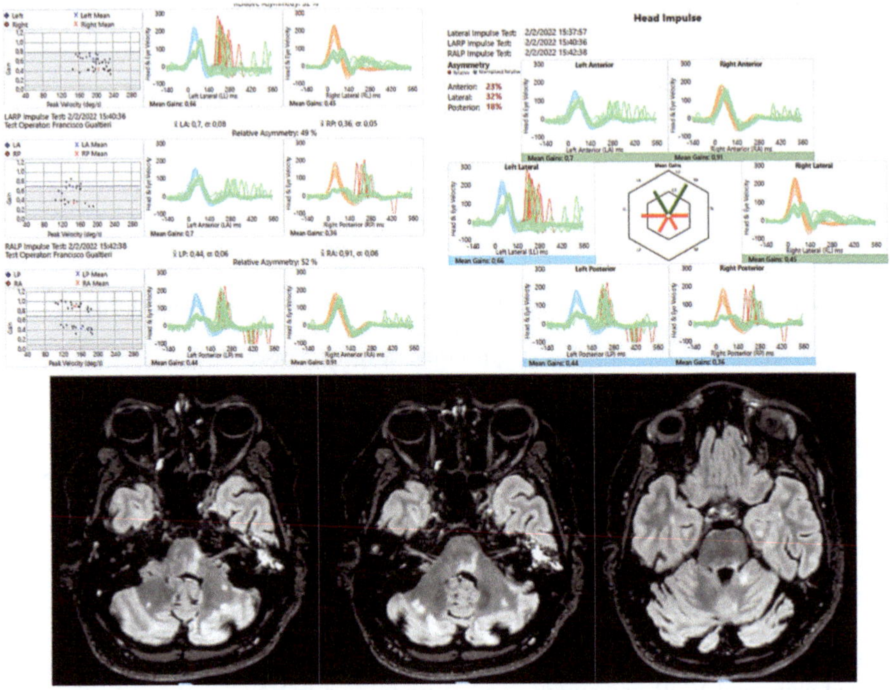

Fig. 15.4 Video head impulse test (vHIT) revealed a bilateral reduction in the gains of the horizontal and posterior semicircular canals. A 51-year-old patient with a 12-year history of RRMS and an EDSS score of 2.0 was evaluated by neurotology due to an acute vestibular syndrome. On examination: right internuclear ophthalmoplegia (INO), impaired vestibulo-ocular reflex (VOR) suppression, bilateral positive head impulse test, altered dynamic visual acuity, and mild ataxic gait. Video head impulse test revealed a bilateral reduction in the gains of the horizontal and posterior semicircular canals with the presence of overt saccades and a deficit in saccade generation to the right (**a**). Brain MRI showed numerous infratentorial lesions (**b**)

Appendix: Vestibular Testing Equipment

Videonystagmography (VNG) and VisualEyes™ 525 (Interacoustics System). The protocol encompasses a comprehensive evaluation of ocular movements: horizontal and vertical smooth pursuit, and saccadic horizontal and vertical eye movement testing. Nystagmus: spontaneous nystagmus, gaze-evoked nystagmus testing, mastoid vibration, and head-shaking maneuvers. Positional testing: postural evaluations in the supine position with head rotation to the right and left, Dix-Hallpike maneuver. Vestibulo-ocular reflex (VOR) suppression.

Video head impulse test (vHIT). The ICS impulse system (Otometrics/Natus) with Otosuite software is employed to perform vHIT. The protocol involves delivering head impulses in three planes: horizontal plane, right-anterior left-posterior (RALP), and left-anterior right-posterior (LARP). Each plane and direction is tested with a minimum of 20 head impulses to ensure reproducibility. Key parameters analyzed include canal gain values; presence, amplitude, configuration, and characteristics of covert and overt saccades.

Table 15.2 Abnormal vHIT results ($n = 16$)

Bilateral dysfunction ($n = 12$ cases)	Unilateral dysfunction ($n = 4$ cases)
Reduced VOR gain: 12 cases	Reduced VOR gain: 3 cases
Overt saccades: 12 cases Covert saccades: 8 cases	Overt saccades: 4 cases Covert saccades: 2 cases
Preserved VOR gain with overt saccades: 1 case	Preserved VOR gain with overt saccades: 1 case
Canal involvement:	Canal involvement:
Horizontal and posterior semicircular canals (CSH-CSP): 5 cases	Horizontal and posterior canals (CSH-CSP): 0 case
Horizontal canal (CSH): 2 cases	Horizontal canal (CSH): 2 cases
Posterior canal (CSP): 2 cases	Posterior canal (CSP): 1 case
All three canals: 3 cases	All three canals: 1 case

Abnormal vHIT findings were identified in 16 MS patients. Each case was categorized based on unilateral or bilateral involvement, with specific emphasis on the presence of saccades and the affected semicircular canal. The observed patterns included: simultaneous horizontal and posterior canal impairment; isolated horizontal or posterior canal dysfunction; triple-canal involvement (superior, horizontal, and posterior canals). These classifications provide a detailed characterization of vestibular deficits in MS, facilitating a more precise understanding of central versus peripheral vestibular dysfunction.

References

1. Cristiano E, Rojas JI. Multiple sclerosis epidemiology in Latin America: an updated survey. Mult Scler J Exp Transl Clin. 2017;3(2):20155217317715050. https://doi.org/10.1177/2055217317715050.
2. Tintoré M, Montalban X, Sastre A, et al. Pathogenesis of multiple sclerosis: insights from clinical and imaging studies. J Neurol. 2018;265(3):579–90. https://doi.org/10.1007/s00415-017-8642-3.
3. Lublin FD, Reingold SC, Cohen JA, et al. Defining the clinical course of multiple sclerosis: the 2013 revisions. Neurology. 2014;83(3):278–86. https://doi.org/10.1212/WNL.0000000000000560.
4. Cochrane GD, Christy JB, Motl RW. Comprehensive clinical assessment of vestibular function in multiple sclerosis. J Neurol Phys Ther. 2021;45(3):228–34. https://doi.org/10.1097/NPT.0000000000000358.
5. Gür E, Binkhamis G, Kluk K. Effects of multiple sclerosis on the audio-vestibular system: a systematic review. BMJ Open. 2022;12:e060540. https://doi.org/10.1136/bmjopen-2021-060540.
6. Yang Y, Wang M, Xu L, et al. Cerebellar and/or brainstem lesions indicate poor prognosis in multiple sclerosis: a systematic review. Front Neurol. 2022;13:874388. https://doi.org/10.3389/fneur.2022.874388.
7. Marsden J, Pavlou M, Dennett R, et al. Vestibular rehabilitation in multiple sclerosis: study protocol for a randomised controlled trial. BMC Neurol. 2020;20:430. https://doi.org/10.1186/s12883-020-01983-y.
8. Pula JH, Newman-Toker DE, Kattah JC. Multiple sclerosis as a cause of the acute vestibular syndrome. J Neurol. https://doi.org/10.1007/s00415-013-6850-1.

9. Russo FY, Ralli M, De Seta D, et al. Autoimmune vertigo: an update on vestibular disorders associated with autoimmune mechanisms. Immunol Res. 2018;66:675–85. https://doi.org/10.1007/s12026-018-9023-6.
10. Kim JA, Jeong IH, Lim YM, Kim KK. Primary position upbeat nystagmus during an acute attack of multiple sclerosis. J Clin Neurol. 2014;10(1):37–41. https://doi.org/10.3988/jcn.2014.10.1.37.
11. Newman-Toker DE, Kattah JC, Alvernia JE, Wang DZ. Normal head impulse test differentiates acute cerebellar strokes from vestibular neuritis. Neurology. 2008;70(24 Pt 2):2378–85.
12. Brevern M, Bertholon P, Brandt T, et al. Benign paroxysmal positional vertigo: diagnostic criteria: consensus document of the Committee for the Classification of vestibular disorders of the Bárány Societyvon. J Vestib Res. 2015;25(3–4):105–17. https://doi.org/10.3233/VES-150553.
13. De Schutter E, Adham ZO, Kattah JC. Central positional vertigo: a clinical-imaging study. Prog Brain Res. 2019;249:345–60.
14. Meshref M, Shaheen A, Elmatboly AM, et al. Central positional vertigo as first initial multiple sclerosis symptom: a case report with systematic review. Clin Case Rep. 2022;10:e06154. https://doi.org/10.1002/ccr3.6154.
15. Frohman EM, Zhang H, Dewey RB, Hawker KS, Racke MK, Frohman TC. Vertigo in MS: utility of positional and particle repositioning maneuvers. Neurology. 2000;55(10):1566–9.
16. Martin CL, Phillips BA, Kilpatrick TJ, et al. Gait and balance impairment in early multiple sclerosis in the absence of clinical disability. Mult Scler. 2006;12(5):620–8.
17. Hebert JR, Corboy JR, Manago MM, Schenkman M. Effects of vestibular rehabilitation on multiple sclerosis–related fatigue and upright postural control: a randomized controlled trial. Phys Ther. 2011;91(8):1166–83. https://doi.org/10.2522/ptj.20100399.
18. Güven H, Bayır Ö, Aytaç E, et al. Vestibular-evoked myogenic potentials, clinical evaluation, and imaging findings in multiple sclerosis. Neurol Sci. 2014;35:221–6. https://doi.org/10.1007/s10072-013-1483-9.
19. Kabel AM, Moaty AS, Afifi KH, Kamh ME. Video head impulse test in multiple sclerosis patients. Menoufia Med J. 2022;35(2):93–100. https://doi.org/10.4103/mmj.mmj_136_21.
20. Koohi N, Mendis S, Lennox A, Whelan D, Kaski D. Video head impulse testing: pitfalls in neurological patients. J Neurol Sci. 2022;442:120417.
21. Choi JY, Kim HJ, Kim JS. Recent advances in head impulse test findings in central vestibular disorders. Neurology. 2018;90(13):602–12.
22. Kim SH, Lee SU, Cho BH, Cho KH, Yu S, Kim BJ, Kim JS. Analyses of head impulse tests in patients with posterior circulation stroke and vestibular neuritis. Neurology. 2023;100(23):e2374–85.
23. Chen L, Halmagyi GM. Central lesions with selective Semicircular Canal involvement mimicking bilateral Vestibulopathy. Front Neurol. 2018;9:264.
24. Kremmyda O, Kirchner H, Glasauer S, Brandt T, Jahn K, Strupp M. False-positive head-impulse test in cerebellar ataxia. Front Neurol. 2012;3:162.
25. Nerrant E, Tilikete C. Ocular motor manifestations of multiple sclerosis. J Neuroophthalmol. 2017;37(3):332–40.
26. Ouattassi N, El Ghoul SEA, Bouchal S, Belahssen MF, Ridal M, El Alami MNEA. Ocular motor impairment in early-stage multiple sclerosis: a video-oculography assessment. Egypt J Otolaryngol. 2023;39(1):8.
27. Serra A, Chisari CG, Matta M. Eye movement abnormalities in multiple sclerosis: pathogenesis, Modeling, and treatment. Front Neurol. 2018;9:31. https://doi.org/10.3389/fneur.2018.00031.
28. Frohman TC, Frohman EM, O'Suilleabhain P, et al. Accuracy of clinical detection of INO in MS: corroboration with quantitative infrared oculography. Neurology. 2003;61(6):848–50. https://doi.org/10.1212/01.WNL.0000085863.54218.72.
29. Tilikete C, Jasse L, Pelisson D, Vukusic S, Durand-Dubief F, Urquizar C, Vighetto A. Acquired pendular nystagmus in multiple sclerosis and oculopalatal tremor. Neurology. 2011;76(19):1650–7.
30. MacMahon H, El Refaie A. The audiovestibular manifestations as early symptoms of multiple sclerosis: a scoping review of the literature. Ir J Med Sci. 2022;191(1):391–400.

Chapter 16
Vestibular Migraine: From Pathophysiology to Treatment

Rosmaninho Seabra Carlos and Rosa Castillo

Keywords Vestibular migraine · Pathophysiology · Vestibular system · Treatment

16.1 Introduction

Vestibular migraine is one of the most common diagnoses in dizziness units all over the world, although some authors believe it is still underdiagnosed. There was some controversy around this clinical entity with several names related to the disease, like migraine associated vertigo, migraine related vestibulopathy, basilar migraine and others until 2012, when the International Committee for Classification of Vestibular Disorders of the Bárány Society and the Migraine Classification Subcommittee of the International Headache Society agreed in a document to establish the diagnostic criteria of this entity which they called **vestibular migraine** [1, 2].

These diagnostic criteria are separated into two levels: definite vestibular migraine and probable vestibular migraine.

16.1.1 Vestibular Migraine (Definite)

A. At least five episodes with vestibular symptoms of moderate or severe intensity, lasting 5 min to 72 h.
B. Current or previous history of migraine with or without aura according to the International Classification of Headache Disorders (ICHD).
C. One or more migraine features with at least 50% of the vestibular episodes:

   (a) Headache with at least two of the following characteristics: one-sided location, pulsating quality, moderate or severe pain intensity, aggravation by routine physical activity.

R. S. Carlos (✉) · R. Castillo
Vertigo Unit, Hospital Cuf Porto/Hospital Luz Setubal, Lisbon, Portugal

P. Ravat et al. (eds.), *Advances in Translational Neuroscience of Vestibular and Eye Movement Disorders*, Contemporary Clinical Neuroscience,
https://doi.org/10.1007/978-3-032-07539-0_16

(b) Photophobia and phonophobia.
(c) Visual aura.

D. Not better accounted for by another vestibular or ICHD diagnosis.

16.1.2 Probable Vestibular Migraine

A. At least five episodes with vestibular symptoms of moderate or severe intensity, lasting 5 min to 72 h.
B. Only one of the criteria B and C for vestibular migraine is fulfilled (migraine history or migraine features during the episode).
C. Not better accounted for by another vestibular or ICHD diagnosis.

It is currently the most common cause of spontaneous recurrent episodic vertigo, affecting 1–2.7% of the population, 11% of patients in a vertigo clinic, and 13% of patients in a headache clinic. The average age of onset is the fifth decade of life, and it is more common in women, where a correlation with the menstrual period and peri-menopausal changes has been demonstrated.

16.2 Pathophysiology

The pathophysiology of the disease is still unknown, but there are several theories, including vascular, neurogenic, genetic factors, and multisensory integration [3–5].

Vascular theory postulates the existence of a reversible vasospasm of the internal auditory artery or its branches, and an infarction of the antero-inferior cerebellar artery (AICA) territory. According to this theory, hypoperfusion of the inner ear due to vessel spasm can be supported by the association with sudden deafness and the increased risk factor for stroke in these patients.

Neurogenic theories are based on the activation and sensitization of the trigeminal vascular system that leads to the release of pro-inflammatory neuropeptides, like substance P and Calcitonin Gene-Related Peptide (CGRP). These neuropeptides, would activate on one hand meninges and intracranial blood vessels and on the other hand the trigeminocervical complex (TCC) and therefore thalamic cortical neurons and cerebral nuclei in the diencephalon and brain stem (like hypothalamus, locus coeruleus) and also the activation of brain areas associated with the processing of nociceptive information as well as areas of the thalamus and the so-called vestibular cortex.

The thalamus and vestibular cortex are essential for the vestibular system in the processing of visual and proprioceptive vestibular input and participation in multi-sensory integration. The thalamus receives input from the trigeminal nerve and is involved in controlling body orientation in space. This cortical thalamic involvement in vestibular migraine has been demonstrated by several neuroimaging tests. This explains that thalamocortical dysrhythmia is caused by the lack of habituation to stimuli. This loss of habituation (repetition of a sensory stimulus causes an increase in the response) occurs in the intercrisis phases of migraine and reverts in crises.

The existence of genetic factors is supported by the recognition of familial occurrence of vestibular migraine. In fact, several genomes have been identified as possibly associated with susceptibility to vestibular migraine, such as the markers rs2651899 in the PRDM16 gene, rs10166942 in the TRPM8 gene, and rs11172113 in the LRP1 gene, between others.

Thus, possible mechanisms involved in the pathogenesis of vestibular migraine would be an abnormal sensory modulation or integration within the thalamocortical network that would result in dizziness and spatial disorientation, a hyperactivity in the trigeminovascular system, and in the nociceptive centres of the brain stem could result in headache. Altered activity of the vestibular system could lead to transient vestibulo-ocular dysfunction or vestibular hypersensitivity associated with migraine features [1].

Although the pathophysiology of migraine is yet unknown, we could state that genetic predisposition combined with behavioural and environmental factors leads to an alteration in cerebral sensory processing, increasing sensory susceptibility. "Normal" sensory information is then perceived as "annoying".

16.3 Treatment [1, 6–8]

The treatment of vestibular migraine is based mainly upon three weapons:

1. Diet and behavioural measures to avoid triggers.
2. Pharmacotherapy with two goals: Control of the crisis attack and avoidance of crisis (prophylactic).
3. Vestibular rehabilitation.

16.4 Diet and Behavioural Measures

- **Stress.**
- **Lack of sleep.**
- Dehydration.
- Weather changes (changes in barometric pressure).
- **Hormonal changes.**
  - **Menstruation.**
- **Foods and drinks.**
  - Chocolate.
  - Red wine.

- Some cheeses.
- Monosodium glutamate.
- Coffee.

16.5 Pharmocotherapy

16.5.1 Control of Crisis Attack

The most important thing in a crisis attack is to exclude the possibility of a stroke.

The drugs used in a crisis attack include analgesics and triptans in order to prevent the allergic symptoms, and antiemetics and vestibuloplegics to control vestibular symptoms like nausea, vomiting, and dizziness/vertigo.

16.5.2 Prophylactic Avoidance of Crisis

When crises are short and occur with a high frequency, it is indicated to perform some prophylactic therapy. This prophylactic therapy includes general measures like enhancing the quality of sleep, a proper diet, and the avoidance of known triggers of the disease (chocolates, wine, some cheeses), and especially the control of stress.

Regarding the drugs used in this case, we would emphasise some of them, like:

Betablockers (propranolol, metoprolol).
Antiepileptics (valproate, topiramate).
Calcium Antagonists (cinnarizine).
Anti-depressants (amitriptyline, venlafaxine).
Anti-serotonergicos (pizotifen).
Anti-HTA (candesartan, lisinopril).
Anti-CGRP antibody (fremanezumab, galcanezumab, eptinezumab).
Anti-CGRP receptor antibody (erenumab).
CGRP receptor antagonist (gepants).

16.6 Vestibular Rehabilitation

Indicated in patients with anxiety, visual dependency, and persistent instability. We should also consider it in patients who are refractory or intolerant to pharmacological treatment and in patients desiring a non-pharmacological approach. It is particularly useful if attacks are very frequent or if there is concurrent persistent postural perceptive dizziness (PPPD) or disabling avoidance behaviours.

References

1. Zanandrea L, Colombo B, Filippi M. Vestibular migraine therapy: update and recent literature review. Audiol Res. 2023;13(5):724–9.
2. Silberstein SD, Holland S, Freitag F, Dodick DW, Argoff C, Ashman E. Evidence-based guideline update: pharmacologic treatment for episodic migraine prevention in adults: report of the quality standards Subcommittee of the American Academy of neurology and the American headache society. Neurology. 2012;78(17):1337–45.
3. Espinosa-Sanchez JM, Lopez-Escamez JA. New insights into pathophysiology of vestibular migraine. Front Neurol. 2015;6:12.
4. Huang TC, Wang SJ, Kheradmand A. Vestibular migraine: an update on current understanding and future directions. Cephalalgia. 2020;40(1):107–21.
5. Morganti LOG, Salmito MC, Duarte JA, Sumi KC, Simões JC, Ganança FF. Vestibular migraine: clinical and epidemiological aspects. Braz J Otorhinolaryngol. 2016;82(04):397–402.
6. Silva VPR, Castro LHM, Calderaro M. Vestibular migraine. Arq Neuropsiquiatr. 2022;80 (5 Suppl 1):232–7.
7. Puledda F, Silva EM, Suwanlaong K, Goadsby PJ. Migraine: from pathophysiology to treatment. J Neurol. 2023;270(7):3654–66.
8. Smyth D, Britton Z, Murdin L, Arshad Q, Kaski D. Vestibular migraine treatment: a comprehensive practical review. Brain. 2022;145(11):3741–54.

References

Chapter 17
Spatial Cognition and Functional Dizziness

Hayo A. Breinbauer

Keywords Functional neurological disorder · Vestibular rehabilitation · Chronic subjective dizziness · PPPD · Cognitive map

17.1 Brief Overview of PPPD

Persistent postural-perceptual dizziness (PPPD) is considered the most common cause of chronic vestibular syndrome [1]. It accounts for 14–22% of all patients in primary care presenting with vertigo, dizziness, or balance issues and over 53% of those reporting these symptoms in tertiary neurotology centers or specialized dizziness clinics [2, 3]. PPPD manifests as non-vertiginous dizziness, unsteadiness, and a sensation of swaying or rocking—symptoms that are present most hours of the day, most days, significantly impacting patients' quality of life [4]. Once PPPD is established, these symptoms are often exacerbated by an upright posture, active or passive movement, and particularly when exposed to visually complex or dynamic environments.

The chronic nature of PPPD can have a profound impact on an individual's daily life, restricting their capacity to engage in routine activities [5]. Impairment may extend to various aspects, affecting work, social interactions, and recreational pursuits [6].

Treatment with individualized vestibular rehabilitation, a moderate dose of a selective serotonin reuptake inhibitor (SSRI) or a serotonin and norepinephrine reuptake inhibitor (SNRI), and cognitive behavioral therapy, alone or in combination, reduces symptom severity by greater than 50% in most patients, although it may take 3–6 months to achieve that improvement [1].

Though its official definition and diagnostic criteria were only established in 2017 by the International Classification of Vestibular Disorders and in 2022 in the

H. A. Breinbauer (✉)
Department of Neuroscience, Universidad de Chile, Santiago, Chile

Facultad de Medicina Clínica Alemana, Universidad del Desarrollo, Santiago, Chile

P. Ravat et al. (eds.), *Advances in Translational Neuroscience of Vestibular and Eye Movement Disorders*, Contemporary Clinical Neuroscience,
https://doi.org/10.1007/978-3-032-07539-0_17

International Classification of Diseases (11th edition), the condition itself is not new. It has been researched extensively since the late 1980s under different labels, now all consolidated under the PPPD umbrella [4]. These precursor entities include phobic postural vertigo, visual vertigo, space motion discomfort, and chronic subjective dizziness [7].

However, there is consensus that the current understanding of its mechanisms is limited [1, 6]. The most current pathophysiological model suggests that a triggering event (often another vestibular disorder, but sometimes a non-vestibular issue like an acute anxiety episode) leads to a cascade of poorly understood functional neural adaptations. These encompass alterations in the processing of sensory information and vestibular responses within the brain, rather than overt structural changes [1, 8, 9].

In this fashion, PPPD may involve functional disruptions at the cortical network level, affecting systems related to motion perception, postural control, locomotion, and spatial orientation [10–25]. Such disruptions could potentially contribute to the characteristic dizziness observed in PPPD, and may also manifest as heightened attention to motion [10, 26], misperception of movement [27], altered posture [28], increased effort in postural control, heightened dependence on visual input [29], and compromised spatial navigation [13, 14], among others [1]. It is important to emphasize that while these associations are suggested, the exact pathophysiological mechanisms underlying these symptoms in PPPD remain to be conclusively determined.

One area of agreement is that PPPD is not associated with definitive structural lesions in either the vestibular organs or the central nervous system. However, the extensive changes in neural function mentioned above have been observed using various techniques such as functional magnetic resonance imaging (fMRI) [16, 18, 19, 30], although, as previously stated, these changes are not yet fully understood. Therefore, there is strong agreement to define PPPD as a functional neurological disorder [1].

17.2 A Functional Neurological Disorder (FND)

Functional neurological disorders (FND) are a group of highly prevalent, costly, and potentially disabling conditions [31]. They encompass motor, sensory, and cognitive aspects, including functional movement disorders (with symptoms such as tremor, dystonia, myoclonus, and gait disorders), functional seizures (psychogenic non-epileptic seizures PNES), functional weakness or paralysis, functional cognitive disorders (memory, attention among others), functional visual loss, and of course, functional dizziness in the paramount form of PPPD [31–34].

These have been termed a "silent epidemic," a "crisis" in neurology, and psychiatry's "blind spot" [34–38]. In recent years, the understanding of these disorders has significantly evolved from their initial description as hysteria disorders

[39], with growing interest in their diagnosis, management, and pathophysiological mechanisms from a neuroscience perspective.

FNDs appear to share several common elements. Clinically, there is a strong call to validate their existence and the subjective experience of patients, emphasizing a positive and inclusive diagnosis rather than considering them a diagnosis of exclusion [38]. Mechanistically, several emerging constructs suggest a common pathophysiology centered around the brain's prediction and inference functions. Alterations in processes such as agency, bodily awareness, interoception, emotion processing dysregulation, salience, and attention are recurrently highlighted [32].

Let us take a moment to explore the central ideas of the predictive brain and active perception, which are highly relevant for understanding FNDs.

Consider an older model of organismal organization to interact with the environment, focusing on the example of maintaining our balance and postural stability. Traditionally, we would have considered that our sensory apparatus (visual, proprioceptive, and vestibular) would capture information from our environment and deliver it to our brain for processing and interpretation. Based on this, a plan of motor responses would be generated and then sent to our muscles to change our body's position, adapting to a changing world.

If we expected this afference-processing-efference circuit to be the indispensable loop for moving through the world, the truth is that we would not be able to stand upright. This circuit would take hundreds of milliseconds to complete a reliable arc. The world around us changes much faster, as does our movement within it. We would always be late, always with errors, and react slowly to the world.

Therefore, it has been necessary to develop mechanisms that allow us to react beforehand and anticipate what is going to happen, enabling us to "live in real-time in the world."

Hence, the concept of the predictive brain, rooted in the predictive coding framework, posits that the brain constantly generates and updates a model of the environment to anticipate sensory input. This model is used to minimize the prediction error, which is the difference between expected and actual sensory input. The brain operates through a hierarchical structure where higher-level cortical areas generate predictions that are matched against incoming sensory data. Discrepancies, or prediction errors, are then propagated back up the hierarchy to refine future predictions [40, 41].

Closely related is the theory of active perception, where perception is not a passive process but an active one involving the continuous generation and testing of hypotheses about the environment. This concept is closely related to the predictive coding framework. In active perception, the brain actively seeks out information to confirm or disconfirm its predictions about the sensory world. This involves not only processing sensory information but also directing attention and motor actions to gather the most relevant data.

The theory emphasizes the role of motor actions and changes in sensory organs as active actors in perception, suggesting that sensory information is often sought out by moving and adapting the sense organs (e.g., eyes, hands) to better explore the environment. Active perception is thus an interplay between sensory input, motor actions, and predictive processes, highlighting the dynamic and interactive nature of perception [42, 43].

These notions apply to the vestibular realm and how balance and the understanding of body position in the world are managed by humans as an organism. In adherence to the principle 'Where there is function, there is potential for dysfunction,' we adhere to the line of thought where the disruption of the predictive brain/ active perception would lead to an FND. Particularly, we posit that, as part of the pathophysiology of PPPD, a disruption of the predictions of the spatial environment (including the position of the body in it) would lead to functional dizziness and the core phenomena of PPPD.

17.3 An Inner Model for an External World

To understand the disruption in predictions in PPPD, we first need to consider that these predictions involve not only the independent sensory inputs from vestibular, visual, and somatosensory systems but also their integrative interaction within an internal multimodal model of the external world. The notion of cognitive-spatial maps and spatial navigation has emerged as a main theoretical framework for understanding this inner model.

In 2014, the Nobel Prize in Physiology or Medicine was awarded to John O'Keefe, May-Britt Moser, and Edvard I. Moser for their groundbreaking discoveries concerning the brain's navigation system, providing profound insights into the neural mechanisms underlying spatial memory and navigation [44–47].

17.3.1 Key Discoveries in Spatial Navigation

- **John O'Keefe's Discovery**: In 1971, John O'Keefe identified "place cells" in the hippocampus of rats, neurons that become active when the animal is in a specific location. This showed that the hippocampus contains a spatial map of the environment, crucial for navigation and memory.
- **Moser and Moser's Contribution**: In 2005, May-Britt and Edvard I. Moser identified "grid cells" in the entorhinal cortex, which create a coordinate system for precise positioning and pathfinding. These cells work with place cells to enable complex spatial navigation and the construction of cognitive maps.

Recent research has identified additional brain areas involved in a vast "spatial navigation network." Neuroscience has traditionally approached the problem of spatial navigation by asking, "Where am I?" "Where do I want to go?" and "How do I get from one place to another?" based on allocentric (external world elements as a reference frame independent of the body) and egocentric cues (perceptions using the body as a reference) [48].

Allocentric cues involve designing a map of the world as seen "from above," where the navigator is a moving point within this map. Visual landmarks play a key role in constructing an allocentric map, making visual information highly relevant for allocentric navigation.

Egocentric cues relate to perceiving how one's body moves in space, where the concept of "path integration" is fundamental. For example, in the classic "triangle completion" task, a person with closed eyes walks three steps forward, turns left, and takes four more steps. They will have a general sense of being about five steps diagonally from the start. To return to the origin, they must estimate how much to turn (angular rotation) and how far to walk (linear displacement), relying heavily on vestibular signals.

Figure 17.1 illustrates allocentric and egocentric cues using as an example a navigation mobile app for vehicular use. It is tempting to simplify the spatial navigation system into a visual-allocentric/vestibular-egocentric dichotomy, since diverse brain structures show preferences for different information types.

However, all available evidence shows that such reductionism is a serious mistake [49]. For example, there are visual cues that are organized egocentrically, such as when I see a bakery and know to turn right to find my friend's house without actually having a map of the streets in my mind. Similarly, egocentric cues are

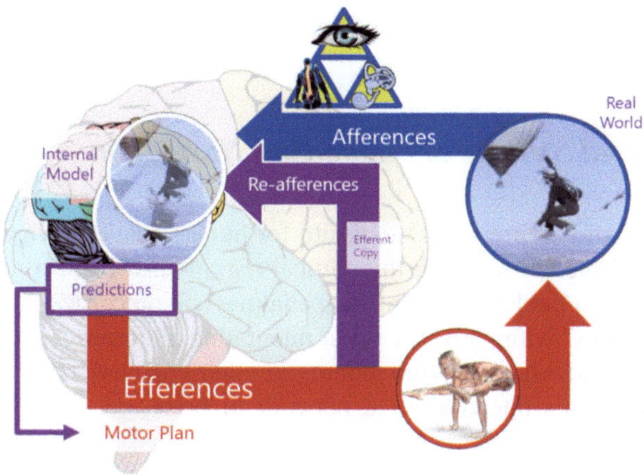

Fig. 17.1 Diagram of the processes for a predictive brain model. Afferent signals from the external world help construct an internal model of the surrounding reality and the body's location within it. Based on this model, a motor plan is proposed and implemented. Long before this motor plan is executed, an efferent copy is used to update the internal model of the surrounding world as a prediction of what is yet to happen. Even before the actual movement is completed, further steps for subsequent motor plans are being reviewed. The internal model (or prediction) is constantly contrasted with new afferent signals from the external world. This allows for a more efficient and faster way of engaging with an ever-changing environment. Not relying on an efferent copy and reafferences to produce predictions, and only depending on afferent signals from the external world, would lead to inadequate delays in adapting and responding to the environment

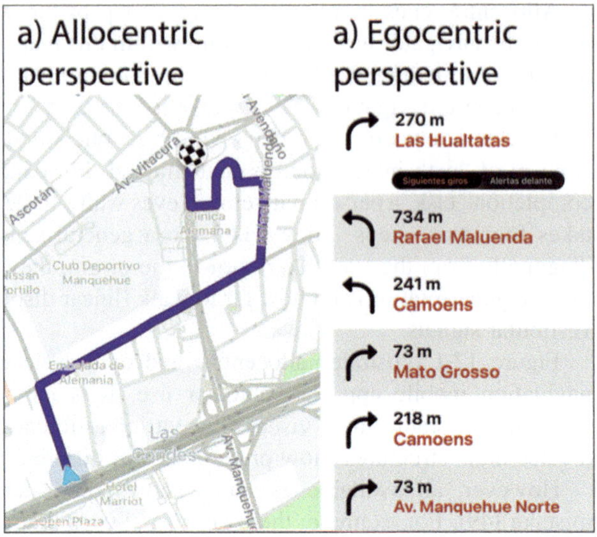

Fig. 17.2 Allocentric vs. egocentric navigation. An example of vehicle navigation apps illustrates the difference between allocentric navigation (**a**), where a cognitive map of the environment references position relative to landmarks, and egocentric navigation (**b**), where decisions (turning left or right) are based on the navigator's perspective as a reference frame ("When I reach this corner, I turn left")

critical for using an allocentric map, as maintaining an awareness of my head's orientation during a zigzagging walk in the forest is crucial for keeping track of where "north" is on the mental map of my surroundings (Fig. 17.2).

17.3.2 Neural Structures Supporting Navigation

The "taxonomy of Aguirre and D'Esposito for human spatial navigation" [50] is a widely accepted model outlining the cortical regions involved in navigation. It is complemented by a model of a "human spatial navigation cortical network," recognizing brain "nodes" with preferences but integrating allocentric, egocentric, visual, and vestibular information, and functioning in other cognitive tasks [49].

Understanding this interconnected network of brain areas helps simplify their roles.

17.3.3 Hippocampus

The hippocampal formation (medial temporal cortex areas, including the hippocampus, parahippocampal, entorhinal, and perirhinal cortices) is central to the navigation network [48]. Place, grid, border, and head direction cells form the basis of the "cognitive map"theory, where the brain constructs an internal, allocentric map of the world [51].

- **Place Cells**: Located predominantly in the hippocampus, they activate when in a specific location. For example, a specific place cell fires at one corner of a bedroom and another at a different room corner. Anatomically, place cells have no topographical relationship to the physical space [52]. The hippocampus "loads" different maps for spatial calculations, organizing hierarchical pseudo-spatial information like workplace hierarchies (e.g., a boss at the "north," colleagues east-west, subordinates south) [48].
- **Border Cells**: Also part of the hippocampus, they define the boundaries of the cognitive map in use [53].
- **Grid Cells**: Located in the entorhinal cortex, grid cells tessellate space into hexagonal units, contributing to spatial magnitude processing [54].
- **Head Direction Cells**: These cells, found in the hippocampus and parahippocampal formation, align with the "north" of the cognitive map, acting like a compass. They are also prevalent in the retrosplenial cortex [45].

17.3.4 Multimodal Integration and Projections

The hippocampal formation receives extensive projections from the visual and vestibular systems. Visual inputs primarily reach through the entorhinal cortex, processing object distances in the cognitive map [55]. Optic flow, encoded in the entorhinal and perirhinal cortices, signals movement within the map, akin to vestibular displacement signals. The parahippocampal cortex identifies visual landmarks in the cognitive map [56].

Vestibular projections are complex and abundant, with at least four main pathways from vestibular nuclei to the hippocampus: a vestibulo-thalamo-cortical pathway (transmitting environmental information via parietal cortices), a "head direction cell" pathway, a theta oscillatory activity pathway (related to memory), and a cerebellar pathway involving spatial learning [57]. Vestibular inputs primarily arrive at anterior hippocampal regions, showing lateral asymmetry with more right hippocampal input [58–60]. Vestibular information is crucial for hippocampal function, as evidenced by atrophy in cases of vestibular failure [61, 62].

17.3.5 Beyond the Hippocampus

While the hippocampus and cognitive maps are central to spatial navigation, they are not the entire process. The Aguirre and D'Esposito taxonomy identifies the posterior parietal cortex, retrosplenial cortex, and lingual gyrus as key components, with the prefrontal cortex now considered integral to the navigation network (Fig. 17.3).

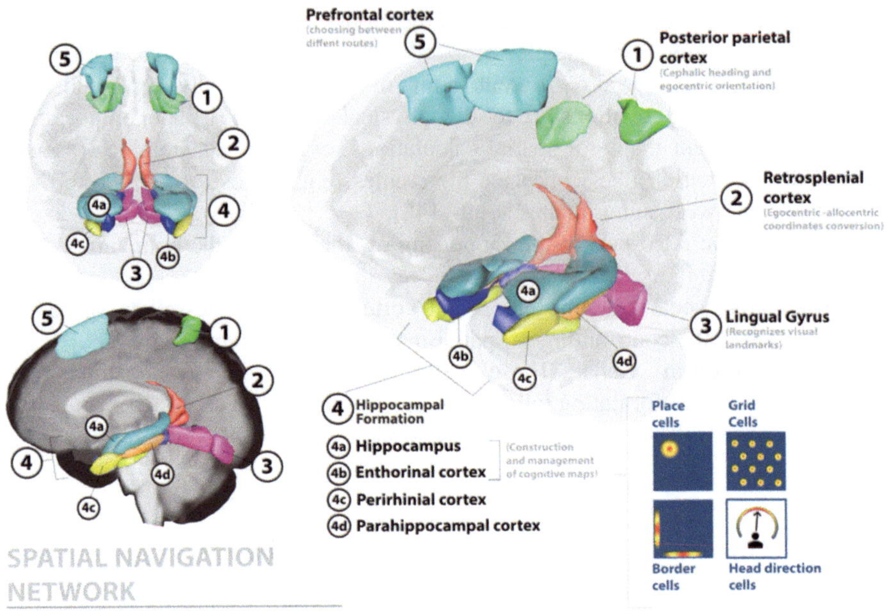

Fig. 17.3 Cortical spatial navigation network. This simplified model shows the main cortical regions involved in spatial navigation, following Aguirre and D'Esposito's taxonomy. The hippocampal formation and posterior parietal cortex receive substantial direct projections from vestibular nuclei. The prefrontal cortex is included, and each area's role is described in the text

17.3.6 Posterior Parietal Cortex and Egocentric Position

The posterior parietal cortex is part of the dorsal visual stream, determining object positions egocentrically ("Where are things relative to me?") for interaction and spatial navigation [63]. It integrates motor and somatosensory information for motor planning and sensory feedback. Vestibular inputs to the ventro-intraparietal region demonstrate the interconnected, redundant nature of the navigation network [57, 64].

17.3.7 Retrosplenial Cortex as Translator

The retrosplenial cortex, rich in head direction cells [46], translates egocentric coordinates into allocentric map views and vice versa [52].

17.3.8 Lingual Gyrus and Landmarks

The lingual gyrus, part of the inferior temporal lobe near the parahippocampal cortex, identifies allocentric landmarks [65]. It categorizes new landmarks and recognizes previously learned ones.

17.3.9 Prefrontal Cortex and Route Selection

The prefrontal cortex, not in the classic Aguirre and D'Esposito taxonomy, plays a role in route selection [66]. It evaluates egocentric and allocentric maps and route options, choosing the best alternative in a recursive, feedback-driven process.

17.3.10 Simplified Navigation Network

Despite caution against reductionism, the navigation network can be summarized as follows [47]:

- **Visual and Vestibular Inputs**: Feed directly into four main nuclei.
- **Posterior Parietal Cortex**: Focuses on egocentric navigation and object interaction.
- **Retrosplenial Cortex**: Converts egocentric data to allocentric coordinates for world modeling.
- **Lingual Gyrus**: Defines and recognizes visual landmarks for anchoring maps.
- **Hippocampal Formation**: Builds, retrieves, and uses cognitive maps with place, grid, border, and head direction cells for internal positioning and route planning.

The **prefrontal cortex** actively participates by demanding and choosing between navigation alternatives.

17.4 More Than Just "Spatial" Navigation and More Than Just "Spatial" Vestibular Computation

While spatial navigation has become a discipline of study in itself, new data suggest that the cognitive maps encoded by the brain contain information beyond merely spatial or physical properties of the world [20, 67]. These cognitive maps are enriched with cognitive and emotional information, such as the perceived risk level of a location or situation.

This "more than just spatial" information can also be observed in the broad range of brain areas and distinct processes involved in vestibular inputs.

17.4.1 Impact of Vestibular Failures on Cognitive Functions

Had this chapter been written one or two decades ago, it would have focused on summarizing the solid evidence showing that patients with peripheral vestibular failures (e.g., vestibular neuritis and bilateral vestibulopathy) not only have difficulties completing spatial navigation tasks but also exhibit atrophy in brain regions dedicated to this function, such as the hippocampus [44, 68–70].

These findings are not surprising, given that spatial navigation functions involve processes to determine the body's position in the world and plan necessary motor actions to reach desired destinations. The vestibular system‘s specialization in linear displacement and angular head movements is crucial for answering "where am I?" while moving from one place to another.

However, in response to recent evidence, it is now known that patients with peripheral vestibular failure show dysfunctions in many other cognitive domains that seem unrelated to vestibular positioning and movement cues. These include impairments in working and short-term memory, attention, executive functions, planning, and processing speed [71–74]. Such alterations are evident not only in visuospatial tests but also in tasks without any visual or spatial cues, such as digit retention, declarative memory, or problem-solving.

There are no significant differences in cognitive impacts between patients with recent versus chronic vestibular failure, suggesting that the loss of higher cognitive functions occurs rapidly after vestibular afference is lost [8]. Some authors emphasize the ubiquitous nature of vestibular signals, which provide the brain with a constant stream of position, velocity, and head rotation signals throughout all human activities. This contrasts with other sensory systems like vision and hearing, which are organized to detect salient stimuli [75]. The vestibular system might provide a constant baseline that allows other sensory modalities to construct a cognitive reality of the individual and their surroundings. A failure in this information flow could weaken this baseline, leading to disorganization in systems accustomed to vestibular afference [64].

Characterizing these cognitive disorders in patients with vestibular dysfunction, understanding their impact on quality of life and daily activities, the pathophysiological processes involved, and potential ad hoc therapeutic measures are areas of ongoing research.

17.4.2 Vestibulo-Cortical Projections

Although our understanding of vestibular cortical networks is incomplete and filled with conjectures, there is sufficient evidence to highlight several unique features of the vestibular system among sensory systems.

Vestibular information ascends to the cortex in a "staircase" arrangement, with multiple relay stations in the brainstem, thalamus, and subcortical pathways. At nearly every station, there is a crossing and recrossing of multiple projections, resulting in significant bilateral representation of any vestibular signal [57]. Projections from the vestibular nuclei reach multiple, widely distributed areas of the cerebral cortex [76]. These connections are proposed to be organized into three major domains [77]:

1. **Sensorimotor Domain**: This domain, predominantly involving projections to parietal cortex areas, relates to processes naturally associated with the vestibular system, with significant integration of somatosensory and visual signals. This includes projections to the navigation network, which will be discussed later. The concept of a parieto-insular vestibular cortex (PIVC), considered a primary vestibular area in some animal models, is not well-defined in humans [78]. Nevertheless, it is part of this first domain.
2. **Autonomic Domain**: Influences functions such as sympathetic responses, cardiovascular regulation, and pain processing.
3. **Cognitive Domain**: Demonstrates relevance in processes like body self-perception, visuospatial processing, decision-making, emotional regulation, perspective-taking, and memory [77].

According to some authors, no cortical regions are exclusively unimodal [66]. While senses like vision and hearing process initial inputs in dedicated primary cortices, which then relay information to other brain areas, the concept of a "vestibular cortex" in humans remains debated. Areas such as the parieto-insular vestibular cortex (PIVC) are proposed as primary regions for vestibular input processing [47]. However, growing evidence suggests that vestibular processing involves a diffuse network, a complex, interwoven set of multimodal networks that directly engage with the vestibular nuclei and project to multiple cortical regions [13, 14, 79–81].

Moreover, evidence indicates that these extensive vestibular networks are not confined to traditional "vestibular domains" like body positioning, movement, and spatial navigation [57, 82, 83]. They also encompass a range of cognitive and autonomic functions, including pain processing, emotional regulation, and executive functions [77, 84].

17.4.3 Multimodal Processing and PPPD

Returning to the issue of functional dizziness and its mechanisms, and considering the previous discussions regarding spatial navigation and the broader spatial and nonspatial cognitive involvement of vestibular signals, we emphasize that constructing a predictive inner model of the external world combines sensory perception, threat evaluation, and cognitive assessment of surrounding space into an integrated multimodal system. This system actively creates context-sensitive (or risk-adaptive) vestibulo-spatial and vestibulo-temporal maps of one's environment [20].

We propose that dysfunctions in the brain's ability to integrate vestibular inputs with other sensory information and cognitive assessments may underlie PPPD. Such a disruption in the integration process could impair the construction of accurate, context-sensitive perceptual maps, which are crucial for interpreting the external world. This disruption may manifest as a persistent mismatch between an error-prone, cognitively enriched perceptual map and the actual sensory afferences, potentially contributing to the persistent dizziness characteristic of PPPD.

17.5 Evidence of Impairment in Spatial Navigation and Cognition in Functional Dizziness

While there is a large body of literature exploring the relationship between vestibular function and cognition and spatial navigation, and growing reports on how these higher-order cerebral functions are disturbed by vestibular damage, few studies specifically address the issue of PPPD, functional dizziness (particularly without peripheral vestibular damage), and spatial-navigation problems.

In 2020 and later in 2023 [13, 14], our group published studies assessing mainly spatial navigation but also other psychocognitive functions in two different cohorts of patients suffering from PPPD. These patients were compared to those with other vestibular disorders but not PPPD and to healthy volunteers as two control groups.

17.5.1 Spatial Navigation Assessment Using the Virtual Morris Water Maze

A version of a virtual Morris water maze (vMWM) test served as our primary tool for assessing spatial navigation capabilities [85]. Originally designed for rodents, this test involves a round pool with visual cues and a hidden platform. Rodents must locate and remember this platform's position, improving their efficiency in reaching it in subsequent trials. Memory impairments, such as those from hippocampal lesions, cause rodents to fail in locating the platform. Adapted virtual versions for humans have been validated to identify memory deficits, including those seen in Alzheimer's patients [86, 87]. The vMWM has also been widely used to assess spatial navigation abilities in individuals with vestibular disorders, such as bilateral vestibulopathy [44, 70].

Figure 17.4 shows an example of the virtual version of our vMWM from the Breinbauer 2020 study [13]. The left panel shows the virtual pool, with the target platform in light blue, the pool boundaries in red, and some of the visual cues in the environment. The right panel displays a graphical representation of results, showing trajectories from a starting point to the hidden target (blue square).

Figure 17.5 presents heat maps summarizing the navigation patterns of all subjects in each group from one of our samples, along with individual trajectories from representative subjects of each category. These exemplify our main findings, which are consistent across different studies with varying samples.

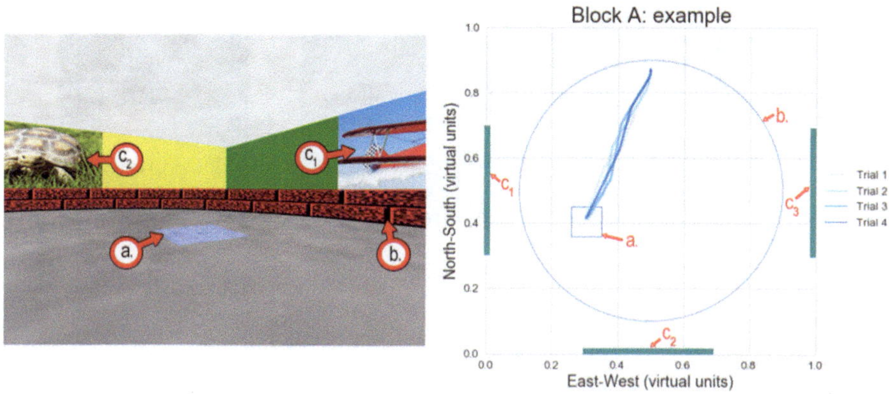

Fig. 17.4 Example of the virtual Morris water maze test. Block A (fixed starting position/visible target). (Left) Screenshot of the virtual environment where the subjects were instructed to navigate toward the blue square on the ground. In the screenshot, the subject was in the northern half of the pool, looking southward and slightly eastward. (Right) Diagram showing the entire virtual environment from above. The four blue lines indicate the path the subject took from his starting point toward the target in each of the four replicate trials. The light blue line shows the first trial, and the dark blue line shows the last trial. In this case, the paths in every trial were almost identical, as the target was visible. (**a**) Square target—visible in this block. (**b**) The boundary of the round pool in which the subjects were allowed to navigate. (**c**) Visual cues on three of the four walls of the virtual environment [13]

When comparing healthy volunteers to patients with vestibular pathology (but not meeting PPPD criteria), the latter showed slightly more difficulty in completing the navigation task. They were able to find the platform similarly to healthy volunteers but with slightly longer trajectories, demonstrating a small increase in the dispersion of frequently visited locations in the virtual pool, although still close to the hidden target. This aligns with reports that vestibular damage can impair navigation systems, consistent with reduced input information.

In stark contrast, subjects with PPPD exhibited significant impairment in spatial navigation, displaying considerable dispersion in the heat map showing the most visited areas of the pool. They had longer, more disorganized, and chaotic trajectories, indicating a lack of systematic or effective search strategies.

Notably, considering the prevalence of vestibular pathologies and peripheral vestibular function values (e.g., video head impulse or vestibular evoked myogenic potentials), the PPPD and non-PPPD vestibular groups are comparable: they exhibit equivalent levels of peripheral vestibular activity. Thus, the significant difference in navigation performance far exceeds what could be attributed solely to diminished vestibular input, suggesting that higher-order cognitive navigation processes are more qualitatively altered (disorganized or chaotic navigation) rather than quantitatively (similar strategies but less efficient).

Figure 17.6 quantitatively describes this behavior with the Cumulative Search Error (CSE) for each group, where a higher value indicates worse navigation. In

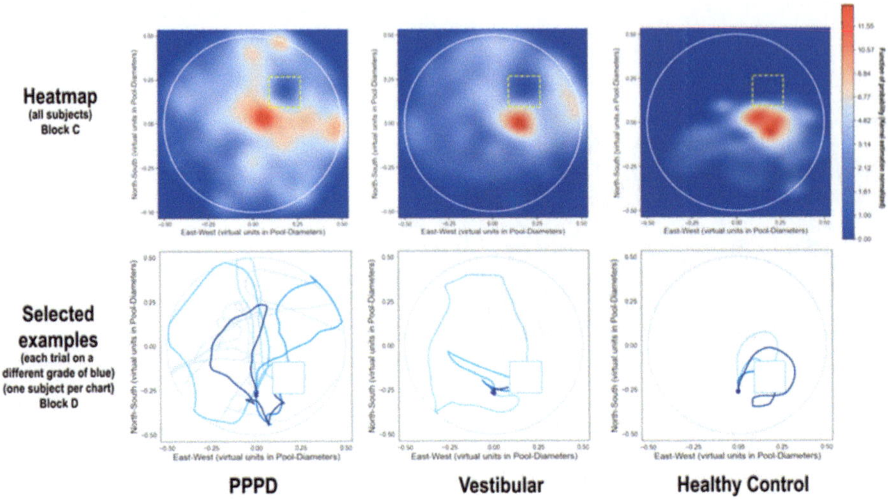

Fig. 17.5 Navigation patterns in the vMWM pool. The upper row showcases density plots derived from kernel estimations, highlighting frequently visited areas during block C navigation, which illustrate each group's overall behavior in relation to the proximity of the hidden target, indicated by yellow dotted squares. Red regions denote the most frequented areas, while blue indicates lesser-visited zones, normalized across 50 probability levels. Healthy controls predominantly navigated near the target, indicated by prevalent deep blue areas, suggesting a low likelihood of finding a control subject elsewhere. Conversely, vestibular patients displayed a broader, yet still target-focused navigation pattern. PPPD patients, while primarily directed toward the target, traversed a more expansive range of areas, with certain regions, particularly near the pool's walls, showing a higher likelihood of PPPD patient presence. The lower row highlights individual navigation paths from selected cases of each group in block D, with varied shades of blue representing each trial's trajectory [14]

various vMWM scenarios, while non-PPPD vestibular patients performed slightly worse than healthy volunteers, PPPD patients performed significantly worse.

vMWM parameters can be adjusted to favor egocentric strategies (starting each trial from the same point, where the target is always "a bit further ahead and to the left relative to myself") or allocentric strategies (starting point and initial gaze orientation are random, so body-relative position is not useful, requiring orientation based on visual cues around the pool). As can be seen on Breinbauer et al. 2023 [14] also shows that in trials favoring allocentric situations, the performance difference between controls and PPPD subjects was greater, suggesting that the main navigation impairment in PPPD involves the brain mechanisms constructing the cognitive inputs needed for allocentric navigation, i.e., allocentric cognitive maps dependent on visual cues.

Figure 17.7 is consistent with this (not yet published in the previously cited reports). These new heat maps do not represent navigation points in Morris Water Maze (MWM) pools but rather fixation points on the computer screen displaying the navigation image. The yellow line represents the pool's horizon level, with visual cues above it. PPPD patients showed greater dispersion of fixation points and a

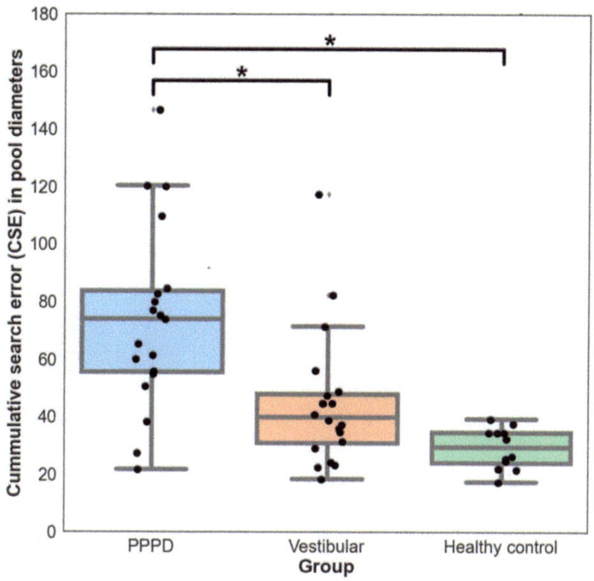

Fig. 17.6 Spatial navigation error per group. Cumulative search error (CSE) describes performance during the virtual Morris water maze (vMWM), thus representing the impairment of spatial navigation skills. When considering all trials of vMWM, the PPPD group showed significantly worse performance than the vestibular and control groups (marked * for *p*-values <0.014). There was no difference between the vestibular and control groups. Black circles represent individual subjects [14]

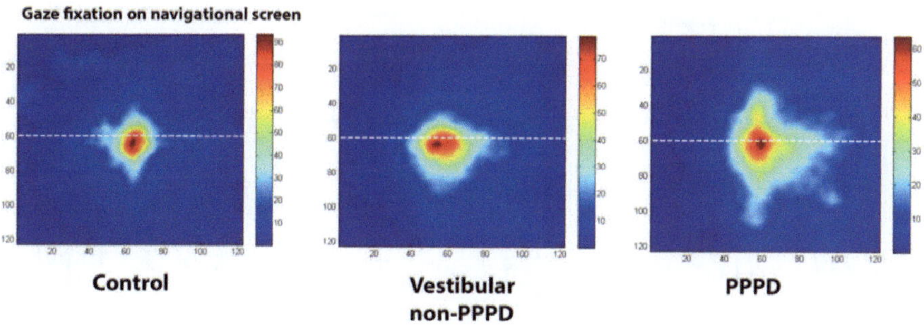

Fig. 17.7 Gaze fixation on the navigational screen. Heat maps/density plots describing locations where gaze was fixed on the computer screen during the virtual Morris water maze task. The yellow line represents the pool's horizon of the pool's rim. Above that line, the visual landmarks can be seen

slightly higher average visual fixation on the screen than the control groups. This may indicate that PPPD subjects employ a more active visual scanning strategy, searching for visual cues to orient themselves in navigation tasks compared to non-PPPD subjects.

17.5.2 Cognitive Testing beyond Spatial Navigation

Figure 17.8 details the performance of each group on cognitive tests not related to spatial navigation. For clarity, certain test scores were inverted so that larger values uniformly indicate more severe impairment. PPPD patients demonstrated greater

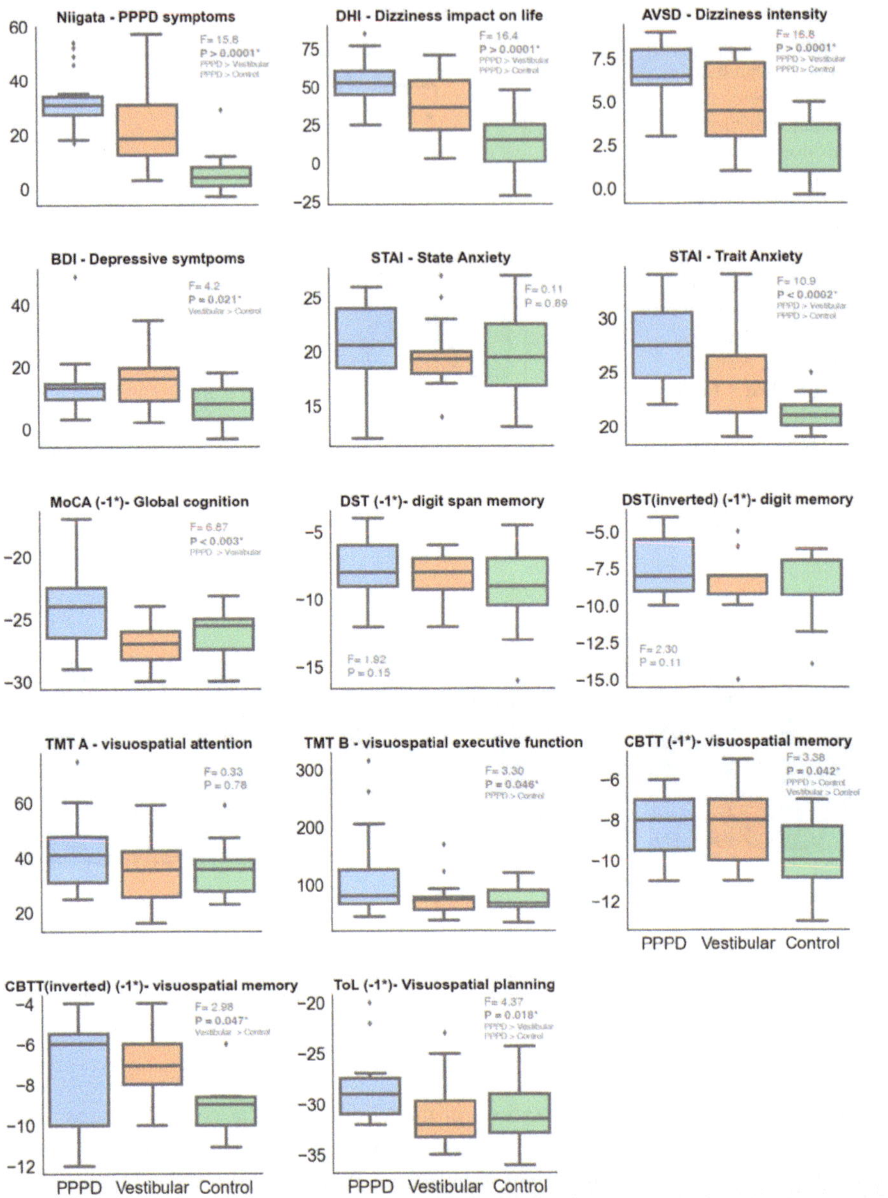

Fig. 17.8 Results from neurocognitive tests. To facilitate interpretation, some test values were inverted (−1*), ensuring that in every subplot, larger values indicate greater impairment or worse performance. Analysis of variance (ANOVA) results are presented for each test. When a significant difference emerged, a post hoc Tukey test was conducted. Specific paired differences are noted when the *p*-value is less than 0.05 [14]

deficits on the Niigata Questionnaire (a recently published validated scale for assessing specifically PPPD symptoms), Dizziness Handicap Inventory (DHI) (dizziness impact), Analogue Visual Scale for Dizziness (AVSD) (dizziness severity), State-Trait Anxiety Inventory (STAI)-Trait (long-term anxiety tendencies), Montreal Cognitive Assessment (MoCA) (global cognition), Trail Making Test B (TMT-B) (executive function in visuospatial planning), Corsi Block-Tapping Task (CBTT) (spatial memory), and Tower of London Test (ToL) (executive function in visuospatial planning) compared to vestibular and healthy controls. There were no observed differences in STAI-State (anxiety at the moment), Digit Span Task (DST) (digit retention/nonspatial memory), or Trail Making Test A (TMT-A) (processing speed on a visuospatial task).

17.5.3 Impact of PPPD on Higher-Demand Cognitive Tests

In higher-demand cognitive tests, PPPD patients' performance was similar to that of their non-PPPD vestibular counterparts in the CBTT, indicating that visuospatial working memory deficits may stem from vestibular loss rather than PPPD alone. However, PPPD patients fared significantly worse on the MoCA (global cognition), TMT-B (executive functions in the visuospatial domain), ToL (visuospatial planning), and spatial navigation measured by CSE during the vMWM. Despite this, all groups showed comparable performance on the DST for nonspatial working memory and the TMT-A for more basic visuospatial skills. These patterns reveal that PPPD's cognitive impact is more pronounced in complex tasks requiring advanced visuospatial processing and executive functions, with the disparity amplifying in tasks with escalating cognitive demands.

17.5.4 Core PPPD Phenomena

A factor analysis of these variables identified a primary factor relating to core PPPD symptoms, significantly comprising the Niigata PPPD score, DHI impact, and AVSD-EVA severity, with these metrics being notably higher in PPPD patients, suggesting a greater perceived disease severity. Another noteworthy point in our data is that PPPD patients demonstrated higher levels of trait anxiety (as measured by STAI-Trait) but not elevated state anxiety (STAI-State) when compared to the control groups. Moreover, trait anxiety emerged in the primary factor previously commented on as a core variable for PPPD, reinforcing the growing view that this trait acts as a predisposing factor for developing PPPD.

While many of the relationships between variables pertain to PPPD severity, impact on life, and known predisposing factors, most of our assessed cognitive variables did not emerge as primary components in this underlying construct. The notable exception is spatial navigation impairment, as gauged by the CSE metric in the vMWM. This cognitive function appears to be intricately linked to the fundamental changes associated with the presence and severity of PPPD in subjects.

17.5.5 Advanced Cognitive Functions

Another main factor emerging from our factor analysis interestingly groups higher-order cognitive functions like TMT-B (visuospatial executive function), ToL (visuospatial planning), and spatial navigation CSE, and is further associated with more complex tasks like the reversed DST (nonspatial memory), reversed CBTT (spatial memory), and MoCA (global cognition), suggesting a trend in PPPD symptomatology. This distinct factor is separate from factor 1's core PPPD symptoms. Although limited by sample size, these findings hint at a broad cognitive impairment affecting both spatial and nonspatial domains in PPPD patients, a hypothesis requiring validation through larger-scale future studies.

17.5.6 Disruption in Higher-Order Cognitive Functions and Its Impact on Cognitive Maps

Our findings suggest that PPPD affects various cognitive functions, particularly in complex tasks involving visuospatial processing, planning, and executive functions, correlating with the spatial navigation deficits at the core of PPPD. This supports models positing that PPPD involves difficulty in creating risk-assessing vestibular spatial-temporal maps. Assessing the integrity or even the construction of these cognitive maps is both conceptually and technically challenging [29, 36, 47]. Nevertheless, we interpret our collective data as supporting the disruption of this map-making process. Key points supporting this concept include:

- Challenges in visuospatial planning and execution in demanding tasks like the TMT-B and ToL, while simpler cognitive functions like memory (assessed by DST and CBTT) remain intact.
- Issues with creating and using allocentric spatial maps, as seen in the vMWM paradigm.
- A link between spatial navigation impairment severity and PPPD, suggesting a correlation with higher cognitive functions, including spatial navigation.

17.5.7 Theoretical Insights from Recent Research

Our understanding of these findings is greatly enhanced by the study conducted by Arshad et al. [20]. Their research, which involved binocular rivalry and vestibular thermal stimulation, introduced an intriguing conceptual model explaining how the brain creates representational maps of the surrounding space-time, based on magnitude estimates that depend on context.

According to their model, functional dizziness involves a complex interaction of factors, including threat assessment, risk tolerance, perceptual bias, and functional changes in space-motion processing and locomotor control. In PPPD patients, there is a reduction in right hemispheric vestibulo-cortical dominance, which affects the cortical processing of brainstem-mediated head velocity signals. This alteration distorts the interhemispheric representations of the magnitudes-space-time continuum, leading to biased dynamic perceptual maps of body position and motion in the physical world. Additionally, an anxious temperament increases threat assessment and decreases risk tolerance. As a result, the usual process of detecting discrepancies between predicted (conscious) and actual (subconscious) movements is disrupted. This disruption elevates unnecessary awareness of these errors to the conscious level, impacting emotional networks. Such changes in brain function explain the primary symptoms of functional dizziness experienced by patients, including a diminished sense of agency over locomotion control.

17.6 Conclusions and Assumptions

The collection of these ideas leads us to consider that the impairment of spatial navigation skills and other cognitive functions supports the notion of PPPD being a broader disorganization of multiple higher-order cortical functions. This dysfunction is not only related to vestibular processing but is also crucial for constructing a reliable internal perceptual map of the external world and the body's position within it. In our proposal, these broad cognitive dysfunctions would affect the reconciliation between predicted perceptual maps (filled with errors) and the reafferences of the actual external world, placing this at the core of PPPD phenomena.

We recognize that the ideas proposed in this chapter are still preliminary and require corroboration in future studies. However, we hope that these lines offer an opportunity to consider concepts and ideas somewhat removed from conventional clinical neuro-otology, such as functional neurological disorders, the predictive brain, active perception, and spatial navigation function. We believe these ideas are not only relevant for understanding the pathophysiology of PPPD but can also lead to the incorporation of notions and tools with a direct influence on the beneficial clinical management of our patients.

References

1. Staab JP. Persistent postural-perceptual dizziness. Neurol Clin. 2023; https://doi.org/10.1016/j.ncl.2023.04.003.
2. Kim HJ, Lee JO, Choi JY, Kim JS. Etiologic distribution of dizziness and vertigo in a referral-based dizziness clinic in South Korea. J Neurol. 2020;267:2252–9. https://doi.org/10.1007/S00415-020-09831-2/METRICS.

3. Staibano P, Lelli D, Tse D. A retrospective analysis of two tertiary care dizziness clinics: a multidisciplinary chronic dizziness clinic and an acute dizziness clinic. J Otolaryngol Head Neck Surg. 2019;48 https://doi.org/10.1186/S40463-019-0336-9.
4. Staab JP, Eckhardt-Henn A, Horii A, et al. Diagnostic criteria for persistent postural-perceptual dizziness (PPPD): consensus document of the committee for the classification of vestibular disorders of the Bárány society. J Vestib Res. 2017;27:191–208. https://doi.org/10.3233/VES-170622.
5. Madrigal J, Olier P, Fonseca M, et al. Quality of life in patients with persistent postural perceptual dizziness: results from a survey in hispanic patients. MOJ Surgery. 2023;11:38–40. https://doi.org/10.15406/MOJS.2023.11.00217.
6. Madrigal J, Herrón-Arango AF, Bedoya MJ, et al. Persistent challenges: a comprehensive review of persistent postural-perceptual dizziness, controversies, and clinical complexities. Cureus. 2024;16 https://doi.org/10.7759/CUREUS.60911.
7. Staab JP. Persistent postural-perceptual dizziness. 2020
8. Dieterich M, Staab JP. Functional dizziness: from phobic postural vertigo and chronic subjective dizziness to persistent postural-perceptual dizziness. Curr Opin Neurol. 2017;30:107–13. https://doi.org/10.1097/WCO.0000000000000417.
9. Dieterich M, Staab JP, Thomas B. Functional (psychogenic) dizziness. Handb Clin Neurol. 2016;139:447–68. https://doi.org/10.1016/B978-0-12-801772-2.00037-0.
10. Cousins S, Kaski D, Cutfield N, et al. Predictors of clinical recovery from vestibular neuritis: a prospective study. Ann Clin Transl Neurol. 2017;4:340–6. https://doi.org/10.1002/acn3.386.
11. Trinidade A, Goebel JA. Persistent postural-perceptual dizziness – a systematic review of the literature for the balance specialist. Otol Neurotol. 2018;39:1291–303. https://doi.org/10.1097/MAO.0000000000002010.
12. Trinidade A, Harman P, Stone J, et al. Assessment of potential risk factors for the development of persistent postural-perceptual dizziness: a case-control pilot study. Front Neurol. 2021;11: 601883. https://doi.org/10.3389/FNEUR.2020.601883/BIBTEX.
13. Breinbauer HA, Contreras MD, Lira JP, et al. Spatial navigation is distinctively impaired in persistent postural perceptual dizziness. Front Neurol. 2020;10 https://doi.org/10.3389/fneur.2019.01361.
14. Breinbauer HA, Arévalo-Romero C, Villarroel K, et al. Functional dizziness as a spatial cognitive dysfunction. Brain Sci. 2023;14:16. https://doi.org/10.3390/BRAINSCI14010016.
15. Yagi C, Kimura A, Horii A. Persistent postural-perceptual dizziness: a functional neuro-otologic disorder. Auris Nasus Larynx. 2024;51:588–98. https://doi.org/10.1016/J.ANL.2023.12.008.
16. Cao Z, Liu X, Ju Y, Zhao X. Neuroimaging studies in persistent postural–perceptual dizziness and related disease: a systematic review. J Neurol. 2022;269:1225–35. https://doi.org/10.1007/s00415-021-10558-x.
17. Passamonti L, Riccelli R, Lacquaniti F, et al. Brain responses to virtual reality visual motion stimulation are affected by neurotic personality traits in patients with persistent postural-perceptual dizziness. J Vestib Res. 2019;28:369–78. https://doi.org/10.3233/VES-190653.
18. Indovina I, Passamonti L, Mucci V, et al. Brain correlates of persistent postural-perceptual dizziness: a review of neuroimaging studies. J Clin Med. 2021;10
19. Castro P, Bancroft MJ, Arshad Q, Kaski D. Persistent postural-perceptual dizziness (PPPD) from brain imaging to behaviour and perception. Brain Sci. 2022;12 https://doi.org/10.3390/BRAINSCI12060753.
20. Arshad Q, Saman Y, Sharif M, et al. Magnitude estimates orchestrate hierarchal construction of context-dependent representational maps for vestibular space and time: theoretical implications for functional dizziness. Front Integr Neurosci. 2022;15
21. Riccelli R, Indovina I, Staab JP, et al. Neuroticism modulates brain visuo-vestibular and anxiety systems during a virtual rollercoaster task. Hum Brain Mapp. 2017; https://doi.org/10.1002/hbm.23411.

22. Nigro S, Indovina I, Riccelli R, et al. Reduced cortical folding in multi-modal vestibular regions in persistent postural perceptual dizziness. Brain Imaging Behav. 2018:1–12. https://doi.org/10.1007/s11682-018-9900-6.
23. Riccelli R, Passamonti L, Toschi N, et al. Altered insular and occipital responses to simulated vertical self-motion in patients with persistent postural-perceptual dizziness. Front Neurol. 2017;8 https://doi.org/10.3389/fneur.2017.00529.
24. Powell G, Derry-Sumner H, Rajenderkumar D, et al. Persistent postural perceptual dizziness is on a spectrum in the general population. Neurology. 2020;94:e1929–38. https://doi.org/10.1212/WNL.0000000000009373.
25. Powell G, Penacchio O, Derry-Sumner H, et al. Visual stress responses to static images are associated with symptoms of persistent postural perceptual dizziness (PPPD). J Vestib Res. 2022;32:69–78. https://doi.org/10.3233/VES-190578.
26. Cousins S, Cutfield NJ, Kaski D, et al. Visual dependency and dizziness after vestibular neuritis. PLoS One. 2014;9 https://doi.org/10.1371/journal.pone.0105426.
27. Yagi C, Morita Y, Kitazawa M, et al. Head roll-tilt subjective visual vertical test in the diagnosis of persistent postural-perceptual dizziness. Otol Neurotol. 2021;42:E1618–24. https://doi.org/10.1097/MAO.0000000000003340.
28. Wuehr M, Pradhan C, Novozhilov S, et al. Inadequate interaction between open- and closed-loop postural control in phobic postural vertigo. J Neurol. 2013;260:1314–23. https://doi.org/10.1007/S00415-012-6797-7/METRICS.
29. Teggi R, Gatti O, Cangiano J, et al. Functional head impulse test with and without optokinetic stimulation in subjects with persistent postural perceptual dizziness (PPPD): preliminary report. Otol Neurotol. 2020;41:E70–5. https://doi.org/10.1097/MAO.0000000000002446.
30. Im JJ, Na S, Jeong H, Chung YA. A review of neuroimaging studies in persistent postural-perceptual dizziness (PPPD). Nucl Med Mol Imaging. 2021;55:53–60.
31. Aybek S, Perez DL. Diagnosis and management of functional neurological disorder. BMJ. 2022;376
32. Finkelstein SA, Popkirov S. Functional neurological disorder: diagnostic pitfalls and differential diagnostic considerations. Neurol Clin; 2023.
33. Jungilligens J, Paredes-Echeverri S, Popkirov S, et al. A new science of emotion: implications for functional neurological disorder. Brain. 2022;145:2648–63.
34. Edwards MJ. Functional neurological disorder: an ethical turning point for neuroscience. Brain. 2019;142:1855–7. https://doi.org/10.1093/BRAIN/AWZ194.
35. Hallett M. Psychogenic movement disorders: a crisis for neurology. Curr Neurol Neurosci Rep. 2006;6:269–71. https://doi.org/10.1007/S11910-006-0015-X.
36. Steinruecke M, Mason I, Keen M, et al. Pain and functional neurological disorder: a systematic review and meta-analysis. J Neurol Neurosurg Psychiatry. 2024:jnnp-2023-332810. https://doi.org/10.1136/JNNP-2023-332810.
37. Cretton A, Brown RJ, Lafrance WC, Aybek S. What does neuroscience tell us about the conversion model of functional neurological disorders? J Neuropsychiatry Clin Neurosci. 2020;32:24–32. https://doi.org/10.1176/APPI.NEUROPSYCH.19040089.
38. Cabreira V, Alty J, Antic S, et al. Perspectives on the diagnosis and management of functional cognitive disorder: an international Delphi study. Eur J Neurol. 2024:e16318. https://doi.org/10.1111/ENE.16318.
39. Stone J, Smyth R, Carson A, et al (2005) Systematic review of misdiagnosis of conversion symptoms and "hysteria." BMJ 331:989–991. https://doi.org/10.1136/bmj.38628.466898.55.
40. Friston K. The free-energy principle: a unified brain theory? Nat Rev Neurosci. 2010;11:127–38. https://doi.org/10.1038/nrn2787.
41. Clark A. Whatever next? Predictive brains, situated agents, and the future of cognitive science. Behav Brain Sci. 2013;36:181–204. https://doi.org/10.1017/S0140525X12000477.
42. Engel AK, Maye A, Kurthen M, König P. Where's the action? The pragmatic turn in cognitive science. Trends Cogn Sci. 2013;17:202–9. https://doi.org/10.1016/j.tics.2013.03.006.

43. O'Regan JK, Noë A. A sensorimotor account of vision and visual consciousness. Behav Brain Sci. 2001;24:939–73. https://doi.org/10.1017/S0140525X01000115.
44. Kremmyda O, Hüfner K, Flanagin VL, et al. Beyond dizziness: virtual navigation, spatial anxiety and hippocampal volume in bilateral Vestibulopathy. Front Hum Neurosci. 2016; https://doi.org/10.3389/fnhum.2016.00139.
45. Hartley T, Lever C, Burgess N, O'Keefe J. Space in the brain: how the hippocampal formation supports spatial cognition. Philos Trans R Soc B Biol Sci. 2014;369
46. King JA, Burgess N, Hartley T, et al. Human hippocampus and viewpoint dependence in spatial memory. Hippocampus. 2002;12:811–20. https://doi.org/10.1002/hipo.10070.
47. Moser EI, Moser M-B, McNaughton BL. Spatial representation in the hippocampal formation: a history. Nat Neurosci. 2017;20:1448–64. https://doi.org/10.1038/nn.4653.
48. Epstein RA, Patai EZ, Julian JB, Spiers HJ. The cognitive map in humans: spatial navigation and beyond. Nat Neurosci; 2017.
49. Ekstrom AD, Arnold AEGF, Iaria G. A critical review of the allocentric spatial representation and its neural underpinnings: toward a network-based perspective. Front Hum Neurosci. 2014;8 https://doi.org/10.3389/fnhum.2014.00803.
50. Aguirre GK, Esposito MD (1999) Topographical disorientation: a synthesis and taxonomy.
51. Bjerknes TL, Dagslott NC, Moser I, Moser M-B Path integration in place cells of developing rats. https://doi.org/10.11582/2018.00003.
52. Krupic J, Bauza M, Burton S, O'Keefe J. Local transformations of the hippocampal cognitive map. Science. 2018;359:1143–6. https://doi.org/10.1126/science.aao4960.
53. Krupic J, Bauza M, Burton S, O'Keefe J. Framing the grid: effect of boundaries on grid cells and navigation. J Physiol. 2016;594:6489–99. https://doi.org/10.1113/JP270607.
54. Moser MB, Rowland DC, Moser EI. Place cells, grid cells, and memory. Cold Spring Harb Perspect Biol. 2015;7 https://doi.org/10.1101/cshperspect.a021808.
55. Miao C, Cao Q, Ito HT, et al. Hippocampal remapping after partial inactivation of the medial entorhinal cortex. Neuron. 2015;88:590–603. https://doi.org/10.1016/j.neuron.2015.09.051.
56. Cornwell BR, Johnson LL, Holroyd T, et al. Human hippocampal and parahippocampal theta during goal-directed spatial navigation predicts performance on a virtual Morris water maze. J Neurosci. 2008;28:5983–90. https://doi.org/10.1523/JNEUROSCI.5001-07.2008.
57. Hitier M, Besnard S, Smith PF. Vestibular pathways involved in cognition. Front Integr Neurosci. 2014;8 https://doi.org/10.3389/fnint.2014.00059.
58. Grady CL. Meta-analytic and functional connectivity evidence from functional magnetic resonance imaging for an anterior to posterior gradient of function along the hippocampal axis. Hippocampus. 2020;30:456–71. https://doi.org/10.1002/hipo.23164.
59. Mao D. Neural correlates of spatial navigation in primate hippocampus. Neurosci Bull. 2023;39:315–27.
60. Theves S, Fernández G, Doeller CF. The hippocampus maps concept space, not feature space. J Neurosci. 2020;40:7318–25. https://doi.org/10.1523/JNEUROSCI.0494-20.2020.
61. Bohbot VD, Copara MS, Gotman J, Ekstrom AD. Low-frequency theta oscillations in the human hippocampus during real-world and virtual navigation. Nat Commun. 2017;8:1–7. https://doi.org/10.1038/ncomms14415.
62. Konishi K, Etchamendy N, Roy S, et al. Decreased functional magnetic resonance imaging activity in the hippocampus in favor of the caudate nucleus in older adults tested in a virtual navigation task. Hippocampus. 2013;23:1005–14. https://doi.org/10.1002/hipo.22181.
63. Goodale MA, Milner AD. Separate visual pathways for perception and action. Trends Neurosci. 1992;15:20–5. https://doi.org/10.1016/0166-2236(92)90344-8.
64. Cullen KE, Wang L. Predictive coding in early vestibular pathways: implications for vestibular cognition. Cogn Neuropsychol. 2020:423–6. https://doi.org/10.1080/02643294.2020.1783222.
65. Nemmi F, Piras F, Péran P, et al. Landmark sequencing and route knowledge: an fMRI study. Cortex. 2013;49:507–19. https://doi.org/10.1016/j.cortex.2011.11.016.
66. Patai EZ, Spiers HJ. The versatile Wayfinder: prefrontal contributions to spatial navigation. Trends Cogn Sci. 2021;25:520–33. https://doi.org/10.1016/J.TICS.2021.02.010.

67. Kaski D, Quadir S, Nigmatullina Y, et al. Temporoparietal encoding of space and time during vestibular-guided orientation. Brain. 2016;139:392–403. https://doi.org/10.1093/brain/awv370.
68. Lucieer FMP, Van Hecke R, van Stiphout L, et al. Bilateral vestibulopathy: beyond imbalance and oscillopsia. J Neurol. 2020;267:241–55. https://doi.org/10.1007/s00415-020-10243-5.
69. Lucieer F, Duijn S, Van Rompaey VV, et al. Full spectrum of reported symptoms of bilateral vestibulopathy needs further investigation-a systematic review. Front Neurol. 2018;9
70. Dobbels B, Mertens G, Gilles A, et al. The virtual Morris water task in 64 patients with bilateral vestibulopathy and the impact of hearing status. Front Neurol. 2020;11:1–12. https://doi.org/10.3389/fneur.2020.00710.
71. Popp P, Wulff M, Finke K, et al. Cognitive deficits in patients with a chronic vestibular failure. J Neurol. 2017;264:554–63. https://doi.org/10.1007/s00415-016-8386-7.
72. Harris LR. Does the vestibular system exert specific or general influences on cognitive processes? Cogn Neuropsychol. 2020;37:430–2. https://doi.org/10.1080/02643294.2020.1785412.
73. Rizk HG, Sharon JD, Lee JA, et al. Cross-sectional analysis of cognitive dysfunction in patients with vestibular disorders. Ear Hear. 2020:1020–7. https://doi.org/10.1097/AUD.0000000000000825.
74. Xia R, Ren J, Li X, et al. Alterations in corticocortical vestibular network functional connectivity are associated with decreased balance ability in elderly individuals with mild cognitive impairment. Brain Sci. 2023;13 https://doi.org/10.3390/brainsci13010063.
75. Agrawal Y. Vestibular cognition: building a framework. Cogn Neuropsychol. 2020;37:421–2. https://doi.org/10.1080/02643294.2020.1782362.
76. Cullen KE. Physiology of central pathways. 1st ed. Elsevier; 2016.
77. Ferrè ER, Haggard P. Vestibular cognition: state-of-the-art and future directions. Cogn Neuropsychol. 2020;37:413–20. https://doi.org/10.1080/02643294.2020.1736018.
78. Zachou A, Bronstein AM. Vestibulo-spatial navigation: pathways and sense of direction. J Neurophysiol. 2023;129:672–84. https://doi.org/10.1152/jn.00422.2022.
79. Do TTN, Lin CT, Gramann K. Human brain dynamics in active spatial navigation. Sci Rep. 2021;11 https://doi.org/10.1038/s41598-021-92246-4.
80. Baumann O, Mattingley JB. Extrahippocampal contributions to spatial navigation in humans: a review of the neuroimaging evidence. Hippocampus. 2021;31:640–57. https://doi.org/10.1002/hipo.23313.
81. Sharma G, Gramann K, Chandra S, et al. Brain connectivity during encoding and retrieval of spatial information: individual differences in navigation skills. Brain Inform. 2017;4:207–17. https://doi.org/10.1007/s40708-017-0066-6.
82. Mast FW, Ellis AW. Internal models, vestibular cognition, and mental imagery: conceptual considerations. Multisens Res. 2015;28:443–60. https://doi.org/10.1163/22134808-00002503.
83. Smith PF. The vestibular system and cognition. Curr Opin Neurol. 2017;30:84–9. https://doi.org/10.1097/WCO.0000000000000403.
84. Huang Y, Zhang X, Tang J, et al. Vestibular cognition assessment system: tablet-based computerized visuospatial abilities test battery. Front Psychol. 2023;14 https://doi.org/10.3389/fpsyg.2023.1095777.
85. Morris R. Developments of a water-maze procedure for studying spatial learning in the rat. J Neurosci Methods. 1984;11:47–60. https://doi.org/10.1016/0165-0270(84)90007-4.
86. Schoenfeld R, Schiffelholz T, Beyer C, et al. Variants of the Morris water maze task to comparatively assess human and rodent place navigation. Neurobiol Learn Mem. 2017;139:117–27. https://doi.org/10.1016/j.nlm.2016.12.022.
87. Wood H. A virtual Morris maze to assess cognitive impairment in Alzheimer disease. Nat Rev Neurol. 2016;12:126. https://doi.org/10.1038/nrneurol.2016.16.

Chapter 18
Causes of Acute-Onset Balance Disorders in Pediatric Patients

Patricia Alejandra Sommerfleck and Gabriela Bettina Grinstein

Keywords Balance disorder · Vertigo · Children · Dizziness · Pediatrics

The consultation of a child with acute-onset symptoms of vertigo, dizziness or imbalance is not common in the emergency department; however, when it does occur, it can present significant challenges in defining a diagnosis and determining a course of action.

The most common disorders are episodic in nature [1], and it is the recurrence of these episodes that captures the parents' attention and leads them to seek specialist advice.

The vestibular system, especially the neural pathway, is not fully developed until early adolescence. In addition, pediatric patients have a significant capacity for adaptation and compensation, due to their greater neural plasticity [2]. Therefore, when vertigo or dizziness appear suddenly and last for hours, immediate medical attention is sought, regardless of the age of the patient.

Consultations for these conditions involve several specialties, mainly general pediatricians, neurologists, otolaryngologists, ophthalmologists, cardiologists, and psychologists. The primary purpose is to detect the condition and establish an accurate diagnosis, and it is important to consider the appropriate complementary studies due to the associated costs and time involved, which could potentially lead to diagnostic delays.

While vertigo and dizziness have a higher incidence in the adult population, various conditions are also prevalent among children and often go undiagnosed. The differential diagnosis of dizziness in children differs from that of adults due to several etiologies that are specific to the pediatric population [3]. Most studies

P. A. Sommerfleck (✉)
Otorhinolaryngology Department, Otoneurology Area, Hospital de Pediatría "Prof. Dr. Juan P. Garrahan", Ciudad Autónoma de Buenos Aires, CABA, Argentina

G. B. Grinstein
Otoneurology Department, Instituto Otorrinolaringólogico Arauz, Ciudad Autónoma de Buenos Aires, CABA, Argentina

P. Ravat et al. (eds.), *Advances in Translational Neuroscience of Vestibular and Eye Movement Disorders*, Contemporary Clinical Neuroscience,
https://doi.org/10.1007/978-3-032-07539-0_18

highlight the challenge of diagnosing this disorder in pediatric patients, with the main reason being their difficulty in expressing it, which is the main cause of underdiagnosis. It is difficult for children, especially those under 4 years of age, to put this sensation into words. Therefore, it is crucial to learn how to inquire effectively, to adapt our language to their mode of expression, and to connect it to everyday situations they are familiar with so that they can describe their "sensation" more easily [4].

There are several reasons why a child may experience dizziness. In this chapter, we focus on the acute causes of vertigo in children.

Vertigo is defined as the subjective sensation of rotational movement, either of the patient themselves or of the surrounding environment. Children often describe it as feeling like the floor or room is spinning, experiencing a sensation of self-rotation, or associating it with being on a merry-go-round or some other spinning element.

On the other hand, the term dizziness represents a general feeling of discomfort and imbalance that is more challenging for the patient to describe. Occasionally, it may be mentioned as a "heavy head feeling," walking on clouds, or feeling like an external force is pushing them. It is a non-specific symptom and may be associated with systemic diseases.

Imbalance is actually a gait disorder that is often multisensory due to involvement of the visual, motor, sensory, or cerebellar pathways. This discomfort disappears when sitting or lying down.

Ataxia is an alteration of posture and control of voluntary movements, secondary to involvement of the cerebellum and cerebellar pathways. The most prominent characteristic of ataxic gait is a broad base of support and motor incoordination, giving the patient a sensation of imminent falling.

These symptoms were assessed through an interview conducted by the healthcare provider who performed the anamnesis.

When considering the causes of acute consultations, differentiation was made based on various forms of presentation [5]:

Acute vestibular syndrome (AVS): Clinical manifestations are characterized by continuous, sudden-onset vertigo, typically lasting hours to days. AVS is often associated with nausea, vomiting, gait instability, nystagmus, and intolerance to head movements. These patients are usually symptomatic at the time of consultation. This presentation suggests a recent and ongoing disorder of the vestibular system. While vestibular neuritis and labyrinthitis are the most common causes of AVS, in the pediatric population, it is important to consider a differential diagnosis, including conditions such as acute postinfectious cerebellitis and complications related to neighboring structures. It is also crucial to determine if AVS appearance followed traumatic brain injury.

Episodic vestibular syndrome (EVS): EVS manifests as a clinical syndrome characterized by transient episodes of vertigo, dizziness, or imbalance, lasting from seconds to hours, often accompanied by features indicative of a temporary disruption of the vestibular system. Additionally, there may be symptoms or signs that suggest a cochlear or central nervous system disorder. EVS typically involves multiple recurrent events, which may be provoked or spontaneous, and may initially occur after the first episode of symptoms. For example: Benign Paroxysmal vertigo in childhood, vestibular migraine, hydrops, and others.

Chronic vestibular syndrome (CVS): Clinically, CVS is characterized as a vertiginous syndrome with a gradual and continuous onset, persisting for weeks, months, or even years. The symptoms may exacerbate or remain constant and overlapping. These symptoms indicate an enduring disorder of the vestibular system.

Approximately 60,000 consultations were conducted at the Department of Ear, Nose, and Throat (ENT) in our high-complexity pediatric hospital during the study period, which was for 7 years (between March 2015 and December 2022). Of these, approximately 2500 consultations were related to balance disorders. Consequently, the incidence of balance disorders accounted for approximately 4.1% of the total consultations in the area of vestibular medicine.

Finally, 720 patients were included to analyze the most frequent causes of AVS: 366 females (50.83%) and 354 males (49.17%), who were followed up and diagnosed.

According to the age classification of the WHO and the American Academy of Pediatrics [6], the patients were distributed as follows: Babies and toddlers: (0–2 years) 54 patients (7.5%); preschoolers: (3–5 years) 118 patients (16.39%); grandschoolers: (6–11 years) 320 patients (44.44%); and teens: (12–18 years) 228 patients (31.67%).

The disorders observed were varied. Figure 18.1 shows the general distribution of the balance disorders.

Based on symptom presentation, these disorders were categorized into AVS, EVS, and CVS, as described above. The most frequent form of presentation were

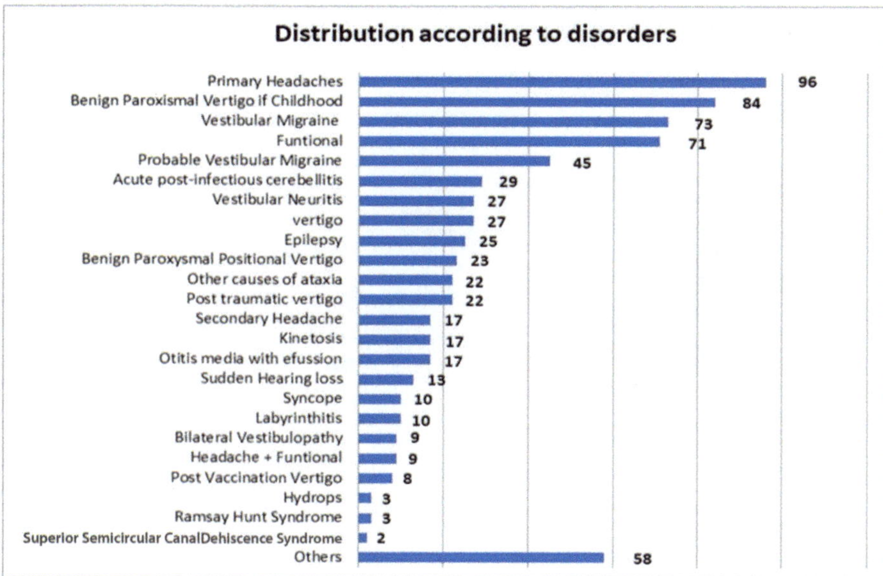

Fig. 18.1 Different diagnoses in this series

episodic disorders (73%—527 patients) it is follow by AVS (18%—129 patients) and CVS (9%—64 patients). In this chapter, the primary focus was on acute-onset disorders, as they demand greater diagnostic certainty and prompt indication of the appropriate treatment.

In this series 129/720 patients (18%) presented with AVS, of whom 61 were girls and 68 were boys. Age distribution was 17/129 patients between 0 and 2 years old; 21/129 patients between 3 and 5 years old; 52/129 patients between 6 and 11 years old; and 39/129 patients between 12 and 17 years old. The Figs. 18.2, 18.3, 18.4, and

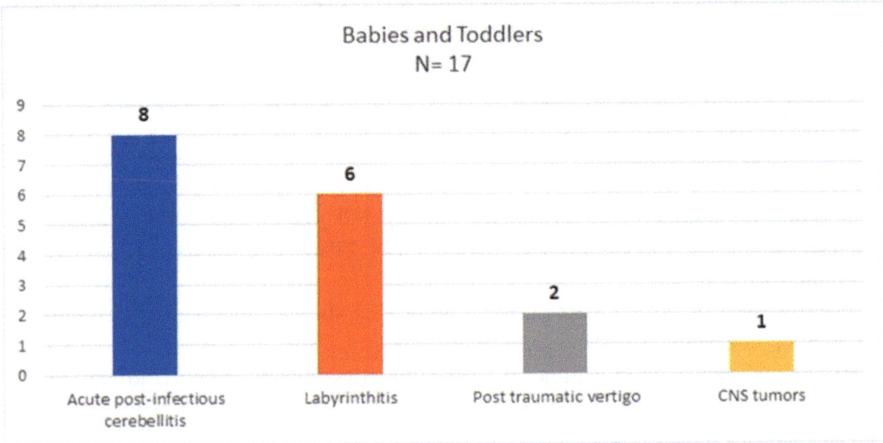

Fig. 18.2 Babies and toddlers: acute presentation in patients aged 0–2 years

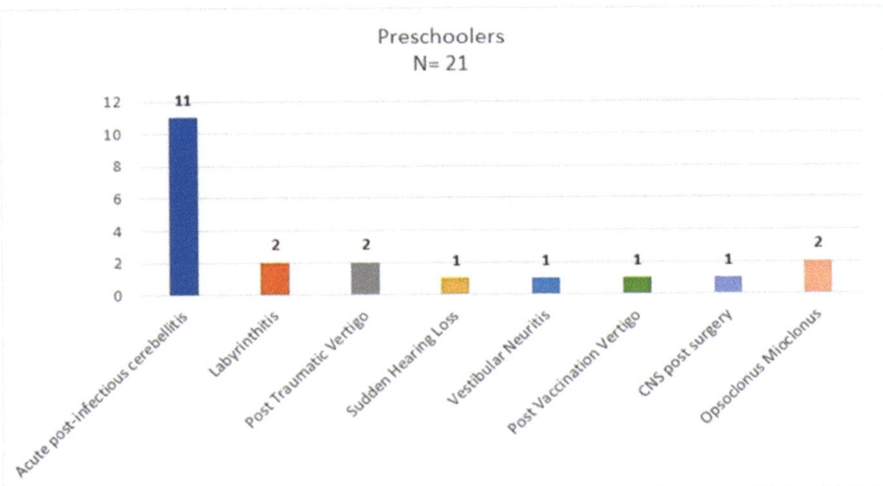

Fig. 18.3 Preschoolers: acute presentation in patients aged 3–5 years

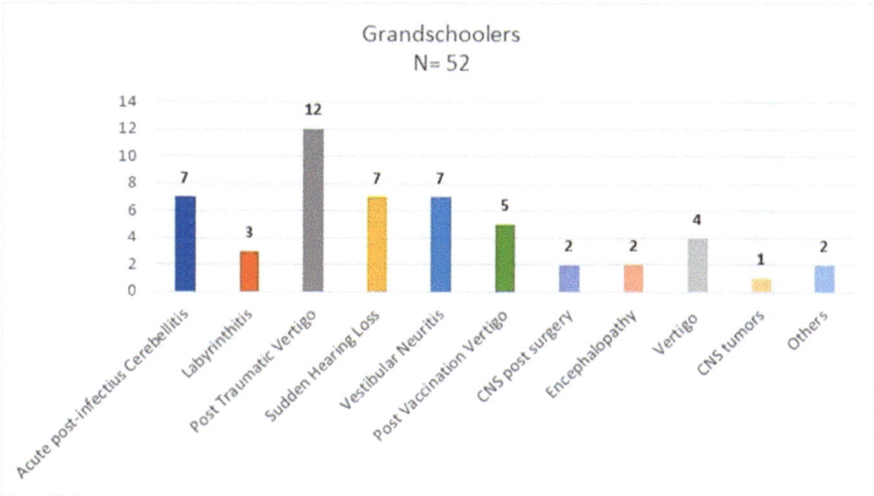

Fig. 18.4 Grandschoolers: acute presentation in patients aged 6–11 years

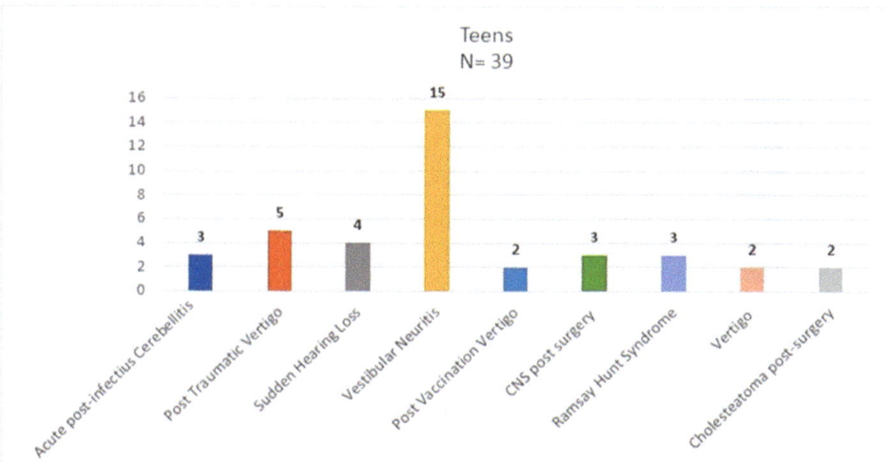

Fig. 18.5 Teens: acute presentation in patients aged 12–17 years

18.5 illustrate the distribution by age range of the disorders with an acute presentation.

Regarding peripheral causes of AVS, the most frequent was vestibular neuritis (23 patients aged 5–14 years—10 males, 13 females), sudden hearing loss (12 patients aged 5–16 years; three of which occurred postCOVID), labyrinthitis as a complication of acute otitis media (AOM) (11 patients aged 1–10 years).

Patients with acute vestibular symptoms due to SARS-CoV-2 virus infection were identified during the pandemic period. They present: four patients vestibular

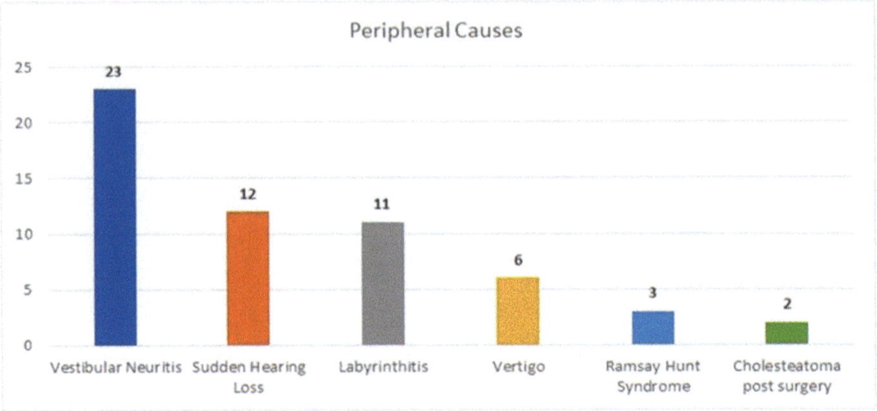

Fig. 18.6 Peripheral causes of acute vestibular disorders

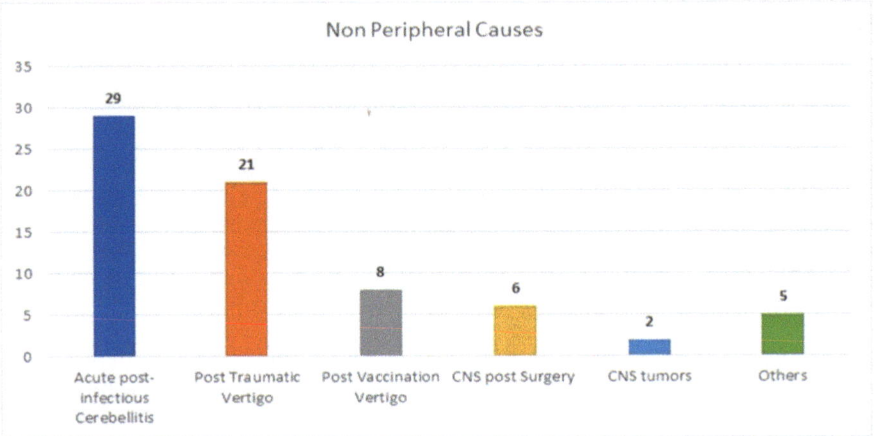

Fig. 18.7 Nonperipheral causes of acute vestibular disorders

neuritis; four patients postcovid with vertigo; three patients with sudden hearing loss; and one patient with acute post-viral cerebellitis.

Figures 18.6 and 18.7 show AVS separated into peripheral and nonperipheral causes.

In this chapter, we present a large series of pediatric patients with balance disorders. The causes and prevalence of acute disorders are highlighted, but it remains important to provide an overview of the typical presentation of a child with dizziness and to consider the main differential diagnoses.

Acute causes are more frequent in patients younger than 5 years of age, and mainly include labyrinthitis as a complication of AOM [7, 8] and acute postinfectious cerebellitis. Benign paroxysmal vertigo of childhood is the primary differential diagnosis in this age group, but its presentation differs; therefore, the interview is a fundamental aspect of the diagnostic process [9].

Labyrinthitis can arise as a complication of AOM. This condition often goes unnoticed in nonambulatory patients, but it should be considered, as more than two-thirds of patients under 18 months old experience an episode of AOM. Typically, this complication resolves with appropriate antibiotic treatment for the otitis. It is important to note that if labyrinthitis is suspected, a subsequent audiological evaluation of the patient should be conducted since it may lead to hearing loss.

Ataxia is a symptom commonly evaluated by child neurologists. It is important to differentiate between ataxias of acute, episodic, or chronic presentation, with the acute forms being the most prevalent.

Acute postinfectious cerebellar ataxia (or acute postinfectious cerebellitis) accounts for 30–50% of cases. It is most commonly associated with viruses such as the varicella zoster virus, Epstein-Barr virus, herpes simplex virus, hepatitis A, legionella, influenza, enterovirus, and others. Ataxia may develop up to 3 weeks after the viral infection, and resolution is usually spontaneous within 2–6 weeks without requiring interventions. Magnetic resonance imaging is usually normal [10].

Balance disturbances that occur after trauma typically present acutely. We have knowledge of the trauma history, and in many cases, they improve spontaneously, or in more severe instances, with the use of symptomatic medication. The prevalence is similar across all age groups, without statistically significant differences. In these cases, we refer to mild trauma where the most appropriate term would be: labyrinthine concussion or posttraumatic labyrinthine failure. This situation results in a temporary disruption of the cochleovestibular receptors [11]. Only one patient in our series experienced severe trauma, resulting in a temporal bone fracture and subsequent hearing loss.

In our experience, acute balance disorders were observed following vaccination. The vaccines involved were those indicated during the SARS-CoV-2 pandemic period. In Argentina, the children received Sinopharm®, but other vaccines included in the mandatory vaccination schedule were also involved, including the Measles, Mumps and Rubeola (MMR) and Human Papillomavirus (HPV) vaccines.

Vestibular neuritis is an acute inflammatory process that primarily affects the superior vestibular nerve (although it can also affect the inferior branch or both branches) unilaterally [12]. It often has a viral origin and typically presents with an acute and prominent onset of vestibular symptoms. This condition predominantly affects young adults and children over 10 years of age. During the SARS-CoV-2 pandemic, there was an observed increase in cases of the disorder following infection with this etiological agent [13–15]. It is frequently associated with other viruses, such as herpes simplex and influenza, among others.

Another condition associated with coronavirus was sudden hearing loss or sudden cochleo-vestibular syndrome. While this latter syndrome is not common in pediatrics, it remains a possible diagnosis. In patients with intense vestibular symptoms (vertigo, nausea, vomiting, instability), hearing loss is usually reported at a later stage once the vestibular symptoms have been controlled. It is important to note that in vestibular neuritis, there is no association with hearing loss.

Episodic disorders are the most common across all age ranges, including conditions such as benign paroxysmal vertigo in childhood, vertigo-associated migraine, vestibular migraine, and disorders with functional or emotional causes [16].

Regarding the disorders with central causes, it is important to differentiate them from epilepsies, which can also occur in all age groups, and from ataxias. As mentioned above, ataxias can have acute, episodic, or chronic presentations, making a differential diagnosis necessary.

Among the causes of chronic vestibular syndrome, it is important to highlight otitis media with effusion, a highly prevalent condition in pediatrics that should either be ruled out or treated.

We emphasize the importance of a thorough interview, including the patient's medical history as well as the physical examination, as we may face a patient with a first episode of an episodic or chronic disorder. Consequently, we stress the importance of continuous monitoring and follow-up of the patient.

In accordance with the classification of the Bárány Society, we show in this chapter the potential diagnoses associated with acute presentations in children. Yilmaz et al. [17] define acute vertigo as a symptom in a patient who presents to the emergency department. The causes of AVS in the pediatric population vary based on age group. In young children, it is important to establish an effective communication channel for accurate symptom interpretation. Moreover, an interdisciplinary approach, including appropriate consultations and complementary studies, is essential to reach a correct diagnosis.

References

1. Davitt M, Delvecchio MT, Aronoff SC. The differential diagnosis of vertigo in children: a systematic review of 2726 cases. Pediatr Emerg Care. 2020;36(8):368–71. https://doi.org/10.1097/PEC.0000000000001281.
2. Cumberworth VL, Patel NN, Rogers W, Kenyon GS. Thematurationof balance in children. J Laryngol Otol. 2007;121(5):449–54. https://doi.org/10.1017/S0022215106004051. Epub 2006 Nov 14.
3. Gruber M, Cohen-Kerem R, Kaminer M, Shupak A. Vertigo in children and adolescents: characteristics and outcome. Sci World J. 2012;2012:109624. https://doi.org/10.1100/2012/109624. Epub 2012 Jan 3. PMID: 22272166; PMCID: PMC3259473.
4. Manzari L, Burgess AM, Macdougall HG, Curthoys IS. Objectivemeasuresof vestibular functionduringan acute vertigoattack in a veryyoungchild. Eur Arch Otorrinolaringol. 2012;269(12):2589–92. https://doi.org/10.1007/s00405-012-2045-0. Epub 2012 May 13.
5. Bisdorff AR, Staab JP, Newman-Toker DE. Overviewofthe international classification of vestibular disorders. Neurol Clin. 2015;33(3):541–50, vii. https://doi.org/10.1016/j.ncl.2015.04.010.
6. https://www.healthychildren.org/english/ages-stages/pages/default.aspx
7. Sommerfleck PA, Bernáldez PC, Hernández CM, Reijtman VR, Lopardo HA. Otitis media aguda: prevalencia de otopatógenos en pacientes de un hospital público [Acute otitis media: prevalenceofearpathogens in patients at a public hospital]. Acta Otorrinolaringol Esp. 2013;64(1):12–6. Spanish. https://doi.org/10.1016/j.otorri.2012.06.005. Epub 2012 Aug 2.

8. Pellegrini S, Gonzalez Macchi ME, Sommerfleck PA, Bernáldez PC. Intratemporalcomplicationsfrom acute otitis media in children: 17 cases in twoyears. Acta Otorrinolaringol Esp. 2012;63(1):21–5. English, Spanish. https://doi.org/10.1016/j.otorri.2011.06.007. Epub 2011 Oct 5.
9. Sommerfleck PS, Pastore CS, Cardozo MS, Quantin LG. Vértigo Paroxístico Benigno de la Infancia y sus diagnósticos diferenciales. Revista de la Federación Argentina de Sociedades de Otorrinolaringología Año. 2021;28(2):23–8.
10. Valle G, Garcia Peñas JJ, Cerisola A. Ataxia aguda. Recomendaciones de la Academia Iberoamericana de Neurología Pediátrica. Capítulo 5. Arroyo H, González G, Campistol J, Editores. Editorial Fundación Garrahan Vol 2. abril 2019. Argentina: CABA.
11. Yang CC, Hua MS, Tu YK, Huang SJ. Early clinical characteristics of patients with persistent post-concussion symptoms: a prospective study. Brain Inj. 2009;23(4):299–306. https://doi.org/10.1080/02699050902788543.
12. Brodsky JR, Cusick BA, Zhou G. Vestibular neuritis in children and adolescents: clinical features and recovery. Int J Pediatr Otorhinolaryngol. 2016;83:104–8. https://doi.org/10.1016/j.ijporl.2016.01.027. Epub 2016 Feb 1.
13. Demir I, Aydin S, Cengiz DU. The short-termeffectof COVID-19 onthecochleovestibularsystem in pediatricpatients. Int J Pediatr Otorhinolaryngol. 2023;164:111406. https://doi.org/10.1016/j.ijporl.2022.111406. Epub 2022 Dec 10. PMID: 36521194; PMCID: PMC9737457.
14. Tannous D, Klepper K. Pediatric COVID-19 vestibular neuritis. Perm J. 2022;26(2):162–5. https://doi.org/10.7812/TPP/21.221. Epub 2022 Jun 20. PMID: 35933661; PMCID: PMC9662244.
15. Giannantonio S, Scorpecci A, Montemurri B, Marsella P. Case of COVID-19-induced vestibular neuritis in a child. BMJ Case Rep. 2021;14(6):e242978. https://doi.org/10.1136/bcr-2021-242978. PMID: 34078625; PMCID: PMC8173285.
16. Sommerfleck PA, González Macchi ME, Weinschelbaum R, De Bagge MD, Bernáldez P, Carmona S. Balance disorders in childhood: mainetiologies according to age. Usefulnessofthe video head impulse test. Int J Pediatr Otorhinolaryngol. 2016;87:148–53. https://doi.org/10.1016/j.ijporl.2016.06.020. Epub 2016 June 7.
17. Yılmaz A, Abseyi SN. Clinical and demographiccharacteristicsofchildren and adolescentswith acute vertigosymptoms: a cross-sectionalstudy. Turk J Med Sci. 2020;50(8):1951–4. https://doi.org/10.3906/sag-2010-228. PMID: 33306337; PMCID: PMC7775689.

Chapter 19
CACNA1A Channelopathies and Associated Eye Movement Disorders

Abhilash Thatikala and Kapil Arya

Keywords Global developmental delay · Channelopathies · Developmental and Epileptic Encephalopathy · P/Q channel · Ataxia

19.1 Introduction

Calcium channels are integral membrane proteins that facilitate the influx of calcium ions (Ca^{2+}) into cells, playing critical roles in various physiological processes, including muscle contraction, neurotransmitter release, and gene expression. They are classified into two main categories based on their activation properties: high voltage-activated (HVA) and low voltage-activated (LVA) channels [1]. P/Q type channel (CaV2.1) belongs to a high-voltage-activated channel.

19.1.1 Calcium Channel Alpha 1A Gene (CACNA1A)

The CACNA1A gene is located on chromosome 19. The CACNA1A gene is bicistronic [2, 3], meaning it encodes two distinct proteins: the calcium channel subunit (α 1A) and a transcription factor known as α 1ACT. The latter is produced through an internal ribosomal entry site (IRES) within the mRNA, allowing for the coordinated expression of both proteins. An IRES is a sequence that allows ribosomes to initiate translation independently of the traditional 5′ cap-dependent mechanism. This means that the ribosome can bind to the IRES and start translating the downstream ORF (in this case, the a1ACT protein) without needing the usual

A. Thatikala
Department of Neurology, University of Arkansas for Medical Sciences, Little Rock, AR, USA
e-mail: athatikala@uams.edu

K. Arya (✉)
Department of Pediatrics, University of Arkansas for Medical Sciences, Little Rock, AR, USA
e-mail: KArya@uams.edu

P. Ravat et al. (eds.), *Advances in Translational Neuroscience of Vestibular and Eye Movement Disorders*, Contemporary Clinical Neuroscience,
https://doi.org/10.1007/978-3-032-07539-0_19

initiation signals. The presence of the IRES allows for the simultaneous expression of both proteins from the same mRNA transcript. This coordinated expression is particularly important in contexts where both proteins need to be produced together to fulfill specific cellular functions, such as in neuronal development and signaling. The a1A subunit is crucial for forming the calcium channel, while a1ACT plays a role in regulating gene expression related to neuronal health and function [4].

19.1.2 Structure of P/Q Calcium Channel (Cav2.1) [5]

(i) **Subunit Composition**:

- The P/Q-type calcium channel is primarily composed of the **α1A subunit**, which is responsible for forming the channel's pore, as shown in Fig. 19.1, and is crucial for its voltage-gated properties. The α1A subunit consists of:
- **N-Terminal Region**: Includes sites for regulatory mutations affecting calcium binding.
- **Four Domains (I–IV)**: Each domain contains six transmembrane segments (S1-S6), with the S4 segment acting as the voltage sensor and the pore loop between S5 and S6 forming the calcium-selective pore.
- **C-Terminal Region**: Plays a role in protein-protein interactions and intracellular signaling.

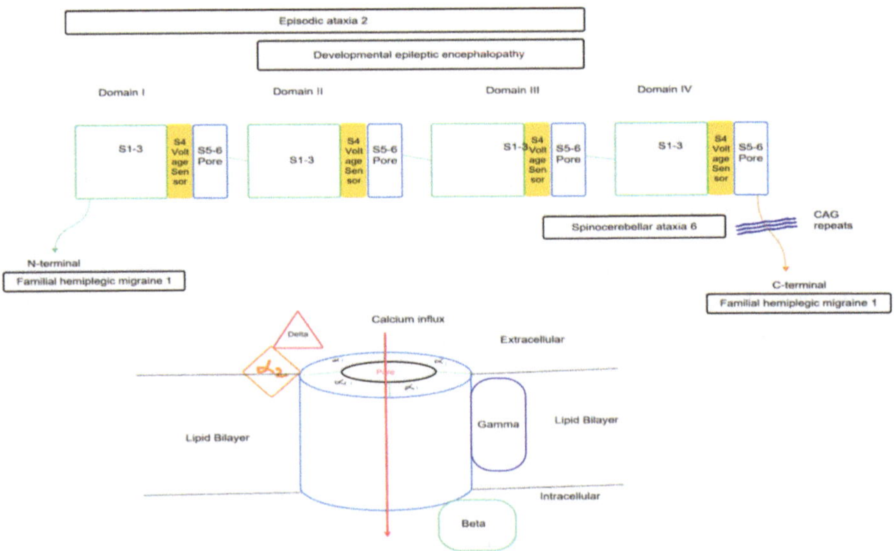

Fig. 19.1 Structure of P/Q channel and various locations of CACNA1A mutations and manifestations

- The channel also associates with auxiliary subunits, including **β**, **α2δ**, and **γ** subunits, which modulate the channel's biophysical properties, surface expression, and pharmacological characteristics.

(ii) **Pore Structure**:

- The pore of the P/Q-type channel, formed by the re-entrant loop between the S5 and S6 segments of each domain, allows for selective permeability to calcium ions (Ca^{2+}) over other ions. The selectivity filter of the channel is primarily determined by a critical glutamic acid residue located in the S6 segments, which contributes to the channel's ability to discriminate between divalent and monovalent cations.

19.1.3 Location of P/Q Calcium Channel [6, 7]

- P/Q-type calcium channels are predominantly expressed in the Central Nervous System (CNS), where they are crucial for neurotransmitter release and synaptic transmission. They are particularly abundant in **Purkinje cells** of the cerebellum and is considered one of the highest density locations for P/Q-type channels. The activation of these channels in Purkinje cells is essential for the release of neurotransmitters that modulate cerebellar output and influence motor control.
- Other regions with significant expression include:
  - **Cerebellar Granule Cells**: These cells also express P/Q-type channels, contributing to cerebellar function.
  - **Hippocampus**: P/Q channels are involved in synaptic plasticity and memory formation.
  - **Brainstem and Spinal Cord**: They are found in various neuronal populations involved in sensory processing and motor control.
  - **Peripheral Nervous System**: P/Q-type are seen in sensory neurons that transmit pain and touch signals.

19.1.4 Function of P/Q Calcium Channel

- P/Q-type channels are activated by membrane depolarization, leading to an influx of calcium ions into the cell. This calcium influx is crucial for various physiological processes, including neurotransmitter release at synapses.
- In Purkinje cells, the activation of P/Q channels triggers the release of neurotransmitters, facilitating synaptic transmission and influencing motor coordination.

19.1.5 Different Types of CACNA1A Mutations and their Effects [6, 8]

(i) **Gain-of-Function (GOF) Variants**: These variants lead to increased calcium ion influx through the channel, enhancing neuronal excitability.

- **Familial Hemiplegic Migraine Type 1 (FHM1)**: Characterized by migraine attacks with aura and associated motor weakness. Mutations are often located in the **N-terminal region** or the **C-terminal tail**, affecting calcium channel gating [9, 10].
- **Developmental and Epileptic Encephalopathy42 (DEE42)**: Severe form of epilepsy with developmental delays. Mutations often occur in **Domains II and III**, disrupting calcium channel function critical for neuronal development.

(ii) **Loss-of-Function (LOF) Variants**: These variants result in reduced or absent channel function, leading to decreased calcium influx.

- **Spinocerebellar Ataxia Type 6 (SCA6):** Progressive cerebellar ataxia associated with **polyglutamine expansions** in the **C-terminal region**.
- **Episodic Ataxia Type 2 (EA2)**: Intermittent attacks of ataxia. Some cases may also involve GOF variants. Mutations are predominantly found in **Domains I-III** [11], impairing calcium influx and neuronal excitability.
- Epilepsy.
- Paroxysmal tonic upzage of infancy (PTU).

(iii) **Phenotypic Variability**: The same genotype can lead to different phenotypes within families, indicating the influence of other genetic or environmental factors.

(iv) **Specific Genotypes Identified.**

(a) **Missense Variants**:

(i) **p.R1352Q**: Associated with severe phenotypes, including hemiplegic migraine and epilepsy [6, 7],
(ii) **p.A713T**: Linked to episodic ataxia type 2 [7],
(iii) **p.G701V**: Linked to EA-2 and hemiplegic migraine [7],
(iv) **p.Y62C**: Associated with severe hemiconvulsion-hemiplegia-epilepsy syndrome [7],

(b) **Nonsense Variants**: These variants lead to premature stop codons, resulting in truncated proteins that typically exhibit LOF. **p.W320***, **p.Y1849***, and **p.R1857*** are known to cause loss of function [7, 12].

(c) **Frameshift Variants**: These variants result from insertions or deletions that shift the reading frame, often leading to LOF.

(i) **p.Q681Rfs*100:** A frameshift variant found in a patient with episodic ataxia type 2 [7],

(ii) **p.S1752Cfs*2:** Is associated with severe phenotypes [7, 12]

(d) **Compound Heterozygous Variants**: Some patients may carry two different mutations in the CACNA1A gene, which can lead to more severe phenotypes. Patients with combinations of missense and truncating mutations have been reported [13].

(e) **CAG Repeat Expansion Disorders:** Spinocerebellar Ataxia Type 6 (SCA6) is caused by an expanded cytosine-adenine-guanine (CAG) repeat within the CACNA1A gene, leading to a polyglutamine tract in the C-terminal domain of the protein [14].

(f) **Splice Site Mutations:** Mutations affecting splice sites can lead to abnormal splicing of the mRNA, resulting in truncated or dysfunctional proteins. These mutations can contribute to the phenotypic variability observed in CACNA1A-related disorders [7].

(g) **Large Deletions and Duplications:** A deletion of the entire exon 35 was identified in a patient, leading to a frameshift mutation [7, 12].

19.2 Disorders Associated with CACNA1A-Related Mutations

The CACNA1A gene is associated with a wide range of neurological disorders, reflecting its critical role in calcium signaling within the nervous system, as shown in Fig. 19.2.

1. Episodic Ataxia Type 2 (EA2): Characterized by recurrent episodes of ataxia, often triggered by stress or physical exertion, and may include other symptoms such as vertigo and migraines [15–23].

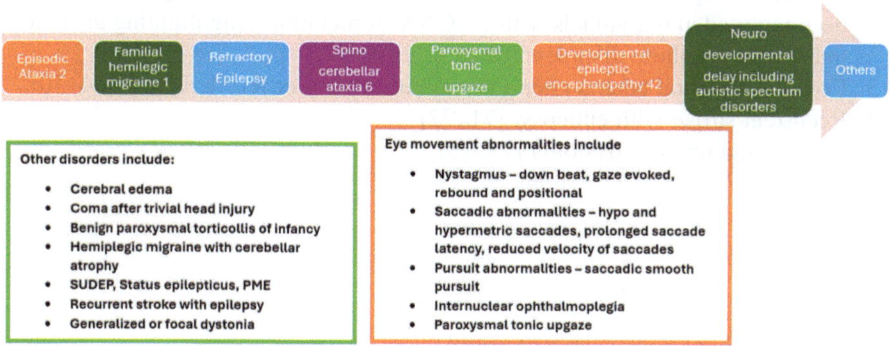

Fig. 19.2 Spectrum of disorders and eye movement abnormalities associated with CACNA1A mutations

2. Familial Hemiplegic Migraine Type 1 (FHM1): A rare form of migraine that can cause temporary paralysis on one side of the body, often accompanied by aura and other neurological symptoms [24–30].
3. Spinocerebellar Ataxia Type 6 (SCA6): A progressive neurodegenerative disorder characterized by ataxia, dysarthria, and other cerebellar symptoms, typically with an onset in mid-adulthood [31–38].
4. Developmental and Epileptic Encephalopathy 42 (DEE42): Severe forms of epilepsy that can present in infancy or early childhood, often associated with global developmental delay and intellectual disability [39]. Some cases are associated with progressive cerebral, cerebellar, and optic nerve atrophy [40–43].
5. Benign Paroxysmal Torticollis of Infancy (BPTI): A movement disorder in infants characterized by recurrent episodes of head tilting and neck stiffness, often associated with a family history of migraines. [40–43]
6. Paroxysmal Tonic Upgaze (PTU): A condition characterized by episodes of sustained upward eye deviation, often accompanied by other neurological symptoms [44–48].
7. Autism Spectrum Disorder (ASD): Some studies have suggested a correlation between CACNA1A mutations and ASD, although the relationship is complex and not fully understood [49–52].
8. Global Developmental Delay (GDD)/ Intellectual Disability (ID):) Many individuals with CACNA1A mutations present with varying degrees of developmental delay, which can affect motor, cognitive, and language skills [6, 53].
9. Progressive Myoclonic Epilepsy: Some mutations have been linked to late-onset forms of myoclonic epilepsy, characterized by myoclonic jerks and progressive neurological decline [54, 55].
10. Severe Hypotonia: Infants with certain CACNA1A mutations may present with severe muscle weakness and hypotonia [56, 57].
11. Refractory Epilepsy: A significant proportion of patients with CACNA1A mutations experience epilepsy that is resistant to standard treatments. [49, 55, 58–61]
12. Sudden Unexpected Death in Epilepsy (SUDEP): Some cases of SUDEP have been reported in individuals with CACNA1A mutations, highlighting the potential severity of these disorders [62, 63].
13. Coma or Cerebral edema after a trivial head injury or febrile illness [64–70].
14. Recurrent stroke with epilepsy. [71, 72]
15. Generalized or focal dystonia [73–75].

19.2.1 Various Eye Movement Abnormalities Related to CACNA1A Mutations

Common Across Disorders Gaze-evoked nystagmus (GEN), impaired smooth pursuit, and downbeat nystagmus (DBN) are shared features in EA2, SCA6, and FHM.

Disease-Specific Patterns

- **EA2**: Features like rebound nystagmus [76] and internuclear ophthalmoplegia (INO) [77] are more specific, as shown in Table 19.1.
- **SCA6**: Positional and head-shaking vertical or perverted nystagmus is pathognomonic. Periodic alternating nystagmus is also described in SCA6 patients. [78, 79]

Table 19.1 Categorization of eye movement abnormalities in CACNA1A-associated disorders

Disorder	Common eye movement abnormalities	Uncommon eye movement abnormalities
Episodic ataxia type 2 (EA2)	Gaze-evoked nystagmus (GEN)	Torsional saccadic palsy
	Downbeat nystagmus (DBN)	Alternating exotropia
	Impaired smooth pursuit	Abduction lag during smooth pursuit (indicative of internuclear ophthalmoplegia (INO))[77]
	Hypermetric or hypometric saccades	Diagonal or side-pocket nystagmus
	Rebound nystagmus	
	Symmetric impairment of optokinetic nystagmus (OKN) and VOR suppression	
Spinocerebellar ataxia type 6 (SCA6)	Saccadic pursuits	Periodic alternating nystagmus
	Reduced smooth pursuit gain	Perverted vertical head-shaking nystagmus
	Slow or dysmetric saccades	Positional nystagmus (apogeotropic or geotropic)
	Square wave jerks (SWJ)	Bow-tie nystagmus in darkness
	DBN, GEN, and rebound nystagmus	
	Vertical saccade velocity reduction	
Hemiplegic migraine (HM)	GEN and DBN	Ocular dipping
	Skew deviation	Coarse nystagmus with simultaneous blinking
	Hypermetric saccades with normal latency	
	Impaired VOR suppression	
Developmental and epileptic encephalopathy 42 (DEE42)	Strabismus	
	Nystagmus	

- **FHM**: Skew deviation [80], ocular dipping [81], and coarse nystagmus are described.
- **Paroxysmal Tonic Upgaze**: Paroxysmal tonic upgaze (PTU) is a rare movement disorder characterized by episodes of sustained upward eye deviation, often accompanied by neck flexion and sometimes truncal ataxia.
- These episodes can occur in clusters and typically manifest in infants, often resolving spontaneously by early childhood. Episodes of conjugate upward eye deviation, which may be sustained and can be associated with down-beating saccades when attempting to look down. Children with PTU often have normal developmental milestones outside of the episodes.
- Brain magnetic resonance imaging (MRI) is typically normal in PTU cases, ruling out structural abnormalities. However, some studies have noted cerebellar atrophy in patients with associated conditions.
- Electroencephalogram (EEG) monitoring may show no electrographic correlate during episodes, further supporting the diagnosis of a movement disorder rather than seizures. PTU was described in mutations in the genes CACNA1A, GRID2, and SEPSECS.
- PTU usually affects children under 2 years of age. Patients usually had benign paroxysmal torticollis before PTU attacks [44, 45, 47, 82, 83].

19.3 Neuroanatomical Substrates Involved in Various Eye Movement Abnormalities

The cerebellum plays a crucial role in controlling eye movements, as shown in Fig. 19.3. Key structures involved include the vestibulocerebellum (flocculus, paraflocculus, nodulus, and uvula), the oculomotor vermis (lobules VI and VII), and fastigial nuclei, as shown in Table 19.2 [84–91].

19.4 Pathophysiology

Purkinje cells are large, GABAergic neurons located in the cerebellar cortex, playing a crucial role in motor coordination and the regulation of cerebellar output. a1ACT acts as a transcription factor that binds to specific DNA sequences, enhancing the expression of genes critical for Purkinje cell development and function. It has been shown to regulate genes such as BTG1, PMCA2, and Granulin gene (GRN), which are involved in neuronal differentiation and signaling. a1ACT promotes neurite outgrowth, which is vital for establishing synaptic connections in developing neurons. This function is crucial for the proper maturation of Purkinje cells [3, 92].

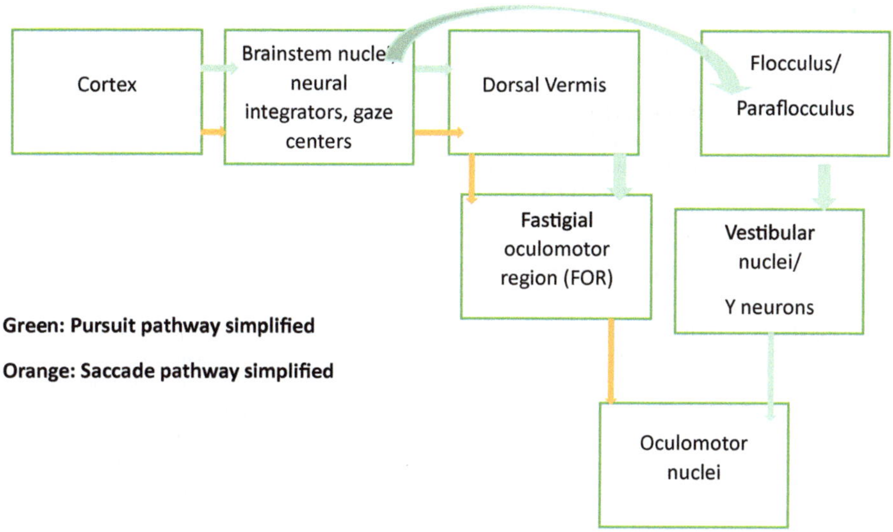

Fig. 19.3 Simplified pathways for saccades and pursuit eye movements

Their function is intricately linked to the activity of various ion channels, including calcium (Ca^{2+}) and potassium (K^+) channels. Purkinje cells exhibit intrinsic pacemaker activity, firing at a baseline rate of approximately 40 Hz in the absence of synaptic inputs. This firing rate can increase to between 40 and 100 Hz during synaptic activity. [93] They integrate excitatory inputs from parallel fibers (which originate from granule cells) and climbing fibers (which originate from the inferior olivary nucleus). The climbing fibers provide strong excitatory input that can generate complex spikes in Purkinje cells. Calcium channels, particularly the P/Q-type (Cav2.1), are crucial for the depolarization of Purkinje cells.

Calcium-activated potassium (KCa) channels are a group of ion channels that link intracellular calcium levels to membrane potential by allowing potassium ions (K^+) to flow out of the cell. This efflux of K^+ hyperpolarizes the cell membrane, reducing excitability and helping regulate neuronal activity, muscle contraction, and secretion processes. There are three different types of KCa channels [93–95]: large-conductance potassium channel (BK), small conductance (SK) channels, and intermediate conductance (IK) channels.

Calcium-activated potassium (KCa) channels, mainly large-conductance (BK) and small-conductance (SK) channels, are critical for repolarizing the membrane after action potentials and generating afterhyperpolarization (AHP), and are closely integrated with P/Q calcium channels [93]. These channels depend on calcium influx to open and regulate neuronal excitability by maintaining the precision of neuronal firing and preventing overexcitation.

Table 19.2 Neuroanatomical substrates involved in various eye movement abnormalities

Lesions	Inputs	Outputs	Findings
Dorsal vermis/ fastigial nucleus (fastigial oculomotor region)	Inferior olivary nucleus (ION), nucleus prepositus hypoglossi (NPH), paramedian pontine reticular formation (PPRF), nucleus reticularis tegmenti pontis (NRTP), vestibular nuclei, dorsolateral pontine nuclei (DLPN)	Caudal fastigial nucleus	Saccadic dysmetria Hypometric saccades in both directions If near FOR—Hypermetric saccades Impaired initiation of smooth pursuit
Uvula/nodulus	Medial and inferior vestibular nuclei, NPH, and ION	Vestibular nuclei	Spontaneous periodic alternating nystagmus Vertical nystagmus during horizontal rotation Downbeat nystagmus after sustained horizontal head-shaking Ocular tilt reaction Impaired downward smooth pursuit Apogeotropic horizontal nystagmus
Flocculus/ paraflocculus	NPH, ION, cell groups of the paramedian tracts (PMT), NRTP, and midbrain reticular formation, vestibular nuclei, DLPN	Ipsilateral superior and medial vestibular nuclei, and the y-group	Impaired tracking of objects leading to saccadic intrusions, either with head still (smooth pursuit) or head moving (VOR suppression) Spontaneous nystagmus Gaze-evoked nystagmus Post-saccadic drift Positional nystagmus (geotropic horizontal nystagmus)

19.4.1 Effects of CACNA1A Mutations on Purkinje Cells [8, 93]

Disrupted Calcium Entry

- CACNA1A mutations reduce calcium current density, limiting the availability of intracellular calcium.

Impaired Activation of KCa Channels

- Reduced calcium entry from dysfunctional P/Q-type channels prevents proper activation of KCa channels, leading to:

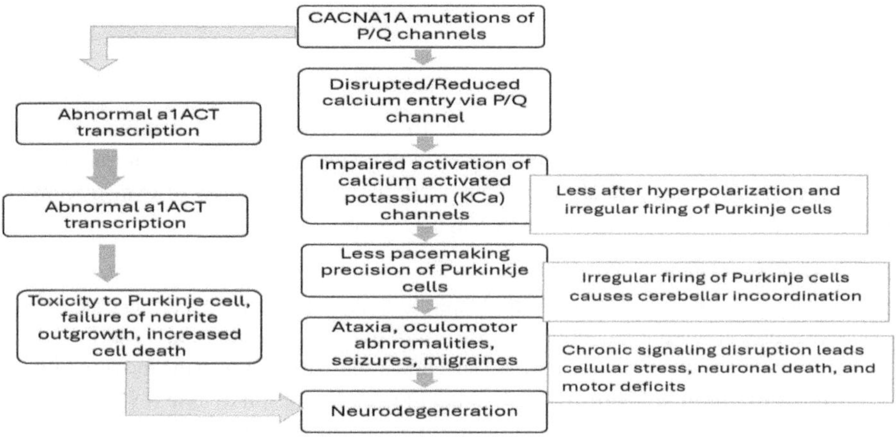

Fig. 19.4 Pathophysiology of CACNA1A mutations

- Decreased afterhyperpolarization (AHP), which normally limits excessive neuronal firing.
- Increased irregularity in Purkinje cell firing patterns is crucial for timing and coordination in cerebellar circuits.

Loss of Pacemaker Precision

- Impaired KCa channel activation disrupts this intrinsic activity, causing irregular and less precise firing. This dysfunction compromises the cerebellum's ability to encode and process motor coordination signals.

Neurodegeneration

- Chronic dysregulation of calcium and potassium signaling in Purkinje cells due to CACNA1A mutations can lead to cellular stress, excitotoxicity, and eventual neuronal death, as shown in Fig. 19.4.
- In SCA6, there is an expansion of the polyglutamine (polyQ) tract in the a1ACT protein which results in failure of neurite outgrowth and is associated with increased cell death in neuronal cultures. The presence of the expanded polyQ tract in also exert toxic effects on cells leading to the degeneration of Purkinje cells and the associated motor dysfunction [92].The exact pathophysiology behind the paroxysmal events or episodic nature is not clearly understood.

19.5 Investigations

The investigation of eye movement abnormalities, especially in the context of CACNA1A mutations, requires a multidisciplinary approach that combines clinical assessments, advanced imaging techniques, and genetic testing. This comprehensive evaluation helps in accurately diagnosing the underlying causes of eye movement disorders and tailoring appropriate management strategies.

Clinical Assessment

- **Neurological Examination**: A thorough neurological examination is essential to assess eye movements, coordination, and overall neurological function. This includes evaluating for nystagmus, saccades, and smooth pursuit movements.
- **Ophthalmological Examination**: A detailed examination to assess visual acuity, eye alignment, and the presence of any ocular abnormalities.

Electrooculography (EOG) This test measures the electrical potential difference between the front and back of the eye, allowing for the assessment of eye movements. It can help identify abnormal eye movement patterns, such as nystagmus.

Video Oculography (VOG) This noninvasive technique uses video cameras to track eye movements in real time. It provides detailed information about the dynamics of eye movements and can help in diagnosing specific types of nystagmus.

Visual Evoked Potentials (VEP) This test measures the electrical activity in the brain in response to visual stimuli. It can help assess the functional integrity of the visual pathways and identify any abnormalities related to eye movement disorders.

Magnetic Resonance Imaging (MRI) Imaging studies can help identify structural abnormalities in the brain, particularly in areas associated with eye movement control, such as the cerebellum and brainstem. MRI can also assess for cerebellar atrophy, which is often associated with CACNA1A mutations.

Electroencephalography (EEG) In cases where seizures are present, EEG can help identify abnormal electrical activity in the brain that may correlate with eye movement abnormalities.

Behavioral Assessments Standardized tests to evaluate cognitive function, developmental milestones, and adaptive behavior can provide context for the impact of eye movement abnormalities on overall functioning.

Functional Imaging This advanced imaging technique can assess brain activity in response to visual stimuli and may provide insights into the neural mechanisms underlying eye movement abnormalities.

Vestibular Function Tests Tests such as caloric testing or vestibular evoked myogenic potentials (VEMPs) can evaluate the vestibular system‘s contribution to eye movement control, particularly in cases of nystagmus.

Genetic Testing

- **Whole Exome Sequencing (WES):** This technique sequences all the protein-coding regions of the genome, allowing for the identification of mutations in the CACNA1A gene, among others. WES is particularly useful for patients with unexplained neurological symptoms, as it can reveal both known and novel mutations.
- **Sanger Sequencing:** A method used to confirm specific mutations identified through WES. It provides a more accurate sequence of the target region. Sanger sequencing is often used to validate the parental origin of identified variants.
- **Targeted Gene Panels:** These panels focus on specific genes known to be associated with particular disorders, including CACNA1A. They are useful for patients with a clinical suspicion of channelopathies or specific neurological conditions.

Functional Studies These studies can provide insights into how specific mutations affect neuronal function and contribute to clinical symptoms.

19.6 Management and Treatment Approaches

19.6.1 Pharmacological Interventions for CACNA1A-Associated Eye Movement Abnormalities

The management of eye movement abnormalities associated with CACNA1A mutations necessitates symptom-specific pharmacological interventions aimed at modulating underlying neuronal excitability, restoring cerebellar function, and alleviating secondary neurological symptoms, as shown in Table 19.3.

19.6.2 Nonpharmacological Interventions for CACNA1A-Associated Eye Movement Abnormalities

Nonpharmacological management of CACNA1A-related eye movement abnormalities involves a multidisciplinary approach that includes vision therapy, vestibular rehabilitation, occupational and physical therapy, patient education, and the use of assistive devices. These strategies aim to improve the quality of life for patients by enhancing their ability to cope with the symptoms associated with their condition. Collaboration among healthcare providers, therapists, and patients is essential for developing an effective management plan tailored to individual needs.

Table 19.3 Pharmacological options for CACNA1A-related disorders

Class of drug	Drug name	Mechanism of action	Clinical applications	Dosage	Adverse effects
Carbonic anhydrase inhibitors	Acetazolamide (ACTZ) [45, 96, 97]	Inhibits carbonic anhydrase, modulating intracellular/extracellular pH, altering neuronal excitability	Preventive treatment of EA2 (50–75% efficacy)	250–1000 mg/day	Nephrolithiasis, paresthesia, muscle stiffness, GastroIntestinal (GI) disturbances
	Sulthiame [98]	Similar to ACTZ but with fewer side effects	Alternative to ACTZ for EA2	50–300 mg/day	Fewer side effects compared to ACTZ
Potassium channel blockers	4-Aminopyridine (4-AP) [99, 100]	Blocks voltage-gated potassium channels, prolonging action potentials, enhancing Purkinje cell excitability	EA2, DBN (reduces slow-phase velocity), SCA6, cerebellar ataxia	Variable	Tremors, dizziness, nervousness in some patients
	3,4-Diaminopyridine (3,4-DAP) [101]	Similar to 4-AP	Similar to 4-AP, including gait stability improvement in cerebellar ataxia	Variable	Similar to 4-AP
Modified amino acids	Acetyl-DL-leucine [98, 102]	Stabilizes neuronal membrane potentials, improves synaptic transmission	EA2, DBN, SCA6; improves balance and spatial orientation	Variable	Minimal reported side effects
SK channel activators	Chlorzoxazone [98, 103, 104]	Activates SK channels, modulating Purkinje cell activity	DBN (improves slow-phase velocity, postural control, and visual acuity)	500 mg three times daily	Sedation, dizziness, mild GastroIntestinal (GI) upset
GABA modulators	Gabapentin	Increases Central Nervous System (CNS) GABA levels, blocks voltage-gated calcium channels, N-methyl-D-aspartate (NMDA) receptor inhibition	Pendular nystagmus, DBN	Up to 2400 mg/day in divided doses	Drowsiness, fatigue, dizziness
	Baclofen [98]	GABA-B receptor agonist; compensates for cerebellar inhibitory loss	DBN; reduces saccadic dysmetria	5 mg three times daily; up to 80 mg/day	Sedation, weakness, fatigue
	Clonazepam	Enhances GABAergic activity	Acquired idiopathic DBN	1.5 mg/day; up to 20 mg/day	Sedation, dependency risk

N-methyl-D-aspartate (NMDA) receptor antagonists	Memantine	Blocks N-methyl-D-aspartate (NMDA) receptors, reduces excitotoxicity and nystagmus amplitude	APN, DBN (improves visual acuity and nystagmus control)	20–40 mg/day in divided doses	Dizziness, headache, confusion
Calcium channel blockers	Verapamil	Blocks voltage-gated calcium channels, reduces neuronal hyperexcitability	Migraine prevention, vestibular symptoms, episodic ataxia	Variable	Bradycardia, fatigue, dizziness
	Flunarizine	Similar to verapamil, with dual benefits for migraines and episodic ataxia	Migraine, episodic ataxia	Variable	Weight gain, fatigue
Anticonvulsants	Lamotrigine [105]	Stabilizes neuronal membranes, reduces glutamate release	Hemiplegic migraines, episodic ataxia, secondary eye movement abnormalities	Variable	Dizziness, rash, potential Stevens-Johnson syndrome
	Valproate [106]	Enhances GABAergic transmission, stabilizes neuronal activity	Seizures, hemiplegic migraines, episodic ataxia	Variable	Weight gain, liver toxicity, tremor
	Levetiracetam [21]	Modulates synaptic neurotransmitter release	Similar applications as valproate and lamotrigine	Variable	Irritability, fatigue
	Topiramate [107]	Enhances GABAergic activity, blocks sodium and calcium channels	Similar to lamotrigine; migraine prophylaxis	Variable	Cognitive slowing, weight loss, paresthesia

Vision Therapy

- Oculomotor exercises designed to improve eye coordination and control can help patients manage their eye movement abnormalities. This may include tracking exercises, saccadic training, and convergence exercises.
- The use of prism lenses can help realign visual input and reduce the symptoms of nystagmus by altering the visual field and helping the brain adapt to abnormal eye movements.

Vestibular Rehabilitation Therapy (VRT)

- VRT focuses on improving balance and reducing dizziness through specific exercises that enhance vestibular function. This can help patients adapt to their eye movement abnormalities and improve overall stability.
- Habituation exercises aim to reduce the sensitivity of the vestibular system to certain movements or positions that trigger symptoms, thereby helping patients cope better with their condition.

Occupational Therapy

- Occupational therapists can work with patients to develop strategies for daily activities that may be affected by eye movement abnormalities. This includes techniques for reading, writing, and using screens.
- Adjusting the environment to reduce visual stressors (e.g., reducing glare, improving lighting) can help patients manage their symptoms more effectively.

Physical Therapy

- Physical therapists can help patients improve their posture and body mechanics, which may alleviate some symptoms associated with eye movement disorders.
- Exercises that enhance overall coordination and motor control can be beneficial, especially for patients with associated ataxia.

Patient Education and Support

- Providing education about the condition and connecting patients with support groups can help them cope with the psychological and social impacts of their eye movement abnormalities.
- Educating family members about the condition can improve support at home and help in understanding the challenges faced by the patient.

Assistive Devices

- Magnifying glasses or other visual aids can assist patients in performing tasks that require detailed vision, such as reading or sewing.
- Utilizing technology such as screen readers or voice recognition software can help patients with significant visual impairments manage daily tasks more effectively.

19.7 Prognosis and Long-Term Impact

The prognosis and long-term impact of CACNA1A mutations are highly variable and depend on the specific clinical phenotype associated with the mutation. While some individuals may experience significant challenges, including refractory epilepsy and profound developmental delays, others may have milder symptoms and better outcomes. Comprehensive care, including medical, educational, and psychosocial support, is essential for improving the quality of life for individuals affected by CACNA1A-related disorders.

Clinical Variability

CACNA1A mutations are associated with a range of neurological disorders, including epilepsy, global developmental delay (GDD), intellectual disability (ID), ataxia, and autism spectrum disorder (ASD). The clinical presentation can vary widely among individuals, even within the same family, due to the complex nature of genotype-phenotype correlations.

Epilepsy

More than half of the patients with CACNA1A-related epilepsy experience refractory seizures, which are difficult to control with standard antiepileptic medications. The most common seizure type associated with these mutations is status epilepticus, particularly linked to gain-of-function variants. The prognosis for patients with refractory epilepsy can be poor, with some individuals experiencing significant morbidity and a reduced quality of life.

Global Developmental Delay and Intellectual Disability

Patients with CACNA1A mutations often present with GDD and ID, with severity ranging from mild to profound. The p.A713T variant, for example, is associated with severe to profound GDD/ID. The long-term impact on cognitive function can be significant, with many patients requiring ongoing support and intervention throughout their lives.

Ataxia and Other Motor Symptoms

Ataxia is a common feature in individuals with CACNA1A mutations, particularly in conditions like episodic ataxia type 2 (EA2) and spinocerebellar ataxia type 6 (SCA6). The prognosis for ataxia can vary; while some patients may experience episodic symptoms, others may have progressive ataxia leading to significant disability. Long-term impacts may include difficulties with balance, coordination, and daily activities.

Mortality Risk

There is an increased risk of early mortality in some patients with severe CACNA1A mutations, particularly those with complex phenotypes involving refractory epilepsy and significant developmental delays. Sudden unexpected death in epilepsy (SUDEP) is a potential risk factor for individuals with severe epilepsy associated with CACNA1A mutations.

Long-Term Impact
The long-term impact of CACNA1A mutations on quality of life can be profound, particularly for individuals with severe GDD/ID, refractory epilepsy, and ataxia. Many patients may require lifelong care and support, impacting both the individuals and their families. Children with CACNA1A mutations often require special educational support and interventions to address their developmental and cognitive challenges. Access to appropriate educational resources and therapies can significantly influence their long-term outcomes. Families of individuals with CACNA1A mutations may face significant psychosocial challenges, including the emotional and financial burden of caring for a child with complex needs. Support groups and counseling can be beneficial for families to cope with these challenges.

19.8 Impact of Early Detection and Intervention

The early detection of CACNA1A mutations plays a crucial role in improving the prognosis and quality of life for affected individuals. By facilitating timely diagnosis, personalized management, and access to resources, early detection can lead to better functional outcomes and enhanced family support. Additionally, it allows for long-term planning and advocacy, ultimately contributing to a more positive trajectory for individuals with CACNA1A-related disorders.

Timely Diagnosis Early identification of CACNA1A mutations allows for a timely diagnosis of associated conditions such as epilepsy, global developmental delay (GDD), intellectual disability (ID), and ataxia. This can lead to more effective management strategies tailored to the specific needs of the patient.

Informed Management Plans With a confirmed diagnosis, healthcare providers can develop personalized management plans that may include pharmacological treatments, nonpharmacological interventions, and supportive therapies. For example, patients with epilepsy may benefit from specific antiepileptic medications, while those with ataxia may require physical therapy.

Monitoring and Intervention Early detection facilitates regular monitoring of developmental milestones and neurological symptoms, allowing for timely interventions that can mitigate the progression of symptoms. For instance, early physical therapy can help improve motor skills in children with ataxia.

Family Counseling and Support Families can receive genetic counseling to understand the implications of the mutation, including inheritance patterns and the potential for other family members to be affected. This knowledge can help families make informed decisions regarding family planning and the management of the condition.

Access to Resources Early diagnosis can connect families with resources such as educational support, therapy services, and community programs designed for individuals with neurodevelopmental disorders. This access can enhance the overall support system for the patient and their family.

19.9 Quality-of-Life Considerations

Improved Functional Outcomes Early intervention and tailored management strategies can lead to improved functional outcomes, including better cognitive, motor, and social skills. This can enhance the individual's ability to participate in daily activities and social interactions.

Reduction in Comorbidities By addressing symptoms early, the risk of developing comorbid conditions, such as anxiety and depression, may be reduced. Individuals with untreated neurological disorders often face additional psychological challenges, which can impact their quality of life.

Enhanced Family Dynamics Families that receive early support and education about CACNA1A-related disorders may experience less stress and anxiety. Understanding the condition can foster a more supportive environment, improving family dynamics and overall well-being.

Long-Term Planning Early detection allows families to plan for the future, including educational and vocational opportunities for the affected individual. This proactive approach can lead to better long-term outcomes and greater independence.

Awareness and Advocacy Increased awareness of CACNA1A mutations can lead to advocacy for research and resources dedicated to these conditions. This can improve the overall landscape for individuals affected by CACNA1A-related disorders, leading to better funding for research and support services.

19.10 Current Research Directions

Current research on CACNA1A mutations is advancing our understanding of the genetic, molecular, and clinical aspects of these disorders. Future directions emphasize the need for personalized approaches, longitudinal studies, and collaborative efforts to improve outcomes for individuals affected by CACNA1A-related conditions. As research progresses, it holds the promise of developing targeted therapies and enhancing the quality of life for affected individuals.

Genotype-Phenotype Correlations Ongoing studies are investigating the relationship between specific CACNA1A mutations and the resulting clinical phenotypes. Research has shown that different mutations can lead to a spectrum of disorders, including epilepsy, global developmental delay (GDD), intellectual disability (ID), and ataxia. For instance, the p.A713T variant is linked with severe to profound

GDD/ID, while other variants are associated with refractory epilepsy and status epilepticus. Understanding these correlations can help predict clinical outcomes and tailor management strategies.

Mechanistic Insights Research is focused on elucidating the cellular and molecular mechanisms by which CACNA1A mutations affect neuronal function. Studies using induced pluripotent stem cell (iPSC)-derived models are being employed to investigate how these mutations lead to synaptic dysfunction and altered neuronal excitability. Investigations into the role of Purkinje cells, which express CACNA1A, are particularly relevant, as these cells are crucial for motor coordination and are affected in various CACNA1A-related disorders.

Animal Models Animal models, particularly mice with targeted mutations in the Cacna1a gene, are being used to study the phenotypic manifestations of CACNA1A mutations. These models help in understanding the pathophysiology of associated disorders and can be used to test potential therapeutic interventions.

Therapeutic Strategies Research is exploring combinatorial therapeutic strategies that may include pharmacological and nonpharmacological approaches to manage symptoms associated with CACNA1A mutations. This includes the potential use of existing medications that target calcium channels and other related pathways. The development of gene therapy approaches to correct or compensate for the effects of CACNA1A mutations is also an area of interest, although this is still in the early stages of research.

19.11 Future Directions

As understanding of the genetic basis of CACNA1A-related disorders improves, there is potential for the development of personalized medicine approaches. This could involve tailoring treatments based on the specific mutation and its associated clinical features. Long-term studies tracking the progression of symptoms in individuals with CACNA1A mutations will be essential for understanding the natural history of these disorders. Such studies can provide insights into the timing of interventions and the effectiveness of various treatment strategies. The implementation of comprehensive genetic screening in patients with unexplained neurological symptoms may lead to the identification of additional CACNA1A mutations. This could enhance the understanding of the mutation spectrum and its clinical implications. Collaborative efforts among researchers, clinicians, and geneticists are crucial for pooling data on CACNA1A mutations. This can facilitate larger studies that improve the statistical power to identify genotype-phenotype correlations and treatment outcomes. Future research should also consider the comorbidities associated with CACNA1A mutations, such as anxiety, depression, and other psychiatric disorders. Understanding these associations can lead to more comprehensive care strategies for affected individuals.

References

1. Mannhold ER, Kubinyi H, Folkers G. Methods and principles in medicinal chemistry.
2. Du X, Wei C, Hejazi Pastor DP, et al. α1ACT is essential for survival and early cerebellar programming in a critical neonatal window. Neuron. 2019;102(4):770–785.e7. https://doi.org/10.1016/j.neuron.2019.02.036.
3. Du X, Wang J, Zhu H, et al. Second cistron in CACNA1A gene encodes a transcription factor mediating cerebellar development and SCA6. Cell. 2013;154(1):118–33. https://doi.org/10.1016/j.cell.2013.05.059.
4. Govek EE, Hatten ME. Tag-team genetics of spinocerebellar ataxia 6. Neuron. 2019;102(4):707–9. https://doi.org/10.1016/j.neuron.2019.04.041.
5. Indelicato E, Boesch S. CACNA1A-related channelopathies: clinical manifestations and treatment options. Handb Exp Pharmacol. 2023;279:227–48. https://doi.org/10.1007/164_2022_625.
6. Kessi M, Chen B, Pang N, et al. The genotype-phenotype correlations of the CACNA1A-related neurodevelopmental disorders: a small case series and literature reviews. Front Mol Neurosci. 2023;16:1222321. https://doi.org/10.3389/fnmol.2023.1222321.
7. Zhang L, Wen Y, Zhang Q, et al. CACNA1A gene variants in eight chinese patients with a wide range of phenotypes. Front Pediatr. 2020;8:577544. https://doi.org/10.3389/fped.2020.577544.
8. Benarroch EE. Neuronal voltage-gated calcium channels: brief overview of their function and clinical implications in neurology. Neurology. 2010;74(16):1310–5. https://doi.org/10.1212/WNL.0b013e3181da364b.
9. Inchauspe CG, Urbano FJ, Di Guilmi MN, et al. Gain of function in FHM-1 Cav2.1 knock-in mice is related to the shape of the action potential. J Neurophysiol. 2010;104(1):291–9. https://doi.org/10.1152/jn.00034.2010.
10. Di Guilmi MN, Wang T, Inchauspe CG, et al. Synaptic gain-of-function effects of mutant Cav2.1 channels in a mouse model of familial hemiplegic migraine are due to increased basal [Ca2+]i. J Neurosci. 2014;34(21):7047–58. https://doi.org/10.1523/JNEUROSCI.2526-13.2014.
11. Bahamonde MI, Serra SA, Drechsel O, et al. A single amino acid deletion (ΔF1502) in the S6 segment of CaV2.1 domain III associated with congenital ataxia increases channel activity and promotes Ca2+ influx. PLoS One. 2015;10(12):e0146035. https://doi.org/10.1371/journal.pone.0146035.
12. Sintas C, Carreño O, Fernàndez-Castillo N, et al. Mutation spectrum in the CACNA1A gene in 49 patients with episodic ataxia. Sci Rep. 2017;7(1):2514. https://doi.org/10.1038/s41598-017-02554-x.
13. Mammadova D, Kraus C, Leis T, et al. Intrafamilial neurological phenotypic variability due to either biallelic or monoallelic pathogenic variants in CACNA1A. Front Neurol. 2024;15:1458109. https://doi.org/10.3389/fneur.2024.1458109.
14. Hommersom MP, van Prooije TH, Pennings M, et al. The complexities of CACNA1A in clinical neurogenetics. J Neurol. 2022;269(6):3094–108. https://doi.org/10.1007/s00415-021-10897-9.
15. Jen J, Kim GW, Baloh RW. Clinical spectrum of episodic ataxia type 2.
16. Idiculla PS, Siddiqui JH. A case of novel CACNA1A mutation causing type 2 episodic ataxia. Neurol Sci. 2021;42(6):2577–8. https://doi.org/10.1007/s10072-021-05059-8.
17. Park D, Kim SH, Lee YJ, Song GJ, Park JS. A novel CACNA1A mutation associated with episodic ataxia 2 presenting with periodic paralysis. Acta Neurol Belg. 2018;118(1):137–9. https://doi.org/10.1007/s13760-018-0895-y.
18. Magis D, Boon E, Coppola G, Daron A, Schoenen J. A novel CACNA1A mutation results in episodic ataxia with migrainous features without headache. Cephalalgia. 2012;32(15):1147–9. https://doi.org/10.1177/0333102412459572.

19. A novel frameshift CACNA1A mutation causing episodic ataxia type 2. 2018. 17.:https://doi.org/10.1007/s12311-018-0931-8.
20. Matsuyama Z, Murase M, Shimizu H, et al. A novel insertion mutation of acetazolamide-responsive episodic ataxia in a Japanese family. J Neurol Sci. 2003;210(1–2):91–3. https://doi.org/10.1016/s0022-510x(03)00008-x.
21. Algahtani H, Shirah B, Algahtani R, Al-Qahtani MH, Abdulkareem AA, Naseer MI. A novel mutation in CACNA1A gene in a Saudi female with episodic ataxia type 2 with no response to acetazolamide or 4-aminopyridine. Intractable Rare Dis Res. 2019;8(1):67–71. https://doi.org/10.5582/irdr.2018.01133.
22. Jen JC, Graves TD, Hess EJ, Hanna MG, Griggs RC, Baloh RW. Primary episodic ataxias: diagnosis, pathogenesis and treatment. Brain. 2007;130(Pt 10):2484–93. https://doi.org/10.1093/brain/awm126.
23. Namekawa M, Takiyama Y, Ueno N, Nishizawa M. A sporadic case of episodic ataxia with nystagmus (EA-2). Rinsho Shinkeigaku. 1998;38(5):446–9.
24. Alfayyadh MM, Maksemous N, Sutherland HG, Lea RA, Griffiths LR. Unravelling the genetic landscape of hemiplegic migraine: exploring innovative strategies and emerging approaches. Genes (Basel). 2024;15(4) https://doi.org/10.3390/genes15040443.
25. Bruun M, Hjermind LE, Thomsen C, et al. Familial hemiplegic migraine type 1 associated with parkinsonism: a case report. Case Rep Neurol. 2015;7(1):84–9. https://doi.org/10.1159/000381827.
26. Sprouse Blum AS, Couperus CJ, Rosen BJ, Haskin-Leahy LF, Shapiro RE. Familial "diplegic" migraine—description of a family with a novel CACNA1A mutation. Headache. 2020;60(3):600–6. https://doi.org/10.1111/head.13741.
27. Nair A, Simonetti M, Birsa N, et al. Familial hemiplegic migraine ca(v)2.1 channel mutation R192Q enhances ATP-gated P2X3 receptor activity of mouse sensory ganglion neurons mediating trigeminal pain. Mol Pain. 2010;6:48. https://doi.org/10.1186/1744-8069-6-48.
28. Garza-López E, González-Ramírez R, Gandini MA, Sandoval A, Felix R. The familial hemiplegic migraine type 1 mutation K1336E affects direct G protein-mediated regulation of neuronal P/Q-type Ca2+ channels. Cephalalgia. 2013;33(6):398–407. https://doi.org/10.1177/0333102412475236.
29. Hansen JM. Familial hemiplegic migraine. Dan Med Bull. 2010;57(9):B4183.
30. Carrera P, Stenirri S, Ferrari M, Battistini S. Familial hemiplegic migraine: a ion channel disorder. Brain Res Bull. 2001;56(3–4):239–41. https://doi.org/10.1016/s0361-9230(01)00570-6.
31. Jodice C, Mantuano E, Veneziano L, et al. Episodic ataxia type 2 (EA2) and spinocerebellar ataxia type 6 (SCA6) due to CAG repeat expansion in the CACNA1A gene on chromosome 19p. Hum Mol Genet. 1997;6(11):1973–8. https://doi.org/10.1093/hmg/6.11.1973.
32. Shimazaki H, Vazifehmand R, Heidari MH, et al. A large family with spinocerebellar ataxia type 6 in Iran: a clinical and genetic study. Arch Iran Med. 2008;11(4):459–62.
33. Takahashi H, Ishikawa K, Tsutsumi T, et al. A clinical and genetic study in a large cohort of patients with spinocerebellar ataxia type 6. J Hum Genet. 2004;49(5):256–64. https://doi.org/10.1007/s10038-004-0142-7.
34. Wang X, Wang H, Xia Y, et al. A neuropathological study at autopsy of early onset spinocerebellar ataxia 6. J Clin Neurosci. 2010;17(6):751–5. https://doi.org/10.1016/j.jocn.2009.10.007.
35. Arpa J, Cuesta A, Cruz-Martínez A, Santiago S, Sarriá J, Palau F. Clinical features and genetic analysis of a Spanish family with spinocerebellar ataxia 6. Acta Neurol Scand. 1999;99(1):43–7. https://doi.org/10.1111/j.1600-0404.1999.tb00656.x.
36. Tonelli A, D'Angelo MG, Salati R, et al. Early onset, non fluctuating spinocerebellar ataxia and a novel missense mutation in CACNA1A gene. J Neurol Sci. 2006;241(1–2):13–7. https://doi.org/10.1016/j.jns.2005.10.007.
37. Globas C, du Montcel ST, Baliko L, et al. Early symptoms in spinocerebellar ataxia type 1, 2, 3, and 6. Mov Disord. 2008;23(15):2232–8. https://doi.org/10.1002/mds.22288.

38. Christova P, Anderson JH, Gomez CM. Impaired eye movements in presymptomatic spinocerebellar ataxia type 6. Arch Neurol. 2008;65(4):530–6. https://doi.org/10.1001/archneur.65.4.530.
39. nature 12439.
40. Vila-Pueyo M, Gené GG, Flotats-Bastardes M, et al. A loss-of-function CACNA1A mutation causing benign paroxysmal torticollis of infancy. Eur J Paediatr Neurol. 2014;18(3):430–3. https://doi.org/10.1016/j.ejpn.2013.12.011.
41. Roubertie A, Echenne B, Leydet J, et al. Benign paroxysmal tonic upgaze, benign paroxysmal torticollis, episodic ataxia and CACNA1A mutation in a family. J Neurol. 2008;255(10):1600–2. https://doi.org/10.1007/s00415-008-0982-8.
42. Giffin NJ, Benton S, Goadsby PJ. Benign paroxysmal torticollis of infancy: four new cases and linkage to CACNA1A mutation. Dev Med Child Neurol. 2002;44(7):490–3. https://doi.org/10.1017/s0012162201002407.
43. Shin M, Douglass LM, Milunsky JM, Rosman NP. The genetics of benign paroxysmal torticollis of infancy: is there an association with mutations in the CACNA1A gene? J Child Neurol. 2016;31(8):1057–61. https://doi.org/10.1177/0883073816636226.
44. Blumkin L, Leshinsky-Silver E, Michelson M, et al. Paroxysmal tonic upward gaze as a presentation of de-novo mutations in CACNA1A. Eur J Paediatr Neurol. 2015;19(3):292–7. https://doi.org/10.1016/j.ejpn.2014.12.018.
45. Lazar SM, Abid F. Pearls & oy-sters: CACNA1A-related paroxysmal tonic upgaze with ataxia responsive to acetazolamide. Neurology. 2024;102(1):e207992. https://doi.org/10.1212/WNL.0000000000207992.
46. Humbertclaude V, Krams B, Nogue E, et al. Benign paroxysmal torticollis, benign paroxysmal vertigo, and benign tonic upward gaze are not benign disorders. Dev Med Child Neurol. 2018;60(12):1256–63. https://doi.org/10.1111/dmcn.13935.
47. Chang MY, Yuen T, Vyas A, Borchert MS. Paroxysmal tonic upgaze (PTU) associated with CACNA1A mutation and gross motor delay. J AAPOS. 2021;25(6):353–353.e1. https://doi.org/10.1016/j.jaapos.2021.04.016.
48. Tantsis EM, Gill D, Griffiths L, et al. Eye movement disorders are an early manifestation of CACNA1A mutations in children. Dev Med Child Neurol. 2016;58(6):639–44. https://doi.org/10.1111/dmcn.13033.
49. Damaj L, Lupien-Meilleur A, Lortie A, et al. CACNA1A haploinsufficiency causes cognitive impairment, autism and epileptic encephalopathy with mild cerebellar symptoms. Eur J Hum Genet. 2015;23(11):1505–12. https://doi.org/10.1038/ejhg.2015.21.
50. Long S, Zhou H, Li S, et al. The clinical and genetic features of co-occurring epilepsy and autism spectrum disorder in Chinese children. Front Neurol. 2019;10:505. https://doi.org/10.3389/fneur.2019.00505.
51. Golshani P. What rodent models teach us about the association of autism and epilepsy. In: Noebels JL, Avoli M, Rogawski MA, Vezzani A, Delgado-Escueta AV, editors. Jasper's basic mechanisms of the epilepsies. 5th ed. Oxford University Press; 2024. p. 1219–32. https://doi.org/10.1093/med/9780197549469.003.0059.
52. Liao X, Li Y. Genetic associations between voltage-gated calcium channels and autism spectrum disorder: a systematic review. Mol Brain. 2020;13(1):96. https://doi.org/10.1186/s13041-020-00634-0.
53. Kessi M, Chen B, Peng J, Yan F, Yang L, Yin F. Calcium channelopathies and intellectual disability: a systematic review. Orphanet J Rare Dis. 2021;16(1):219. https://doi.org/10.1186/s13023-021-01850-0.
54. Lv Y, Wang Z, Liu C, Cui L. Identification of a novel CACNA1A mutation in a Chinese family with autosomal recessive progressive myoclonic epilepsy. Neuropsychiatr Dis Treat. 2017;13:2631–6. https://doi.org/10.2147/NDT.S145774.
55. Hayashida T, Saito Y, Ishii A, et al. CACNA1A-related early-onset encephalopathy with myoclonic epilepsy: a case report. Brain Dev. 2018;40(2):130–3. https://doi.org/10.1016/j.braindev.2017.08.006.

56. Kramer AA, Bennett DF, Barañano KW, Bannister RA. A neurodevelopmental disorder caused by a dysfunctional CACNA1A allele. eNeurologicalSci. 2023;31:100456. https://doi.org/10.1016/j.ensci.2023.100456.
57. Natiq A, Elalaoui SC, Miesch S, et al. A new case of de novo 19p13.2p13.12 deletion in a girl with overgrowth and severe developmental delay. Mol Cytogenet. 2014;7:40. https://doi.org/10.1186/1755-8166-7-40.
58. Romero VI, Sáenz S, Arias-Almeida B, DiCapua D, Hosomichi K. AARS and CACNA1A mutations: diagnostic insights into a case report of uncommon epileptic encephalopathy phenotypes in two siblings. Front Neurol. 2024;15:1376643. https://doi.org/10.3389/fneur.2024.1376643.
59. Mitta N, Menon RN, McTague A, et al. Genotype-phenotype correlates of infantile-onset developmental & epileptic encephalopathy syndromes in South India: a single Centre experience. Epilepsy Res. 2020;166:106398. https://doi.org/10.1016/j.eplepsyres.2020.106398.
60. Liu J, Tong L, Song S, et al. Novel and de novo mutations in pediatric refractory epilepsy. Mol Brain. 2018;11(1):48. https://doi.org/10.1186/s13041-018-0392-5.
61. Bolte KN, Assaf M, Zach T, Peche S. Two children with early-onset strokes and intractable epilepsy, both with CACNA1A mutations. Child Neurol Open. 2022;9:2329048X221094977. https://doi.org/10.1177/2329048X221094977.
62. Coll M, Allegue C, Partemi S, et al. Genetic investigation of sudden unexpected death in epilepsy cohort by panel target resequencing. Int J Legal Med. 2016;130(2):331–9. https://doi.org/10.1007/s00414-015-1269-0.
63. Jansen NA, Schenke M, Voskuyl RA, Thijs RD, van den Maagdenberg AMJM, Tolner EA. Apnea associated with brainstem seizures in Cacna1a(S218L) mice is caused by medullary spreading depolarization. J Neurosci. 2019;39(48):9633–44. https://doi.org/10.1523/JNEUROSCI.1713-19.2019.
64. Curtain RP, Smith RL, Ovcaric M, Griffiths LR. Minor head trauma-induced sporadic hemiplegic migraine coma. Pediatr Neurol. 2006;34(4):329–32. https://doi.org/10.1016/j.pediatrneurol.2005.08.033.
65. Kors EE, Terwindt GM, Vermeulen FL, et al. Delayed cerebral edema and fatal coma after minor head trauma: role of the CACNA1A calcium channel subunit gene and relationship with familial hemiplegic migraine. Ann Neurol. 2001;49(6):753–60. https://doi.org/10.1002/ana.1031.
66. McClain IJ, Tsai MA, Guardia CF 3rd. Prolonged postictal coma in hemiplegic migraine: a case report. Headache. 2019;59(9):1641–2. https://doi.org/10.1111/head.13606.
67. Wada T, Kobayashi N, Takahashi Y, Aoki T, Watanabe T, Saitoh S. Wide clinical variability in a family with a CACNA1A T666m mutation: hemiplegic migraine, coma, and progressive ataxia. Pediatr Neurol. 2002;26(1):47–50. https://doi.org/10.1016/s0887-8994(01)00371-x.
68. Camia F, Pisciotta L, Morana G, et al. Combined early treatment in hemiplegic attacks related to CACNA1A encephalopathy with brain oedema: blocking the cascade? Cephalalgia. 2017;37(12):1202–6. https://doi.org/10.1177/0333102416668655.
69. Vahedi K, Denier C, Ducros A, et al. CACNA1A gene de novo mutation causing hemiplegic migraine, coma, and cerebellar atrophy. Neurology. 2000;55(7):1040–2. https://doi.org/10.1212/wnl.55.7.1040.
70. Gauquelin L, Hawkins C, Tam EWY, Miller SP, Yoon G. Pearls & oy-sters: fatal brain edema is a rare complication of severe CACNA1A-related disorder. Neurology. 2020;94(14):631–4. https://doi.org/10.1212/WNL.0000000000009223.
71. Gudenkauf FJ, Azamian MS, Hunter JV, Nayak A, Lalani SR. A novel CACNA1A variant in a child with early stroke and intractable epilepsy. Mol Genet Genomic Med. 2020;8(10):e1383. https://doi.org/10.1002/mgg3.1383.
72. Knierim E, Leisle L, Wagner C, et al. Recurrent stroke due to a novel voltage sensor mutation in Cav2.1 responds to verapamil. Stroke. 2011;42(2):e14–7. https://doi.org/10.1161/STROKEAHA.110.600023.

73. Alshareet M, Alakkas A, Alsinaidi OA, Bawazeer S, Peer-Zada AA. Novel de novo heterozygous CACNA1A gene variant in generalised dystonia: a case report. BMJ Neurol Open. 2024;6(1):e000710. https://doi.org/10.1136/bmjno-2024-000710.
74. Fletcher CF, Tottene A, Lennon VA, et al. Dystonia and cerebellar atrophy in Cacna1a null mice lacking P/Q calcium channel activity. FASEB J. 2001;15(7):1288–90. https://doi.org/10.1096/fj.00-0562fje.
75. Fuerte-Hortigón A, Pérez-Noguera R, Dotor García-Soto J, Royo Boronat I, Álvarez de Andrés S, García-Moreno JM. Novel CACNA1A variant may cause cervical dystonia and cerebellar ataxia syndrome. J Neurol Sci. 2020;415:116909. https://doi.org/10.1016/j.jns.2020.116909.
76. Kim HJ, Kim JS, Choi JH, Shin JH, Choi KD, Zee DS. Rebound upbeat nystagmus after lateral gaze in episodic ataxia type 2. Cerebellum. 2014;13(3):411–3. https://doi.org/10.1007/s12311-014-0547-6.
77. Rucker JC, Jen J, Stahl JS, Natesan N, Baloh RW, Leigh RJ. Internuclear ophthalmoparesis in episodic ataxia type 2. Ann N Y Acad Sci. 2005;1039:571–4. https://doi.org/10.1196/annals.1325.070.
78. Hashimoto T, Sasaki O, Yoshida K, Takei Y, ichi, Ikeda S ichi. Periodic alternating nystagmus and rebound nystagmus in spinocerebellar ataxia type 6. Mov Disord. 2003;18(10):1201–4. https://doi.org/10.1002/mds.10511.
79. Colen CB, Ketko A, George E, Van Stavern GP. Periodic alternating nystagmus and periodic alternating skew deviation in spinocerebellar ataxia type 6. J Neuroophthalmol. 2008;28(4):287–8. https://doi.org/10.1097/WNO.0b013e318183bf5a.
80. Yadegari S. Skew deviation and ocular tilt reaction as presenting feature of CACNA1A mutation. J AAPOS. 2023;27(3):166–9. https://doi.org/10.1016/j.jaapos.2023.03.006.
81. Ocular dipping in a patient with hemiplegic migraine. 55.; 2018. https://doi.org/10.3928/01913913-20171129-01.
82. Quade A, Thiel A, Kurth I, et al. Paroxysmal tonic upgaze: a heterogeneous clinical condition responsive to carbonic anhydrase inhibition. Eur J Paediatr Neurol. 2020;25:181–6. https://doi.org/10.1016/j.ejpn.2019.11.002.
83. Massuyama BK, Tonholo Silva TY, Gambirasio BG, Pedroso JL, Barsottini OGP. Paroxysmal tonic upward gaze: a clinical clue for CACNA1A-related disorders. Mov Disord Clin Pract. 2023;10(8):1225–7. https://doi.org/10.1002/mdc3.13809.
84. Kheradmand A, Zee DS. Cerebellum and ocular motor control. Front Neur. 2011:2. https://doi.org/10.3389/fneur.2011.00053.
85. Ye BS, Kim YD, Nam HS, Lee HS, Nam CM, Heo JH. Clinical manifestations of cerebellar infarction according to specific lobular involvement. Cerebellum. 2010;9(4):571–9. https://doi.org/10.1007/s12311-010-0200-y.
86. Walker MF, Tian J, Shan X, Tamargo RJ, Ying H, Zee DS. Lesions of the cerebellar nodulus and uvula impair downward pursuit. J Neurophysiol. 2008;100(4):1813–23. https://doi.org/10.1152/jn.01193.2007.
87. Straube A, Deubel H, Ditterich J, Eggert T. Cerebellar lesions impair rapid saccade amplitude adaptation. Neurology. 2001;57(11):2105–8. https://doi.org/10.1212/WNL.57.11.2105.
88. Baier B, Stoeter P, Dieterich M. Anatomical correlates of ocular motor deficits in cerebellar lesions. Brain. 2009;132(8):2114–24. https://doi.org/10.1093/brain/awp165.
89. Buttner U, Straube A, Spuler A. Saccadic dysmetria and "intact" smooth pursuit eye movements after bilateral deep cerebellar nuclei lesions. J Neurol Neurosurg Psychiatry. 1994;57(7):832–4. https://doi.org/10.1136/jnnp.57.7.832.
90. Brodsky MC, Donahue SP, Vaphiades M, Brandt T. Skew deviation revisited. Surv Ophthalmol. 2006;51(2):105–28. https://doi.org/10.1016/j.survophthal.2005.12.008.
91. Hashimoto T, Sasaki O, Yoshida K, Takei Y, Ikeda S. Periodic alternating nystagmus and rebound nystagmus in spinocerebellar ataxia type 6. Mov Disord. 2003;18(10):1201–4. https://doi.org/10.1002/mds.10511.

92. α1ACT Is essential for survival and early cerebellar programming in a critical neonatal window. 102.; 2019. https://doi.org/10.1016/j.neuron.2019.02.036.
93. Benarroch E. What is the role of potassium channels in ataxia? Neurology. 2021;97(20): 938–41. https://doi.org/10.1212/WNL.0000000000012870.
94. Womack MD, Chevez C, Khodakhah K. Calcium-activated potassium channels are selectively coupled to P/Q-type calcium channels in cerebellar purkinje neurons. J Neurosci. 2004;24(40): 8818–22. https://doi.org/10.1523/JNEUROSCI.2915-04.2004.
95. Bozzola E, Savasta S, Peruzzi C, Bozzola M, Bona G. Spinocerebellar ataxias in infancy: pathogenesis of potassium and calcium channels' diseases, clinical features and therapeutical approach. Minerva Pediatr. 2007;59(2):149–56.
96. Harno H, Hirvonen T, Kaunisto MA, et al. Acetazolamide improves neurotological abnormalities in a family with episodic ataxia type 2 (EA-2). J Neurol. 2004;251(2):232–4. https://doi.org/10.1007/s00415-004-0299-1.
97. Kotagal V. Acetazolamide-responsive ataxia. Semin Neurol. 2013;32(05):533–7. https://doi.org/10.1055/s-0033-1334475.
98. Rucker JC. Current treatment of nystagmus. Curr Treat Options Neurol. 2005;7(1):69–77. https://doi.org/10.1007/s11940-005-0008-0.
99. Schniepp R, Wuehr M, Ackl N, et al. 4-Aminopyridine improves gait variability in cerebellar ataxia due to CACNA 1A mutation. J Neurol. 2011;258(9):1708–11. https://doi.org/10.1007/s00415-011-5987-z.
100. Strupp M, Kalla R, Claassen J, et al. A randomized trial of 4-aminopyridine in EA2 and related familial episodic ataxias. Neurology. 2011;77(3):269–75. https://doi.org/10.1212/WNL.0b013e318225ab07.
101. Strupp M, Kalla R, Glasauer S, et al. Aminopyridines for the treatment of cerebellar and ocular motor disorders. In: Progress in brain research, vol. 171. Elsevier; 2008. p. 535–41. https://doi.org/10.1016/S0079-6123(08)00676-6.
102. Schniepp R, Strupp M, Wuehr M, et al. Acetyl-DL-leucine improves gait variability in patients with cerebellar ataxia-a case series. Cerebellum Ataxias. 2016;3:8. https://doi.org/10.1186/s40673-016-0046-2.
103. McLean RJ, Gottlob I. The pharmacological treatment of nystagmus: a review. Expert Opin Pharmacother. 2009;10(11):1805–16. https://doi.org/10.1517/14656560902978446.
104. Alviña K, Khodakhah K. K_{Ca} channels as therapeutic targets in episodic ataxia type-2. J Neurosci. 2010;30(21):7249–57. https://doi.org/10.1523/JNEUROSCI.6341-09.2010.
105. Byers HM, Beatty CW, Hahn SH, Gospe SMJ. Dramatic response after lamotrigine in a patient with epileptic encephalopathy and a De NovoCACNA1A variant. Pediatr Neurol. 2016;60: 79–82. https://doi.org/10.1016/j.pediatrneurol.2016.03.012.
106. Scoggan KA, Friedman JH, Bulman DE. CACNA1A mutation in a EA-2 patient responsive to acetazolamide and valproic acid. Can J Neurol Sci. 2006;33(1):68–72. https://doi.org/10.1017/s0317167100004728.
107. González-Mingot C, López-Ortega R, Brieva-Ruiz L. The efficacy of combining topiramate and 4-aminopyridine to reduce relapses and interictal progression in two cases of episodic ataxia type 2. Neurol Sci. 2022;43(8):5099–101. https://doi.org/10.1007/s10072-022-06144-2.

Chapter 20
Treatment of Central Positional Nystagmus

João Lemos

Keywords Vestibulocerebellar pathways · Cerebellar disinhibition · Continuous vestibular monitoring · Positional nystagmus

20.1 Introduction

Central positional nystagmus (CPN) is characterized by nystagmus precipitated by changes in head position with respect to gravity and is caused by cerebellar and/or brainstem disorders involving the central vestibulo-cerebellar pathways. The latter disorders are believed to interfere with the normal integration of semicircular canal-related signals by the cerebellar nodulus, uvula and/or tonsil, which on its turn leads to an erroneous estimation of head tilt and/or eye position coordinates and ultimately manifests as CPN [1–4]. CPN can be paroxysmal (less than 1 min) or persistent, and is frequently downbeating or upbeating in head hanging positions and/or apogeotropic or geotropic in lateral supine positions. CPN can be symptomatic and associated with intense vertigo, oscillopsia and dysautonomia, particularly its paroxysmal form [1, 2, 5, 6]. Moreover, CPN is frequently associated with additional central ocular motor and/or neurological signs, which by themselves can cause additional symptoms (e.g. diplopia, dysarthria, ataxia) [7]. Among its common aetiologies, CPN is seen in 15% of patients with degenerative/genetic cerebellar ataxia, up to 100% of vestibular migraine patients during an attack, up to ~50% of patients with infratentorial stroke, 20% to ~80% of patients with infratentorial tumours, and in ~6% of patients with multiple sclerosis [8–14]. Auto-immune (paraneoplastic and non-paraneoplastic) ataxias and toxicity account for the remaining cases [7]. Treatments in symptomatic CPN might be directed towards the underlying disorder (e.g. intravenous methylprednisolone), the underlying CPN mechanism (e.g. aminopyridines) and/or be used for alleviating clinical symptoms (e.g. antiemetics, vestibular suppressants) [15]. In the following sections, CPN

J. Lemos (✉)
Neurology of Vision and Balance Unit, Coimbra University Hospital Centre, Coimbra, Portugal

P. Ravat et al. (eds.), *Advances in Translational Neuroscience of Vestibular and Eye Movement Disorders*, Contemporary Clinical Neuroscience,
https://doi.org/10.1007/978-3-032-07539-0_20

treatment will be discussed separately for each disorder. Importantly, despite its high prevalence in several neurological disorders, CPN treatment and outcome has been scarcely reported in the literature [15].

20.2 Structural Lesions

Strategic lesions causing CPN involve either the cerebellar nodulus, uvula, tonsil and/or fibres connecting the latter structures with the vestibular nuclei in the cerebellar peduncles, lateral medulla, and/or dorsolateral region of the fourth ventricle. Cerebellar tumours, multiple sclerosis and ischemic stroke are the common culprits in this context. Importantly, particularly in multiple sclerosis patients, CPN can manifest either acutely, usually associated with spontaneous nystagmus in the seated position (i.e. acute spontaneous and positional vestibular syndrome), or manifest as a chronic disorder with no clear acute precipitant [7].

20.2.1 Tumour

Infratentorial (mostly cerebellar) tumours can cause CPN associated with vertigo and/or posturally evoked vomiting, which may be fully or partially improved after tumoural resection [16, 17] and post-operative radiation [18–20], albeit patients might be left with new ocular motor and vestibular deficits attributable to the surgical procedure itself [18, 20]. The above treatments might improve CPN by alleviating the compression of critical structures and/or pathways within the central vestibulo-cerebellar network. In two patients with a cerebellar astrocytoma causing paroxysmal downbeating and upbeating CPN, respectively, surgical resection and radiation led to resolution of CPN [20]. In contrast, in another patient with a cerebellar hemangioblastoma, paroxysmal downbeating CPN did not resolve after tumour resection [21]. Pharmacological treatment directed to the putative underlying mechanism of CPN is another option for tumour-related CPN, particularly in patients where surgery had no effect on CPN. Accordingly, aminopyridines have been used anecdotally, showing inconsistent benefit [22, 23]. 3,4-diaminopyridine (DAP) was used in a patient with paroxysmal downbeating CPN after surgical removal of a cerebellar multi-nodular dysembroplastic neuroepithelial tumour. There was only a modest effect on CPN velocity 90 min after ingestion, and its long term effect was not evaluated [23]. On the other hand, the use of 4-aminopyridine (4AP) seemed to show a clear benefit in paroxysmal downbeating CPN associated with a cerebellar vermis hemangioblastoma [22]. Aminopyridines, by blocking potassium channels and increasing Purkinje cells excitability [24], may partially restore an uvulo-nodular deficit and reduce the asymmetric charge of the velocity storage [22, 23].

20.2.2 Stroke

Infratentorial stroke may cause CPN [10, 13, 25]. Importantly, CPN and positional vertigo may go unnoticed during the acute phase as they might be overshadowed by concurrent and prominent spontaneous nystagmus and vertigo. Still, unlike spontaneous nystagmus, CPN may persist over time in this context and become progressively noticeable, particularly downbeating CPN [10]. Reports on the efficacy of treatment on stroke-related CPN are scarce. Oral clonazepam was shown to improve positional vertigo and intensity of paroxysmal downbeating and apogeotropic CPN associated with an intraventricular haemorrhage [26]. An initial trial of oral baclofen failed due to drug intolerance. Clonazepam might act on CPN by enhancing gamma-aminobutyric acid (GABA)ergic input from the cerebellar nodulus/uvula to the vestibular nuclei, this way precluding an exaggerated post-acceleratory secondary response by disinhibited irregular vestibular afferents [2, 26]. In contrast, neither the use of baclofen or 4AP had any effect on another patient with paroxysmal downbeating CPN associated with lateral medullary ischemia (Fig. 20.1; Patient 18 from reference [10]).

20.2.3 Multiple Sclerosis

Similarly to what occurs in infratentorial stroke, CPN in MS can present only acutely during a relapse, or remain chronically [10, 27, 28]. Intravenous methylprednisolone for 5 days has abolished persistent torsional CPN in right head hanging position, and persistent downbeating CPN in lateral supine in two unrelated patients with MS vestibular relapses [27, 28]. In the latter case, oral clonazepam has been added [28]. Methylprednisolone might have improved CPN by exerting its anti-inflammatory effect, ultimately leading to the restoring of the otolithic connections between the cerebellum and the vestibular nuclei [29]. Of note, in MS, two particular mechanisms, i.e. transversely spreading ephaptic activation of axons and ion channel dysfunction in partially demyelinated lesions seem to be associated with paroxysmal events (i.e. brainstem paroxysms, trigeminal neuralgia, etc.) [30, 31]. Thus, some forms of CPN in MS patients might theoretically respond to antiepileptic drugs (e.g. carbamazepine, phenytoin, lamotrigine, lacosamide, levetiracetam), which have been shown to improve paroxysmal events in MS [30–32]. Interestingly, in one MS patient, oral phenobarbital dramatically improved subjective positional vertigo associated with paroxysmal downbeating CPN in head hanging positions, 2 years after right pontomedullary demyelination. Still, this was not accompanied by a relevant change in CPN intensity. Previous trials of 4AP and baclofen improved neither positional vertigo or CPN. (Fig. 20.2; Patient 21 from reference [10]).

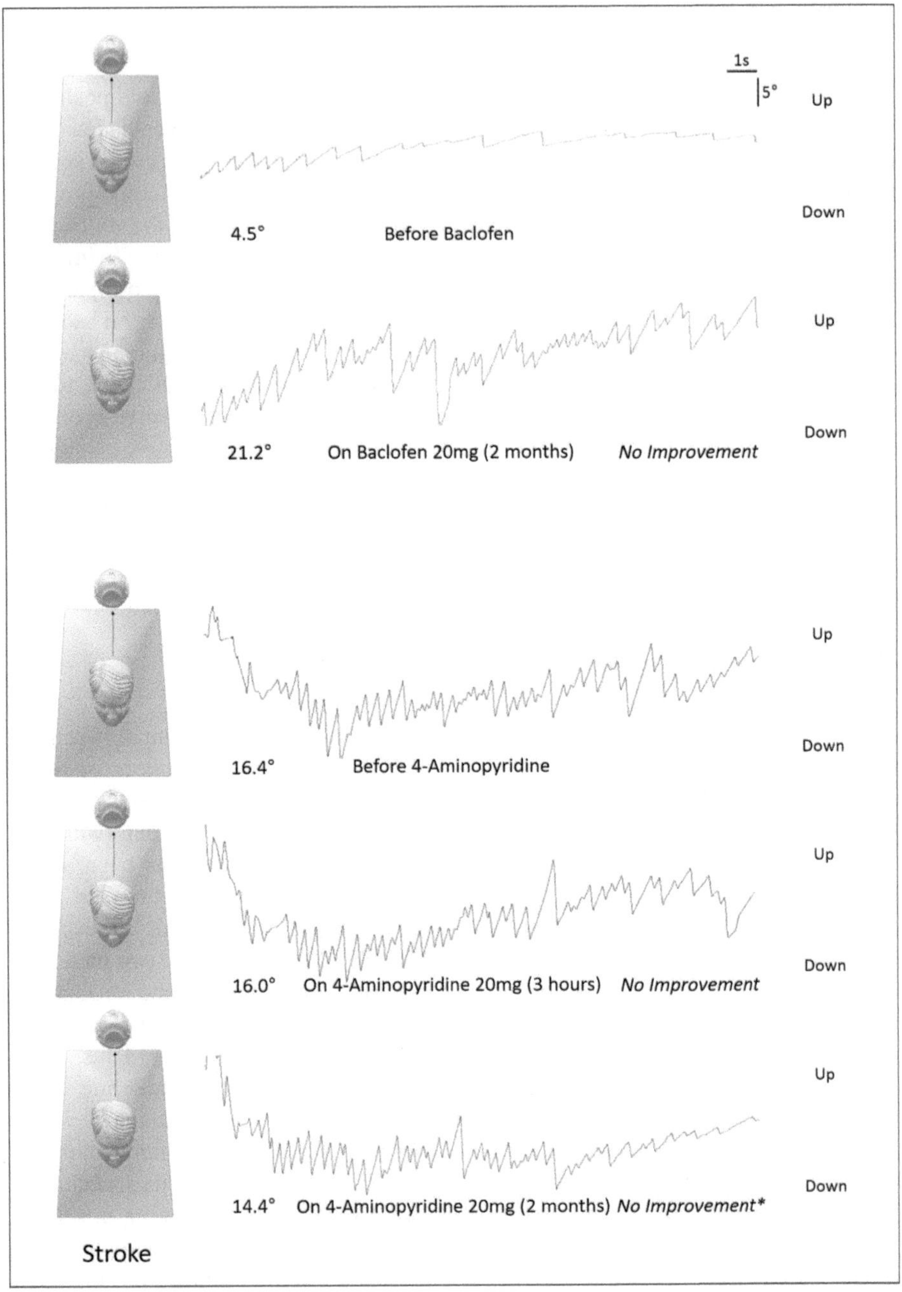

Fig. 20.1 Positional testing and its representation in recordings
Mean slow phase velocity of positional nystagmus and patient's subjective change of positional vertigo with intervention (improvement vs. no improvement) is shown in the bottom left and right corner of each recording, respectively

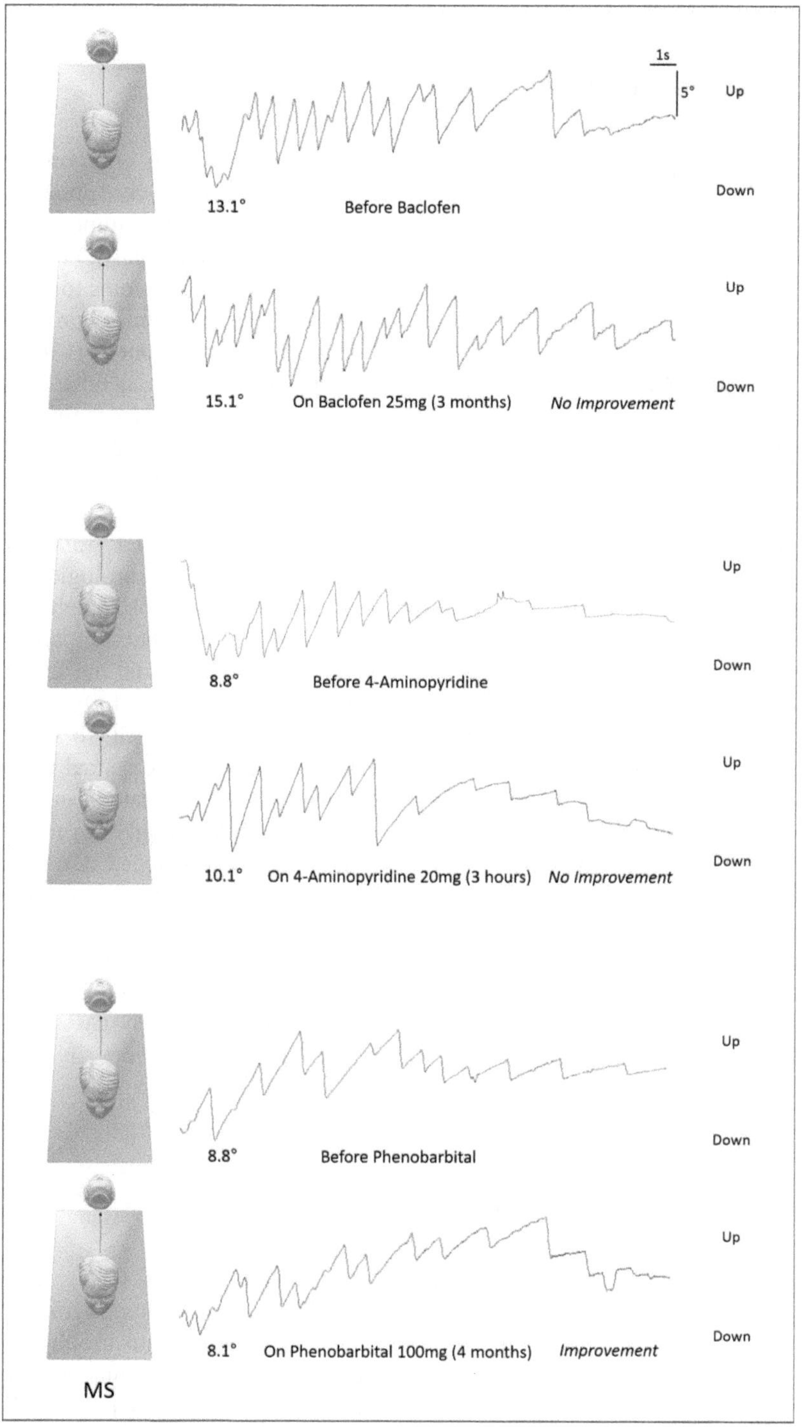

Fig. 20.2 Positional testing
Mean slow phase velocity of positional nystagmus and patient's subjective change of positional vertigo with intervention (improvement vs. no improvement) is shown in the bottom left and right corner of each recording, respectively. *MS* multiple sclerosis

20.3 Auto-Immune Ataxia

Immune-mediated cerebellar ataxias (IMCAs) respond to immunotherapy (e.g. steroids, immunoglobulins, immunosuppressants such as mycophenolate, plasmapheresis and/or rituximab), in addition to a gluten-free diet or tumour-directed therapy [33]. Importantly, IMCAs can be associated (i.e. paraneoplastic) or not with a tumour. In paraneoplastic IMCAs, CPN response to treatment seems to be highly heterogeneous, may or may not accompany tumour response, and its prognosis is obviously dependent on the overall survival dictated by the tumour [34–36]. Since paraneoplastic CPN constitutes an immunological response against neuronal antigens expressed both by an underlying tumour and the cerebellum, both tumour-directed therapies (e.g. resection, chemotherapy, radiotherapy, other) and immunotherapies (e.g. steroids, immunoglobulin, plasmapheresis, cyclophosphamide, rituximab) have been tried in this context [37]. In one case, 4 weeks after a chondrosarcoma resection, paroxysmal torsional CPN during right supine disappeared. However, 6 weeks later, CPN, among other neurological signs, progressed inexorably, showing no response to chemotherapy, plasmapheresis, or oral baclofen [34]. In another patient with Hodgkin lymphoma, paroxysmal downbeating CPN in supine, prone and head-hanging positions did not respond to chemotherapy or radiotherapy, but the tumour itself fully remitted [35]. Non-paraneoplastic IMCAs include anti-GAD Ab- and anti-GQ1b Ab-related ataxias, among others. In one patient with anti-GAD ataxia, paroxysmal upbeating CPN in supine and head-hanging positions and geotropic CPN in lateral supine fully responded to intravenous immunoglobulin for 5 days. Interestingly, an initial trial with oral baclofen also abated CPN, but the drug had to be stopped due to gastric intolerance [38]. In the above case, baclofen might have restored nodulus/uvula GABAergic inhibitory output over the vestibular nuclei, this way abating CPN [38]. Likewise, immunoglobulin use might have restored the same inhibitory output by exerting an anti-inflammatory effect [38, 39]. In anti-GQ1b Ab ataxia, no treatment outcome has been reported for CPN. Still, in the CPN cases detailed, spontaneous remission was observed in all cases over months, without any treatment intervention [40, 41].

20.4 Degenerative and Genetic Ataxia

Some degenerative and genetic ataxias may manifest with CPN, including spinocerebellar ataxia type 6 (SCA6), familial hemiplegic migraine type 1 (FHM1), episodic ataxia type 2 (EA2), and multiple system atrophy (MSA) [9]. Notably, SCA6, FHM1, and EA2 share the presence of mutations in the brain-specific P/Q type Ca^{2+} channel alpha1 subunit gene, CACNA1A [42]. While the latter channelopathies are known to manifest with ictal episodes of ataxia (SCA6, EA2) or hemiplegia/headache (FHM1), CPN on the other hand, when present, is usually in the context of progressive cerebellar ataxia. Thus, CPN in the above genetic ataxias probably represents a manifestation of cerebellar atrophy and neuronal loss [43]. Acetazolamide has been used in the above conditions, mostly to reduce ictal events, as it aids in stabilizing transient abnormal ion channel function, through its acidification of pH [44]. This might explain why acetazolamide is less effective in resolving permanent/non-ictal CPN and ataxia than paroxysmal/ictal episodes [45, 46]. As an example, in three SCA6 patients with inter-ictal downbeating CPN and ictal episodes of ataxia (in two of the patients), the use of acetazolamide markedly reduced the ictal episodes, while showing no benefit on CPN [45]. Likewise, in a FHM1 patient, acetazolamide improved headache, but not downbeating CPN [46]. Still, in another FHM1 patient, inter-ictal downbeating CPN, slowly resolved after 14 months of treatment with acetazolamide [43]. Whether the latter finding constitutes spontaneous resolution of CPN, occasionally seen in genetic ataxia, or reflects a definite treatment response is difficult to appreciate in a single case report [47]. Finally, in episodic ataxias including EA1 and EA2, while downbeating CPN has often been described in EA2, and both acetazolamide and 4AP have been shown to reduce ictal events up to 60%, there are no reports evaluating CPN response to treatment in EA2 [48, 49]. Instead, inter-ictal downbeating CPN has been reported for the first time in EA1 in one patient [50]. Here, acetazolamide completely abolished CPN, with its effect persisting after 6 years (Fig. 20.3). In SCA3, CPN is also rare [51]. In one such patient, we

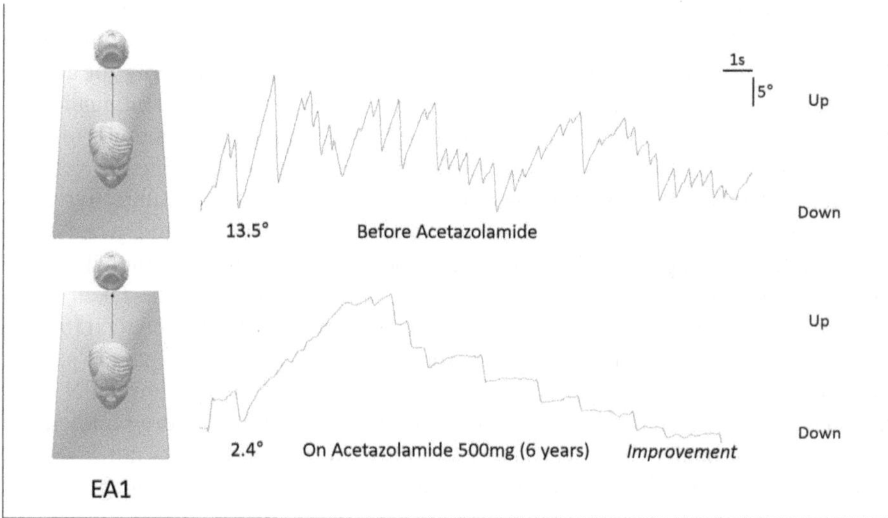

Fig. 20.3 Positional testing
Mean slow phase velocity of positional nystagmus and patient's subjective change of positional vertigo with intervention (improvement vs. no improvement) are shown in the bottom left and right corner of each recording, respectively. *EA1* episodic ataxia type 1

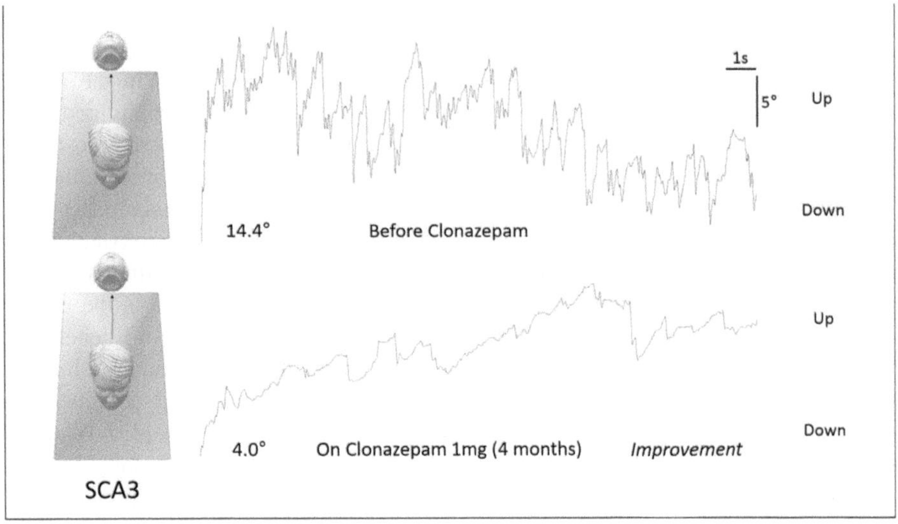

Fig. 20.4 Positional testing
Mean slow phase velocity of positional nystagmus and patient's subjective change of positional vertigo with intervention (improvement vs. no improvement) are shown in the bottom left and right corner of each recording, respectively. *SCA3* spinocerebellar ataxia type 3

were able to fully abort symptomatic downbeating CPN with the use of oral clonazepam, with its effect persisting over time (Fig. 20.4). Recently, in a heterogeneous group of 15 patients with progressive ataxia, including SCA6, SCA1, SCA 2, SCA 17, and MSA, among others, the use of oral baclofen was associated with a ~ 40% decrement of downbeating CPN intensity after taking the drug for at least ~1 week, and a re-increase of CPN intensity ~2 weeks after discontinuation of baclofen. Baclofen might have enhanced the inhibitory action of the vestibulocerebellum on the vestibular nuclei, by specifically inhibiting the velocity storage mechanism of the vestibulo-ocular reflex [52]. Importantly, subjective positional vertigo also improved and persisted, even during the period when CPN reemerged, after drug discontinuation [52]. Therefore, subjective positional vertigo might not fully correlate with CPN intensity, and careful should be taken when interpreting subjective versus objective outcomes of treatment in this context [52]. In the above series, it was nevertheless surprising to see the overall patient's good tolerability to oral baclofen 30 mg per day, when in most of the previous CPN cases reported so far, baclofen had to be stopped at lower dosages due to drug intolerance [26, 38]. Additionally, its effect and tolerability after 1 month was not assessed, leaving unanswered if such approach is feasible in the long-term management of patients with chronic degenerative ataxia.

20.5 Vestibular Migraine

Vestibular migraine (VM) attacks often present with positional vertigo and CPN, which is usually mild, persistent and with variable direction [53]. Given the lack of strong evidence to help selecting an effective treatment for improving VM in the long-term, patients are usually treated with drugs commonly used in the preventive treatment of migraine (e.g. betablockers, antidepressants, antiepileptics, calcium channel blockers, etc.). Propranolol and venlafaxine seem to show slightly more benefit than other drugs [54, 55]. Studies specifically addressing long term response of CPN and positional vertigo attacks to preventive treatment in VM are rare. Still, a few underpowered uncontrolled studies have showed that Vestibular migraine (VM) patients with predominant positional vertigo and CPN seem to respond to preventive treatment, by improving their attacks' frequency and/or severity [56–58]. However, conclusions about the long-term effects of these drugs should be carefully weighted, taking into account the natural course of VM, which can spontaneously vary over time [59, 60]. Prophylactic drugs in VM may act on one or more of the following putative pathomechanisms: spreading depression, vasoconstriction, ion channel defect, sterile inflammation, release of neuropeptides, reduced threshold for trigeminovascular stimulation, and/or locus coeruleus and dorsal raphe nucleus hypersensitization, all ultimately affecting the cerebellar nodulus, uvula and tonsil, and/or the vestibular nuclei [11, 61–63]. On the other hand, abortive treatment in a VM attack could theoretically dampen ictal CPN and positional vertigo more quickly. Unfortunately, treatment studies in this setting are lacking and the role of triptans and related drugs in CPN attacks in VM is largely unknown. Interestingly, the use of non-invasive vagus nerve stimulation (NVNS) in a patient with a VM attack, dramatically aborted persistent right beating CPN and vertigo during, 15 min after stimulation [64]. NVNS might have modulated the neuronal activity in the vestibular nuclei, vestibulocerebellum and their connections by stimulating vagal fibres which travel to brainstem loci that host trigemino-vestibulovagal connections in this case [64, 65].

20.6 Toxicity

Drugs acting on the cerebellar nodulus, uvula and/or tonsil, may cause CPN [66–68]. These include antiepileptic (pregabalin, lamotrigine) and anti-arrhythmic drugs (e.g. amiodarone), among others. Usually, discontinuation or reduction of the offending drug has been sufficient to dampen CPN over time, albeit clonazepam and midazolam had been added in one case report [66–68].

Vestibular suppressants (e.g. dimenhydrinate, flunarizine, cinnarizine, promethazine) and antiemetics (e.g. ondasetron, domperidone, metoclopramide) may temporarily improve CPN-related symptoms [16], but should be otherwise used only for short periods of time, taking into account their possible role in delaying vestibular adaptation [69].

Table 20.1 summarizes CPN treatment.

Table 20.1 Treatment of central positional nystagmus

Treatment targeting CPN mechanism	Clonazepam Baclofen 3,4-Diaminopyridine, 4-aminopyridine Carbamazepine[a], phenytoin[a], lamotrigine[a], lacosamide[a], levetiracetam[a,b], acetazolamide[a,b]
Treatment of the underlying disorder	*Tumour* Tumour directed-therapy (surgical removal, radiotherapy, chemotherapy, etc.) *Multiple sclerosis* Intravenous methylprednisolone *Autoimmune ataxia* Intravenous methylprednisolone, dexamethasone Plasmapheresis Intravenous immunoglobulin Mycophenolate, cyclosporine, cyclophosphamide Rituximab *Vestibular migraine* Zolmitriptan, sumatriptan, eletriptan (acutely) Propranolol, bisoprolol, metoprolol, verapamil, amlodipine, flunarizine, cinnarizine, lomerizine, valproic acid, topiramate, lamotrigine, acetazolamide, amitriptyline, nortriptyline, venlafaxine, paroxetine, duloxetine, escitalopram, di-hydroergotamine *Toxicity* Offending drug discontinuation or reduction
Vestibular suppressants and antiemetics	Meclizine, dimenhydrinate, scopolamine, diazepam, lorazepam, clonazepam, flunarizine, cinnarizine, betahistine, haloperidol, promethazine Ondansetron, domperidone, metoclopramide

CPN central positional nystagmus
[a]Carbamazepine, phenytoin, lamotrigine, lacosamide, levetiracetam, and acetazolamide have been used in paroxysmal events, mostly in multiple sclerosis
[b]Acetazolamide and levetiracetam use has mostly been described in patients with familial hemiplegic migraine type 1, spinocerebellar ataxia type 6 and episodic ataxia type 2

20.7 Conclusions

Currently, CPN treatment is based on underpowered data coming from single case reports and uncontrolled small series with heterogeneous outcomes' assessment. Nevertheless, there is a suggestion that drugs hypothetically targeting CPN mechanism, i.e. abnormally enhanced canal responses during positioning due to cerebellar disinhibition, including clonazepam, baclofen, and aminopyridines, might reduce/dampen CPN. Whether this directly translates into a reduction of the associated subjective positional vertigo needs to be addressed in future studies, since a discrepancy between subjective and objective outcomes has been clearly shown in this context. Therefore, in future studies, reports on CPN treatment should provide data for both types of outcomes. Moreover, studies addressing long-term progression of

CPN, with and without treatment, are lacking. Until then, it becomes difficult to contextualize a single measurement in time showing CPN improvement hours, days or weeks after an intervention. Continuous vestibular monitoring during daily activities might become a promising tool in CPN treatment trials, as it has the potential to provide a long-term, objective recording of eye movements under more natural conditions, data which could eventually better correlate with patients' symptoms [70]. In summary, prospective randomized multicentric treatment trials using standard assessment for measuring objective and subjective outcomes, and including different CPN populations, are highly needed to better support treatment strategies in CPN.

References

1. Choi J-Y, Glasauer S, Kim JH, Zee DS, Kim J-S. Characteristics and mechanism of apogeotropic central positional nystagmus. Brain. 2018;141:762–75.
2. Choi J-Y, Kim JH, Kim HJ, Glasauer S, Kim J-S. Central paroxysmal positional nystagmus: characteristics and possible mechanisms. Neurology. 2015;84:2238–46.
3. Glasauer S, Dieterich M, Brandt T. Central positional nystagmus simulated by a mathematical ocular motor model of otolith-dependent modification of listing's plane. J Neurophysiol. 2001;86:1546–54.
4. Choi SY, Jang JY, Oh EH, Choi JH, Park JY, Lee SH, Choi KD. Persistent geotropic positional nystagmus in unilateral cerebellar lesions. Neurology. 2018;91:e1053–7.
5. Büttner U, Helmchen C, Brandt T. Diagnostic criteria for central versus peripheral positioning nystagmus and vertigo: a review. Acta Otolaryngol. 1999;119:1–5.
6. Choi JY, Kim JS. Central positional nystagmus: characteristics and model-based explanations. Prog Brain Res. 2019;249:211–25.
7. Lemos J, Strupp M. Central positional nystagmus: an update. J Neurol. 2021; https://doi.org/10.1007/S00415-021-10852-8.
8. Polensek SH, Tusa RJ. Nystagmus during attacks of vestibular migraine: an aid in diagnosis. Audiol Neurotol. 2010;15:241–6.
9. Feil K, Strobl R, Schindler A, Krafczyk S, Goldschagg N, Frenzel C, Glaser M, Schöberl F, Zwergal A, Strupp M. What is behind cerebellar vertigo and dizziness? Cerebellum. 2019;18: 320–32.
10. Lemos J, Martins AI, Duque C, Pimentel S, Nunes C, Gonçalves AF. Positional testing in acute vestibular syndrome: a transversal and longitudinal study. Otol Neurotol. 2018; https://doi.org/10.1097/MAO.0000000000002067.
11. Young AS, Lechner C, Bradshaw AP, MacDougall HG, Black DA, Michael Halmagyi G, Welgampola MS. Capturing acute vertigo: a vestibular event monitor. Neurology. 2019;92: E2743–53.
12. Nylen C. Statistical analysis of oto-neurological symptomatology in supra- and subtentorial tumours. Acta Otolaryngol. 1939;27:24–31.
13. Lee SU, Choi JY, Kim HJ, Park JJ, Zee DS, Kim JS. Impaired tilt suppression of post-rotatory nystagmus and cross-coupled head-shaking nystagmus in cerebellar lesions: image mapping study. Cerebellum. 2017;16:95–102.
14. Thomsen J, Zilstorff K, Johnsen NJ. Positional nystagmus of the persistent type. ORL. 1978;40: 86–91.
15. Martins AI, Jorge A, Lemos J. Central positional nystagmus. Curr Treat Options Neurol. 2022;24:453–84.

16. Drachman DA, Diamond ER, Hart CW. Posturally – evoked vomiting; association with posterior fossa lesions. Ann Otol Rhinol Laryngol. 1977;86:97–101.
17. Chan T, Logan P, Eustace P. Intermittent downbeat nystagmus secondary to vermian arachnoid cyst with associated obstructive hydrocephalus. J Clin Neuroophthalmol. 1991;11:293–6.
18. Barber HO. Positional nystagmus. Otolaryngol Neck Surg. 1984;92:649–55.
19. Shoman N, Longridge N. Cerebellar vermis lesions and tumours of the fourth ventricle in patients with positional and positioning vertigo and nystagmus. J Laryngol Otol. 2007;121: 166–9.
20. Gregorius FK, Crandall PH, Baloh RW. Positional vertigo with cerebellar astrocytoma. Surg Neurol. 1976;6:283–6.
21. Lea J, Lechner C, Halmagyi GM, Welgampola MS. Not so benign positional vertigo: paroxysmal downbeat nystagmus from a superior cerebellar peduncle neoplasm. Otol Neurotol. 2014; https://doi.org/10.1097/MAO.0000000000000245.
22. Kremmyda O, Zwergal A, La Fougère C, Brandt T, Jahn K, Strupp M. 4-Aminopyridine suppresses positional nystagmus caused by cerebellar vermis lesion. J Neurol. 2013;260:321–3.
23. Helmchen C, Gottschalk S, Sander T, Trillenberg P, Rambold H, Sprenger A. Beneficial effects of 3,4-diaminopyridine on positioning downbeat nystagmus in a circumscribed uvulo-nodular lesion. J Neurol. 2007;254:1126–8.
24. Etzion Y, Grossman Y. Highly 4-aminopyridine sensitive delayed rectifier current modulates the excitability of Guinea pig cerebellar Purkinje cells. Exp Brain Res. 2001;1394(139):419–25.
25. Sakata E, Ohtsu K, Itoh Y. Positional nystagmus of benign paroxysmal type (BPPN) due to cerebellar vermis lesions. Acta Otolaryngol Suppl. 1991;481:254–7.
26. Oh EH, Kim H, Choi SY, Choi KD, Choi JH. Paroxysmal central positional nystagmus responsive to clonazepam. J Neurol. 2022;269:1028–31.
27. Anagnostou E, Varaki K, Anastasopoulos D. A minute demyelinating lesion causing acute positional vertigo. J Neurol Sci. 2008;266:187–9.
28. Anagnostou E, Mandellos D, Limbitaki G, Papadimitriou A, Anastasopoulos D. Positional nystagmus and vertigo due to a solitary brachium conjunctivum plaque. J Neurol Neurosurg Psychiatry. 2006;77:790–2.
29. Meier CA. Mechanisms of immunosuppression by glucocorticoids. Eur J Endocrinol. 1996;134:50.
30. Ostermann PO, Westerberg CE. Paroxysmal attacks in multiple sclerosis. Brain. 1975;98:189–202.
31. Espir ML, Millac P. Treatment of paroxysmal disorders in multiple sclerosis with carbamazepine (Tegretol). J Neurol Neurosurg Psychiatry. 1970;33:528.
32. Ciampi E, Uribe-San-Martín R, Godoy-Santín J, Cruz JP, Cárcamo-Rodríguez C, Juri C. Secondary paroxysmal dyskinesia in multiple sclerosis: clinical-radiological features and treatment. Case report of seven patients. Mult Scler. 2017;23:1791–5.
33. Mitoma H, Manto M, Hadjivassiliou M. Immune-mediated cerebellar ataxias: clinical diagnosis and treatment based on immunological and physiological mechanisms. J Mov Disord. 2021;14: 10–28.
34. Kearsley JH, Johnson P, Halmagyi GM. Paraneoplastic cerebellar disease: remission with excision of the primary tumor. Arch Neurol. 1985;42:1208–10.
35. Eggers SDZ, Pittock SJ, Shepard NT, Habermann TM, Neff BA, Klebig RR. Positional periodic alternating vertical nystagmus with PCA-Tr antibodies in Hodgkin lymphoma. Neurology. 2012;78:1800–2.
36. Tafur AJ, Baumann Kreuzinger LM, Quevedo F. 28-year-old man with severe vertigo. Mayo Clin Proc. 2008;83:1070–3.
37. Wray SH, Dalmau J, Chen A, King S, Leigh RJ. Paraneoplastic disorders of eye movements. Ann N Y Acad Sci. 2011;1233:279–84.
38. Martins AI, Carvalho JN, Amorim AM, Geraldo A, Eggenberger E, Lemos J. Disabling central paroxysmal positioning upbeat nystagmus and vertigo associated with the presence of anti–glutamic acid decarboxylase antibodies. J Neuro-Ophthalmol. 2017;38:1.

39. Gelfand EW. Intravenous immune globulin in autoimmune and inflammatory diseases. N Engl J Med. 2012;367:2015–25.
40. Jeong S-H, Nam J, Kwon MJ, Kim JK, Kim JS. Nystagmus and ataxia associated with antiganglioside antibodies. J Neuro-Ophthalmol. 2011;31:326–30.
41. Lee SU, Kim HJ, Choi JY, Kim JK, Kim JS. Acute vestibular syndrome associated with anti-GQ1b antibody. Neurology. 2019;93:e1085–92.
42. Ophoff RA, Terwindt GM, Vergouwe MN, et al. Familial hemiplegic migraine and episodic ataxia type-2 are caused by mutations in the Ca2+ channel gene CACNL1A4. Cell. 1996;87: 543–52.
43. Suzuki M, Fujiwara K, Tsubuku T, Yabe I, Sasaki H, Fukuda S. Time course of downbeat positioning nystagmus in familial hemiplegic migraine type 1 treated with acetazolamide. J Neurol Sci. 2016;368:206–8.
44. Strupp M, Zwergal A, Brandt T. Episodic ataxia type 2. Neurotherapeutics. 2007;4:267–73.
45. Jen JC, Yue Q, Karrim J, Nelson SF, Baloh RW. Spinocerebellar ataxia type 6 with positional vertigo and acetazolamide responsive episodic ataxia. J Neurol Neurosurg Psychiatry. 1998;65: 565–8.
46. Yabe I, Kitagawa M, Suzuki Y, et al. Downbeat positioning nystagmus is a common clinical feature despite variable phenotypes in an FHM1 family. J Neurol. 2008;255:1541–4.
47. Yu-Wai-Man P, Gorman G, Bateman DE, Leigh RJ, Chinnery PF. Vertigo and vestibular abnormalities in spinocerebellar ataxia type 6. J Neurol. 2009;256:78–82.
48. Ling X, Kim HJ, Lee JH, Choi JY, Yang X, Kim JS. Positioning velocity matters in central paroxysmal positional vertigo: implication for the mechanism. Front Neurol. 2020;11:591602.
49. Muth C, Teufel J, Schöls L, Synofzik M, Franke C, Timmann D, Mansmann U, Strupp M. Fampridine and acetazolamide in EA2 and related familial EA. Neurol Clin Pract. 2021;11:e438–46.
50. Jorge A, Melancia D, Figueiredo C, Galego O, Oliveira J, Martins AI, Lemos J. Downbeat nystagmus in episodic ataxia type 1 associated with a novel KCNA1 mutation. Mov Disord. 2022;37:430–2.
51. Kim JS, Kim JS, Youn J, Seo D-W, Jeong Y, Kang J-H, Park JH, Cho JW. Ocular motor characteristics of different subtypes of spinocerebellar ataxia: distinguishing features. Mov Disord. 2013;28:1271–7.
52. Yun SY, Lee JH, Kim HJ, Choi J, Kim JS. Effects of baclofen on central paroxysmal positional downbeat nystagmus. Cerebellum. 2024; https://doi.org/10.1007/S12311-024-01684-Z.
53. Young AS, Nham B, Bradshaw AP, Calic Z, Pogson JM, D'Souza M, Halmagyi GM, Welgampola MS. Clinical, oculographic, and vestibular test characteristics of vestibular migraine. Cephalalgia. 2021;41:1039.
54. Byun YJ, Levy DA, Nguyen SA, Brennan E, Rizk HG. Treatment of vestibular migraine: a systematic review and meta-analysis. Laryngoscope. 2021;131:186–94.
55. Wang F, Wang J, Cao Y, Xu Z. Serotonin-norepinephrine reuptake inhibitors for the prevention of migraine and vestibular migraine: a systematic review and meta-analysis. Reg Anesth Pain Med. 2020;45:323–30.
56. Beh SC. Horizontal direction-changing positional nystagmus and vertigo: a case of vestibular migraine masquerading as Horizontal Canal BPPV. Headache. 2018; https://doi.org/10.1111/head.13356.
57. Roberts RA, Gans RE, Kastner AH. Differentiation of migrainous positional vertigo (MPV) from horizontal canal benign paroxysmal positional vertigo (HC-BPPV). Int J Audiol. 2006; https://doi.org/10.1080/14992020500429658.
58. Mahrous MM. Vestibular migraine and benign paroxysmal positional vertigo, close presentation dilemma. Acta Otolaryngol. 2020; https://doi.org/10.1080/00016489.2020.1770857.
59. Cutrer FM, Baloh RW. Migraine-associated dizziness. Headache J Head Face Pain. 1992; https://doi.org/10.1111/j.1526-4610.1992.hed3206300.x.
60. Neuhauser H, Radtke A, Von Brevern M, Lempert T. Zolmitriptan for treatment of migrainous vertigo: a pilot randomized placebo-controlled trial. Neurology. 2003;60:882–3.

61. von Brevern M, Zeise D, Neuhauser H, Clarke AH, Lempert T. Acute migrainous vertigo: clinical and oculographic findings. Brain. 2004;128:365–74.
62. Von Brevern M, Radtke A, Clarke AH, Lempert T. Migrainous vertigo presenting as episodic positional vertigo. Neurology. 2004;62:469–72.
63. King S, Priesol AJ, Davidi SE, Merfeld DM, Ehtemam F, Lewis RF. Self-motion perception is sensitized in vestibular migraine: pathophysiologic and clinical implications. Sci Rep. 2019; https://doi.org/10.1038/s41598-019-50803-y.
64. Beh SC. Nystagmus and vertigo in acute vestibular migraine attacks: response to non-invasive Vagus nerve stimulation. Otol Neurotol. 2021;42:e233–6.
65. Henssen DJHA, Derks B, van Doorn M, Verhoogt N, Cappellen V, van Walsum AM, Staats P, Vissers K. Vagus nerve stimulation for primary headache disorders: an anatomical review to explain a clinical phenomenon. Cephalalgia. 2019; https://doi.org/10.1177/0333102419833076.
66. Choi JY, Park YM, Woo YS, Kim SU, Jung JM, Kwon DY. Perverted head-shaking and positional downbeat nystagmus in pregabalin intoxication. J Neurol Sci. 2014;337:243–4.
67. Oh S-Y, Kim JS, Lee YH, Lee AY, Kim J, Kim JM. Downbeat, positional, and perverted head-shaking nystagmus associated with lamotrigine toxicity. J Clin Neurol. 2006;2:283.
68. Arbusow V, Strupp M, Brandt T. Amiodarone-induced severe prolonged headpositional vertigo and vomiting. Neurology. 1998;51:917.
69. Scholtz AW, Steindl R, Burchardi N, Bognar-Steinberg I, Baumann W. Comparison of the therapeutic efficacy of a fixed low-dose combination of cinnarizine and dimenhydrinate with betahistine in vestibular neuritis: a randomized, double-blind, non-inferiority study. Clin Drug Investig. 2012;32:387–99.
70. Phillips JS, Newman J, Cox S. Towards providing an automated approach to differentiating the nystagmus of Ménière's disease, vestibular migraine, and benign paroxysmal positional vertigo. Otol Neurotol. 2021;42:890–6.

Chapter 21
The Five S's of Dysfunctional Saccades: Slow, Sticky, Sloppy, Stunted, and Spontaneous

Janet C. Rucker

Keywords Saccades · Supranuclear gaze palsy · Ocular motor apraxia · Saccadic intrusions · Opsoclonus

21.1 Introduction

Saccades are fast conjugate eye movements that rapidly shift gaze to bring salient visual targets onto the fovea, the portion of the retina with the highest density of photoreceptors and the highest visual acuity. Saccades are examined clinically by asking the patient to move their eyes rapidly upon command between two visual targets separated horizontally and then vertically, noting initiation, speed, trajectory, and accuracy. To optimize visual clarity, vision is suppressed during saccades; thus, for optimal visual function, saccades must be brief (typically less than 100 msec), fast (up to 600 deg./sec), and land accurately on target (Fig. 21.1a) [1]. Peak saccadic velocity and saccade duration follow a main sequence by which saccadic velocity increases linearly with saccade amplitude up to a point of velocity saturation for large saccades. The initiation of a saccade requires a neural burst, or pulse, from the excitatory burst neurons (EBN) in the brainstem to the ocular motoneurons that is transmitted to generate high-frequency phasic activity in agonist extraocular muscles. Simultaneously, the antagonist muscles relax due to input from inhibitory burst neurons (IBN) located in the medullary reticular formation. For horizontal and vertical saccades, respectively, the pulse originates in EBN located in the paramedian pontine reticular formation (PPRF) and the rostral interstitial medial longitudinal fasciculus (RIMLF) [2, 3]. Following the pulse, a tonic signal output

Supplementary Information The online version contains supplementary material available at https://doi.org/10.1007/978-3-032-07539-0_21.

J. C. Rucker (✉)
Departments of Neurology and Ophthalmology, NYU Grossman School of Medicine, New York, NY, USA
e-mail: Janet.rucker@nyulangone.org

P. Ravat et al. (eds.), *Advances in Translational Neuroscience of Vestibular and Eye Movement Disorders*, Contemporary Clinical Neuroscience,
https://doi.org/10.1007/978-3-032-07539-0_21

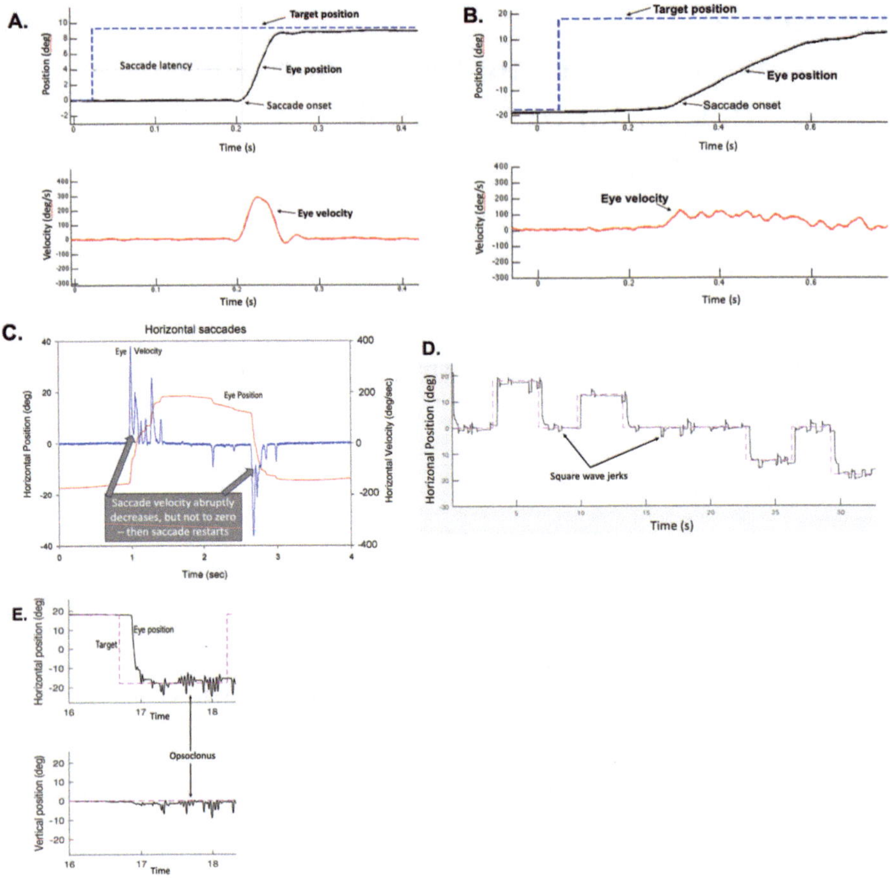

Fig. 21.1 Examples of saccade profiles obtained with infrared oculography. (**A**). **Normal.** Position and velocity traces of a single 10-degree horizontal saccade in a normal healthy individual. The light gray vertical line represents saccade onset. (**B**). **Slow.** Position and velocity traces of a single 40-degree horizontal saccade that is very slow and fails to reach the target in a patient with dysfunction of excitatory burst neurons for horizontal saccades in the paramedian pontine reticular formation. (**C**). **Stunted.** Position and velocity traces of a nearly 40-degree saccade in a patient with late-onset Tay-Sachs disease. The eye position trace shows hypometric, multistep saccades. The eye velocity trace shows saccades with "discrete decelerations," sometimes to a stop and sometimes with transient decline, but not to zero velocity, followed by continuation of the saccade. (**D**). **Spontaneous: square wave jerks.** Horizontal position trace with spontaneous square wave jerks occurring intermittently. Note the small saccades that remove the eye from and return it to fixation, with a brief intersaccadic interval between the pairs of saccades comprising the square wave jerks. (**E**). **Spontaneous: opsoclonus.** A burst of opsoclonus after a 40-degree leftward saccade. Note the back-to-back small saccades in the horizontal (top) and vertical (bottom) eye position traces and the lack of an intersaccadic interval between spontaneous saccades.
By convention, upward deflections in position traces represent rightward movement in horizontal traces or upward movement in vertical traces. Hatched lines represent the visual target

called the step is required to maintain the eyes in the new position. The step is generated by neural integrators located, for horizontal and vertical saccades, respectively, in the nucleus prepositus hypoglossi (NPH)/medial vestibular nucleus (MVN) in the medulla and the interstitial nucleus of Cajal (INC) in the midbrain. At all times other than during a saccade, EBN are inhibited by omnipause neurons (OPN) in the raphe interpositus (RIP) in the caudal pons [4]. At the end of a saccade, the OPN resume firing. The dysfunction of each of the above neural components leads to a variety of localizable clinical and subclinical saccadic disorders. A schematic of five S's of saccadic dysfunction is presented here (Fig. 21.1, Video 21.1). See references for more comprehensive coverage of each type of disorder.

21.2 Slow Saccades

Orbital disease, including thyroid eye disease and extraocular myopathies such as chronic progressive external ophthalmoplegia (CPEO), can cause slowing of saccades. However, in these scenarios, the localization and diagnosis are usually evident from other examination features. Thyroid eye disease causes external diagnostic clues, including proptosis, chemosis, eyelid retraction, and periorbital edema. CPEO is accompanied by bilateral ptosis with impaired levator palpebrae superioris excursion, along with ophthalmoplegia, with which eye movement range is not improved by testing of vestibulo-ocular reflexes. CPEO may occur in isolation or accompanied by other neurological signs of mitochondrial disease, such as systemic myopathy. Cranial neuropathies also cause slowing of saccades in the direction of action of the denervated muscles. The detection of saccadic slowing, for example, during abduction of one eye, suggests sixth nerve dysfunction rather than an alternate diagnosis, such as ocular myasthenia gravis.

Saccadic slowing of adducting saccades, called adduction lag, occurs with internuclear ophthalmoplegia (INO) due to a lesion of the medial longitudinal fasciculus (MLF). The MLF interneurons originate in the abducens nucleus, which contains two populations of neurons: motoneurons destined for the ipsilateral lateral rectus and interneurons that decussate and ascend in the contralateral MLF to allow for conjugate horizontal gaze shifts. Additional signs of INO include impaired adduction of the ipsilesional eye that may or may not be overcome by convergence and compensatory abducting nystagmus in the contralesional eye. These deficits are often enhanced during saccade testing and, with a subtle INO, slowing of adducting saccades may be the only finding [5]. Saccades place a greater demand on the motoneuron than pursuit movements, and the demyelinated or infarcted MLF is less equipped to meet the high demands of saccadic movements.

Saccadic slowing due to EBN pathology is a characteristic finding in supranuclear gaze palsies of brainstem origin (Fig. 21.1b) [6, 7]. Slowing of horizontal saccades occurs with PPRF EBN lesions; slowing of vertical saccades, with RIMLF EBN lesions. The hallmark feature of a supranuclear gaze palsy is ophthalmoplegia that is overcome by vestibulo-ocular reflexes. More subtle supranuclear gaze palsies may

Table 21.1 Differential diagnosis of the most common causes of slow saccades and supranuclear brainstem saccadic gaze palsy

Neurodegenrative
Progressive supranuclear palsy (V > H)
Huntington's disease
Corticobasal syndrome (later in disease course)
Vascular
Midbrain (V) or pontine (H) infarction
Following cardiac/aortic surgery
Paraneoplastic/automimmune
Ma, GAD, GlyR, and IgLON5 antibodies
Demyelinating disease
Genetic/metabolic
Niemann-pick
Gaucher type 3
Kufor-Rakeb
Wilson's disease
Spinocerebellar ataxia type 2 (most common, but also types 1, 3, 6, 7, and 28)
Leigh syndrome
Prion disease
Creutzfeld-Jacob disease
Infection
Whipple's disease

Abbreviations: *H* horizontal, *V* vertical

manifest as isolated saccadic slowing [8]. Checking optokinetic nystagmus (OKN) is also helpful and will often reveal absent quick phases, since OKN quick phases are generated by EBN. Saccade trajectory may be impaired if a supranuclear brainstem saccadic gaze palsy affects horizontal or vertical EBN alone or asymmetrically, resulting in "round the houses" sign—a curved saccadic trajectory. For example, with a vertical supranuclear gaze palsy, saccades between two vertically placed visual targets may demonstrate a horizontal component leading to a curved trajectory of the saccade. Brainstem stroke is the most common cause of acute-onset supranuclear gaze palsy. With insidious onset leading to chronic dysfunction, progressive supranuclear palsy is a common cause of vertical supranuclear gaze palsy. See Table 21.1 for a list of classic etiologies of slow saccades due to supranuclear saccadic gaze palsy [9].

Clinical assessment of saccadic speed may fail to accurately differentiate slow from normal saccades. On clinical examination, horizontal saccades are more accurately judged as normal and vertical saccades as slow. The accuracy of detection of conjugate saccade slowing and of adducting saccade slowing in INO can be enhanced by infrared-oculographic recordings with direct measurement of saccadic velocities [10, 11].

21.3 Sticky Saccades

Sticky saccades refer to the saccadic characteristics of ocular motor apraxia from bilateral frontal or parietal cortical eye field lesions [12]. The cortical eye fields project to the superior colliculus, which transforms the cortical spatially-encoded saccade command to a temporally-encoded motor command transmitted to the EBN. Ocular motor apraxia causes impaired initiation of voluntary eye movements, including saccades and smooth pursuit, but spares reflexive eye movements such as OKN. Saccades have prolonged latency; latency can be shortened by enhancing the salience of the visual target (e.g., using money as a saccade target, rather than the examiner's finger). Once initiated, the velocity of saccades is normal. Patients will often utilize head thrusting or blinking to facilitate saccade initiation.

21.4 Sloppy Saccades

Saccades must be highly accurate to place the small fovea onto objects of visual interest. Saccades in normal, healthy individuals may fall slightly short of the target and require a corrective saccade to bring the eye to the target [13]. Pathologic saccadic dysmetria includes both hypometric and hypermetric saccades. Dysmetric saccades are classically due to dysfunction of the cerebellar dorsal vermis and caudal fastigial nucleus, though hypometric saccades are less specifically localizing than hypermetric saccades. Bilateral dorsal vermis lesions cause hypometric saccades, whereas bilateral fastigial nucleus lesions result in hypermetric saccades [14].

21.5 Stunted Saccades

Interruption, or stunting, of a saccade in progress is a disorder of termination of saccades that have been properly initiated and, thus, is not related to an EBN abnormality. OPN resume firing at the end of a saccade, but, in many OPN, this occurs after the saccade has already been terminated. It has been proposed that saccade termination may occur through a "choke" signal sent to the IBN contralateral to saccade direction by the caudal fastigial nucleus. These contralateral IBN then "choke" the "drive" signal from the EBN ipsilateral to saccade direction [15]. However, OPN stimulation in nonhuman primates during a saccade does cause an abrupt saccade deceleration, followed by continuation of the saccade when OPN stimulation stops [16]. Transient decelerations mid-saccade, leading to hypometric, multistep saccades that are undetectable on clinical examination but detectable with ocular motor recordings, have been well-studied in late-onset Tay-Sachs (Fig. 21.1c) [17]. Such transient decelerations consist of an abrupt decline in saccadic velocity mid-saccade, but not to zero (often remaining above 50 deg/sec).

Neuropathologic analysis of OPN and the cerebellar dorsal vermis, which is important in saccade control, in late-onset Tay-Sachs revealed involvement of both structures, with diffuse gliosis and vesicular inclusions [18]. Neuromimetic modeling studies showed that lesioning of OPN does lead to abrupt deceleration of saccades, but saccades do not resume following the deceleration; thus, premature OPN reactivation alone does not replicate the stunted saccade behavior [19]. In contrast, failure of the caudal fastigial nucleus inhibition replicates the stunted saccade behavior, leading to the hypothesis that stunted saccades in late-onset Tay-Sachs are due to fastigial nucleus dysregulation, which leads to transient early IBN firing that causes transient decelerations in saccadic velocity mid-saccade [19, 20].

21.6 Spontaneous Saccades

Abnormal spontaneous saccades that intrude upon fixation are saccadic intrusions [21]. Saccadic intrusions are divided into two categories based on whether the saccades making up the intrusion have a brief pause, or an intersaccadic interval, between subsequent saccades. Examples of saccadic intrusions with an intersaccadic interval are square wave jerks and macrosaccadic oscillations. Examples without an intersaccadic interval are ocular flutter and opsoclonus.

Square wave jerks are composed of pairs of small saccades (usually 0.5–5 degrees; when larger than 5 degrees they are termed macro-square wave jerks) that remove the eye from and return the eye to central fixation, with a brief pause of 200 msec on average between the two small saccades (Fig. 21.1d). These are often normal in healthy older individuals. Frequent, larger square wave jerks are typically pathologic and especially common in certain disease states, such as progressive supranuclear palsy and Friedreich ataxia [22, 23]. Square wave jerks arise from dysfunction of structures involved in fixation, such as the superior colliculus and its inputs. Occasionally, they occur in runs of back-to-back square wave jerks called square wave oscillations that can mimic ocular flutter clinically. Macrosaccadic oscillations also have intersaccadic intervals between a series of abnormal spontaneous saccades, but this type of saccadic intrusion differs from square wave jerks. These oscillations consist of a series of spontaneous saccades that cross the midline, gradually building and then reducing in amplitude. They are a manifestation of dysfunction in the ocular motor vermis and caudal fastigial nucleus involved in saccadic accuracy.

Ocular flutter and opsoclonus are spontaneous, high-frequency, back-to-back saccades that intrude upon fixation without an intersaccadic interval. The saccades are ocular flutter when they are limited to the horizontal plane and opsoclonus when they occur in multiple planes (e.g., horizontal, vertical, or torsional) (Fig. 21.1e). Unlike square wave jerks, which are typically asymptomatic, ocular flutter and opsoclonus usually cause oscillopsia, a perception of visual motion. These pathologic eye movements occur when brainstem or cerebellar disease leads to a pause or delay in reactivation of OPN that unleashes the EBN, which are prone to oscillate. This EBN propensity to oscillate is due to two inherent properties: (1) post-inhibitory rebound firing in EBN facilitates saccade acceleration upon EBN release from OPN inhibition [24, 25] and (2) positive feedback loops created by mutual IBN inhibition [26]. Ocular flutter and opsoclonus are usually caused by either parainfectious states or by paraneoplastic etiologies, though they may occur with other etiologies also, such as trauma and uremia [27, 28].

References

1. Leigh RJ, Zee DS. The neurology of eye movements. New York: Oxford University Press; 2015.
2. Horn AK, Büttner-Ennever JA. Premotor neurons for vertical eye movements in the rostral mesencephalon of monkey and human: histologic identification by parvalbumin immunostaining. J Comp Neurol. 1998;392:413–27.
3. Horn AK, Büttner-Ennever JA, Suzuki Y, Henn V. Histological identification of premotor neurons for horizontal saccades in monkey and man by parvalbumin immunostaining. J Comp Neurol. 1995;359:350–63.
4. Büttner-Ennever JA, Cohen B, Pause M, Fries W. Raphe nucleus of the pons containing omnipause neurons of the oculomotor system in the monkey, and its homologue in man. J Comp Neurol. 1988;267:307–21.
5. Crane TB, Yee RD, Baloh RW, Hepler RS. Analysis of characteristic eye movement abnormalities in internuclear ophthalmoplegia. Arch Ophthalmol. 1983;101:206–10.
6. Jensen K, Shaikh AG. Slow saccades. In: Shaikh A, Ghasia F, editors. Annals in translational neuroscience of eye movement disorders. Berlin: Springer; 2019.
7. Rucker JC, Hudson T, Rizzo JR. Translational neurology of slow saccades. In: Shaikh A, Ghasia F, editors. Annals in translational neuroscience of eye movement disorders. Berlin: Springer; 2019.
8. Hardwick A, Rucker JC, Cohen ML, Friedland RP, Gustaw-Rothenberg K, Riley DE, Leigh RJ. Evolution of oculomotor and clinical findings in autopsy-proven Richardson syndrome. Neurology. 2009;73:2122–4.
9. Lloyd-Smith Sequeira A, Rizzo JR, Rucker JC. Clinical approach to Supranuclear brainstem saccadic gaze palsies. Front Neurol. 2017;8:429.
10. Frohman TC, Frohman EM, O'suilleabhain P, Salter A, Dewey RB Jr, Hogan N, Galetta S, Lee AG, Straumann D, Noseworthy J, Zee D, Corbett J, Corboy J, Rivera VM, Kramer PD. Accuracy of clinical detection of INO in MS: corroboration with quantitative infrared oculography. Neurology. 2003;61:848–50.
11. Grossman SN, Calix R, Hudson T, Rizzo JR, Selesnick I, Frucht S, Galetta SL, Balcer LJ, Rucker JC. Accuracy of clinical versus oculographic detection of pathological saccadic slowing. J Neurol Sci. 2022;442:120436.
12. Tilikete C, Matthieu PR. Ocular Motor Apraxia. In: Shaikh A, Ghasia F, editors. Annals in translational neuroscience of eye movement disorders. Berlin: Springer Nature; 2019.
13. Pelisson D, Prablanc C. Kinematics of centrifugal and centripetal saccadic eye movements in man. Vis Res. 1988;28:87–94.
14. Shemesh AA, Zee DS. Eye movement disorders and the cerebellum. J Clin Neurophysiol. 2019;36:405–14.
15. Optican LM, Quaia C. Distributed model of collicular and cerebellar function during saccades. Ann N Y Acad Sci. 2002;956:164–77.

16. Keller EL. Control of saccadic eye movements by midline brainstem neurons. In: Baker R, Berthoz A, editors. Control of gaze by brainstem neurons. Amsterdam: Elsevier; 1977.
17. Rucker JC, Shapiro BE, Han YH, Kumar AN, Garbutt S, Keller EL, Leigh RJ. Neuro-ophthalmology of late-onset Tay-Sachs disease (LOTS). Neurology. 2004;63:1918–26.
18. Rucker JC, Leigh RJ, Optican LM, Keller EL, Bu Ttner-Ennever JA. Ocular motor anatomy in a case of interrupted saccades. Prog Brain Res. 2008;171:563–6.
19. Optican LM, Rucker JC, Keller EL, Leigh RJ. Mechanism of interrupted saccades in patients with late-onset Tay-Sachs disease. Prog Brain Res. 2008;171:567–70.
20. Rucker JC, Ying SH, Moore W, Optican LM, Buttner-Ennever J, Keller EL, Shapiro BE, Leigh RJ. Do brainstem omnipause neurons terminate saccades? Ann N Y Acad Sci. 2011;1233:48–57.
21. Rucker JC. Nystagmus and saccadic intrusions. Continuum (Minneap Minn). 2019;25:1376–400.
22. Otero-Millan J, Serra A, Leigh RJ, Troncoso XG, Macknik SL, Martinez-Conde S. Distinctive features of saccadic intrusions and microsaccades in progressive supranuclear palsy. J Neurosci. 2011;31:4379–87.
23. Shallo-Hoffmann J, Petersen J, Muhlendyck H. How normal are "normal" square wave jerks? Invest Ophthalmol Vis Sci. 1989;30:1009–11.
24. Miura K, Optican LM. Membrane channel properties of premotor excitatory burst neurons may underlie saccade slowing after lesions of omnipause neurons. J Comput Neurosci. 2006;20:25–41.
25. Shaikh AG, Ramat S, Optican LM, Miura K, Leigh RJ, Zee DS. Saccadic burst cell membrane dysfunction is responsible for saccadic oscillations. J Neuroophthalmol. 2008;28:329–36.
26. Ramat S, Leigh RJ, Zee DS, Optican LM. Ocular oscillations generated by coupling of brainstem excitatory and inhibitory saccadic burst neurons. Exp Brain Res. 2005;160:89–106.
27. Changa AR, Rucker JC, Drummond PS. Opsoclonus in Uremia with resolution after Hemodialysis. J Neuroophthalmol. 2022;42:e448–9.
28. Rizzo JR, Hudson TE, Sequeira AJ, Dai W, Chaudhry Y, Martone J, Zee DS, Optican LM, Balcer LJ, Galetta SL, Rucker JC. Eye position-dependent opsoclonus in mild traumatic brain injury. Prog Brain Res. 2019;249:65–78.

Chapter 22
Navigating Eye Movement Challenges in Parkinson's Disease: Insights and Solutions

Ibada Chowdhary, Inaya Chowdhary, Fatema F. Ghasia, and Aasef G. Shaikh

Keywords Saccades · Vergence · Spatial navigation · Motor and perceptual dysfunction · Deep brain stimulation

22.1 Introduction

Parkinson's disease (PD), the second most common neurodegenerative disorder, affects around ten million people worldwide [18]. Visual impairments in PD are more widespread than often recognized. PD patients score significantly lower on the composite visual function questionnaire (VFQ) compared to healthy controls, with notable deficits in ocular motor functions essential for visual scanning [2]. These visual deficits in PD can stem from various causes. Difficulty in quickly shifting gaze (saccades) disrupts reading and scanning of visual scenes [2–4, 45, 74]. Additionally, the impaired simultaneous movement of both eyes in opposite directions (vergence) leads to compromised depth perception and spatial navigation challenges in PD [9, 13]. These vergence abnormalities can cause misalignment of the eyes, resulting

I. Chowdhary · I. Chowdhary
Daroff-Dell'Osso Ocular Motility Laboratory, Louis Stokes Cleveland VA Medical Center, Cleveland, OH, USA

F. F. Ghasia
Daroff-Dell'Osso Ocular Motility Laboratory, Louis Stokes Cleveland VA Medical Center, Cleveland, OH, USA

Cole Eye Institute, Cleveland Clinic, Cleveland, OH, USA

A. G. Shaikh (✉)
Neurological Institute, University Hospitals Neurology Service, Louis Stokes Cleveland, VA Medical Center, Department of Neurology, Case Western Reserve University, Cleveland, OH, USA
e-mail: Axs848@case.edu

P. Ravat et al. (eds.), *Advances in Translational Neuroscience of Vestibular and Eye Movement Disorders*, Contemporary Clinical Neuroscience,
https://doi.org/10.1007/978-3-032-07539-0_22

in strabismus and abnormal fusion of visual signals, leading to blurred or double vision (diplopia). Up to one-third of PD patients experience diplopia, and 23% report tired eyes or blurred vision during reading [45, 74]. Vergence issues in PD include increased latency, reduced speed, lower gain, and reliance on alternative eye movements to achieve a three-dimensional gaze shift [22, 23, 27]. This review aims to summarize the objective features of abnormal eye movements in PD and explore their underlying mechanisms. The following sections will be organized by individual eye movement subclasses, the objective assessment of their abnormalities, and their mechanistic supports.

22.2 Gaze Holding in Parkinson's Disease

Gaze-holding abnormalities in PD include excessive fixational eye movements like microsaccades, drifts after microsaccades, and controversial persistent pendular eye movements. The pathophysiology of microsaccades is directly linked to abnormal basal ganglia function, while pervasive pendular movements are considered a physiological response (the vestibulo-ocular reflex) to tremor-related oscillations [7, 8, 21, 38–40, 62].

Fixational saccades are miniature eye movements essential for counteracting fading [47, 48, 73]. Square wave jerks, related to fixational saccades, can impair visual clarity when excessive [1]. Notably, square waves are prominent in PD but also occur in other Parkinsonian disorders [5, 7, 8, 17, 55, 66, 67, 78].

Figure 1 illustrates abnormal gaze holding in a PD patient compared to a healthy subject, plotting the angular position vector of horizontal eye position against time. The healthy control's data is shown in Fig. 22.1a, while Fig. 22.1b shows the PD patient's results. The traces' excursions (indicated by arrows in Fig. 22.1) depict fixational saccades; certain excursions are followed by returning saccades, constituting a square wave jerk (closed arrow, Fig. 22.1). PD exhibits larger amplitudes of fixational saccades than healthy controls, often linked with return saccades that form square wave jerks.

Fixational saccades are crucial markers of PD pathophysiology, with the basal ganglia implicated in their abnormality alongside square wave jerks in PD. This effect is believed to manifest through connections with the superior colliculus and the cerebellum, based on the hypothesis that the substantia nigra pars reticulata (SNr) exerts excessive inhibitory control over the superior colliculus [29–33]. The superior colliculus influences the downstream burst generators necessary for saccade generation. Furthermore, the subthalamus, connected to the cerebellum via precerebellar neurons in the pons, modulates the saccade matrix, with abnormal subthalamic activity potentially affecting microsaccades and square waves. The impact of subthalamic (STN) deep brain stimulation (DBS) on microsaccades is examined

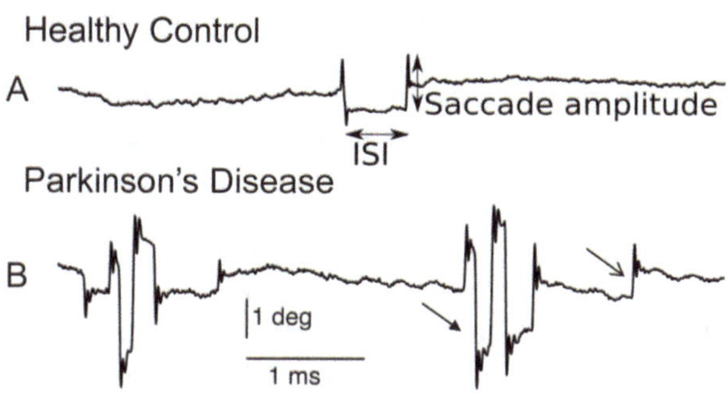

Fig. 22.1 Eye position traces from a healthy control (**a**) and a PD patient (**b**). Horizontal and vertical eye positions are shown. Metrics used: saccade amplitude and intersaccadic interval (ISI). Microsaccades (open arrow) and square waves (closed arrow) are indicated in panel B. ISI: intersaccadic interval; deg.: degrees; ms: milliseconds. (Modified from Beylergil et al. [8])

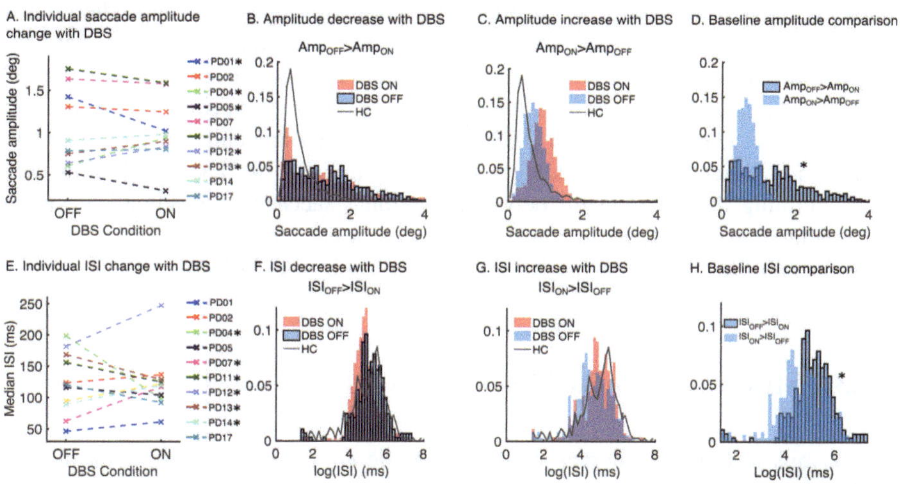

Fig. 22.2 Summary of microsaccade analyses. (**a**) Median saccade amplitudes in DBS-off and DBS-on conditions. Asterisks indicate significant changes in six patients. (**b**) Normalized amplitude histograms of microsaccades showing significant increase due to DBS. (**c**) Normalized amplitude histograms showing significant decrease due to DBS. (**d**) Normalized baseline amplitudes of the two response groups. (**e**) Median ISI durations in DBS-off and DBS-on conditions. Asterisks indicate significant changes in six patients. (**f**) Normalized ISI histograms showing DBS-related increase. (**g**) Normalized ISI histograms showing DBS-related decrease. (**h**) Normalized baseline ISI of the two response groups. (Modified from Beylergil et al. [8])

through reversible interventions that demonstrate effects on the kinematic properties of fixational saccades and square waves.

Figure 22.2 compares the amplitude of fixational saccades in healthy controls, PD patients with STN DBS turned on, and those with DBS off. Participants had an average age of 69 ± 8.1 years (age-matched controls) and an average disease duration of 13.1 ± 4.9 years. The effects of DBS on the same individual are shown in Fig. 22.2a. Generally, healthy controls exhibit smaller fixational saccade amplitudes compared to those with PD (Fig. 22.2b). The influence of STN DBS on fixational saccade amplitude varies; in some patients, DBS significantly decreases amplitudes, while in others, it results in a notable increase (Fig. 22.2b, c). PD patients with smaller baseline fixational saccade amplitudes and variance in gaze holding tend to show increased amplitudes with subthalamic DBS, whereas patients with larger baseline amplitudes may see a reduction.

The intersaccadic intervals (ISIs) between consecutive fixational saccades are also influenced by PD and altered by STN DBS (Fig. 22.2e). Collectively, ISIs are shorter in PD patients relative to healthy controls, further modulated by DBS. Again, two distinct groups of PD patients emerge: one where STN DBS decreases ISIs (Fig. 22.2f) and another where ISIs are prolonged with DBS (Fig. 22.2g). For those PD patients whose ISIs increase with DBS, there tends to be a higher frequency of microsaccades at baseline, and vice versa for those whose ISIs decrease (Fig. 22.2h). Furthermore, PD patients exhibit more frequent oblique saccades when STN DBS is active compared to when it is off; this corresponds to the same group where DBS amplifies the microsaccade amplitude, which is not observed in the group experiencing reduced microsaccade amplitudes. As fixational saccade amplitudes increase, they frequently become part of a square wave jerk. The ratio of square wave jerks to fixational saccades changes with DBS, with STN DBS enhancing this ratio in nearly 70% of PD patients while reducing it in the remaining 30%. The coupling mechanism defining the relationship between fixational saccades and square waves is documented in the literature [54–56], demonstrating strong coupling in healthy controls and PD patients with DBS on, but not in PD with DBS off.

22.3 Strabismus in Parkinson's Disease

Strabismus in PD often comes with nonspecific complaints like diplopia, blurred vision, or reading difficulties. For example, in a study of 37 PD patients, about one-fourth reported tired eyes or blurred vision while reading, whereas only 7% experienced diplopia (9 out of 37 patients reported tired eyes or blurred vision, and 3 out of 37 reported diplopia). Other studies have shown that up to 20% of PD patients can experience diplopia, with all participants in a study of 44 PD patients with diplopia presenting with convergence insufficiency [45, 74]. The occurrence of strabismus in PD may be linked to dopamine's crucial role in the vergence pathway; disruptions in this mechanism can lead to vergence impairment and resultant strabismus.

Convergence insufficiency in PD patients can improve with levodopa or deep brain stimulation (DBS) in conjunction with levodopa [2, 57]. In a study involving 27 PD patients, there was a noted increase in convergence amplitude during the "on" medication phase compared to the "off" phase. Although there was an improvement during the "on" medication phase, the amplitude remained lower than that of healthy controls. Many PD patients exhibited considerable exotropia at near vision, but there was no significant difference in mean exodeviation or ocular ductions during medication on or off periods [2]. The fluctuation of convergence ability throughout the day presents a significant challenge for the ophthalmic management of PD patients, adversely affecting their quality of life. This fluctuation highlights the necessity of considering medication timing during ophthalmological evaluations in PD patients.

Eye deviation is a recognized side effect of subthalamic DBS [68]. Transient diplopia is common among patients undergoing DBS surgery, but this deficit typically resolves after adjusting stimulation parameters [41, 75]. The mechanism behind DBS-induced diplopia is related to electrical overflow affecting structures adjacent to the subthalamic nucleus (STN), such as the corticospinal and corticobulbar tracts, as they traverse the internal capsule, lateral to the STN. It is not uncommon for incorrectly placed STN DBS to influence the suprabulbar fibers of the extraocular motor nerves or their nuclei, which run alongside the red nucleus and may be impacted by implants placed too medially [68]. A case of hypertropia leading to vertical diplopia was reported as a side effect of DBS implantation, though this was attributed to hemorrhaging at the implantation site rather than the stimulation itself [53]. In summary, strabismus and diplopia are recognized side effects of DBS that warrant careful monitoring in the postoperative period to prevent negative impacts on the quality of life for PD patients.

Evaluating the characteristics of strabismus and vergence insufficiency in PD provides vital insights into disease pathophysiology and may reveal potential biomarkers for disease progression. While strabismus and vergence insufficiency alone are neither necessary nor sufficient for diagnosing the disease, their correlation with symptom severity could serve as meaningful progression markers. For instance, a study involving 39 PD patients found a correlation between binocular abnormalities and disease severity, indicating that convergence insufficiency's response to conventional PD treatments underscores its relationship with overall disease pathophysiology and symptomatology.

22.4 Vergence in Parkinson's Disease

PD significantly impacts the perception of spatial dimensions and depth, primarily facilitated by two types of synchronous eye movements in opposite directions—vergence. Vergence movements are categorized into convergence (shifting focus from distant to near) and divergence (shifting focus from near to far). Both types are coupled with accommodation to maintain the proper focal length of the intraocular lens, ensuring sharp image projection on the retina (blur prevention). Consequently, two triggers initiate vergence: disparity in image depth and blur on the retina [15, 35, 42, 43, 58]. Vergence and accommodation are cross-coupled (illustrated in Fig. 22.3), allowing accommodation to be activated by either a change in disparity or a change in blur [76].

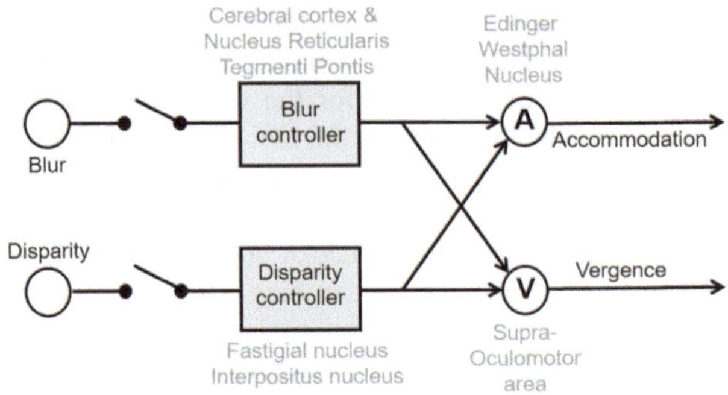

Fig. 22.3 Schematic model of disparity- and blur-driven vergence system. Blur and disparity controllers are upstream, while nodes "A" and "V" are downstream. Blur controller: cerebral cortex and nucleus reticularis tegmenti pontis. Disparity controller: fastigial and interpositus nucleus. Node A: Edinger–Westphal nucleus. Node V: supraoculomotor nucleus. (Modified from Gupta et al. [22, 23])

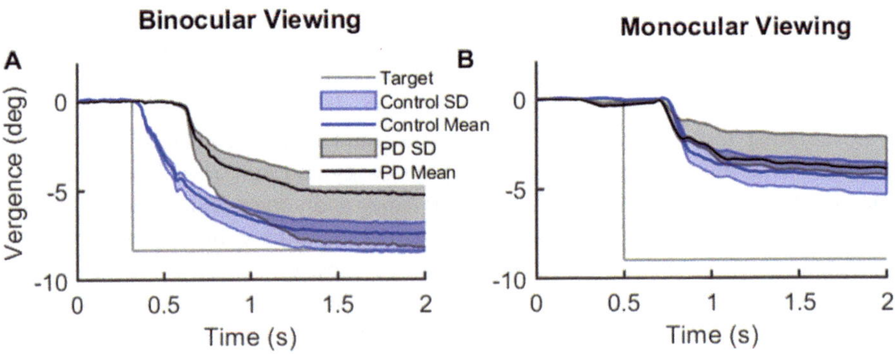

Fig. 22.4 Comparison of blur- and disparity-driven vergence in PD patients and healthy controls. (**a**) Disparity-driven vergence. (**b**) Blur-driven vergence. Black line: PD patients; blue line: controls. Light shades: standard deviation. (Modified from Gupta et al. [22, 23])

Several neural substrates coordinate accommodation and vergence, transmitting motor information generally after the cross-coupling between the two processes has occurred (e.g., nodes "A" and "V" in Fig. 22.3). The downstream motor signal responds equally to blur- or disparity-driven vergence. In contrast, sensory or sensorimotor signals (i.e., controllers of blur or disparity, Fig. 22.3) originate upstream and respond differently to disparities versus blur-driven signals. PD patients appear to exhibit blur-driven vergence functionality comparable to healthy age-matched controls [27, 28, 59], whereas disparity-driven vergence shows substantial impairment in PD compared to healthy individuals [22, 23].

Figure 22.4a illustrates the eye position trajectory and standard deviation around the mean during convergence movements in PD patients and healthy controls (viewing binocularly, thus triggering disparity-driven vergence). Unlike healthy controls, PD patients demonstrate significant limitations in generating

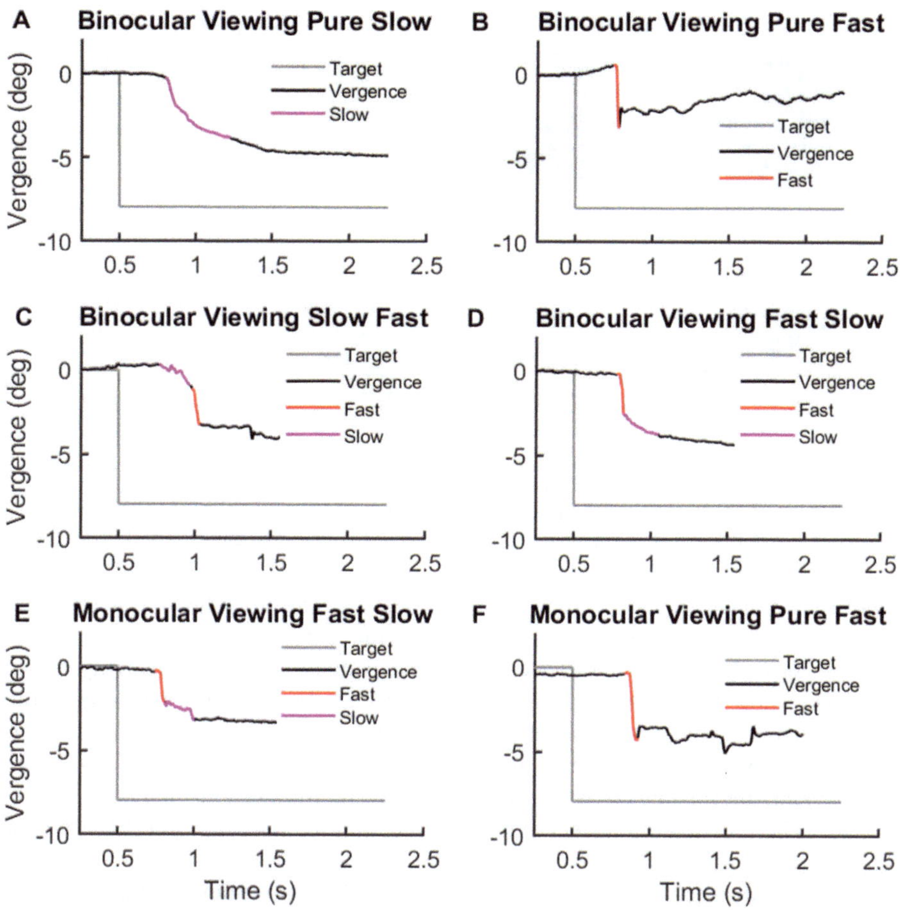

Fig. 22.5 Gaze-shifting strategies to compensate for vergence deficits in PD patients. Each panel shows one strategy. Gray line: desired target position. (Modified from Gupta et al. [22, 23])

disparity-driven vergence. Figure. 22.4b depicts the same convergence task conducted under monocular viewing, thereby initiating blur-driven vergence. Notably, blur-driven vergence is also impaired in both PD patients and healthy controls, indicating that such impairment is not exclusive to PD pathophysiology.

The fundamental abnormality resulting in impaired vergence within PD leads to four compensatory strategies, as depicted in Fig. 22.5. One strategy encompasses a "pure-slow" vergence gaze shift (Fig. 22.5a). The second includes saccades and is labeled "pure-fast" (Fig. 22.5b). The third and fourth strategies entail combining slow and fast movements, either slow followed by fast or vice versa (Fig. 22.5c, d). In contrast to disparity-driven vergence, blur-driven vergence is characterized by only two strategies: fast followed by slow (Fig. 22.5e) or pure-fast (Fig. 22.5f).

Figures 22.6a, b, e, and f summarizes the prevalence of the four strategies for disparity-driven vergence in PD patients and healthy controls. As expected,

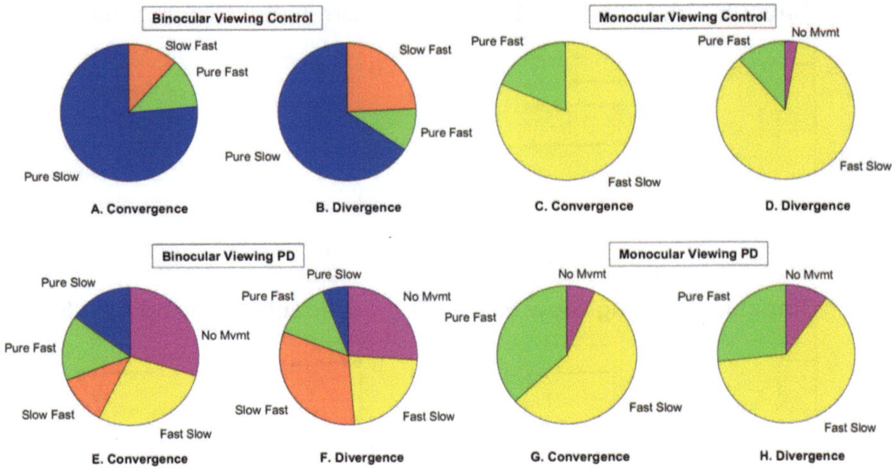

Fig. 22.6 Gaze-shift strategies in PD patients and healthy controls, grouped by convergence and divergence under binocular and monocular conditions. (Modified from Gupta et al. [22, 23])

pure-slow movements are most common in healthy controls, followed by slow-fast and pure-fast movements. In contrast, vergence movements were absent or preceded by fast saccades in most instances when PD patients attempted disparity-driven vergence; slow-fast or pure-fast strategies were also observed. The pure-slow strategy, which is the majority in healthy controls, was much less frequent in PD patients (Fig. 22.6a, b, e, f). Under monocular viewing conditions, depicting the blur-driven component, prominent fast-slow movements were followed by pure-fast movements. The distribution was comparable in healthy controls and PD patients (Fig. 22.6c, d, g, h).

PD not only affects the ability to make vergence eye movements but also impacts the latency with which vergence is made. In addition to increased vergence latency, the velocity is also decreased [22, 23, 27, 28]. It has also been shown that decreasing the vergence velocity gain correlates with the propensity for generating compensatory fast, saccadic eye movements [22, 23]. These findings are consistent with previous studies in nonhuman primates, which described separate brain areas controlling divergence and convergence and noted the significant role that the midbrain supraoculomotor area plays in controlling vergence movements [11, 12]. The mesencephalic reticular formation, involved in mediating vergence velocity, is complemented by a separate group of convergence burst cells located in the dorsal mesencephalic region rostral to the superior colliculus [49]. It is possible that a more robust neural network is in place for mediating convergence eye movements, enabling them to compensate for motor insufficiency seen in PD.

22.5 Saccades in Parkinson's Disease

PD affects saccades made to novel visual targets (i.e., visually guided saccades), but even more so, saccades to remembered locations (i.e., memory-guided saccades). Impaired visually guided saccades may affect patients' ability to read seamlessly and precisely scan their visual environment. PD patients also have a greater anti-saccadic (i.e., inhibition of reflexive saccades) error rate compared to healthy individuals [3, 61]. Impairment of visually guided saccades and anti-saccades is present early in PD, and both have been associated with cognitive dysfunction in PD [3, 27, 28, 70]. The study of saccadic eye movements in PD offers a prototype to examine the physiology of complex motor and perceptual phenomena. Such studies often elucidate the etiologies of motor abnormalities in physiologically simpler circuits of visually guided saccades. By delineating such pathophysiology, we can better understand the complex motor and perceptual phenomenon of movement disorders in PD.

The superior colliculus is the critical structure for visuomotor coordination; its malfunction leads to abnormal saccade scaling, as seen in PD [71, 72]. Visually guided saccades in PD are hypometric in the setting of increased inhibition of the superior colliculus through the abnormal basal ganglia output via SNr; this pathway also affects pre-attentional visual processing and saccade execution [6, 29–34, 46, 77]. There is a reduced fMRI activation of the frontal eye fields in PD patients in the preparatory phase of the visually guided saccade [36, 37, 60]. It was suggested that the underlying physiology might involve the superior colliculus, which sends corollary information to the frontal ocular motor and occipital visual regions at the onset of eye movement [36, 37, 69]. It was also suggested that interruption or malprogramming of saccade size and curvature, as seen in PD and progressive supranuclear palsy (though to a smaller extent), could be partially due to an abnormality affecting the brainstem burst generators.

An example of a visually guided saccade from a PD patient and a healthy control is illustrated in Fig. 22.7. The saccadic eye movement in the healthy control is uninterrupted, as is evident from a single peak in the saccade velocity (Fig. 22.7a, b). The trajectory of the shown saccade from the healthy control is straight (Fig. 22.7c). Unlike the healthy control, the PD patient has an interruption in the ongoing saccade (Fig. 22.7d). The interruption is prominent in the vertical axis and is clearly shown in the plotted velocity trace (Fig. 22.7e). The saccade is not only interrupted, but it also has a curved trajectory (Fig. 22.7f). The saccade depicted in Fig. 22.7d–f is a vertical saccade. The interruptions are also seen in horizontal saccades (Fig. 22.7g, h), but their curvature irregularity is less pronounced (Fig. 22.7i).

The three types of abnormalities in visually guided saccades in PD subjects—frequent interruption of the ongoing saccade, irregular saccade trajectory resulting in longer time to complete the movement, and slowing of segmented saccade velocity—can be explained by the maladaptation in burst generation and colliculus network. Timely saccade initiation requires transient cessation of the tonic GABAergic inhibition from the SNr neurons on the superior colliculus [16, 19,

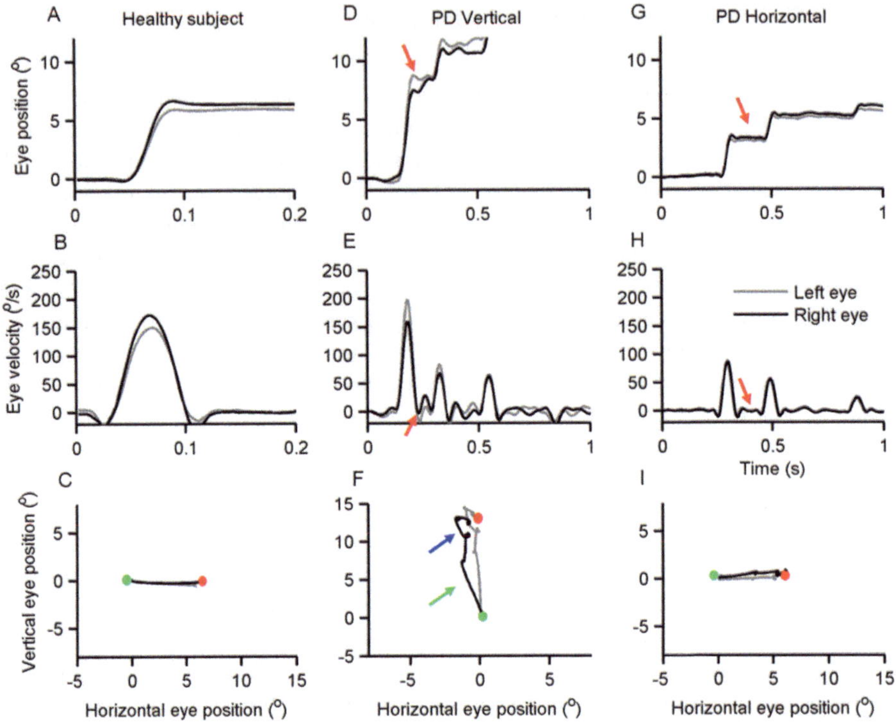

Fig. 22.7 Examples of visually guided saccades from a healthy subject and a PD patient. Top row: eye position over time. Black line: right eye; gray line: left eye. Red arrows: interruptions. Middle row: eye velocity over time. Red arrows: interruptions. Bottom row: saccade trajectories. Green dot: start point; red dot: stop point. Arrows: curvature. (Modified from Shaikh and Ghasia [63])

26, 29–32]. Removal of such inhibition by injecting muscimol in the SNr causes saccadic intrusions and contralateral spontaneous saccades [33, 77]. In contrast, the electrical stimulation of SNr causes decreased latency and hypometria of visually guided saccades [6]. PD is expected to disinhibit SNr, prematurely interrupting the ongoing saccades by imposing phasic inhibition to the superior colliculus. This theory, however, predicts equal impairment in horizontal and vertical saccades and saccade velocity always reaching zero during the break (i.e., complete cessation of the eye movements during saccades). However, the data in PD patients are inconsistent with such a prediction (Fig. 7), hence an alternative explanation is needed [63].

The abnormal saccades in PD may be due to pathology in the burst generators. The saccade signals from the cerebral cortex follow two pathways: the nucleus reticularis tegmenti pontis (NRTP) of the pontine reticular formation, and the superior colliculus [44]. The projections of the NRTP go to the oculomotor vermis (OMV), located in lobules five to seven of the cerebellum [44]. OMV projects to the caudal fastigial nucleus (the fastigial oculomotor region, FOR) via GABAergic inhibition. The FOR projects to the omnidirectional pause neurons (OPN) in the nucleus raphe interpositus of the midline pons, and inhibitory and excitatory burst neurons (IBNs and EBNs, respectively) [44]. The choke in external inhibition triggers an abrupt peak in firing, known as the post-inhibitory rebound (PIR), necessary for high saccade velocity [14, 50, 64–67]. The abrupt cessation of the OPN activity and spike in the burst neuron firing are correlated with saccade velocity and acceleration [14, 50, 64–67]. The caveat of this theory is that the OPNs equally affect horizontal and vertical saccades via their influence on NRTP; therefore, an OPN abnormality alone does not account for the asymmetric involvement of horizontal and vertical saccades. However, burst neurons for horizontal and vertical saccades are located in different anatomical regions; hence impaired excitation of vertical and horizontal saccades generators can result in differences in abnormal matrices of horizontal and vertical saccades in PD.

If stimulated during the saccade, the superior colliculus fixation zone can cause directional changes in the ongoing eye movements [20, 51, 52]. The redirected saccade path depends upon the coordinates of the stimulated site in the collicular fixation zone [20]. Idiosyncrasy and irregularity in the saccade trajectory in PD may be explained by the premature activation of superior colliculus fixation zones [63]. It is possible that PD first affects EBNs and IBNs for vertical saccades, and a degenerative lesion of IBNs causes an asynchronous, irregular discharge pattern during an ongoing vertical saccade [63]. This discharge pattern of IBNs during an ongoing vertical saccade is inferred as a command to stop the eye movement, hence the abrupt interruption of the ongoing saccade. It is then forwarded into a feedback loop to the superior colliculus, leading to superior colliculus fixation zone stimulation. Such unwanted activation of the superior colliculus leads to redirected saccades in an arbitrary direction. These predictions underlying the pathophysiology of abnormal visually guided saccades are supported by the histology literature in PD, which suggests the involvement of the mesencephalic reticular formation in the area for vertical saccades [10, 24, 25]. In other words, it was predicted that the irregularities and slowing, prominently seen in the vertical saccades in PD, could be due to impaired function of EBNs and IBNs leading to maladaptive feedback causing the premature activation of the superior colliculus [63].

22.6 Summary

In PD, eye movement issues include abnormal fixational saccades, strabismus, impaired convergence, and visually guided saccades. These abnormalities are linked to the basal ganglia and superior colliculus. Deep brain stimulation (DBS) can reset these movements, highlighting the basal ganglia's role. Strabismus and convergence issues suggest the importance of dopamine, with improvements seen with levodopa and DBS. These eye movement problems correlate with disease severity and could serve as markers for progression. PD patients show more convergence impairment and different adaptive strategies compared to healthy individuals. They also have increased latency and decreased velocity in vergence movements, controlled by the midbrain. Monitoring eye deviation post-DBS is important. Visually guided saccades in PD are often interrupted and slow, involving the superior colliculus and other brain areas. Overall, eye movement abnormalities in PD provide insights into the disease and may help in monitoring its progression.

Acknowledgments Grants from the US Department of Veterans Affairs SPiRE I21-RX003878-01 (Shaikh), The American Academy of Neurology (Shaikh), American Parkinson's Disease Association George C Cotzias Memorial Fellowship (Shaikh), Dystonia Medical Research Foundation Research (Shaikh), and philanthropic funds to the Department of Neurology at University Hospitals (The Allan Woll Fund). Shaikh also has Penni and Steven Weinberg Chair in Brain Health.

References

1. Abadi RV, Gowen E. Characteristics of saccadic intrusions. Vis Res. 2004;44(23):2675–90.
2. Almer Z, Klein KS, Marsh L, Gerstenhaber M, Repka MX. Ocular motor and sensory function in Parkinson's disease. Ophthalmology. 2012;119(1):178–82.
3. Antoniades CA, Demeyere N, Kennard C, Humphreys GW, Hu MT. Antisaccades and executive dysfunction in early drug-naive Parkinson's disease: the discovery study. Mov Disord. 2015;30(6):843–7.
4. Armstrong RA. Oculo-visual dysfunction in Parkinson's disease. J Parkinsons Dis. 2015;5(4): 715–26.
5. Averbuch-Heller L, Stahl JS, Hlavin ML, Leigh RJ. Square-wave jerks induced by pallidotomy in parkinsonian patients. Neurology. 1999;52(1):185–8.
6. Basso MA, Liu P. Context-dependent effects of substantia nigra stimulation on eye movements. J Neurophysiol. 2007;97(6):4129–42.
7. Beylergil SB, Murray J, Noecker AM, Gupta P, Kilbane C, McIntyre CC, Ghasia FF, Shaikh AG. Temporal patterns of spontaneous Fixational eye movements: the influence of basal ganglia. J Neuroophthalmol. 2022;42(1):45–55.
8. Beylergil SB, Murray J, Noecker AM, Gupta P, Kilbane C, McIntyre CC, Shaikh AG, Ghasia FF. Effects of subthalamic deep brain stimulation on fixational eye movements in Parkinson's disease. J Comput Neurosci. 2021;49(3):345–56.
9. Bloem BR, Hausdorff JM, Visser JE, Giladi N. Falls and freezing of gait in Parkinson's disease: a review of two interconnected, episodic phenomena. Mov Disord. 2004;19(8):871–84.
10. Braak H, Del Tredici K, Rub U, de Vos RA, Jansen Steur EN, Braak E. Staging of brain pathology related to sporadic Parkinson's disease. Neurobiol Aging. 2003;24(2):197–211.
11. Das VE. Cells in the supraoculomotor area in monkeys with strabismus show activity related to the strabismus angle. Ann N Y Acad Sci. 2011;1233:85–90.
12. Das VE. Responses of cells in the midbrain near-response area in monkeys with strabismus. Invest Ophthalmol Vis Sci. 2012;53(7):3858–64.
13. Davidsdottir S, Wagenaar R, Young D, Cronin-Golomb A. Impact of optic flow perception and egocentric coordinates on veering in Parkinson's disease. Brain. 2008;131(Pt 11):2882–93.
14. Enderle J, Engelken EJ. Simulation of oculomotor post-inhibitory rebound burst firing using a Hodgkin-Huxley model of a neuron. Biomed Sci Instrum. 1995;31:53–8.
15. Fincham EF, Walton J. The reciprocal actions of accommodation and convergence. J Physiol. 1957;137(3):488–508.
16. Fisher RS, Buchwald NA, Hull CD, Levine MS. The GABAergic striatonigral neurons of the cat: demonstration by double peroxidase labeling. Brain Res. 1986;398(1):148–56.

17. Fisk JD, Goodale MA, Burkhart G, Barnett HJ. Progressive supranuclear palsy: the relationship between ocular motor dysfunction and psychological test performance. Neurology. 1982;32(7): 698–705.
18. Foundation PS. 2018. Statistics. 2023 from https://www.parkinson.org/understanding-parkinsons/statistics.
19. Francois C, Percheron G, Yelnik J. Localization of nigrostriatal, nigrothalamic and nigrotectal neurons in ventricular coordinates in macaques. Neuroscience. 1984;13(1):61–76.
20. Gandhi NJ, Keller EL. Comparison of saccades perturbed by stimulation of the rostral superior colliculus, the caudal superior colliculus, and the omnipause neuron region. J Neurophysiol. 1999;82(6):3236–53.
21. Gitchel GT, Wetzel PA, Baron MS. Pervasive ocular tremor in patients with Parkinson disease. Arch Neurol. 2012;69(8):1011–7.
22. Gupta P, Beylergil S, Murray J, Jacobs J, Kilbane C, Shaikh AG, Ghasia FF. Effects of Parkinson disease on blur-driven and disparity-driven Vergence eye movements. J Neuroophthalmol. 2021a;41(4):442–51.
23. Gupta P, Beylergil S, Murray J, Kilbane C, Ghasia FF, Shaikh AG. Computational models to delineate 3D gaze-shift strategies in Parkinson's disease. J Neural Eng. 2021b;18(4):0460a5.
24. Halliday GM, Del Tredici K, Braak H. Critical appraisal of brain pathology staging related to presymptomatic and symptomatic cases of sporadic Parkinson's disease. J Neural Transm. 2006;(Suppl(70)):99–103.
25. Halliday GM, Li YW, Blumbergs PC, Joh TH, Cotton RG, Howe PR, Blessing WW, Geffen LB. Neuropathology of immunohistochemically identified brainstem neurons in Parkinson's disease. Ann Neurol. 1990;27(4):373–85.
26. Handel A, Glimcher PW. Quantitative analysis of substantia nigra pars reticulata activity during a visually guided saccade task. J Neurophysiol. 1999;82(6):3458–75.
27. Hanuska J, Bonnet C, Rusz J, Sieger T, Jech R, Rivaud-Pechoux S, Vidailhet M, Gaymard B, Ruzicka E. Fast vergence eye movements are disrupted in Parkinson's disease: a video-oculography study. Parkinsonism Relat Disord. 2015;21(7):797–9.
28. Hanuska J, Rusz J, Bezdicek O, Ulmanova O, Bonnet C, Dusek P, Ibarburu V, Nikolai T, Sieger T, Sonka K, Ruzicka E. Eye movements in idiopathic rapid eye movement sleep behaviour disorder: high antisaccade error rate reflects prefrontal cortex dysfunction. J Sleep Res. 2019;28(5):e12742.
29. Hikosaka O, Wurtz RH. Effects on eye movements of a GABA agonist and antagonist injected into monkey superior colliculus. Brain Res. 1983a;272(2):368–72.
30. Hikosaka O, Wurtz RH. Visual and oculomotor functions of monkey substantia nigra pars reticulata. I. Relation of visual and auditory responses to saccades. J Neurophysiol. 1983b;49 (5):1230–53.
31. Hikosaka O, Wurtz RH. Visual and oculomotor functions of monkey substantia nigra pars reticulata. III. Memory-contingent visual and saccade responses. J Neurophysiol. 1983c;49(5): 1268–84.
32. Hikosaka O, Wurtz RH. Visual and oculomotor functions of monkey substantia nigra pars reticulata. IV. Relation of substantia nigra to superior colliculus. J Neurophysiol. 1983d;49: 1285–301.
33. Hikosaka O, Wurtz RH. Modification of saccadic eye movements by GABA-related substances. I. Effect of muscimol and bicuculline in monkey superior colliculus. J Neurophysiol. 1985;53(1):266–91.
34. Hikosaka O, Wurtz RH. Saccadic eye movements following injection of lidocaine into the superior colliculus. Exp Brain Res. 1986;61(3):531–9.
35. Hung GK, Semmlow JL. Static behavior of accommodation and vergence: computer simulation of an interactive dual-feedback system. IEEE Trans Biomed Eng. 1980;27(8):439–47.
36. Javaid MA, Amassian V, Glazman S, Fesharaki A, Stefanov D, Bodis-Wollner I. Cortical control of voluntary saccades in Parkinson's disease and pre-emptive perception. Parkinsonism Relat Disord. 2012;18(Suppl 1):S100–3.

37. Javaid MA, Weeden J, Flom P, Avitable M, Glazman S, Bodis-Wollner I. Perisaccadic gamma modulation in Parkinson disease patients and healthy subjects. Clin EEG Neurosci. 2010;41(2): 94–101.
38. Kaski D, Bronstein AM. Ocular tremor in Parkinson's disease: discussion, debate, and controversy. Front Neurol. 2017;8:134.
39. Kaski D, Saifee TA, Buckwell D, Bronstein AM. Ocular tremor in Parkinson's disease is due to head oscillation. Mov Disord. 2013a;28(4):534–7.
40. Kaski D, Saifee TA, Buckwell D, Bronstein AM. Reply to letter–Eye oscillations in Parkinson's disease relate to the vestibulo-ocular reflex. Mov Disord. 2013b;28(6):845–6.
41. Krack P, Fraix V, Mendes A, Benabid AL, Pollak P. Postoperative management of subthalamic nucleus stimulation for Parkinson's disease. Mov Disord. 2002;17(Suppl 3):S188–97.
42. Krishnan VV, Phillips S, Stark L. Frequency analysis of accommodation, accommodative vergence and disparity vergence. Vis Res. 1973;13(8):1545–54.
43. Krishnan VV, Stark L. A heuristic model for the human vergence eye movement system. IEEE Trans Biomed Eng. 1977;24(1):44–9.
44. Leigh RJ, Zee DS. The neurology of eye movements. Oxford University Press; 2015.
45. Lepore FE. Parkinson's disease and diplopia. Neuro-Ophthalmology. 2006;30(2–3):37–40.
46. Lieb K, Brucker S, Bach M, Els T, Lucking CH, Greenlee MW. Impairment in preattentive visual processing in patients with Parkinson's disease. Brain. 1999;122(Pt 2):303–13.
47. Martinez-Conde S. Fixational eye movements in normal and pathological vision. Prog Brain Res. 2006;154:151–76.
48. Martinez-Conde S, Macknik SL, Troncoso XG, Dyar TA. Microsaccades counteract visual fading during fixation. Neuron. 2006;49(2):297–305.
49. Mays LE, Porter JD, Gamlin PD, Tello CA. Neural control of vergence eye movements: neurons encoding vergence velocity. J Neurophysiol. 1986;56(4):1007–21.
50. Miura K, Optican LM. Membrane channel properties of premotor excitatory burst neurons may underlie saccade slowing after lesions of omnipause neurons. J Comput Neurosci. 2006;20(1): 25–41.
51. Munoz DP, Wurtz RH. Fixation cells in monkey superior colliculus. I. Characteristics of cell discharge. J Neurophysiol. 1993a;70(2):559–75.
52. Munoz DP, Wurtz RH. Fixation cells in monkey superior colliculus. II. Reversible activation and deactivation. J Neurophysiol. 1993b;70(2):576–89.
53. Ortiz-Perez S, Sanchez-Dalmau B, Molina J, Adan A, Candela S, Rumia J. Ocular tilt reaction as a delayed complication of deep brain stimulation for Parkinson disease. J Neuroophthalmol. 2009;29(4):286–8.
54. Otero-Millan J, Macknik SL, Serra A, Leigh RJ, Martinez-Conde S. Triggering mechanisms in microsaccade and saccade generation: a novel proposal. Ann N Y Acad Sci. 2011a;1233:107–16.
55. Otero-Millan J, Schneider R, Leigh RJ, Macknik SL, Martinez-Conde S. Saccades during attempted fixation in parkinsonian disorders and recessive ataxia: from microsaccades to square-wave jerks. PLoS One. 2013;8(3):e58535.
56. Otero-Millan J, Serra A, Leigh RJ, Troncoso XG, Macknik SL, Martinez-Conde S. Distinctive features of saccadic intrusions and microsaccades in progressive supranuclear palsy. J Neurosci. 2011b;31(12):4379–87.
57. Racette BA, Gokden M, Tychsen L, Perlmutter JS. Convergence insufficiency in idiopathic Parkinson's disease responsive to levodopa. Strabismus. 1999;7(3):169–74.
58. Rashbass C, Westheimer G. Independence of conjugate and disjunctive eye movements. J Physiol. 1961;159:361–4.
59. Repka MX, Claro MC, Loupe DN, Reich SG. Ocular motility in Parkinson's disease. J Pediatr Ophthalmol Strabismus. 1996;33(3):144–7.
60. Rieger JW, Kim A, Argyelan M, Farber M, Glazman S, Liebeskind M, Meyer T, Bodis-Wollner I. Cortical functional anatomy of voluntary saccades in Parkinson disease. Clin EEG Neurosci. 2008;39(4):169–74.

61. Rivaud-Pechoux S, Vidailhet M, Brandel JP, Gaymard B. Mixing pro- and antisaccades in patients with parkinsonian syndromes. Brain. 2007;130(Pt 1):256–64.
62. Saifee TA, Kaski D, Buckwell D, Bronstein AM. Tremor of the eyes in Parkinson's disease: merely a measure of the head movement. Parkinsonism Relat Disord. 2014;20(12):1447–8.
63. Shaikh AG, Ghasia FF. Saccades in Parkinson's disease: Hypometric, slow, and maladaptive. Prog Brain Res. 2019;249:81–94.
64. Shaikh AG, Miura K, Optican LM, Ramat S, Leigh RJ, Zee DS. A new familial disease of saccadic oscillations and limb tremor provides clues to mechanisms of common tremor disorders. Brain. 2007;130(Pt 11):3020–31.
65. Shaikh AG, Wong AL, Optican LM, Miura K, Solomon D, Zee DS. Sustained eye closure slows saccades. Vis Res. 2010;50(17):1665–75.
66. Shaikh AG, Xu-Wilson M, Grill S, Zee DS. Staircase' square-wave jerks in early Parkinson's disease. Br J Ophthalmol. 2011a;95(5):705–9.
67. Shaikh AG, Zee DS, Optican LM, Miura K, Ramat S, Leigh RJ. The effects of ion channel blockers validate the conductance-based model of saccadic oscillations. Ann N Y Acad Sci. 2011b;1233:58–63.
68. Shields DC, Gorgulho A, Behnke E, Malkasian D, DeSalles AA. Contralateral conjugate eye deviation during deep brain stimulation of the subthalamic nucleus. J Neurosurg. 2007;107(1): 37–42.
69. Sommer MA, Wurtz RH. Visual perception and corollary discharge. Perception. 2008;37(3): 408–18.
70. Stuart S, Lawson RA, Yarnall AJ, Nell J, Alcock L, Duncan GW, Khoo TK, Barker RA, Rochester L, Burn DJ, I.-P. s. group. Pro-saccades predict cognitive decline in Parkinson's disease: ICICLE-PD. Mov Disord. 2019;34(11):1690–8.
71. Terao Y, Fukuda H, Ugawa Y, Hikosaka O. New perspectives on the pathophysiology of Parkinson's disease as assessed by saccade performance: a clinical review. Clin Neurophysiol. 2013;124(8):1491–506.
72. Terao Y, Fukuda H, Yugeta A, Hikosaka O, Nomura Y, Segawa M, Hanajima R, Tsuji S, Ugawa Y. Initiation and inhibitory control of saccades with the progression of Parkinson's disease - changes in three major drives converging on the superior colliculus. Neuropsychologia. 2011;49(7):1794–806.
73. Troncoso XG, Macknik SL, Martinez-Conde S. Microsaccades counteract perceptual filling-in. J Vis. 2008;8(14):15 11–9.
74. Urwyler P, Nef T, Killen A, Collerton D, Thomas A, Burn D, McKeith I, Mosimann UP. Visual complaints and visual hallucinations in Parkinson's disease. Parkinsonism Relat Disord. 2014;20(3):318–22.
75. Wang Y, Liu ZY, Dou WC, Ma WB, Wang RZ, Guo Y. Application of preoperative CT/MRI image fusion in target positioning for deep brain stimulation. Chin Med Sci J. 2016;31(3): 161–7.
76. Westheimer G. Amphetamine, barbiturates, and accommodation-convergence. Arch Ophthalmol. 1963;70:830–6.
77. Wurtz RH, Hikosaka O. Role of the basal ganglia in the initiation of saccadic eye movements. Prog Brain Res. 1986;64:175–90.
78. Zhou H, Wang X, Ma D, Jiang Y, Li F, Sun Y, Chen J, Sun W, Pinkhardt EH, Landwehrmeyer B, Ludolph A, Zhang L, Zhao G, Wang Z. The differential diagnostic value of a battery of oculomotor evaluation in Parkinson's disease and multiple system atrophy. Brain Behav. 2021;11(7):e02184.

[illegible]

Chapter 23
Dorsal Midbrain (Parinaud) Syndrome with INO and Mild Bilateral INO, Two Case Report and a Literature Review: The Spectrum of Oculomotor Abnormalities in Medial Longitudinal Fasciculus Lesions

Carlos Palomino-Diaz and Jorge C. Kattah

Keywords MLF · Vestibulo-ocular Reflex · WEMINO · WEBINO · Excitatory burst neurons · Omnipause neurons

23.1 Introduction

The medial longitudinal fasciculus (MLF) is a bundle of paired, highly specialized, myelinated tracts that travel in a dorsomedial direction, near the midline (close to the fourth ventricle), through the dorsal pons and midbrain tegmentum [3]. This tract originates from interneurons in the VIth nerve nucleus (abducens nucleus), containing both ascending and descending axons that travel along the brainstem and terminate in the interstitial nucleus of Cajal (INC) and the rostral interstitial nucleus of the medial longitudinal fasciculus (riMLF), while also projecting to the oculomotor nucleus and trochlear nucleus. The MLF runs ventral to the central gray matter and extends caudally toward the spinal cord. In contrast, the dorsal longitudinal fasciculus is a periventricular pathway originating from the nuclei of the posterior medial hypothalamic region and extending to the inferior medullary region. It contains thin ascending and descending fibers that connect the forebrain with visceral autonomic centers in the lower brainstem and does not play any role in ocular motor function [4].

The MLF contains axons responsible for horizontal gaze: the internuclear neurons of the abducens nucleus project to the contralateral medial rectus motor neurons of the oculomotor nucleus. Vestibular projections within the MLF mediate the vertical

C. Palomino-Diaz (✉)
Department of Neurology, Hospital Lima Este Vitarte, Lima, Lima, Peru
e-mail: cpalomino@cientifica.edu.pe

J. C. Kattah
Peoria, IL, USA

P. Ravat et al. (eds.), *Advances in Translational Neuroscience of Vestibular and Eye Movement Disorders*, Contemporary Clinical Neuroscience,
https://doi.org/10.1007/978-3-032-07539-0_23

and torsional vestibulo-ocular reflex (VOR). The riMLF contains burst neurons that generate vertical gaze, and the INC serves as the vertical gaze integrator. These structures are involved in generating slow tracking, vertical VOR, gaze holding, and utricular projections [5, 6].

In addition to the MLF, the motor neurons of the medial rectus subnucleus of the oculomotor complex also receive direct projections from the position-vestibular-pause (PVP) neurons of the ipsilateral medial vestibular nucleus through the ascending tract of Deiters (ATD). The functional significance of the ATD is not entirely clear, but it is thought to play a role in the horizontal vestibulo-ocular reflex. Lesions in the MLF can cause adduction paralysis for conjugate movements on the side of the lesion; however, adduction may still be possible due to direct convergence signals to the medial rectus. This has led to the designation of the ATD as an extra-MLF pathway [5, 6].

23.2 Smooth Pursuit

The position signal from a target of interest, originating from the visual cortex (areas V1, V2, V3), the medial temporal (MT), and superior medial temporal (MST) areas, projects to visual-motor areas such as the frontal eye field (FEF) and supplemental eye field (SEF). FEF is known to play a role in predictive pursuit and contribute to the dynamic gain control of smooth pursuit [7, 8]. Both the MT and the MST, as well as the FEF, project directly to the dorsal lateral pontine nucleus (DLPN). For horizontal movements, these projections cross the first decussation to the contralateral ventral flocculus, ventral paraflocculus, dorsal vermis, and fastigial nucleus. Pursuit fibers then reach the contralateral abducens nucleus (second decussation) to project to the ipsilateral lateral rectus and the contralateral medial rectus through the MLF [9].

For vertical movements, however, the projections reach the rostral reticular nucleus of the pontine tegmentum (NRTP), which then projects to the Y group (a group of cells in the inferior cerebellar peduncle). These signals are sent to the oculomotor and trochlear nuclei via the superior cerebellar peduncle and the ventral crossed tegmental tract (VTT), as well as the INC and the dentate nucleus of the cerebellum [5, 9].

23.3 Saccadic System

The circuit of saccadic eye movements corresponds to a complex system involving a descending pathway from the frontal eye field (FEF) to premotor areas in the caudal reticular nucleus of the pontine tegmentum (cNRTP) and the riMLF in the brainstem. The cNRTP receives input from the superior colliculus (SC), both of which carry trigger signals for the initiation of saccades. These structures, in turn, project to the cerebellum, including the dorsal vermis, fastigial nucleus, flocculus, and ventral paraflocculus [6, 9].

The FEF and SC also project directly to the PPRF (Paramedian Pontine Reticular Formation) and riMLF in the brainstem. For horizontal saccadic movements, the PPRF receives bilateral projections from the FEF and contralateral projections from the SC. The PPRF then projects to abducens motor neurons and internuclear neurons of the abducens, which send axons to the contralateral medial rectus subnucleus via the MLF [6, 9].

For vertical upward saccades, excitatory burst neurons (EBNs) located in the riMLF project to the ipsilateral superior rectus and inferior oblique subnucleus of the oculomotor nucleus via the MLF. Collateral axons cross into the oculomotor nucleus to innervate the contralateral superior rectus muscle. For downward saccades, the EBNs project through the MLF to the ipsilateral inferior rectus subnucleus of the oculomotor nucleus and to the ipsilateral trochlear nucleus. The axons decussate to innervate the superior oblique muscle of the opposite eye [6, 9, 10].

Two classes of neurons in the brainstem are critical for the generation of saccades, burst neurons and omnipause neurons. Burst neurons are divided into excitatory (EBN) and inhibitory (IBN) types. Excitatory burst neurons are further categorized into medium-lead and long-lead neurons. Medium-lead EBN generate the premotor command for saccades by activating motoneurons that control agonist muscles. They fire at high frequencies before and during saccades but remain silent during fixation and other eye movements. These neurons are located in the nucleus reticularis pontis caudalis within the PPRF for horizontal saccades and in the riMLF for vertical and torsional saccades. Long-lead EBN, found in the rostral PPRF and mesencephalic reticular formation, activate medium-lead EBN and inhibit omnipause neurons to release their tonic inhibition on saccade generation. Inhibitory burst neurons (IBN) inhibit motoneurons controlling antagonist muscles and are active just before and during saccades. They are located in the nucleus paragigantocellularis dorsalis in the PPRF for horizontal saccades and in the interstitial nucleus of Cajal (and possibly riMLF) for vertical and torsional saccades [5, 6, 9].

Omnipause neurons tonically inhibit medium-lead burst neurons during fixation and smooth eye movements. When a saccade is initiated, omnipause neurons are inhibited (likely by long-lead burst neurons), allowing medium-lead EBN to generate saccades. These neurons are located in the nucleus raphe interpositus within the Paramedian Pontine Reticular Formation (PPRF). The neural integrator responsible for integrating the velocity-to-position command for the eye during horizontal saccades is the prepositus hypoglossal complex and medial vestibular nucleus (NPH-MVN). In vertical saccades, this function is performed by the interstitial nucleus of Cajal (INC) [6, 9, 10].

However, there are other structures involved in this process. Another pathway comprises the caudate basal ganglia system, the pars reticulata of the substantia nigra (SNPr), the subthalamic nucleus, and the amygdala, which act depending on the context, expectations, memory, reward, and emotional situation of the scenario [6, 11].

23.4 Semicircular Projections

The medial longitudinal fasciculus (MLF) serves as the primary communication pathway for motor neurons involved in eye movements generated by vestibular signals. If the eye movements are compensatory for angular head movements mediated by the semicircular canals (SCC), they are referred to as the rotational vestibular-ocular reflex (R-VOR). Conversely, if the eye movements are compensatory for linear head acceleration, they are mediated by the otolith organs, specifically the utricle and the saccule [1].

During a cephalic rotation to the left, the left lateral semicircular canal (LSCC) is activated to maintain straightforward gaze and stabilize vision. Axons of the left LSCC neurons project to the ipsilateral vestibular nucleus, primarily the medial vestibular nucleus (MVN), through two ascending excitatory vestibular pathways [1, 12].

The first pathway involves neurons in the vestibular nucleus activated by the LSCC. These neurons send signals through the brainstem to reach the contralateral abducens nucleus (VI). This signal connects to the right lateral rectus muscle, causing the right eye to abduct. Simultaneously, another signal crosses to the opposite side via the MLF. These crossed interneurons then project to the medial rectus subnucleus of the oculomotor nucleus which controls the left medial rectus muscle, responsible for moving the left eye inward (adduction) [1, 12].

The second pathway originates from the ventrolateral vestibular nucleus and travels via the ascending tract of Deiters (ATD). This pathway projects directly to the medial rectus motor neurons, bypassing the MLF [1, 12] (Fig. 23.1).

The semicircular canals detect angular acceleration, while the otolithic organs (the saccule and utricle) detect linear acceleration. The neural pathways of the vestibular nuclei responsible for otolithic-ocular reflexes are not as clearly defined as those for semicircular canal reflexes. Unlike the well-established three-neuron arc for semicircular canal reflexes, there is limited evidence for a direct disynaptic pathway involving the otolithic organs, the vestibular nucleus, and the ocular motor neurons of the oculomotor and the trochlear cranial nerves via the MLF. In response to the cephalic tilt, an ocular counter-rolling reflex is produced, mediated by the activation of the cyclovertical ocular muscles. For example, during a left head tilt, the eyes counter-roll in the opposite (right) direction. This response involves the coordinated innervation of specific ocular muscles on each side: the superior rectus and inferior oblique muscles on one side, and the inferior rectus and superior oblique muscles on the contralateral side. These muscles work synergistically to generate the torsional eye movements necessary for stabilizing gaze during head tilts [13].

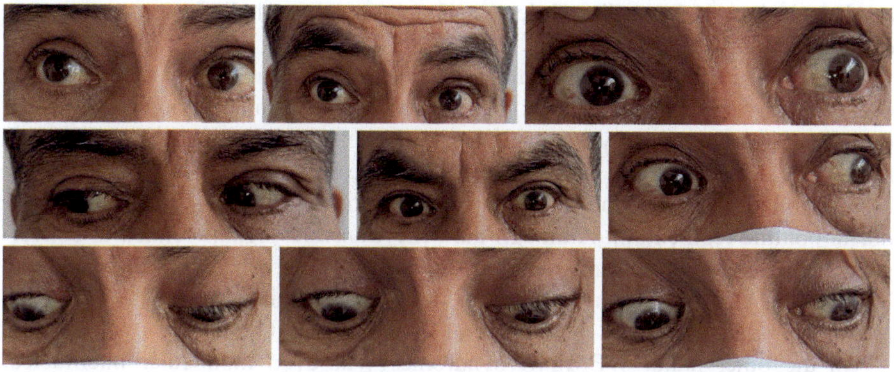

Fig. 23.1 Structures involved in saccadic movements, slow tracking, and ascending projections of the semicircular canals, specifically the lateral semicircular canal (LSC) during left cephalic rotation. Axons from the right vestibular nucleus (receiving input from the ipsilateral LSC) project to the ipsilateral medial rectus subnucleus via the ascending tract of Deiters (ATD). These axons also project to the contralateral abducens nucleus, which sends signals via the medial longitudinal fasciculus (MLF) to the contralateral medial rectus subnucleus and ipsilateral lateral rectus nucleus. For the generation of horizontal saccades, excitatory burst neurons reside in the paramedian pontine reticular formation (PPRF), while velocity and eye position commands are integrated by the nucleus prepositus hypoglossi-medial vestibular nucleus (NPH-MVN) complex

23.5 Internuclear Ophthalmoplegia

A lesion of the medial longitudinal fasciculus (MLF) causes the classic syndrome called internuclear ophthalmoplegia (INO). This condition is characterized by the paralysis of adduction on the side of the lesion, with nystagmus in the contralateral eye during abduction. During recovery, the nystagmus often becomes bilateral but remains more pronounced in amplitude in the contralateral eye (the eye in abduction). Vergence pathways are typically spared, so convergence remains intact in most patients despite adduction weakness during lateral gaze [1].

Unilateral INO is frequently accompanied by a skew deviation, a vertical misalignment of the eyes caused by lesions in the prenuclear otolithic inputs. These inputs carry gravitropic signals from the vestibular system, leading to the vertical misalignment [6, 14].

Skew deviation is characteristically associated with ocular torsion: the elevated eye exhibits intorsion (visible only through fundus photography), while the downward eye exhibits extorsion. Both eyes cyclotort to the same side, a hallmark of skew torsion [15, 16]. This condition must be differentiated from trochlear palsy, where the elevated eye is extorted. Skew deviation is often accompanied by other neurological signs. When combined with ocular torsion and head tilt, it forms a clinical triad called the ocular tilt reaction (OTR). OTR is characterized by head tilt, ocular counter-roll (with extorsion in the lower eye and intorsion in the higher eye), and can help localize central lesions. Peripheral and caudal pontomedullary lesions may result in ipsilateral head tilt, while rostral pontomesencephalic lesions may cause contralateral head tilt [6, 17].

In the ocular tilt reaction, the pathological head tilt is ipsilateral to the hypotropic eye, reflecting the underlying pathological mechanism. Ocular torsion occurs when the upper pole of both eyes rotates in the same direction as the head tilt (the hypotropic eye is extorted, and the hypertropic eye is intorted) [6].

Bilateral INO occurs due to a bilateral lesion of both MLFs, typically near the dorsal pontine tegmentum. Although the MLF transmits signals related to convergence, unilateral MLF lesions usually do not affect convergence. However, bilateral MLF lesions can impair convergence, resulting in exotropia in the primary position, a condition known as "wall-eyed bilateral INO" (WEBINO) [6, 18].

On the other hand, WEMINO syndrome (Wall-eyed monocular internuclear ophthalmoplegia) is a rare variant of internuclear ophthalmoplegia (INO) with ipsilateral exotropia whose etiology is unknown. One hypothesis suggests that abnormal vestibular signals that reach the medial rectus nucleus induce an asymmetry of muscle tone and the consequent ocular abduction toward the injured side [19]. It has been proposed that Wall-Eyed Monocular Internuclear Ophthalmoplegia (WEMINO) results from discrete unilateral damage to burst-tonic fibers in the MLF. In contrast, orthotropic INO requires more extensive damage to extrafascicular pathways (lateral, ventral, and caudal to the MLF), which carry inhibitory signals to the medial rectus subnucleus [20].

Contralateral exotropia, though rarely reported, appears to be a frequent finding at the onset of INO. The mechanism is not fully understood, but some authors attribute it to overexcitation of the contralateral pontine paramedian reticular formation [21]. This condition is known as paralytic pontine exotropia (PPE) or nonparalytic pontine exotropia (NPPE).

23.6 Case 1

We present the case of a 71-year-old hypertensive male who experienced sudden and spontaneous symptoms of dizziness, spinning vertigo, unsteadiness, and horizontal diplopia. One week prior to admission, he had also developed weakness and paresthesia on the left side of his body.

On oculomotor examination, primary gaze revealed downward deviation of both eyes, with paralysis of right adduction and abducting nystagmus on left gaze, consistent with right internuclear ophthalmoplegia (INO) (characterized by adduction deficits and bilateral abducting nystagmus). The patient also exhibited severe restriction of upgaze in both eyes and impaired convergence (Fig. 23.2). Attempted upgaze triggered convergence–retraction nystagmus. Pupillary examination showed both pupils at 3 mm, reactive to light.

Neurological examination revealed mild left-sided brachiocrural hemiparesis and superficial and nociceptive hypoesthesia. No other cranial nerve, sensory, motor, or cerebellar symptoms were observed.

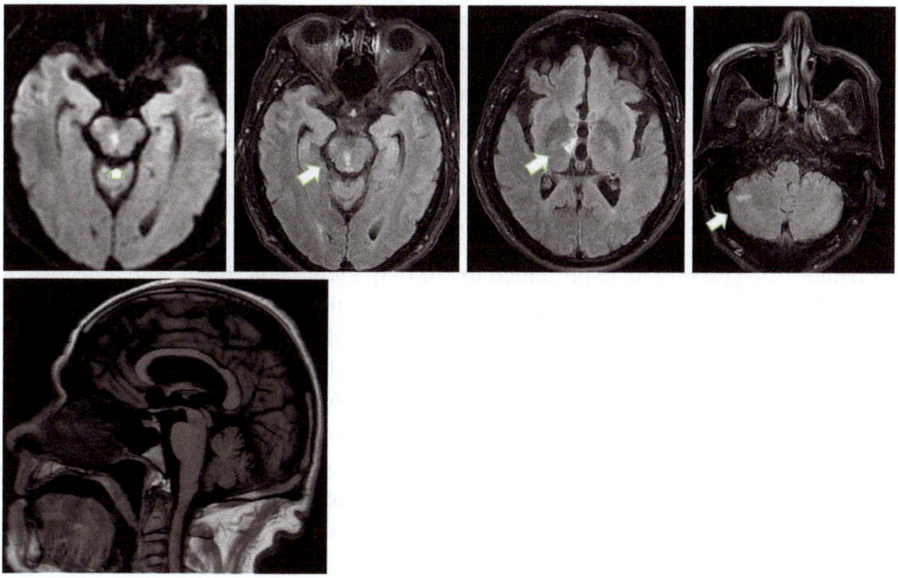

Fig. 23.2 Photographs of the nine eye positions in different gaze directions. Note the presence of right internuclear ophthalmoplegia (INO), characterized by the right eye's inability to adduct beyond the midline, accompanied by abduction nystagmus in the right eye and marked limitation of upgaze

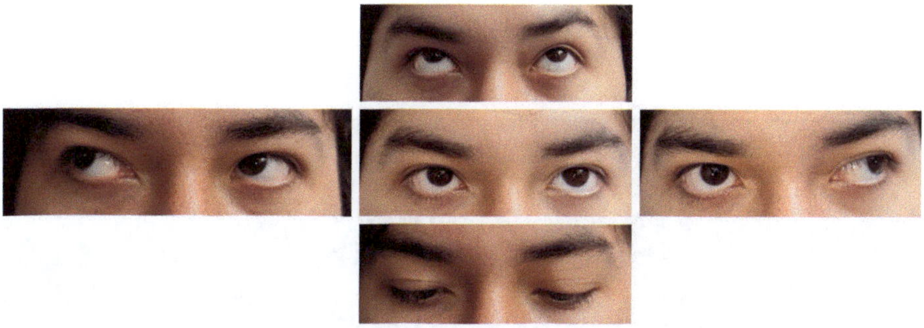

Fig. 23.3 MRI brain images, including axial FLAIR and diffusion-weighted imaging (DWI) sequences, show a hyperintense lesion (arrow) at the left dorsal tegmentum of the pontomesencephalic junction, extending ventrally into the right cerebellar hemisphere

Brain magnetic resonance imaging (MRI) demonstrated lesions affecting the right superior cerebellar peduncle, the MLF, and the right paramedian rostral thalamus. These findings were consistent with a right subacute thalamic infarction and an acute dorsal tegmentum infarction at the pontomesencephalic junction, involving the right anteromedial and posterior arterial territories of the brainstem (Fig. 23.3).

During hospitalization, the patient was found to have atrial fibrillation and was treated with rivaroxaban and antihypertensive agents. One month later, his diplopia and upgaze palsy showed significant improvement.

23.7 Case 2

A healthy 24-year-old male presented to our outpatient department with a history of blurred vision and diplopia. Neurological examination was unremarkable, with no evidence of cranial nerve, sensory, motor, or cerebellar symptoms.

On oculomotor examination, forward gaze revealed that the ocular position was fixed at the midline. The examination showed mild bilateral internuclear ophthalmoplegia (INO), characterized by adduction deficits and bilateral abducting nystagmus. Specifically, there was impaired adduction of the right eye with abduction nystagmus in the left eye, as well as mild impaired adduction of the left eye with abduction nystagmus in the right eye. Convergence was normal, and vertical eye movements were unaffected (Fig. 23.4).

MRI revealed multiple lesions in the cortical, subcortical, and periventricular white matter of the left anterior frontal and right parietal regions, bilateral centrum semiovale, and the right medial pontine area. These findings were strongly suggestive of a demyelinating disorder, most likely multiple sclerosis (Fig. 23.5). Spinal cord MRI was normal.

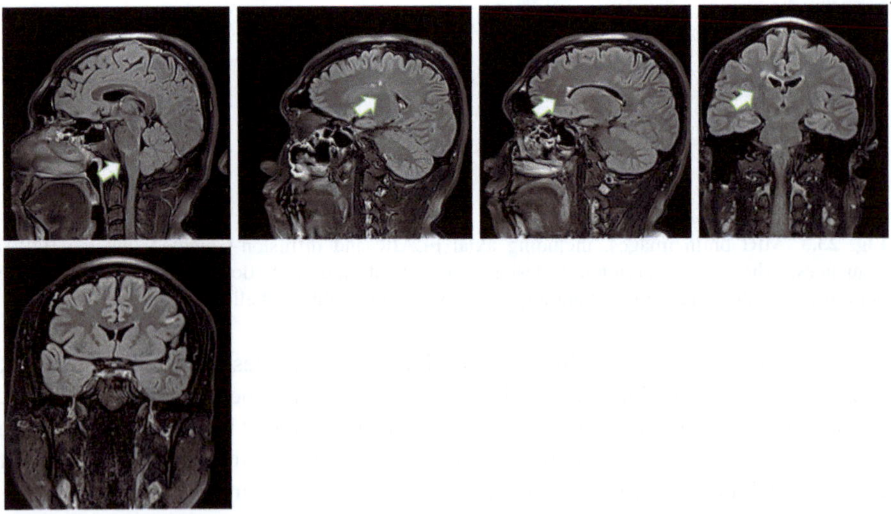

Fig. 23.4 Photographs of eye positions in different gaze directions during the initial attack. Note bilateral internuclear ophthalmoplegia (INO), predominantly on the right side

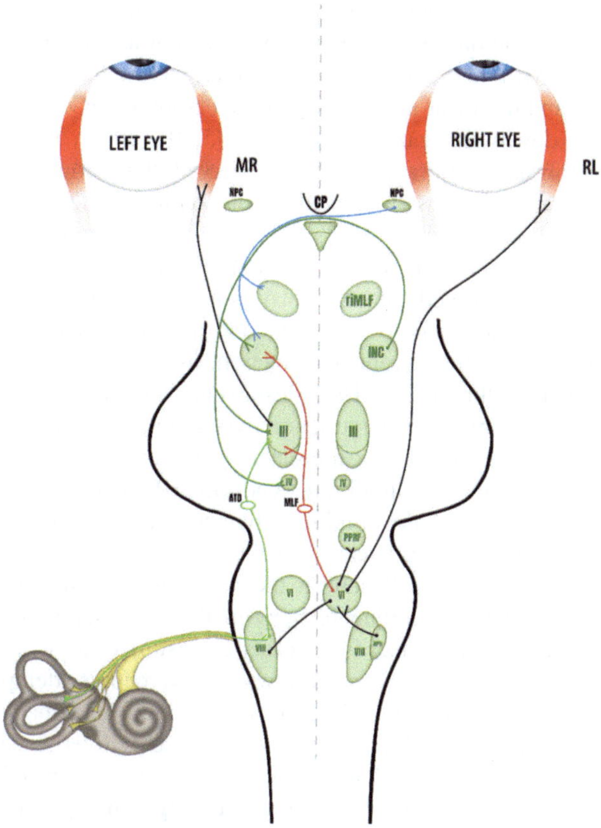

Fig. 23.5 Sagittal and coronal T2-weighted enhanced brain MRI during the acute phase. Abnormal signal intensity lesions are visible in the cortical, subcortical, and periventricular white matter of the left anterior frontal and right parietal regions, bilateral centrum semiovale, and the paramedian portion (arrow) of the upper pontine tegmentum

The patient was initially treated with interferon Beta-1a. After one relapse, he was switched to natalizumab. One month after treatment, his diplopia and extraocular movements showed almost complete recovery.

23.8 Discussion

Internuclear ophthalmoplegia (INO) caused by lesions of the medial longitudinal fasciculus (MLF) is often associated with heterogeneous oculomotor disorders. The most frequent causes of INO include stroke, demyelinating diseases, hemorrhages, tumors, and other conditions.

In the first case, the finding of INO was associated with upgaze palsy, which involves lesions in the mesencephalic pathways responsible for vertical gaze. Key structures involved include the riMLF, the INC, and the posterior commissure (PC). The riMLF contains burst neurons for both upward and downward gaze, while the INC acts as a neural integrator, providing ocular position signals to maintain upward gaze. Upgaze commands cross in the PC and reach the superior rectus subnucleus of the oculomotor nerve (third nerve). In contrast, axons for downward gaze arise from the riMLF and INC, projecting directly to the ipsilateral and contralateral oculomotor and trochlear nuclei without traversing the PC. It is believed that the projections for downward gaze are stronger [5, 9], which may explain why upward gaze paralysis is more frequently observed than downward gaze paralysis, even in unilateral lesions of the posterior commissure pathway. Additionally, vestibular signals affecting the VTT impair upward eye movement drives, leading to downward deviation, as seen in our first patient.

INO with vertical gaze palsy is rarely reported and typically involves lesions at the mesencephalic-pontine junction and the dorsomedial tegmentum of the combined pontomesencephalic junction, as well as the paramedian tegmentum of the dorsal midbrain [22]. These lesions are often coupled with vergence abnormalities, such as convergence–retraction nystagmus. This saccadic disorder, characterized by asynchronous and opposite saccadic movements during attempted upward gaze, is a hallmark of dorsal midbrain syndrome (Parinaud syndrome).

Three key clinical findings characterize Parinaud syndrome (PS). First is limited conjugate upgaze, because of the involvement of the riMLF, INC, and Posterior commissure (CP). Second is the convergence–retraction nystagmus, which is characterized by disconjugate horizontal jerk nystagmus, where both eyes converge quickly due to a saccadic disorder consisting of asynchronous and opposite saccadic movements when a vertical upward movement is attempted, cataloged as a vergence anomaly caused by damage to the supranuclear fibers which normally inhibit the neurons responsible for convergence and divergence. The third is a pupillary disorder, which is characterized by the impairment of constriction in response to light, while pupillary constriction during convergence and accommodation is preserved. The pretectal nucleus sends a signal to the ipsilateral and contralateral Edinger–Westphal nucleus via the posterior commissure, which contributes to securing the consensual reflex [22, 23].

To our knowledge, the association of Parinaud syndrome (PS) with internuclear ophthalmoplegia (INO) has not been previously reported. In our case, pupillary disorders were absent, which we presume is due to the lesion not involving the Edinger–Westphal nuclei or because of the alternative pathways for the pupillary light reflex that bypass the pretectal circuit [23]. Additionally, our patient presented with an ischemic lesion in the rostral thalamic region, which explains the left-sided paresthesia and mild paresis.

It is well-established that demyelinating diseases are among the most frequent causes of INO, particularly when bilateral and occurring in young patients. In our first case, we observed bilateral right-dominant INO due to lesions affecting both the medial longitudinal fasciculi (MLF), with greater involvement on the right side. The lesion was localized to the pons, and this case represented the initial presentation leading to the diagnosis of multiple sclerosis.

The complex anatomy of the brainstem, particularly the course of the MLF, gives rise to diverse oculomotor abnormalities, posing a diagnostic challenge for clinicians. Our patients exhibited significant oculomotor alterations, including INO accompanied by disturbances in vertical eye movements due to involvement of mesencephalic structures such as the INC, riMLF and PC. These structures are closely associated with the origin of the MLF. Demyelinating lesions remain a predominant cause of such presentations, especially in young patients.

Despite extensive studies and theories on oculomotor pathways, a thorough understanding of these oculomotor signs is crucial for achieving accurate clinical–radiological correlations. This knowledge aids in precise lesion localization and consideration of relevant etiologies, ultimately guiding appropriate diagnosis and management.

References

1. Frohman TC, Galetta S, Fox R, Solomon D, Straumann D, Filippi M, et al. Pearls & oy-sters: the medial longitudinal fasciculus in ocular motor physiology. Neurology. 2008;70(17):e57–67.
2. Johkura K, Kudo Y, Amano Y, Kikyo H, Imazeki R, Amari K, et al. Gaze palsy and exotropia in internuclear ophthalmoplegia. J Neurol Sci. 2015;353(1–2):158–60.
3. Fiester P, Baig SA, Patel J, Rao D. An anatomic, imaging, and clinical review of the medial longitudinal fasciculus. J Clin Imaging Sci. 2020;10:83.
4. Standring S, Elsevier C. Gray's anatomy: the anatomical basis of clinical practice. Forty-second edition. ed. New York: Elsevier; 2021.
5. Leigh RJ, Zee DS. Neural Basis for Conjugate Eye Movements. In: The Neurology of Eye Movements. Oxford University Press; 2015. p. 386–473.
6. Wray SH. Eye movement disorders in clinical practice. Oxford University Press; 2014.
7. Ono S, Mustari MJ. Smooth pursuit-related information processing in frontal eye field neurons that project to the NRTP. Cereb Cortex. 2009;19(5):1186–97.
8. Leigh RJ, Zee DS. Smooth; visual tracking and fixation. In: The Neurology of Eye Movements. Oxford University Press; 2015. p. 0.
9. Wong A. Eye Movement Disorders. USA: Oxford University Press; 2008.
10. Leigh RJ, Zee DS. The neurology of eye movements. Oxford University Press; 2015.
11. Maeda K, Inoue KI, Kunimatsu J, Takada M, Hikosaka O. Primate Amygdalo-Nigral pathway for boosting oculomotor action in motivating situations. iScience. 2020;23(6):101194.
12. Aw ST, Chen L, Todd MJ, Barnett MH, Halmagyi GM. Vestibulo-ocular reflex deficits with medial longitudinal fasciculus lesions. J Neurol. 2017;264(10):2119–29.
13. Brodsky MC, Donahue SP, Vaphiades M, Brandt T. Skew deviation revisited. Surv Ophthalmol. 2006;51(2):105–28.
14. Leigh RJ, Zee DS. Diagnosis of peripheral ocular motor palsies and strabismus. In: The Neurology of Eye Movements. Oxford University Press; 2015. p. 0.
15. Furman JM, Lempert T. Handbook of clinical neurology, (3rd series) Neuro-Otology, vol. 137; 2016.
16. Urban PP, Caplan LR. Brainstem disorders. Springer; 2011.
17. Chang T-P, Zee DS, Kheradmand A. Chapter 2 – Technological advances in testing the dizzy patient: the bedside examination is still the key to successful diagnosis. In: Gleason AT, Kesser BW, editors. Dizziness and vertigo across the lifespan. Elsevier; 2019. p. 9–30.
18. Leigh RJ, Zee DS. Disorders of ocular motility due to disease of the brainstem; cerebellum, and diencephalon. In: The neurology of eye movements. Oxford University Press; 2015. p. 0.
19. Ceballos-Lizarraga R, Palomino-Díaz C, Romero-Figueroa JÁ. Wall-eyed monocular Internuclear Ophthalmoplegia (WEMINO) and Millard-Gubler syndromes in a patient with isolated pontine infarction: topographic, oculomotor, and radiological analysis of two very uncommon conditions. Case Rep Neurol. 2019;11(2):230–7.

20. Johnston JL, Sharpe JA. Wall-eyed Internuclear Ophthalmoplegia: history and hypothesis. Canadian J Neurol Scis / J Canadien des Sci Neurol. 2024:1–9.
21. Haider AS. Unilateral internuclear ophthalmoplegia, strabismus and transient torsional nystagmus in focal pontine infarction. BMJ Case Rep. 2016;2016:bcr2016216503.
22. Wai YZ, Ng QX, Lim TH, Lim LT. A rare case of unilateral Cogan's anterior internuclear ophthalmoplegia, upgaze palsy and ataxia caused by dorsal tegmentum lesion at pontomesencephalic junction. BMC Ophthalmol. 2021;21(1):105.
23. Ortiz JF, Eissa-Garces A, Ruxmohan S, Cuenca V, Kaur M, Fabara SP, et al. Understand Parinaud's Syn Brain Sci. 2021;11:11.

Chapter 24
Covert Corrective Saccades in Acute Unilateral Vestibular Patients Induced by Predictable and Unpredictable Automated Head Impulses: A LATER Model Analysis

Stefano Ramat, Hassen Kerkeni, David S. Zee, Athanasia Korda, Miranda Morrison, and Georgios Mantokoudis

Keywords Linear Approach to Threshold with Ergodic Rate (LATER) · Saccade latency · Saccadic correction · Vestibulopathy · Compensatory response

24.1 Introduction[1]

The vestibulo-ocular reflex (VOR) is responsible for visual stabilization during head rotations while viewing a stationary target, which it accomplishes by producing equal and opposite eye rotations through a three-neuron arc. The first afferents from the excited semicircular canal project to the ipsilateral vestibular nuclei (VN), which project to the contralateral abducens nuclei, in turn causing the contraction of the contralateral lateral rectus while also exciting the ipsilateral oculomotor nuclei that innervate the ipsilateral medial rectus and hence produce a conjugate, compensatory eye rotation.

[1] Some of the material presented in this work was published in Kerkeni et al. [18].

S. Ramat (✉)
Department of Electrical, Computer and Biomedical Engineering, University of Pavia, Pavia, Italy
e-mail: stefano.ramat@unipv.it

H. Kerkeni
Department of Neurology, Inselspital, Bern University Hospital, University of Bern, Bern, Switzerland

D. S. Zee
Department of Neurology, Johns Hopkins University School of Medicine, Baltimore, MD, USA

A. Korda · M. Morrison · G. Mantokoudis
Department of Otorhinolaryngology, Head and Neck Surgery, Inselspital, Bern University Hospital, University of Bern, Bern, Switzerland

P. Ravat et al. (eds.), *Advances in Translational Neuroscience of Vestibular and Eye Movement Disorders*, Contemporary Clinical Neuroscience,
https://doi.org/10.1007/978-3-032-07539-0_24

If the vestibular system is impaired, or head rotations exceed the capabilities of the VOR [1, 2], the brain rapidly learns to produce corrective saccades in the same direction of the deficient response.

Vestibular function is commonly assessed using the head impulse test (HIT) [3] in which abrupt, small amplitude head rotations are imposed on the head of the patient while fixating on a stationary target. The paradigm allows testing the VOR at high accelerations and frequencies, and unveils vestibular dysfunction reflected in a corrective saccade. It is also effective in chronic unilateral vestibular patients, which may have compensated the deficient VOR at low accelerations because the intensity of the head impulse toward the affected side drives the healthy side to inhibitory cutoff, thus revealing the affected side.

These corrective saccades can be triggered while the head is still moving (covert saccades) or when the head has stopped rotating (overt saccades). Covert saccades can appear with latencies as low as ~70 ms [2, 4] and were shown to be functionally useful to improve visual performance in dynamic conditions [5, 6].

While overt saccades can usually be detected during the bedside head impulse test by the clinician, covert saccades are frequently revealed only by recording eye movements, e.g., using the video head impulse test. When both eye and head movements are recorded—for instance, using a video-oculography recording system that typically embeds a gyroscope—a further indicator of VOR dysfunction is its gain, i.e., the ratio of eye movement to head movement. A gain value, as assessed during the head impulse test, below 0.7 [7, 8] or 0.8 [9, 10] is typically considered pathological, and in the case of unilateral vestibular patients, it is expected to be asymmetrical.

Although several hypotheses have been proposed for explaining the triggering of covert saccades, including neck proprioception [11, 12], residual labyrinthine function, or other clues such as somatosensory information from the pressure of the examiner's hands, the mechanisms determining the timing and amplitude of covert saccades are still poorly understood. Studies have shown that covert saccades are triggered sooner when head rotations are active [13, 14] as well as when passive head rotations are predictable [15, 16]. Previous studies by our group have shown that the amplitude of a covert corrective saccade compensates for the gaze error expected at the saccade end, accounting for the head displacement occurring during the saccade rather than merely correcting the gaze error present when the saccade is initiated, and that covert saccades triggered during active head movements are more precise than during passive, unpredictable ones [17].

On the other hand, though, what triggers a corrective saccade remains largely unknown, and previous hypotheses on the existence of a threshold gaze error value [17] do not satisfactorily explain the experimental results.

Here, we set out to understand how the brain adapts to vestibular loss, estimates head velocity with deficient labyrinthine function, and uses such information to trigger covert corrective saccades.

We carried out a study at the Bern University Hospital using an automated head impulse test device and used it to deliver repeatable stimuli to patients with acute unilateral vestibular neuritis, to reduce the confounding effects of stimulus variability [18]. We measured the latencies of the first corrective saccades in response to predictable and unpredictable impulses. We studied VOR gains and saccade amplitudes, but mostly focused on the latencies of covert saccades. To interpret the distribution of the latencies of covert saccades and to speculate on the decision process that led to their triggering, we used the LATER model of motor response latency. We hypothesized that decision-making processes rely on a build-up of activity toward a threshold of confidence, triggering a response when the threshold is reached. This modeling approach considers the decision to make a corrective saccade as a process in which the brain estimates the likelihood, presumably based on vestibular signals, that the compensation of the head movement will require a corrective saccade to refixate the target, and when enough evidence in support of this hypothesis has been gathered, the saccade is triggered.

24.2 Materials and Methods

We considered 14 patients with acute vestibular neuritis who were recruited from the emergency department of the Bern University Hospital and used the aHIT device [19] to deliver head impulses in the horizontal plane while they viewed a stationary target.

Head impulses were either 150°/s peak head velocity with 10° amplitude or 250°/s with 20° amplitude. Peak angular acceleration was always $3000°/s^2$. The duration of the slow and fast head impulses was therefore different and averaged 130 ms in the first group and 190 ms in the second.

The movement of the right eye was recorded at 220 Hz using an EyeSeeCam video-oculography device, and the gyroscope signals from the incorporated Inertial Measurement Unit (IMU) were used to compute head angular velocity.

We developed one paradigm in which the head rotation direction, peak velocity, amplitude, and timing were predictable and used it to deliver ten cycles, consisting of one rotation from center to lateral followed by one rotation back to center, with an interval between head impulses of 1.2 s. Such a paradigm is referred to as "predictable" in our analysis. In the other paradigm, the timing, peak velocity, and amplitude of the head impulses were unpredictable, and we delivered ten cycles to each patient. This paradigm is referred to as "unpredictable" in our analysis (Fig. 24.1).

We considered the first zero-crossing of head velocity after its peak as the end of each head impulse and defined those corrective saccades beginning before that time as covert saccades. Saccades beginning after the zero-crossing were instead considered as overt saccades. This definition of overt and covert saccades, based on their occurrence relative to head motion, lacks physiological significance, as head impulses can vary in duration. Instead, it reflects the practical challenge encountered in clinical examinations when trying to identify saccadic corrections that occur during head rotation (covert saccades). Conversely, the definition of covert corrective saccades can be viewed as an approximation of the corrections made during the open-loop phase of the oculomotor response. Indeed, from a physiological

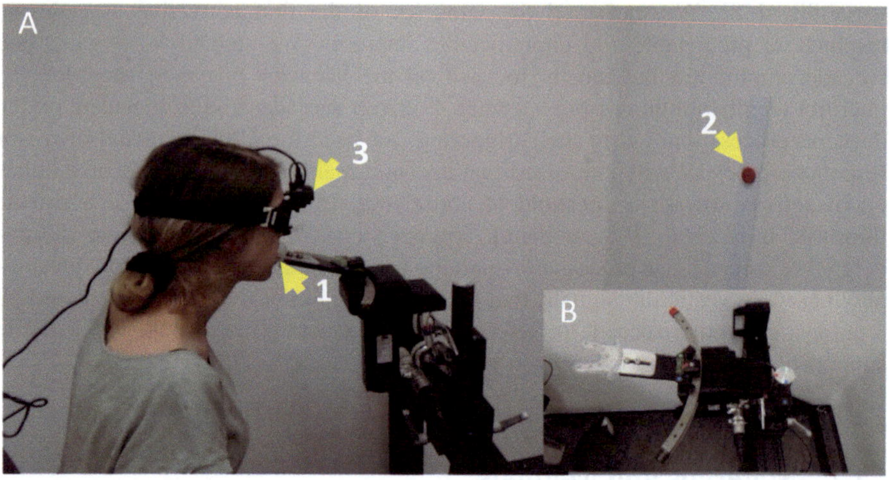

Fig. 24.1 (**a**) Patient on the bite block (1) of the aHIT™, fixating a fixation target (2). The eyes and head movements are recorded with a EyeSeeCam video-oculography system (3). (**b**) automated head impulse device (aHIT™)

standpoint, one aspect of scientific interest is the understanding of how the brain triggers nonvisually guided corrective saccades, which are thought to be related to mechanisms triggering covert saccades during head movements. A threshold latency of 150 ms, roughly corresponding to the average duration of correctly imposed manual head impulses, has also been considered the maximum latency for nonvisually guided covert saccades [17]. Choosing this criterion would not have significantly changed the number of saccadic corrections considered covert in this work, as it would have led to discarding only four corrections compared to using the zero-crossing criterion used in our analysis.

Gaze error was computed as the sum of head position and eye position, since the fixating target centered in front of the subject corresponds to a gaze angle of 0 degrees. We measured gaze error at saccade onset, at saccade end, and at the end of the head movement. The gain of the VOR response was computed as the ratio of the average of the desaccaded eye velocity to that of head velocity, computed between the time of peak acceleration and the time of peak velocity of the head [17].

The distribution of covert saccade latencies across the population of subjects was analyzed using the LATER model of response time [20], which considers decision-making processes triggered by a stimulus as resulting from a build-up of activity toward a threshold of confidence, S_T. Such activity is hypothesized to rise linearly from its initial value representing a prior in Bayesian terms, S_O, with a rate r, which is drawn from a Gaussian distribution with mean μ and standard deviation σ. The resulting reaction time is therefore $T = (S_T - S_O)/r$. With the LATER approach, saccade latencies are considered in terms of their reciprocal, and in this representation, the skewed distribution of latencies becomes Gaussian. The cumulative histogram of reciprocal latencies, plotted using a nonlinear probit scale for the vertical axis and the reciprocal of saccade latency on the horizontal axis, building what is commonly called a reciprobit plot [21], is a useful visualization. Indeed, with this representation, the Gaussian distribution of reciprocal latencies becomes a straight line, allowing one to intuitively verify whether the distribution of reciprocal latencies is normal and obtain a visual estimate of the two free parameters of the model, μ and σ. The model was previously used to explain response latencies in several experimental contexts, from eye movements to manual responses in controls, as well as to quantify the progress of neurodegenerative diseases [22].

24.3 Results

We compared the gain of the VOR in response to the predictable and unpredictable stimuli toward both the affected and unaffected sides. While there were no statistically significant differences toward the unaffected side, the gain in response to unpredictable stimuli toward the affected side was significantly ($p < 0.001$) lower (0.39 ± 0.22) than with the predictable ones (0.47 ± 0.28).

The prevalence of trials with covert corrective saccades was higher in predictable (63.5%) than in unpredictable (51.5%) head impulses.

An example of the eye movements (eye angular velocity) recorded in two representative patients during the predictable and the unpredictable paradigms is shown in Fig. 24.2, showing both a higher VOR gain and a higher number of corrective saccades in the responses to the predictable paradigm.

As expected, given the higher VOR gain and lower latency, we found that the gaze error at the onset of the first covert corrective saccade was significantly lower ($p < 0.001$) in predictable ($6.7 \pm 4.5°$) than in unpredictable ($8.7 \pm 4.2°$) trials. Similarly, gaze error at the time the head stopped, i.e., the first zero-crossing of head velocity, was lower with predictable head impulses, and the same was true for gaze error at the end of the first covert corrective saccade (Fig. 24.3).

The amplitude of covert saccades is highly correlated with the total error, i.e., the difference between head position at saccade end and eye position at saccade onset, in response to both predictable and unpredictable head impulses (Fig. 24.4), whereas the correlation with gaze error at saccade onset is significantly lower (not shown).

Covert saccades toward the pathological side have a lower mean latency (104 ± 27 ms) in predictable head impulses than in unpredictable head impulses (114 ± 25 ms), whereas the latencies of the first corrective saccades are not different toward the healthy side.

24.3.1 Saccade Latency

Besides the significant difference in saccade latencies in response to predictable and unpredictable head impulses captured by the Analysis of Variance (ANOVA) test of

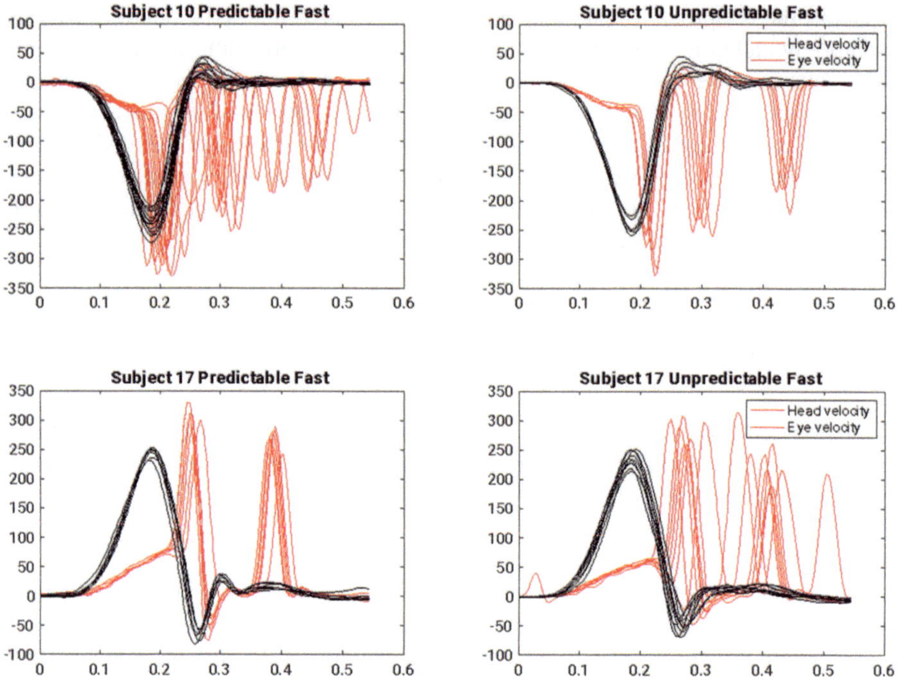

Fig. 24.2 Responses to automatic head impulses toward the affected side in two representative patients. Left column, responses to predictable trials. Right column, responses to unpredictable trials

variances ($p < 0.001$), the distributions of saccade latencies are different in the two paradigms (Fig. 24.5a). Both distributions show that early covert saccades, with a latency lower than 90 ms, are more frequent with predictable head impulses.

As explained in the Methods section, we chose to analyze covert saccade latencies using the LATER model and built a reciprobit plot of their cumulative distributions, shown in Fig. 24.5b.

Using this representation of the data, the cumulative distribution of latencies from both predictable and unpredictable head impulses appeared to be best fit by two straight lines, suggesting two Gaussian distributions of reciprocal latencies. In fact, their best fit showed the typical broken-line plot observed by Carpenter with visually evoked saccades [21], suggesting the existence of two decision mechanisms: one responsible for the longer-latency covert corrections and one responsible for the early covert saccades.

Earlier covert saccades are represented by the shallower lines in Fig. 24.5b, as they are less frequent in both paradigms than the longer-latency ones.

Different from early saccades to visual targets, we found that with vestibular stimuli they may occur sooner, with a latency lower than 100 ms. In predictive trials, early covert saccades accounted for almost 30% of covert corrective saccades, as indicated by the y-coordinate of the intersection point of the two straight lines, which

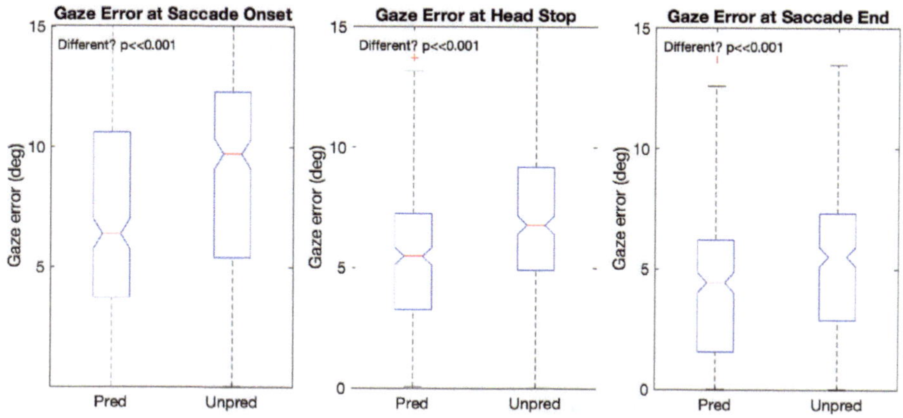

Fig. 24.3 Left panel: Boxplot of gaze error at the time of covert saccade onset in predictable and unpredictable head impulses. The red line represents the median of the distribution. The bottom and top lines represent the 25th and 75th percentiles, respectively, while the whiskers represent the extremes of the distribution. Center panel: Boxplot of gaze error at the time the head stops (the first zero-crossing of head velocity after head impulse onset) in both predictable and unpredictable head impulses containing a covert saccade. Right panel: Boxplot of gaze error at the time of covert saccade end

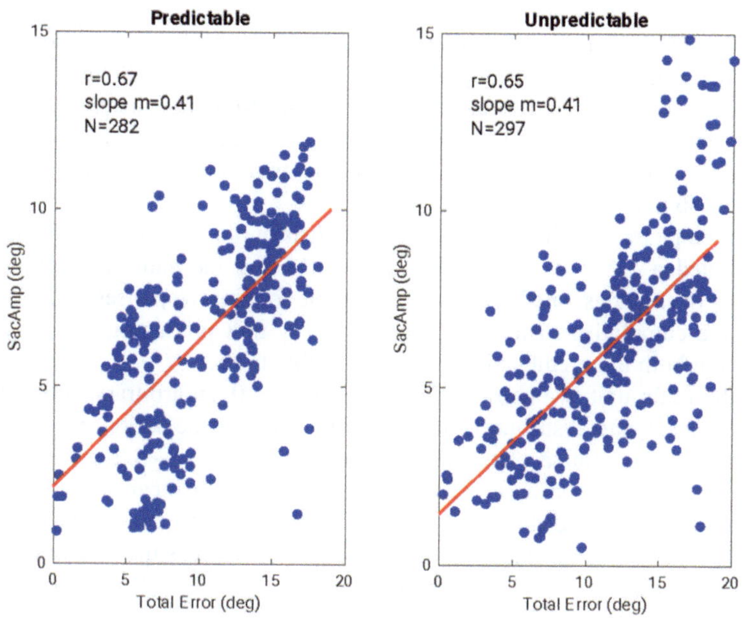

Fig. 24.4 Saccade amplitude versus total error in predictable (left) and unpredictable (right) head impulses. *r*: correlation coefficient; *m:* slope of the interpolating line. Total error is the difference, in degrees, between head position at saccade end and eye position at saccade onset

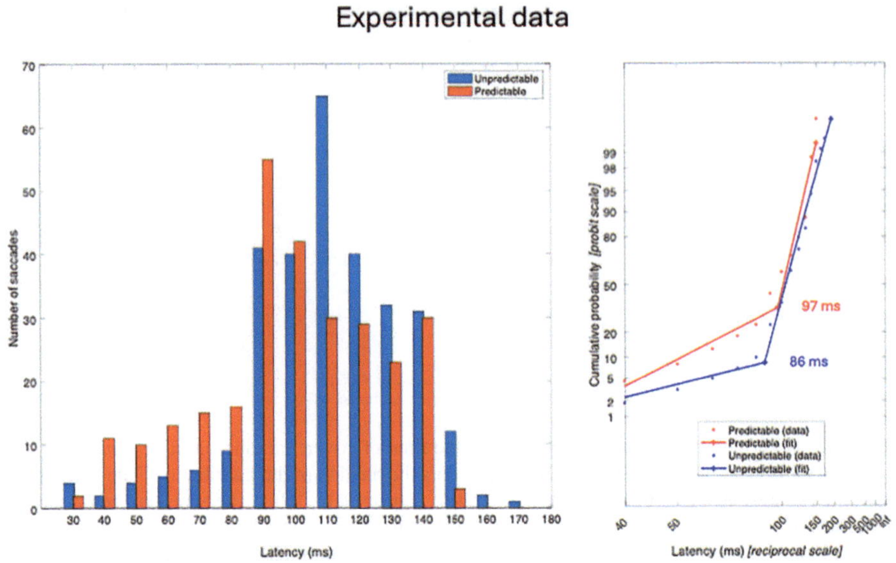

Fig. 24.5 Distribution of covert saccades' latencies in experimental data. Left panel: Histograms of latencies grouped in 10 ms bins for the unpredictable (blue bars) and the predictable (red bars) conditions. Right panel: Reciprobit plot of the cumulative distributions of reciprocal latencies in the unpredictable (blue line) and the predictable (red line) conditions

occured at 97 ms. Indeed, in predictable head impulse trials, early covert saccades were more frequent than in unpredictable ones; therefore, their distribution was steeper (red line compared to blue line in Fig. 24.5b). During unpredictable trials, instead, early covert saccades accounted for less than 10% of the cumulative distribution, and the intersection between the two lines (distributions) occured sooner, at 86 ms.

To attempt to explain the experimental distribution of saccade latencies, we implemented a LATER model with two competing mechanisms leading to the triggering of saccades: one Main unit responsible for most responses, i.e., the steeper part of the cumulative distribution, and an Early unit producing the early covert saccades in both the predictable and unpredictable trials.

For the Main unit, we chose an initial state $S_0 = 0$, an arbitrary threshold $S_T = 1$, and a mean rate distribution $\mu = 9$ with standard deviation $\sigma = 1.1$, which produced decisions for saccades with a mean latency of $(S_T - S_O)/\mu = 110$ ms. The Early unit was modeled as a Gaussian distribution with mean $\mu_E = 0$, standard deviation $\sigma_E = 9.2$, and threshold $S_{TE} = 1$. During unpredictable trials, the initial state for this decision mechanism was also $S_{0E} = 0$, whereas the greater confidence related to predictable trials was modeled as a higher initial state, i.e. a higher prior, $S'_{0E} = 0.3$ (Fig. 24.6).

The model simulation produced the distributions in Fig. 24.7, with the usual pattern of distribution on the left side and the reciprobit plot on the right.

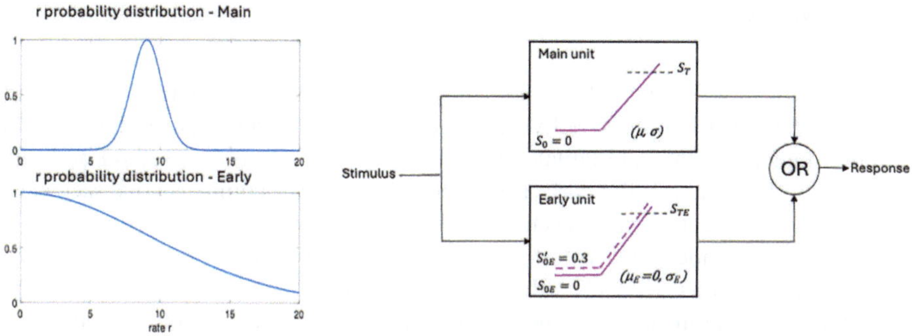

Fig. 24.6 The two decision mechanisms LATER model of covert corrective saccades' generation. Left panel: Probability distribution of the rate *r* of the rising decision signal for the Main (top, μ, σ) and the Early (bottom, μ_E, σ_E) units used in the model simulations. Right panel: The model with the two competing decision units triggering the covert corrective saccades showing the initial states S_O and S_{OE} and the corresponding thresholds S_T and S_{TE} that are set to 1 for all simulations. In the simulations of unpredictable trials, both initial states are set to 0, while for the simulation of predictable trials, the higher confidence in the vestibular input signal is modeled by a higher *prior* for the Early distribution, which is represented by an initial state $S'_{OE} = 0.3$

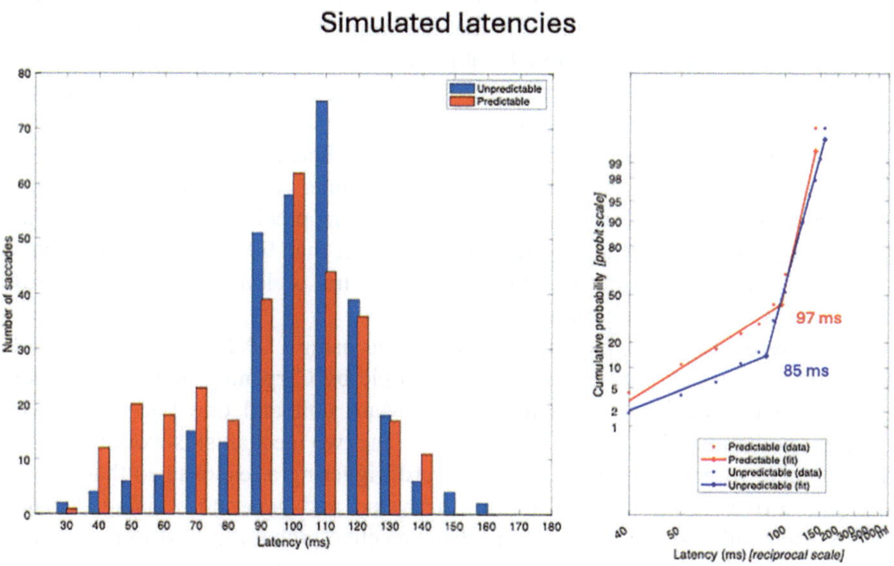

Fig. 24.7 Distribution of simulated covert saccades' latencies. Panel A: Histograms of latencies grouped in 10 ms bins for the unpredictable (blue bars) and the predictable (red bars) conditions. Panel B: Reciprobit plot of cumulative distributions of reciprocal latencies in the unpredictable (blue line) and the predictable (red line) conditions

24.4 Discussion

In this work, we have studied the compensatory response to predictable and unpredictable head impulses, delivered by an automatic head impulse device, in a group of patients with acute unilateral vestibular. Our analysis focused on both the VOR slow phase and on the early triggering of corrective compensatory saccades, i.e., the covert corrective saccades that are triggered before the head stops rotating, a time point that we identified with the first zero-crossing of head velocity.

As in our previous study [13], the amplitude of the corrective saccades appeared to take into account the expected head movement occurring during the saccade, as it correlates better with the total error, i.e., the difference between head position at saccade end and eye position at saccade onset, than with the error at saccade onset. This suggests that an estimate of the head's angular displacement was used to compute saccade amplitude.

Similar to previous studies on the translational VOR [23] and to the findings on the angular VOR in chronic unilateral patients [24], we found that knowing the direction of the upcoming head movement influences the compensatory vestibular response, which presents with higher gain compared to when the rotation direction is unknown.

A higher VOR gain also led to a lower gaze error during head rotation. However, in response to predictable head impulses, we found earlier and more frequent corrective saccades compared to the responses to unpredictable HITs. In fact, our experimental data showed that the gaze error at the onset of the first corrective saccade was significantly smaller and that the proportion of early covert saccades is higher in predictable than in unpredictable trials.

Hence, the idea that corrective saccades are triggered when the estimated gaze error reaches a threshold value [25] does not explain the experimental data from this study, since it would imply later saccades in the predictable condition because gaze error grows more slowly than in the unpredictable one, yet we found the opposite.

Furthermore, in both conditions, and more so in predictable trials, a relevant number of corrective saccades were triggered earlier than 90 ms from head impulse onset, but most corrections had latencies of 90 ms or longer, and their overall distributions are far from normal.

We therefore tried to explain these distributions using a LATER model of saccade latencies, inspired by previous work on the topic by Carpenter, and built a model comprising two decision mechanisms: one Main unit and one Early unit, with different probability distributions of the rate *r*. The two mechanisms work in parallel and compete with each other in a race toward triggering a response so that when a stimulus arrives, both mechanisms are activated and generate a rising decision signal, each with a slope *r* drawn from the respective probability distribution, rising toward the threshold. The first signal reaching it will be the one triggering the saccade. The Main unit is responsible for triggering most covert saccades and has a probability distribution with mean $\mu = 9$ with standard deviation 1.1, while the Early unit has a mean $\mu = 0$ with standard deviation 9.2. For simulating covert saccades' latencies during unpredictable head impulses both units have an initial state $S_0 = 0$ and a decision threshold $S_T = 1$. Simulations of this model produced an overall distribution of saccade latencies that replicated quite accurately the experimental data recorded in unpredictable head impulses. Figure 24.6 shows the distribution of latencies resulting from one such simulation.

Here, we suggest that the change between the unpredictable and the predictable head impulses is represented by a change in the prior probability accorded to the Early mechanism, S_{OE}. Indeed, choosing an initial state $S_{OE} = 0.3$ for modeling a higher confidence accorded to the Early unit during predictable head impulses, reduced the amount of information that this unit needs to accumulate to reach the decision threshold to 0.7, providing a higher chance for its decision signal to reach the triggering threshold first.

Simulations of this model replicated quite precisely the distributions of saccade latencies in terms of the cumulative amount of early covert corrective saccades and the intersection of the two lines representing the two underlying distributions, in both predictable and unpredictable trials.

More specifically, the increase in the initial state S_{OE} between the unpredictable and the predictable trials appears to explain the two different cumulative distributions of covert saccade latencies and the lower median gaze error at the onset of the first covert corrective saccade, since the Main unit behaves identically in the two conditions, as found in the predictable vs. the unpredictable trials. In other words, the model accounts for the apparent two different gaze error thresholds for triggering the first covert corrective saccade (Fig. 24.3, left panel) by increasing the confidence level of the Early mechanism during predictable trials.

24.5 Conclusions

In this study, we used an automated device to impose passive head impulses on acute unilateral vestibular patients looking at a stationary target. We used two different paradigms: one in which the direction, timing, peak velocity, and amplitude of the head movement were predictable, and one in which these factors were unpredictable. Our results show that cognitive factors, such as the predictability of the speed, timing, and direction of head rotation, influenced the gain of the VOR, which is increased when the passive head movement parameters are predictable, and also affect the timing of covert saccades. In the predictable condition, covert saccades may be triggered more frequently and with a lower mean latency than in the unpredictable one.

Here, we focused our analysis on covert corrective saccades, which we defined as those occurring before the first zero-crossing of head velocity, and used a LATER approach to analyze their latency distributions in the two conditions. The reciprobit plot of covert saccades' cumulative distributions hinted at the existence of two subtypes of covert saccades: those occurring earlier and those occurring later than 90–100 ms. This led us to model the data using two competing raise-to-threshold decision mechanisms competing with each other, similarly to how Carpenter and Williams modeled saccades to sudden visual targets appearing with predefined statistics [21]. We hypothesized that the two subtypes of saccades are generated by a Main unit, responsible for most covert saccades, producing saccades with a mean latency of 110 ms, and an Early unit that can produce, although less likely, shorter-latency saccades. In simulating the unpredictable head impulse condition, both units had the same initial state set at zero and the decision threshold set at one (arbitrary units). To reproduce the higher ratio of early covert saccades in the predictable condition trials, we assumed that, when the features of the head impulse are predictable, the Early decision mechanism would be prioritized by assigning it a higher initial state (0.3), thereby reducing the amount of information that needs to be accumulated to trigger the saccade. This change in the initial state reflects a higher confidence assigned to the Early unit, a change in the *prior* in Bayesian terms. Overall, considering the changes for the predictable trials, the model accounts well for the distribution of covert saccade latencies in both the unpredictable and the predictable conditions. Further experiments, using different patterns of predictive information about head motion, for example, timing versus direction, timing versus amplitude, or timing versus speed, can be used to test our model. The ultimate goal will be to discover which circuits within the brain engage these predictive capabilities, which help overcome impaired vestibular function.

References

1. Peng GCY, Zee DS, Minor LB. Phase-plane analysis of gaze stabilization to high acceleration head thrusts: a continuum across Normal subjects and patients with loss of vestibular function. J Neurophysiol. 2003;91:1763–81. https://doi.org/10.1152/jn.00611.2002.
2. Tian J, Crane BT, Demer JL. Vestibular catch-up saccades in labyrinthine deficiency. Exp Brain Res. 2000;131:448–57.
3. Halmagyi GM, Curthoys IS. A clinical sign of canal paresis. Arch Neurol. 1988;45:737–9. https://doi.org/10.1001/archneur.1988.00520310043015.
4. Weber KP, Aw ST, Todd MJ, McGarvie LA, Curthoys IS, Halmagyi GM. Head impulse test in unilateral vestibular loss: vestibulo-ocular reflex and catch-up saccades. Neurology. 2008;70: 454–63. https://doi.org/10.1212/01.wnl.0000299117.48935.2e.
5. Ramaioli C, Colagiorgio P, Sağlam M, Heuser F, Schneider E, Ramat S, Lehnen N. The effect of Vestibulo-ocular reflex deficits and covert saccades on dynamic vision in opioid-induced vestibular dysfunction. PLoS One. 2014;9:e110322. https://doi.org/10.1371/journal.pone.0110322.
6. Hermann R, Pelisson D, Dumas O, Urquizar C, Truy E, Tilikete C. Are covert saccade functionally relevant in vestibular hypofunction? Cerebellum. 2018;17(3):300–7.
7. Machner B, Erber K, Choi JH, Trillenberg P, Sprenger A, Helmchen C. Usability of the head impulse test in routine clinical practice in the emergency department to differentiate vestibular neuritis from stroke. Eur J Neurol. 2021;28:1737–44. https://doi.org/10.1111/ENE.14707.
8. Mantokoudis G, Tehrani ASS, Wozniak A, Eibenberger K, Kattah JC, Guede CI, Zee DS, Newman-Toker DE. VOR gain by head impulse video-oculography differentiates acute vestibular neuritis from stroke. Otol Neurotol. 2015:36. https://doi.org/10.1097/MAO.0000000000000638.

9. McGarvie LA, MacDougall HG, Halmagyi GM, Burgess AM, Weber KP, Curthoys IS. The video head impulse test (vHIT) of semicircular canal function – age-dependent normative values of VOR gain in healthy subjects. Front Neurol. 2015:6. https://doi.org/10.3389/FNEUR.2015.00154.
10. Starkov D, Strupp M, Pleshkov M, Kingma H, van de Berg R. Diagnosing vestibular hypofunction: an update. J Neurol. 2021;268:377–85. https://doi.org/10.1007/S00415-020-10139-4.
11. Macdougall HG, Curthoys IS. Plasticity during vestibular compensation: the role of saccades. Front Neurol. 2012;3:21. https://doi.org/10.3389/fneur.2012.00021.
12. Iwasaki S, Kamogashira T, Fujimoto C, Kabaya K, Kinoshita M, Yamasoba T. The role of neck input in producing corrective saccades in the head impulse test. Front Neurol. 2022:13. https://doi.org/10.3389/fneur.2022.881411.
13. Peng GCY, Minor LB, Zee DS. Gaze position corrective eye movements in normal subjects and in patients with vestibular deficits. Ann N Y Acad Sci. 2005;1039:337–48. https://doi.org/10.1196/annals.1325.032.
14. Black RA, Halmagyi GM, Thurtell MJ, Todd MJ, Curthoys IS. The active head-impulse test in unilateral peripheral vestibulopathy. Arch Neurol. 2005;62:290–3. https://doi.org/10.1001/archneur.62.2.290.
15. Schubert MC, Mantokoudis G, Xie L, Agrawal Y. Acute VOR gain differences for outward vs. inward head impulses. J Vestib Res Equilib Orientat. 2014;24 https://doi.org/10.3233/VES-140523.
16. Mantokoudis G, Agrawal Y, Newman-Toker DE, Xie L, Saber Tehrani AS, Wong A, Schubert MC. Compensatory saccades benefit from prediction during head impulse testing in early recovery from vestibular deafferentation. Eur Arch Otorrinolaringol. 2016; https://doi.org/10.1007/s00405-015-3685-7.
17. Colagiorgio P, Versino M, Colnaghi S, Quaglieri S, Manfrin M, Zamaro E, Mantokoudis G, Zee DS, Ramat S. New insights into vestibular-saccade interaction based on covert corrective saccades in patients with unilateral vestibular deficits. J Neurophysiol. 2017:117. https://doi.org/10.1152/jn.00864.2016.
18. Kerkeni H, Zee DS, Korda A, Morrison M, Mantokoudis G, Ramat S. Corrective saccades in acute vestibular neuritis: studying the role of prediction with automated passively induced head impulses. J Neurophysiol. 2023:129. https://doi.org/10.1152/jn.00382.2022.
19. Tan GX, Schoo DP, Della Santina CC, Rahman MA, Valentin Contreras NS, Sun C-H, Chiang B. Automated head motion system improves reliability and lessens operator dependence for head impulse testing of vestibular reflexes. IEEE Int Symp Med Meas Appl Proc. 2017;2017: 94–9. https://doi.org/10.1109/MeMeA.2017.7985856.
20. Carpenter RHS. Oculomotor procrastination. In: Eye movements. Routledge; 1981. p. 237–46. https://doi.org/10.4324/9781315437415-19.
21. Carpenter RHS, Williams MLL. Neural computation of log likelihood in control of saccadic eye movements. Nature. 1995;377:59–62. https://doi.org/10.1038/377059a0.
22. Noorani I, Carpenter RHS. The LATER model of reaction time and decision. Neurosci Biobehav Rev. 2016;64:229–51. https://doi.org/10.1016/J.NEUBIOREV.2016.02.018.
23. Ramat S, Straumann D, Zee DS. Interaural translational VOR: suppression, enhancement, and cognitive control. J Neurophysiol. 2005;94:2391–402. https://doi.org/10.1152/jn.01328.2004.
24. Lee SH, Newman-Toker DE, Zee DS, Schubert MC. Compensatory saccade differences between outward versus inward head impulses in chronic unilateral vestibular hypofunction. J Clin Neurosci. 2014:21. https://doi.org/10.1016/j.jocn.2014.01.024.
25. Colagiorgio P, Versino M, Colnaghi S, Quaglieri S, Manfrin M, Zamaro E, Mantokoudis G, Zee DS, Ramat S. New insights into vestibular-saccade interaction based on covert corrective saccades in patients with unilateral vestibular deficits. J Neurophysiol. 2017;117:2324–38. https://doi.org/10.1152/jn.00864.2016.

Chapter 25
Neurobiology of Saccadic Neurons in Monkeys

Anja Horn, Ümit Mayadali, Kim Suza Kuhn, Michael Abspacher, and Aasef G. Shaikh

Keywords Ion channels · Burst neurons · Omnipause neurons

25.1 Premotor Saccadic Network

Saccades are brief (less than 100 ms) and very fast (up to 500°/s) eye movements that move the eyes between fixation targets. They serve to bring an object of interest into focus on the fovea [1, 2]. For many years, the basic saccadic circuitry, with firing characteristics of the crucial neuronal groups, has been investigated by recording and anatomical tract-tracing studies in different animal species, including nonhuman primates [2–5].

The immediate premotor command for saccade initiation is provided by the interaction of different sets of neurons within the brainstem referred to the saccadic burst generator. A high-frequency burst of activity is transmitted from excitatory burst neurons (EBNs) to motoneurons, inducing a phasic activity (pulse) in the corresponding extraocular muscles to overcome the viscosity of the orbital tissues [4]. At the new eye position, the motoneurons exhibit a new level of tonic activity (step) to keep the eye in place. This input is transmitted by premotor neural integrator cells as a position signal calculated from the incoming velocity signal of premotor EBNs [2, 6]. At the same time, inhibitory burst neurons (IBNs) targeting the motoneurons of the antagonistic eye muscles prevent them from contracting. During non-saccadic eye movements, including fixation, burst neurons are tonically inhibited by glycinergic omnipause neurons (OPNs) [7, 8]. For initiation of a

A. Horn (✉) · Ü. Mayadali · K. S. Kuhn · M. Abspacher
Institute of Anatomy, Faculty of Medicine, LMU Munich, Munich, Germany
e-mail: Anja.Bochtler@med.uni-muenchen.de; lncs@springer.com; kimsuza@freenet.de

A. G. Shaikh
Neurological Institute, University Hospitals Neurology Service, Louis Stokes Cleveland, VA Medical Center, Department of Neurology, Case Western Reserve University, Cleveland, OH, USA
e-mail: aasef.shaikh@case.edu

P. Ravat et al. (eds.), *Advances in Translational Neuroscience of Vestibular and Eye Movement Disorders*, Contemporary Clinical Neuroscience,
https://doi.org/10.1007/978-3-032-07539-0_25

saccade, OPNs are inhibited, which releases the inhibition from burst neurons and allows the activation or inhibition of the respective motoneurons. Thus, the OPNs have a key function—to maintain sustained inhibition of EBNs and IBNs when steady fixation is required.

The knowledge of the saccadic premotor circuitry helps to understand the saccadic symptoms in patients with vascular or structural lesions [for review: 2]. Similar symptoms are observed in monkeys after chemical lesions, for example, of the PPRF, resulting in slowed/abolished horizontal saccades [4, 9, 10], or of the rostral interstitial nucleus of the medial longitudinal fascicle (RIMLF), resulting in slowed/abolished vertical saccades [11]. However, the immediate cause for saccadic disorders in metabolic, autoimmune, or neurodegenerative diseases affecting the whole brain is not clear [12–14]. MRI scans in patients may help in the diagnosis of the disease, but do not highlight the malfunctioning neurons. Even in clinicopathological postmortem studies, the observed eye movement deficits cannot always be correlated with the expected loss or morphological damage of functional cell groups, for example, the OPNs in cases of opsoclonus [15], but may rather result from alterations of the membrane biology or extracellular matrix, as found in a case with saccadic palsy after aorta surgery [16].

There is growing evidence that the dysfunction of ion channels contributes significantly to the pathophysiology in many neurodegenerative diseases [17, 18], including those associated with saccadic disorders [19]. Based on the successful treatment of eye movement symptoms in vestibular disorders by drugs that act on transmitter-receptors or specific ion channels [20, 21], the intrinsic properties of neurons have been incorporated in recent, more advanced conductance-based neuromimetic models of the saccadic system [22]. Hypotheses have been put forward that the intrinsic properties of saccadic circuit neurons, e.g., alterations of ion channels or transmitter-receptors, may be a mechanism for the observed saccadic disorders rather than pure cell loss [22–27].

The firing pattern and the overall excitability of neurons are determined by various factors. This includes the neuron size (surface of the membrane), the intracellular calcium concentration and type of calcium buffers (calcium-binding proteins), and the ion channel expression on the neuronal membrane, along with the transmitter inputs they receive and transmitter-receptors they express [28–30]. In addition, the extracellular matrix formed by molecules released from neurons and glial cells in the extracellular space may contribute to firing characteristics [31]. A highly condensed form of extracellular matrix molecules known as perineuronal nets enwrap the somata and proximal dendrites of highly active neurons, such as those of the auditory system and saccadic system considered here [32–34].

In addition to sodium channels, voltage-gated potassium channel subunits (Kv) are the major regulators of a neurons' ability to return to the resting membrane potential, enabling the fast-firing capacity of neurons as required for the bursting behavior of premotor burst neurons [35]. Low-voltage-activated cation channels govern effective recovery from hyperpolarized voltages by post-inhibitory rebound and are provided by T-type calcium channels (Cav3) and hyperpolarization-activated cyclic nucleotide-gated (HCN) channels [36, 37]. Accordingly, neuromimetic models of the saccadic system based on clinical and experimental data proposed the expression of these channels in premotor burst neurons [22, 24, 38] and successfully explained the cause of irregular saccades in several disorders as dysregulation of bursts by the alterations in the expression of these channels [25, 39].

Here we review findings of neuron properties in the premotor saccadic circuit in monkey including transmitter inputs and those ion channels, which are particularly involved in fast firing and post-inhibitory rebound (PIR) [40], and whose malfunction may result in saccadic disorders in human [24, 41].

25.2 Premotor Saccadic Burst Neurons

25.2.1 Excitatory Burst Neurons (EBN)

EBNs for vertical saccades are located in the wing-shaped rostral interstitial nucleus of the medial longitudinal fascicle (RIMLF) in the rostral mesencephalon (Fig. 25.1A), whereas EBNs for horizontal saccades lie in the nucleus reticularis pontis caudalis (NRPC) within the paramedian pontine reticular formation (PPRF) (Fig. 25.2A) [42, 43].

Calcium-Binding Proteins and Perineuronal Nets All the EBNs in RIMLF and PPRF express strong immunoreactivity for the calcium-binding protein parvalbumin and nonphosphorylated neurofilaments labeled by SMI32-antibody, and their cell bodies are enwrapped by strong perineuronal nets (PNNs) (Figs. 25.1B, 25.2B) [34, 44, 45].

In the RIMLF, only the EBNs for upward saccades contain an additional calcium-binding protein—calretinin (CR) (Fig. 25.1B, arrows) [34, 46–48]. This implies differences between upgaze and downgaze burst neurons, despite their similar firing patterns [49, 50]. More recent studies of the synapses in the cochlear nucleus suggest the distinct Ca-buffering capacity of parvalbumin and calretinin under different activity levels. With its higher affinity to Ca, parvalbumin helps to maintain low Ca levels at rest or irregular firing, whereas CR takes over at higher firing rates due to cooperative Ca ion binding [51, 52]. Although the exact function of the variable expression of parvalbumin and calretinin in saccadic burst neurons is unknown, it suggests the necessity of kinetically different Ca binding proteins in up-burst and down-burst neurons, which may cause differential vulnerability resulting in distinct forms of vertical gaze palsy in neurodegenerative diseases [2, 53, 54].

Ion Channels The voltage-gated potassium channels Kv1.1, Kv3.1b and KCC2, which enable the fast firing of neurons, are present in the EBNs for vertical saccades in the RIMLF. The expression pattern of the potassium channels is similar for EBNs for upward or downward saccades, which is consistent with similar burst behavior and firing rates in upgaze and downgaze EBNs [48–50]. An identical pattern for Kv1.1 and Kv3.1b channels is present for the EBNs in the PPRF (Horn, unpublished observations), which can be identified as medium-sized parvalbumin- or SMI32-

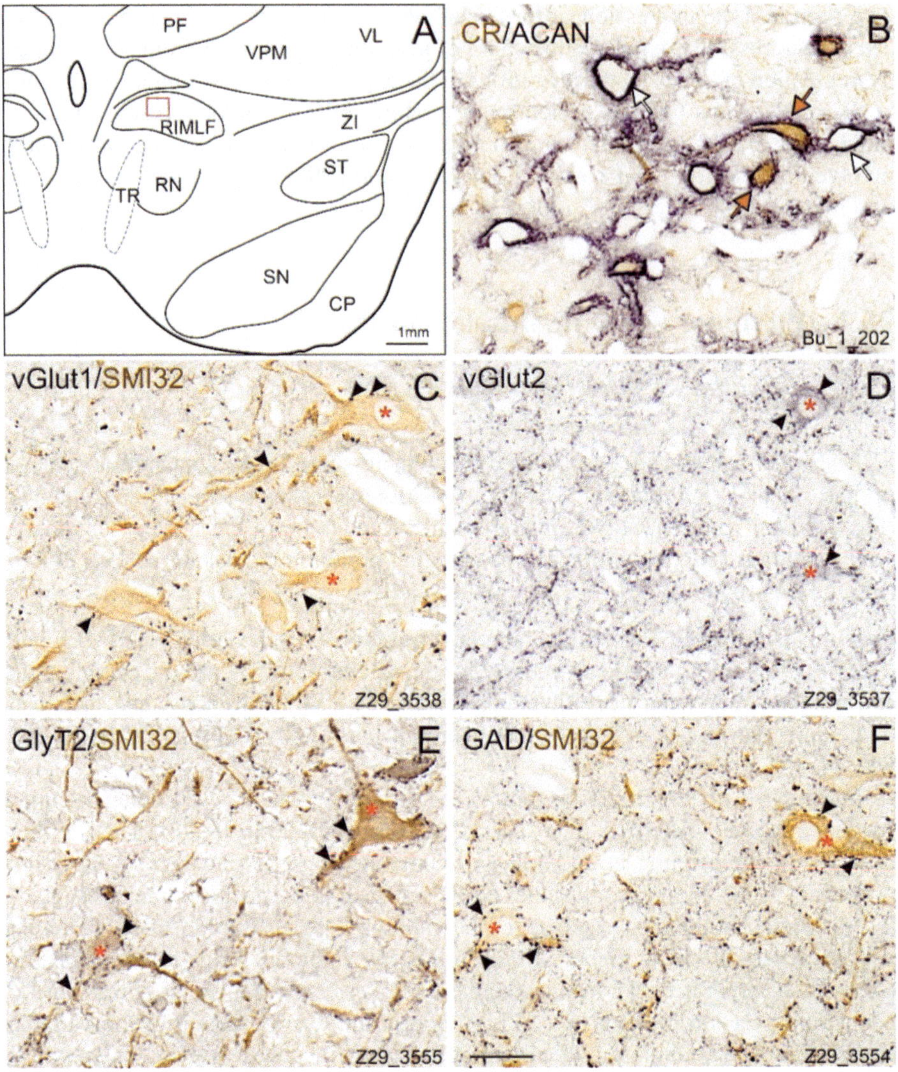

Fig. 25.1 Vertical excitatory burst neurons (EBN) in transverse sections of the rostral interstitial nucleus of the medial longitudinal fascicle (RIMLF) in monkeys. (**A**) Drawing with the area (red rectangle) for detailed views shown in (**B-F**); Detail of calretinin (CR)-positive upward EBNs (red arrows) and CR-negative downward EBNs (white arrows), both ensheathed by strong perineuronal nets (black); (**C**, **D**) Neighboring 5 μm sections stained with SMI32 to identify EBNs (red asterisk) combined with vesicular glutamate transporter 1 (vGlut1) or only vGlut2 (arrowheads); (**E**, **F**) Neighboring 5 μm sections stained with neurofilament antibody SMI32 (brown) and either glycine transporter 2 (GlyT2, black) (**E**) or glutamate decarboxylase (GAD) (black) (**F**), indicating glycinergic and GABAergic inputs. Red asterisks point to identical neurons in neighboring sections. Scale bar: 50 μm, applies to **B**–**F**

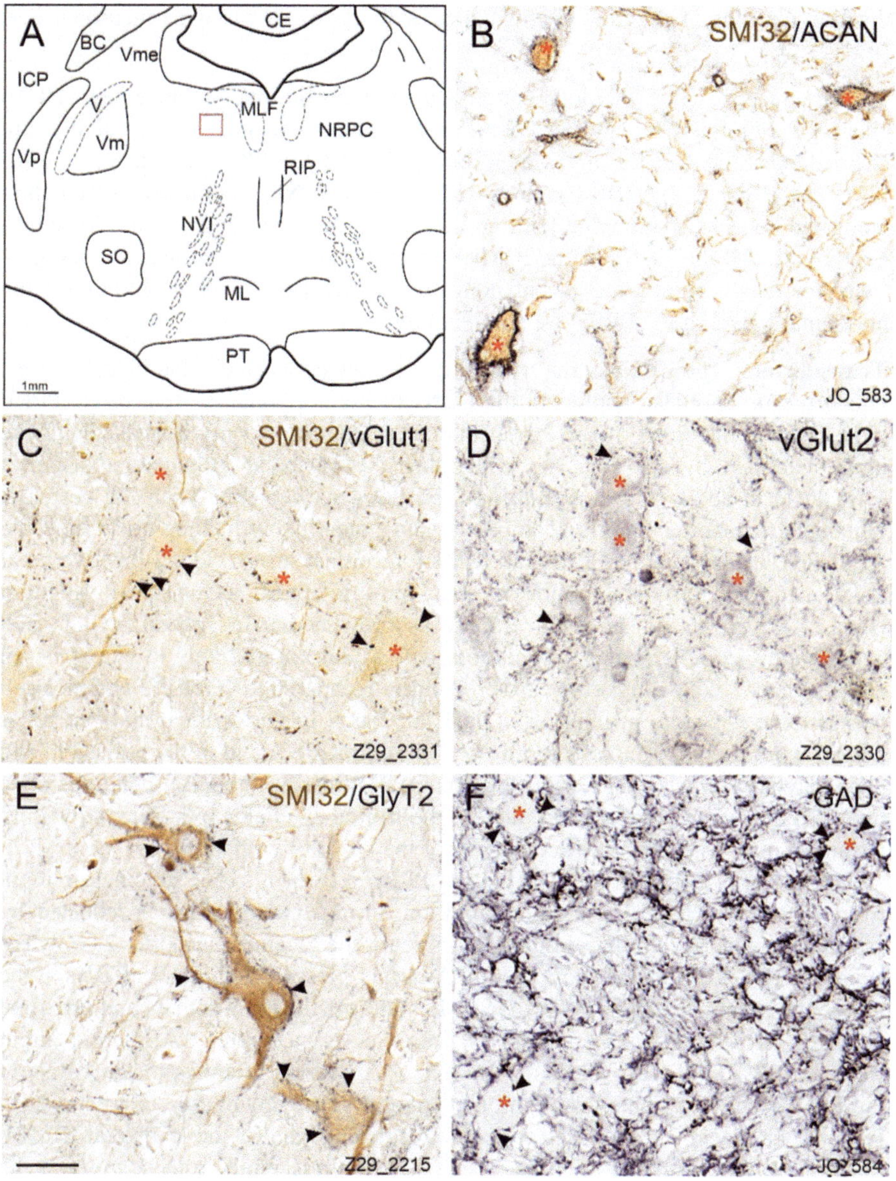

Fig. 25.2 Horizontal excitatory burst neurons (EBN) in transverse sections of the caudal pontine reticular nucleus (NRPC) in monkeys. (**A**) Location of NRPC with the area (red rectangle) for details shown in **B**–**F**. (**B**) SMI32-positive EBNs are enwrapped by perineuronal nets (black). (**C**, **D**) SMI32-positive EBNs (red asterisk) with glutamate inputs by vesicular glutamate transporter 1 (vGlut1) or vGlut2 (arrowheads) in neighboring 5 μm sections; (**E**, **F**) SMI32-positive EBNs with strong glycine transporter 2 (GlyT2) and glutamate decarboxylase (GAD) input (black, arrows). Red asterisks in F indicate the same EBNs of adjacent section in **B**. Scale bar: 50 μm, applies to **B**–**F**

positive neurons in the NRPC enwrapped with aggrecan-based perineuronal nets (Fig. 25.2B).

The presence of hyperpolarization-activated cyclic nucleotide–gated (HCN) channel subunits HCN1 and HCN2 and low-voltage-activated T-type calcium ion channels Cav3.1, Cav3.2, and Cav3.3 are proposed for PIR behavior in neurons [23]. The EBNs in the RIMLF show HCN1- and HCN2-subunit immunolabeling along the neuronal membrane with no differences between upgaze and downgaze EBNs [48]. All the EBNs of the RIMLF express moderate to strong immunoreactivity for Cav3.2 and Cav3.3, but none for Cav3.1 [48]. Similar studies for the horizontal EBNs in the PPRF are still pending.

Transmitters The EBNs in the RIMLF use glutamate as an excitatory transmitter, which presumably also applies to the EBNs of the horizontal system in the PPRF [55]. The motoneurons of the extraocular muscles receive only a weak glutamate input mediated by the synapses containing the vesicular glutamate transporter 1 (vGlut1) in addition to a strong input by the synapses containing the vesicular glutamate transporter 2 (vGlut2) [56], the latter most likely include those from the RIMLF. The EBNs in the PPRF and RIMLF receive a strong vGlut2-positive input to somata and proximal dendrites, but only a sparse vGlut1-positive input (Fig. 25.1C, D arrowheads; Fig. 25.2C, D arrowheads). In general, vGlut1 and vGlut2 show complementary expression, with vGlut2 more likely to be expressed at the synapses with a higher release probability compared to vGlut1 [57]. Since cell bodies expressing the mRNA for vGlut1 are found in the cerebral and cerebellar cortex, whereas those expressing mRNA for vGlut2 are found in the brainstem and deep cerebellar nuclei [57], vGlut1 terminals confined to the EBN dendrites arise most likely from cortical sources such as the frontal eye fields [58, 59], and vGlut2 terminals may arise from the deep layers of the superior colliculus [60].

Furthermore, EBNs in the RIMLF and PPRF receive a strong glycinergic input from OPNs, which provides their inhibition during fixation. This is reflected by associated synaptic terminals immunoreactive for the glycine transporter 2 (vGlyT2) (Figs. 25.1E, 25.2E arrowheads) [8, 60]. A further glycinergic input for EBNs of the horizontal system arises from IBNs [61]. In addition, the EBN somata and proximal dendrites in the RIMLF and PPRF receive a moderate GABAergic input (Figs. 25.1D; 25.2D arrowheads) [60]. The GABAergic afferents to the EBNs in the RIMLF may originate from inhibitory nonpremotor downgaze burst neurons in INC, which have been proposed to send a feedback signal to the ipsilateral RIMLF [50, 62]. EBNs in the PPRF may receive GABAergic input from neurons in the central mesencephalic reticular formation [63].

25.2.2 Inhibitory Burst Neurons (EBN)

In primates, the premotor IBNs of the vertical saccadic system are located within the interstitial nucleus of Cajal (INC) [64], whereas those for horizontal saccades lie in the dorsal paragigantocellular nucleus (PGD) within the medullary reticular formation (Fig. 25.3A) [65, 66]. Recent studies suggest that the IBNs not only inhibit the motoneurons of antagonistic eye muscles and their contralateral counterparts but may also provide the inhibition of omnipause neurons enabling saccade initiation [61, 67]. The reciprocal projections between the IBNs of both sides could make the circuit prone to oscillations as found in ocular flutter or opsoclonus [22].

Calcium-Binding Proteins and Perineuronal Nets As saccadic EBNs, the IBNs contain parvalbumin and they are ensheathed by well-developed PNNs (Fig. 25.3B) [44, 48]. In INC and PGD, additional populations of CR-positive neurons ensheathed by PNNs are present, indicating highly active neurons. In INC, premotor CR-positive neurons were shown to be excitatory, and they may represent burst and burst-tonic neurons involved in upgaze [47, 62]. The function of a similar CR-positive neuron population with PNNs and parvalbumin intermingled with IBNs in the PGD is not clear yet. They may represent other bursting neurons, e.g., long-lead burst neurons or gaze-related neurons [48, 66, 68].

Ion Channels The glycinergic IBNs for horizontal saccades in the PGD show the same profile of voltage-gated potassium channels Kv1.1, Kv3.1b, and KCC2 as those observed for horizontal and vertical EBNs [48]. This includes qualitatively similar HCN1 and HCN2 labeling patterns as those for EBNs. Unlike EBNs in RIMLF, few IBNs in the PGD show weak somatic Cav3.1 labeling, whereas Cav3.2 and Cav3.3 channels are expressed in all putative IBNs [48]. The ion channel profile of presumed GABAergic IBNs of the vertical saccadic system in the INC remains to be studied.

Transmitters IBNs of the vertical saccadic system in INC use GABA as a transmitter, whereas IBNs for horizontal saccades in PGD are glycinergic [69, 70]. As with EBNs, the IBNs of the horizontal saccadic system receive a strong vGlyT2 input, which derives from glycinergic OPNs, but also from contralateral IBNs (Fig. 25.3C arrowheads) [8, 61, 71]. This reciprocal inhibitory connection between the IBNs of both sides is considered as the cause of why this circuit is prone to oscillations [72]. Whether similar reciprocal projections exist between the GABAergic IBNs of both INCs remains to be proved [67].

As with the EBNs, the IBNs receive a considerable glutamatergic input mainly via vGlut2-positive synapses located at somata and proximal dendrites, whereas only a few vGlut1-positive terminals are found (Fig. 25.3E, F arrowheads). One major source of glutamatergic afferent inputs to the IBNs of the vertical and horizontal saccadic system may arise from EBNs in the RIMLF and PPRF, respectively [42, 55]. These glutamatergic inputs are most likely mediated via vGlut2-positive terminals, which are located in high numbers on somata and dendrites of burst neurons—similar to the motoneurons of the extraocular muscles (Fig. 25.3F) [56]. Similarly, the excitatory projection from the contralateral superior colliculus may provide another source of vGlut2-positive terminals at IBNs [64]. As argued above, the vGlut1 input may represent cortical input.

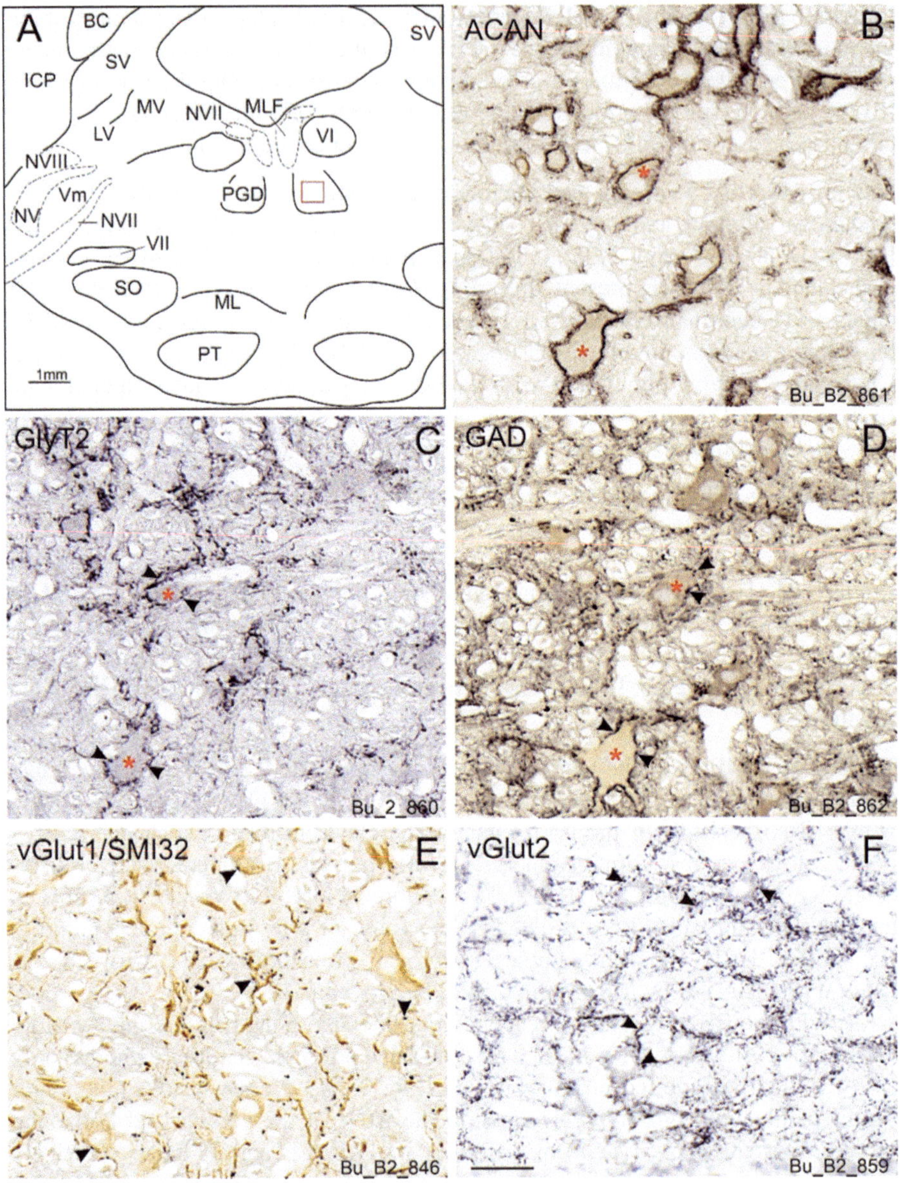

Fig. 25.3 Horizontal inhibitory burst neurons (IBN) in transverse brainstem sections in monkeys. (**A**) Location of the dorsal paragigantocellular nucleus (PGD) and area (red rectangle) with details shown in **B-F**. **B**, **C**, **D**: Neighboring 5 μm sections showing IBNs with perineuronal nets (**B**), which receive a strong glycine transporter 2 (vGlyT2) and strong glutamate decarboxylase (GAD) input (**C**, **D**, arrowheads). Red asterisks in **B-D** indicate the same IBNs in 5 μm neighboring sections. (**E**, **F**) SMI32-positive IBNs with weak input by vesicular glutamate transporter 1-positive terminals (vGlut1), but strong vGlut2-positive input (**F**, **G**, arrowheads). Scale bar: 50 μm, applies to **B**–**F**

25.3 Omnipause Neurons

Saccadic omnipause neurons (OPN) appear as two vertical columns of medium-sized neurons with their long axis oriented horizontally at the midline of the caudal pons. They form a cytoarchitectural entity located between the raphe nuclei, hence being termed the nucleus raphe interpositus (RIP) (Fig. 25.4A, B) [73].

Calcium-Binding Proteins and Perineuronal Nets OPNs contain high levels of parvalbumin and nonphosphorylated neurofilaments, and they are enwrapped by prominent PNNs (Fig. 25.4B). CR is not expressed in the OPN somata, but confined to numerous terminals associated with the OPNs' membrane, indicating a strong CR-positive input [34].

Ion Channels The voltage-gated potassium channels Kv1.1, Kv3.1b and KCC2 are strongly expressed in OPNs [74], and they contain the calcium channels Cav3.2 and Cav3.3 (Fig. 25.4C, D).

Transmitters Saccadic OPNs use glycine as a transmitter, by which the saccadic EBNs and IBNs are inhibited during fixation [8, 71]. Further, the OPN somata are contacted by numerous GlyT2-positive puncta, reflecting a strong glycinergic input most likely from IBNs in the PGD, which may provide the trigger signal for saccade initiation (Fig. 25.4E, arrowheads) [8, 64, 71]. Another strong inhibitory input is represented by numerous GAD-positive terminals outlining the somata and proximal dendrites of OPNs (Fig. 25.4F, arrowheads). This strong GABAergic input to OPNs arises at least in part not only from the neurons in the central mesencephalic reticular formation [63] but possibly also from vertical IBNs in INC as indicated by recent recording studies [67].

The OPNs receive a glutamatergic input mainly to the dendrites, as shown with glutamate antibodies [8]. As in IBNs and EBNs, glutamatergic afferents appear as distinct terminals containing either vGlut1 or vGlut2, with vGlut1 confined to dendrites and the numerically more frequent vGlut2 terminals located on somata and dendrites (Fig. 25.4G, H). As stated above for glutamatergic inputs to burst neurons [57], vGlut1 terminals arise most likely from neurons in the frontal and supplementary eye fields [58, 59], and vGlut2 terminals from neurons in the deep layers of the rostral superior colliculus, which excite OPNs during fixation [64, 75, 76], as well as from neurons in the fastigial nucleus [77].

25.4 Conclusion

The findings in nonhuman primate tissue presented here build a bridge between the histochemical and electrophysiological properties of distinct neuron types of the saccadic circuitry. The ion channel profile of neurons consisting of fast voltage-gated potassium channels, HCN, and Cav3 channels enables fast firing and its resumption after inhibition, known as post-inhibitory rebound. Their expression, along with the

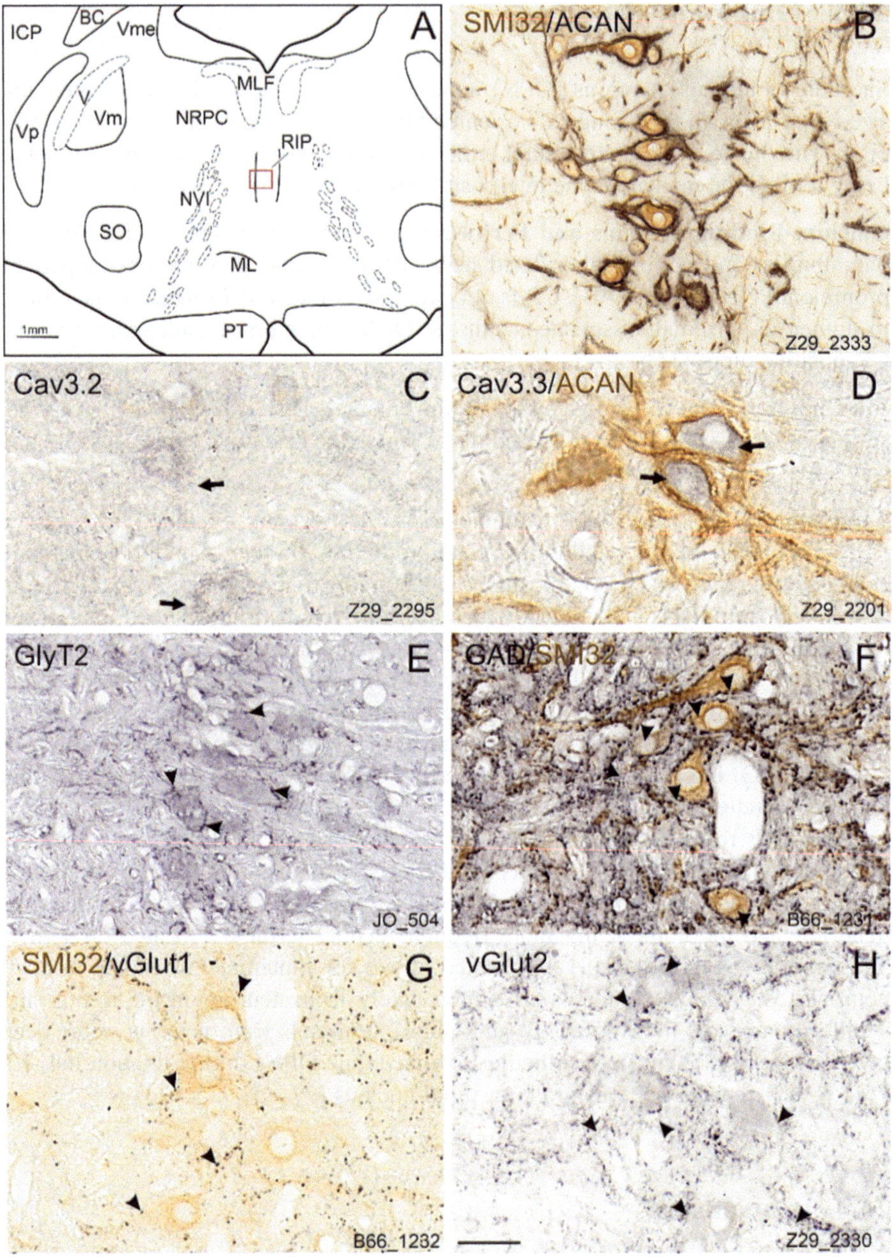

Fig. 25.4 Omnipause neurons (OPNs) in the nucleus raphe interpositus (RIP) in monkeys. (**A**) Area for details in **B–F** are indicated by red rectangle; (**B**) SMI32-positive (brown) OPNs ensheathed by perineuronal nets. (**C, D**) Granular Cav3.2- and more homogenous Cav3.3-immunolabeling (black) in OPNs (**D**). (**E, F**) OPNs with strong supply by glycine transporter 2 (GlyT2) and glutamate decarboxylase (GAD)-positive terminals (**E, F**, black, arrowheads). (**G, H**) SMI32-positive OPNs with vesicular glutamate transporter 1 (vGlut1)- and vGlut2-positive terminals (**G, H**, black arrowheads). Scale bar: 50 μm, applies to **B–F**

presence of the calcium-binding protein parvalbumin and strong perineuronal nets in saccadic IBNs, EBNs, and OPNs, demonstrates in monkeys that the neurons of the saccadic burst generator contain the molecular machinery required for fast firing and post-inhibitory rebound as requested by proposed models [40]. The identification of the molecular signature of neurons involved in saccade generation will allow the refinement of existing neuromimetic models of saccadic disorders with realistic parameters to develop therapeutic strategies [25]. Furthermore, the data will serve as a basis for studies with corresponding investigations in human control cases and those with saccade disorders, such as opsoclonus.

References

1. Horn A, Straka H. Functional organization of extraocular motoneurons and eye muscles. Ann Rev Vis Sci. 2021;7:793–825.
2. Leigh J, Zee D. The neurology of eye movements. 5th ed. Oxford/New York: Oxford University Press; 2015.
3. Moschovakis A, Scudder C, Highstein S. The microscopic anatomy and physiology of the mammalian saccadic system. Prog Neurobiol. 1996;50(2–3):133–254.
4. Sparks D. The brainstem control of saccadic eye movements. Nat Rev Neurosci. 2002;3(12): 952–64.
5. Takahashi M, Shinoda Y. Brain stem neural circuits of horizontal and vertical saccade systems and their frame of reference. Neurosci. 2018;392:281–328.
6. Moschovakis AK. The neural integrators of the mammalian saccadic system. Front Biosci. 1997;2:552–77.
7. Keller E. Participation of medial pontine reticular formation in eye movement generation in monkey. J Neurophysiol. 1974;37:316–32.
8. Horn A, Büttner-Ennever J, Wahle P, Reichenberger I. Neurotransmitter profile of saccadic omnipause neurons in nucleus raphe interpositus. J Neurosci. 1994;14(4):2032–46.
9. Henn V, Lang W, Hepp K, Reisine H. Experimental gaze palsies in monkeys and their relation to human pathology. Brain. 1984;107(2):619–36.
10. Barton E, Nelson J, Gandhi N, Sparks D. Effects of partial lidocaine inactivation of the paramedian pontine reticular formation on saccades of macaques. J Neurophysiol. 2003;90 (1):372–86.
11. Suzuki Y, Büttner-Ennever J, Straumann D, Hepp K, Hess M, Henn V. Deficits in torsional and vertical rapid eye movements and shift of listing's plane after uni- and bilateral lesions of the rostral interstitial nucleus of the medial longitudinal fasciculus. Exp Brain Res. 1995;106:215–32.
12. Eghbali A, Hassan S, Seehra G, FitzGibbon E, Sidransky E. Ophthalmological findings in Gaucher disease. Mol Genet Metab. 2019;127(1):23–7.
13. Shaikh A, Factor S, Juncos J. Saccades in progressive supranuclear palsy – maladapted, irregular, curved, and slow. Mov Disord Clin Pract. 2017;4(5):671–81.
14. Sekar A, Panouillères M, Kaski D. Detecting abnormal eye movements in patients with neurodegenerative diseases - current insights. Eye Brain. 2024;16:3–16.
15. Ridley A, Kennard C, Scholtz C, Büttner-Ennever J, Summers B, Turnbull A. Omnipause neurons in two cases of opsoclonus associated with oat cell carcinoma of the lung. Brain. 1987;110(6):1699–709.
16. Eggers S, Horn A, Roeber S, Härtig W, Nair G, Reich D, Leigh R. Saccadic palsy following cardiac surgery: possible role of perineuronal nets. PLoS One. 2015;10(7):e0132075.

17. Roselli F, Livrea P, E J. Voltage-gated sodium channel blockers as immunomodulators. Recent Pat CNS Drug Discov. 2006;1(1):83–91.
18. Kumar P, Kumar D, Jha S, Jha N, Ambasta R. Ion channels in neurological disorders. Adv Protein Chem Struct Biol. 2016;103:97–136.
19. Anderson T, MacAskill M. Eye movements in patients with neurodegenerative disorders. Nat Rev Neurol. 2013;9(2):74–85.
20. Soto E, Vega R. Neuropharmacology of vestibular system disorders. Curr Neuropharmacol. 2010;8(1):26–40.
21. Zwergal A, Strupp M, Brandt T. Advances in pharmacotherapy of vestibular and ocular motor disorders. Exp Opin on Pharmacother. 2019;20(10):1267–76.
22. Shaikh A, Ramat S, Optican L, Miura K, Leigh R, Zee D. Saccadic burst cell membrane dysfunction is responsible for saccadic oscillations. J Neuroophthalmol. 2008;28(4):329–36.
23. Miura K, Optican L. Membrane channel properties of premotor excitatory burst neurons may underlie saccade slowing after lesions of omnipause neurons. J Comput Neurosci. 2006;20:25–41.
24. Ramat S, Leigh R, Zee D, Shaikh A, Optican L. Applying saccade models to account for oscillations. Prog Brain Res. 2008;171:123–30.
25. Shaikh A. Saccadic oscillations – membrane, model, and medicine. Expert Rev Ophthalmol. 2012;7(5):481–6.
26. Shaikh A, Wilmot G. Opsoclonus in a patient with increased titers of anti-GAD antibody provides proof for the conductance-based model of saccadic oscillations. J Neurol Sci. 2016;362:169–73.
27. Optican L, Pretegiani E. What stops a saccade? Philos Trans R Soc Lond Ser B Biol Sci. 2017;372:1718.
28. Catterall W. The molecular basis of neuronal excitability. Science. 1984;223:653–61.
29. Catterall WA. Voltage-gated calcium channels. Cold Spring Harb Perspect Biol. 2011;3(8): a003947.
30. Albéri L, Lintas A, Kretz R, Schwaller B, Villa AE. The calcium-binding protein parvalbumin modulates the firing 1 properties of the reticular thalamic nucleus bursting neurons. J Neurophysiol. 2013;109(11):2827–41.
31. Wingert J, Sorg B. Impact of perineuronal nets on electrophysiology of parvalbumin interneurons, principal neurons, and brain oscillations: a review. Front Synaptic Neurosci. 2021;13: 673210.
32. Brückner G, Brauer K, Härtig W, Wolff J, Rickmann M, Derouiche A, Delpech B, Girard N, Oertel W, Reichenbach A. Perineuronal nets provide a polyanionic, glia-associated form of microenvironment around certain neurons in many parts of the rat brain. Glia. 1993;8(3): 183–200.
33. Sonntag M, Blosa M, Schmidt S, Rübsamen R, Morawski M. Perineuronal nets in the auditory system. Hear Res. 2015;329:21–32.
34. Horn A, Brückner G, Härtig W, Messoudi A. Saccadic omnipause and burst neurons in monkey and human are ensheathed by perineuronal nets but differ in their expression of calcium-binding proteins. J Comp Neurol. 2003;455(3):341–52.
35. Kaczmarek L, Zhang Y. Kv3 channels: enablers of rapid firing, neurotransmitter release, and neuronal endurance. Physiol Rev. 2017;97(4):1431–68.
36. Kodama T, Gittis A, Shin M, Kelleher K, Kolkman K, McElvain L, Lam M, du Lac S. Graded coexpression of ion channel, neurofilament, and synaptic genes in fast-spiking vestibular nucleus neurons. J Neurosci. 2020;40(3):496–508.
37. Mishra P, Narayanan R. The enigmatic HCN channels: a cellular neurophysiology perspective. Proteins. 2025;93(1):72–92.
38. Daye P, Optican L, Roze E, Gaymard B, Pouget P. Neuromimetic model of saccades for localizing deficits in an atypical eye-movement pathology. J Transl Med. 2013;11:125.

39. Shaikh A, Miura K, Optican L, Ramat S, Leigh R, Zee D. A new familial disease of saccadic oscillations and limb tremor provides clues to mechanisms of common tremor disorders. Brain. 2007;130(Pt 11):3020–31.
40. Enderle J, Engelken E. Simulation of oculomotor post-inhibitory rebound burst firing using a Hodgkin-Huxley model of a neuron. Biomed Sci Instrum. 1995;31:53–8.
41. Shaikh A, Zee D, Optican L, Miura K, Ramat S, Leigh R. The effects of ion channel blockers validate the conductance-based model of saccadic oscillations. Ann N Y Acad Sci. 2011;1233:58–63.
42. Strassman A, Highstein S, McCrea R. Anatomy and physiology of saccadic burst neurons in the alert squirrel monkey. I. Excitatory burst neurons. J Comp Neurol. 1986;249:337–57.
43. Horn A. The reticular formation. Prog Brain Res. 2006;151:127–55.
44. Horn A, Büttner-Ennever J, Suzuki Y, Henn V. Histological identification of premotor neurons for horizontal saccades in monkey and man by parvalbumin immunostaining. J Comp Neurol. 1995;359(2):350–63.
45. Horn A, Büttner-Ennever J. Premotor neurons for vertical eye movements in the rostral mesencephalon of monkey and human: histologic identification by parvalbumin immunostaining. J Comp Neurol. 1998;392(4):413–27.
46. Adamczyk C, Strupp M, Jahn K, Horn A. Calretinin as a marker for premotor neurons involved in upgaze in human brainstem. Front Neuroanat. 2015;9:153.
47. Ahlfeld J, Mustari M, Horn A. Sources of calretinin inputs to motoneurons of extraocular muscles involved in upgaze. Ann N Y Acad Sci. 2011;1233:91–9.
48. Mayadali Ü, Lienbacher K, Shaikh A, Horn A. Saccadic premotor burst neurons and histochemical correlates of their firing patterns in rhesus monkey. J Neurol Sci. 2022;439:120328.
49. Moschovakis A, Scudder C, Highstein S. Structure of the primate oculomotor burst generator. I. Medium-lead burst neurons with upward on-directions. J Neurophysiol. 1991;65(2):203–17.
50. Moschovakis A, Scudder C, Highstein S, Warren J. Structure of the primate oculomotor burst generator. II. Medium-lead burst neurons with downward on-directions. J Neurophysiol. 1991;65(2):218–29.
51. Schwaller B. Cytosolic Ca^{2+} buffers are inherently Ca^{2+} signal modulators. Cold Spring Harb Perspect Biol. 2020;12(1):a035543.
52. Zhang C, Wang M, Lin S, Xie R. Calretinin-expressing synapses show improved synaptic efficacy with reduced asynchronous release during high-rate activity. J Neurosci. 2022;42(13):2729–42.
53. Salari M, Alikhani A, Hojjati Pour F, Etemadifar M. Supranuclear vertical gaze palsy in movement disorders. Neuro-Ophthalmol. 2024:1–18.
54. Salsano E, Umeh C, Rufa A, Pareyson D, Zee D. Vertical supranuclear gaze palsy in Niemann-pick type C disease. Neurol Sci. 2012;33(6):1225–32.
55. Spencer R, Wang S. Immunohistochemical localization of neurotransmitters utilized by neurons in the rostral interstitial nucleus of the medial longitudinal fasciculus (riMLF) that project to the oculomotor and trochlear nuclei in the cat. J Comp Neurol. 1996;366(1):134–48.
56. Zeeh C, Mustari M, Hess B, Horn A. Transmitter inputs to different motoneuron subgroups in the oculomotor and trochlear nucleus in monkey. Front Neuroanat. 2015;9:95.
57. Fremeau R, Troyer M, Pahner I, Nygaard G, Tran C, Reimer R, Bellocchio E, Fortin D, Storm-Mathisen J, Edwards R. The expression of vesicular glutamate transporters defines two classes of excitatory synapse. Neuron. 2001;31:247–60.
58. Schnyder H, Reisine H, Hepp K, Henn V. Frontal eye field projection to the paramedian pontine reticular formation traced with wheat germ agglutinin in the monkey. Brain Res. 1985;329(1–2):151–60.
59. Lynch J, Tian J. Cortico-cortical networks and cortico-subcortical loops for the higher control of eye movements. Prog Brain Res. 2006;151:461–501.
60. Horn A, Büttner-Ennever J. Brainstem circuits controlling lid-eye coordination in monkey. Prog Brain Res. 2008;171:87–95.

61. Strassman A, Highstein S, McCrea R. Anatomy and physiology of saccadic burst neurons in the alert squirrel monkey. II. Inhibitory burst neurons. J Comp Neurol. 1986;249:358–80.
62. Helmchen C, Rambold H, Büttner U. Saccade-related burst neurons with torsional and vertical on-directions in the interstitial nucleus of Cajal of the alert monkey. Exp Brain Res. 1996;112: 63–78.
63. Wang N, Perkins E, Zhou L, Warren S, May P. Anatomical evidence that the superior colliculus controls saccades through central mesencephalic reticular fomation gating of omnipause neuron activity. J Neurosci. 2013;33:16285–96.
64. Takahashi M, Sugiuchi Y, Na J, Shinoda Y. Brainstem circuits triggering saccades and fixation. J Neurosci. 2022;42:789–803.
65. Yoshida K, McCrea R, Berthoz A, Vidal P. Morphological and physiological characteristics of inhibitory burst neurons controlling horizontal rapid eye movements in the alert cat. J Neurophysiol. 1982;48(3):761–84.
66. Scudder C, Fuchs A, Lanager T. Characteristics and functional identification of saccadic inhibitory burst neurons in the alert monkey. J Neurophysiol. 1988;59(5):1430–54.
67. Takahashi M, Sugiuchi Y, Shinoda Y. Brainstem neural circuits triggering vertical saccades and fixation. J Neurosci. 2024;44(1):e1657232023.
68. Whittington D, Lestienne F, Bizzi E. Behavior of preoculomotor burst neurons during eye-head coordination. Exp Brain Res. 1984;55(2):215–22.
69. Spencer R, Wenthold R, Baker R. Evidence for glycine as an inhibitory neurotransmitter of vestibular, reticular, and prepositus hypoglossi neurons that project to the cat abducens nucleus. J Neurosci. 1989;9(8):2718–36.
70. Horn A, Helmchen C, Wahle P. GABAergic neurons in the rostral mesencephalon of the macaque monkey that control vertical eye movements. Ann N Y Acad Sci. 2003;1004:19–28.
71. Kanda T, Iwamoto Y, Yoshida K, Shimazu H. Glycinergic inputs cause the pause of pontine omnipause neurons during saccades. Neurosci Lett. 2007;413:16–20.
72. Ramat S, Leigh R, Zee D, Optican L. Ocular oscillations generated by coupling of brainstem excitatory and inhibitory saccadic burst neurons. Exp Brain Res. 2005;160(1):89–106.
73. Büttner-Ennever J, Cohen B, Pause M, Fries W. Raphe nucleus of the pons containing omnipause neurons of the oculomotor system in the monkey, and its homologue in man. J Comp Neurol. 1988;267(3):307–21.
74. Mayadali Ü, Lienbacher K, Mustari M, Strupp M, Horn A. Potassium channels in omnipause neurons. Prog Brain Res. 2019;249:117–23.
75. Munoz D, Wurtz R. Fixation cells in monkey superior colliculus. II. Reversible activation and deactivation. J Neurophysiol. 1993;70(2):576–89.
76. Büttner-Ennever J, Horn A, Henn V, Cohen B. Projections from the superior colliculus motor map to omnipause neurons in monkey. J Comp Neurol. 1999;413(1):55–67.
77. Sugita S, Noda H. Pathways and terminations of axons arising in the fastigial oculomotor region of macaque monkeys. Neurosci Res. 1991;10:118.

Chapter 26
Strabismus in Parkinson's Disease: Assessing Binocular Misalignment and the Role of Subthalamic Nucleus Stimulation During Near Fixation

Palak Gupta, Camilla Kilbane, Aasef G. Shaikh, and Fatema F. Ghasia

Keywords Role of STN in visual alignment · Eye alignment · Binocular viewing · Visual deficits in PD

26.1 Introduction

Strabismus, the misalignment of the eyes, is a common ocular motor presentation in PD. Strabismus can be attributed to both convergence insufficiency and a more intricate subcortical deficit specific to PD and can cause double vision in about one-third of PD patients [1, 4, 14, 22, 25, 29, 32, 34, 41]. Vergence insufficiency, the inability of the two eyes to coordinate in response to shifting the gaze from far to near or vice versa, is also common in PD [2, 3, 25, 31, 41]. Such inability can lead to a number of quality-of-life issues, such as difficulty reading, reaching near-located

P. Gupta
Department of Biomedical Engineering, Case Western Reserve University, Cleveland, OH, USA

C. Kilbane
Department of Neurology, University Hospitals Cleveland Medical Center, Cleveland, OH, USA

Department of Neurology, Case Western Reserve University, Cleveland, OH, USA

A. G. Shaikh (✉)
Neurological Institute, University Hospitals Neurology Service, Louis Stokes Cleveland, VA Medical Center, Department of Neurology, Case Western Reserve University, Cleveland, OH, USA
e-mail: Axs848@case.edu

F. F. Ghasia
Cole Eye Institute, Cleveland Clinic, Cleveland, OH, USA

P. Ravat et al. (eds.), *Advances in Translational Neuroscience of Vestibular and Eye Movement Disorders*, Contemporary Clinical Neuroscience,
https://doi.org/10.1007/978-3-032-07539-0_26

objects, or viewing closely located objects. Strabismus and vergence insufficiency are even more evident on instrumented measures, as noted in our previous study where 55% of patients had increased eye misalignment under binocular viewing conditions (i.e., tropia). Nearly all participants had impaired eye alignment under monocular viewing (i.e., phoria) and had vergence insufficiency [14]. The abnormal binocular control translates to lower visual health in PD with significant reductions in visual quality-of-life survey scores, particularly for near vision [1, 4, 15, 29, 33, 37, 43]. The impact of these visual deficits extends beyond mere discomfort, significantly contributing to the higher prevalence of depression and anxiety commonly observed in PD [16, 45]. Considered collectively, these reports highlight the imperative need for more detailed mechanistic investigations, directed toward effective interventions.

This study provides an objective demonstration of the strabismus phenomenology during convergence in PD and provide insights into its potential treatment with STN-DBS. We address two unanswered questions:

(a) Which aspects of fixational stability and related strabismus is affected in PD?
(b) Can STN-DBS alleviate binocular misalignment that contributes to strabismus in PD?

These questions provide an approach examining the effects of PD on strabismus, whether STN is involved in such pathophysiology, and how STN-DBS modulates the pathological phenomenon. The concept is motivated by the premise that the dorsal STN and the white-matter pathways, such as zona incerta that resides at the dorsal margin of the STN, communicate with the precerebellar pontine neurons influencing the cerebellar outflow. The cerebellar outflow modulates the supra-oculomotor area (SOA) within the brainstem, the region where the strabismus-angle-sensitive neurons are located [8, 9].

26.2 Methods

Fixation stability and eye misalignment were examined for targets at distances of 30 cm and 244 cm in 33 PD participants and 10 healthy controls. Fixation stability was measured using the bivariate contour ellipse (BCEA), which includes 95% of fixation points. Eye position data were filtered with a moving-average filter to remove fast, low-amplitude eye movements. The angles of strabismus or eye alignment deviations were then calculated from the filtered eye position data after eliminating rapid movements in each plane.

Since monocular calibration was performed at 55 cm while eye alignment data were assessed at 30 cm and 244 cm, we used an approach to determine actual and expected eye positions as follows. Under monocular viewing conditions, we assessed the viewing eye's position to establish empirical or baseline positions as reference points. Figure 26.1a, b show eye positions for monocular viewing with the

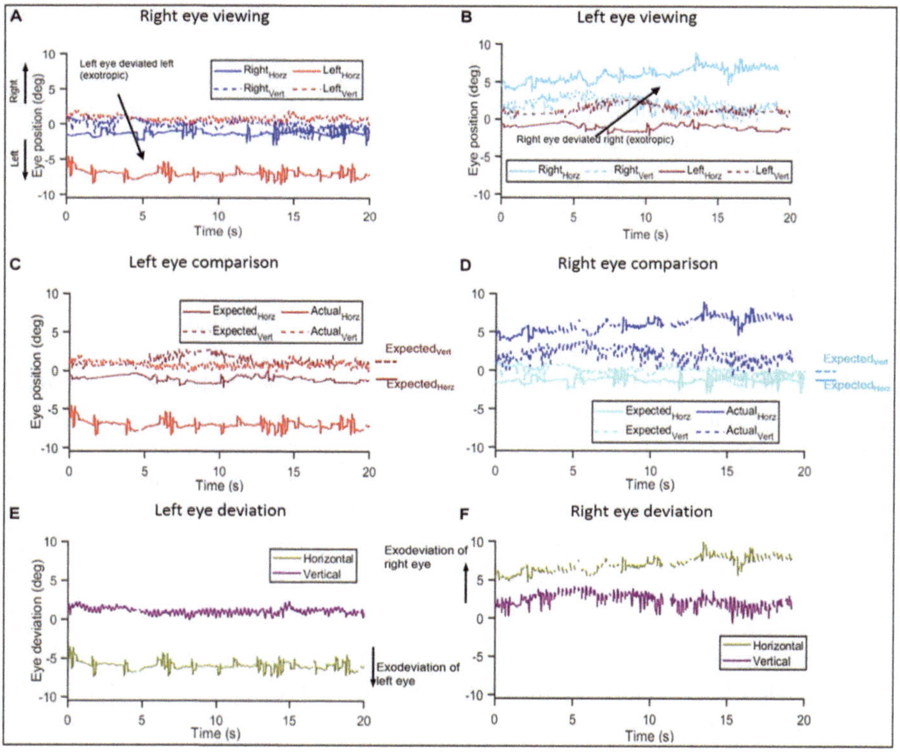

Fig. 26.1 Methodology of eye deviation analysis: (**a**, **b**) show the time series of right and left eye horizontal and vertical eye position data from a Parkinson's disease subject collected under right (**a**) and left (**b**) eye viewing conditions while viewing a target. Methodology of eye deviation analysis: (**a**, **b**) show the time series of right and left eye horizontal and vertical eye position data from a Parkinson's disease subject collected under right (**a**) and left (**b**) eye viewing conditions while viewing a target in primary position for 20 s. Notice that in the right eye viewing condition (**a**), the left eye is deviated to the left and during the left eye viewing (**b**), the right eye is deviated to the right indicative of an exodeviation. The viewing eye data obtained under monocular viewing is designated as the expected eye position, i.e., the right eye position data [blue trace in (**a**)] under right eye viewing is designated as right eye expected eye position data and vice versa, whereas the nonviewing eye data obtained under monocular viewing is the actual eye position data, i.e., nonviewing right eye data [cyan trace in (**b**)] obtained under left eye viewing is designated as actual eye position data of the right eye and vice versa. (**c, d**) show the respective "expected" (dark blue and dark red traces) and "actual" eye position traces (light blue and light red) for the right (**c**) and left eye (**d**). Annotations on the side show median values for expected horizontal and expected vertical eye positions to use in the calculation of the angle of deviation. The eye deviation was measured by comparing the position of each eye with itself when it was viewing vs. nonviewing (i.e., expected and actual eye positions, respectively). (**e, f**) show the computed difference in horizontal and vertical planes between the "expected" and "actual" eye positions for the right (**e**) and left eyes (**f**)

right eye (Fig. 26.1a—expected eye position data for the right eye) and left eye (Fig. 26.1b—expected eye position data for the left eye). The median horizontal and vertical eye positions of the viewing eye were calculated and designated as the expected eye position data (Figs. 26.1c, d). Thus, the expected position of the right eye was derived from the right eye trace during right eye viewing, and the expected position of the left eye was obtained from the left eye trace during left eye viewing.

To compute eye deviation, we defined the actual eye positions for both monocular and binocular viewing as follows:

1. Monocular viewing: Eye deviation was calculated by comparing the eye position data of the nonviewing eye (actual eye position) to the median expected eye position data obtained under monocular viewing.
2. Binocular viewing: For each trial, we determined whether each eye fixated on the target by evaluating the differences between the expected and actual eye position data for each eye. The eye that deviated for a longer duration during the trial was designated as the non-fixating eye. Eye deviation under binocular viewing was then obtained by calculating the difference between the expected median eye position data (from monocular viewing) and the actual eye position data of the non-fixating eye (from binocular viewing) for each participant.

The 95% lower and upper bounds and the span of deviation were computed under both binocular and monocular viewing conditions for each subject. To establish a normal range benchmark, we measured the 95th percentile of eye deviation in healthy participants, which was 3.5° horizontally and 2° vertically (referred to as the threshold window).

Eye deviation (the difference calculated between the expected and actual eye positions) underwent further filtering to eliminate rapid movements such as saccadic intrusions. The cleaned eye deviation data, devoid of rapid-movement artifacts, were clustered using Density-Based Spatial Clustering and Application with Noise (DBSCAN). This method enables an objective assessment of the control of eye deviation in intermittent strabismus observed in PD subjects. DBSCAN employs minimum density level estimation with a radius determined interactively to be 0.05° (epsilon) and a core-point threshold potential determined adaptively to be proportional to twice the expected number of points given an even distribution of points within the 95% BCEA approximation or a minimum number of points present in a verified fixational eye movement segment, whichever is greater. The clustering algorithm parameters (epsilon and minimum number of points) were chosen to ensure that eye trajectories arising from rapid movements with quick changes in eye alignment were not sufficiently dense to join all clusters together, while also avoiding a large number of small clusters in the neighborhood. Clusters with horizontal and vertical means within the normative ranges (threshold window) were labeled as "good" or well-aligned, while all others were labeled as "bad" clusters, henceforth referred to as "misaligned clusters."

Figure 26.2 illustrates the application of DBSCAN and the clustering approach in analyzing eye deviation under binocular viewing in a PD subject. Figures 26.2a–c

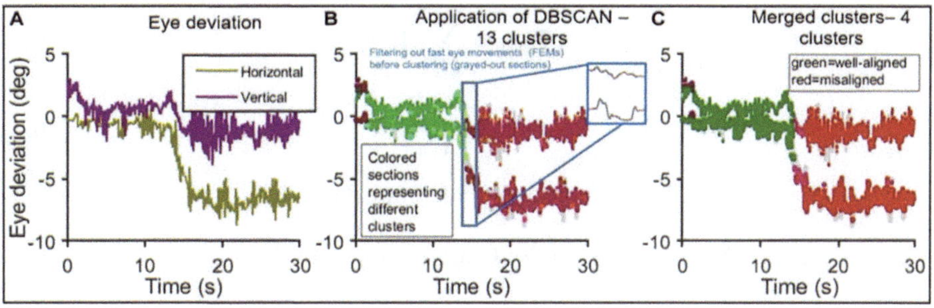

Fig. 26.2 Application of DBSCAN: (**a**–**c**) shows the time series [x-axis: time(s); y-axis: eye deviation (deg)]; (**a**) shows the computed difference between expected and actual eye positions in horizontal and vertical planes; (**b**) shows the initial application of DBSCA N resulting in five well-aligned and eight misaligned clusters. Inset shows the instance of filtering out rapid eye movements, in this case, an SWJ; (**c**) shows a single well-aligned super-cluster. Misaligned clusters with horizontal means <2° and vertical means <1° apart are combined. End result: 1 super-cluster of good binocular alignment and three misaligned super-clusters

outline the steps involved in clustering time series data. Figure 26.2a displays horizontal and vertical eye deviation obtained after calculating the difference between expected and actual eye positions in horizontal and vertical planes, respectively, as outlined in Fig. 26.1. On the initial application of DBSCAN, 13 clusters were identified (Fig. 26.2b). Of these, five clusters are depicted in varying shades of green, indicating well-aligned eyes, while eight clusters are shown in varying shades of red, indicating misaligned eyes (outside the determined threshold window—3.5° horizontally and 2° vertically). Data points not part of any cluster, i.e., not density-reachable from any core points, are depicted in gray. Well-aligned and misaligned clusters were merged into super-clusters using the following criteria: (1) all well-aligned clusters were combined into one super-cluster and (2) misaligned clusters with both horizontal means <2° and vertical means <1° apart were merged (Fig. 26.2c). The total percentage of time duration when both eyes were well-aligned (within the threshold window) is referred to as "average good time %."

To further understand the changes in eye deviation under different viewing conditions, the time-based weighted mean of misaligned clusters ($Mean_{Weighted}$) was calculated. Using Eq. 26.1, composite values were calculated for each of the misaligned super-clusters (n) means ($Composite_n$). Subsequently, the time duration associated with each of these misaligned super-clusters (Tn) was calculated as a fraction (FT_n) of the total time (T_{total}), as described in Eq. 26.2. The cluster composite ($ClusterComposite_n$) was calculated based on the time spent for each of the clusters, as described in Eq. 26.3. Finally, the weighted mean of all misaligned super-clusters was determined using Eq. 26.4.

$$T_{total} = \sum_{n=1}^{n=N} T_n \tag{26.1}$$

$$FT_n = \frac{T_n}{T_{total}} \tag{26.2}$$

$$\text{ClusterComposite}_n = \text{Composite}_n * FT_n \tag{26.3}$$

$$\text{Mean}_{\text{Weighted}} = \left(\sum \text{ClusterComposite}_n\right) / T_{total} \tag{26.4}$$

Additionally, for both pilot and main study data, to objectively compare the angular difference (strabismus or deviation) across different participants, we parsed the eye position signal within 2 millisecond epochs. Then we measured strabismus angles for each 2 ms epochs. We then utilized the cumulative distribution function (ECDF), a statistical tool to visualize and analyze the distribution of the strabismus angle dataset. The ECDF is a nonparametric estimator of the cumulative distribution function (CDF) associated with variables, in this case strabismus angles. It is calculated by first sorting the data in ascending order and then computing the cumulative sum of the sorted values. The resultant sequence of cumulative sums is then normalized by dividing each of the sums by the total number of data points. The normalized cumulative sum represents the ECDF and provides a visual representation of the data distribution without making assumptions for the underlying probability distribution. In addition, we quantified the strabismus angle over time (fixational stability), computing the maximum eye position difference between the two eyes across the maximum time data was collected, and the percent time during which the eyes were aligned, maintaining fusion.

Statistical Analysis The statistical analysis was conducted using MATLAB and SPSS. The Mann–Whitney U-test was employed to assess the age of study participants (control and PD). Due to normality violation, the nonparametric Kruskal–Wallis H-test, followed by Dunn's multiple comparisons tests for all eye alignment parameters with statistically significant results was performed. CDF results were interpreted with KS statistics. A significance level of $p < 0.05$ was defined for all tests. The Kruskal–Wallis H-test was utilized to compare disease severity via the Unified Parkinson's Disease Rating Scale (UPDRS) scores across PD groups.

26.3 Results

Our study aimed to examine the hypothesis that PD patients have significantly increased ocular misalignment compared to healthy controls, and that the misalignment can be influenced by modulation of the STN.

Convergent (Near Target) Fixation We analyzed the results of the eye alignment task, comparing horizontal and vertical eye alignment among control subjects and PD group 1 ($n = 15$), group 2 ($n = 10$), and group 3 ($n = 8$). Figures 26.3a–d

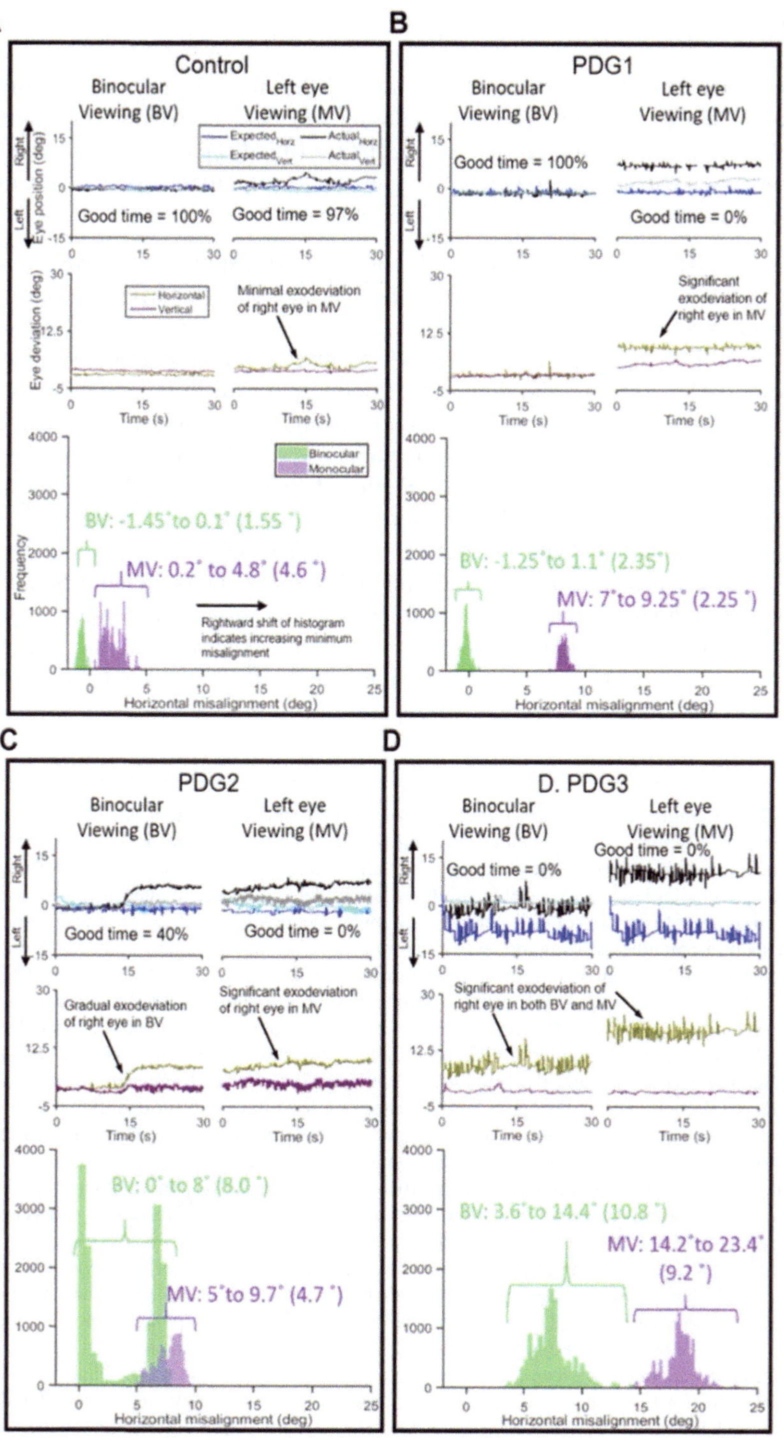

Fig. 26.3 Examples of horizontal and vertical eye positions obtained from the right eye during a 30 s epoch under left eye monocular and binocular viewing conditions: From a control subject in binocular viewing [(**a**)-left top panel)] and monocular viewing [(**a**)-right top panel)], eye positions

present: the actual and expected positions of the nonfixing eye during binocular viewing and of the nonviewing eye during monocular viewing (top panel); the resulting horizontal and vertical eye deviation (middle panel); and the histogram of horizontal eye deviation with 95% upper and lower bounds, showing the range of eye deviation from a healthy control (bottom panel).

Characterization of Eye Movements In the control subjects, the eyes were well-aligned under binocular viewing and in the majority of monocular viewing instances, with horizontal mean deviation less than 3.5° and vertical mean deviation less than 2°. PD subjects in group 1 exhibited eye alignment similar to controls in both eyes viewing (BV), but with the right eye deviating outward in monocular viewing by approximately 8° (exophoria). Histograms in Fig. 26.3b indicated comparable eye deviation bounds to controls under binocular viewing but increased horizontal eye deviation under monocular viewing. Figure 26.3c depicted a PD patient in group 2 initially displaying normal eye alignment in both eyes viewing (BV), which deteriorated gradually with the right eye drifting to the right by 8°, indicative of exotropia. This patient showed increasing rightward deviation in monocular viewing (MV) throughout the recording period. Histograms showed intermittent horizontal eye deviation in binocular viewing with increased exodeviation compared to controls under monocular viewing (MV). Figure 26.3d illustrated a PD patient in group

Fig. 26.3 (continued) maintained fairly close together under binocular viewing with minimal increase in difference (exodeviation) between actual and expected eye positions observed under monocular viewing; (**a**) (middle panel): eye deviation, i.e., difference in eye positions for binocular viewing (right) and monocular viewing (left) in horizontal and vertical planes in control subject; (**a**) (bottom panel): histogram of difference of horizontal eye positions (95% lower and upper bounds with span in parenthesis) showing less spread in binocular viewing than monocular viewing with the majority of data points falling within the threshold window of 3.5°). PDG1 subject: (**b**) (left top panel): binocular viewing—eye positions maintained close together; (**b**) (right top panel): monocular viewing—significant difference (7°–9.25° horizontal) between actual and expected eye position; (**b**) (middle panel): minimal difference in eye positions for binocular viewing (right) and a significantly larger difference in monocular viewing (left) in horizontal and vertical planes in PDG1; (**b**) (bottom panel): histogram of difference of horizontal eye positions showing constrained spread in binocular viewing and relatively larger spread in monocular viewing. PDG2 subject: (**c**) (left top panel): binocular viewing—eye positions maintained close together initially with the right eye deviating up to 8° starting around the halfway point; (**c**) (right top panel): monocular viewing—gradually increasing rightward deviation in horizontal plane; (**c**) (middle panel): difference in eye positions for binocular viewing (right) and monocular viewing (left) in horizontal and vertical planes in PDG2 showing significantly higher deviation; (**c**) (bottom panel): histogram of difference of eye positions showing large spread in both binocular viewing and monocular viewing with increased span under binocular viewing. PDG3 subject: (**d**) (left top panel): binocular viewing—large difference in eye positions throughout; (**d**) (right top panel): monocular viewing—large angle deviation in horizontal between actual and expected eye position; (**d**) (middle panel): difference in eye positions for binocular viewing (right) and monocular viewing (left) in horizontal and vertical planes and horizontal eye deviation in PDG3; (**d**) (bottom panel): histogram of the horizontal difference of eye positions showing extremely large spread in both binocular and monocular viewing with increased span under both binocular and monocular viewing

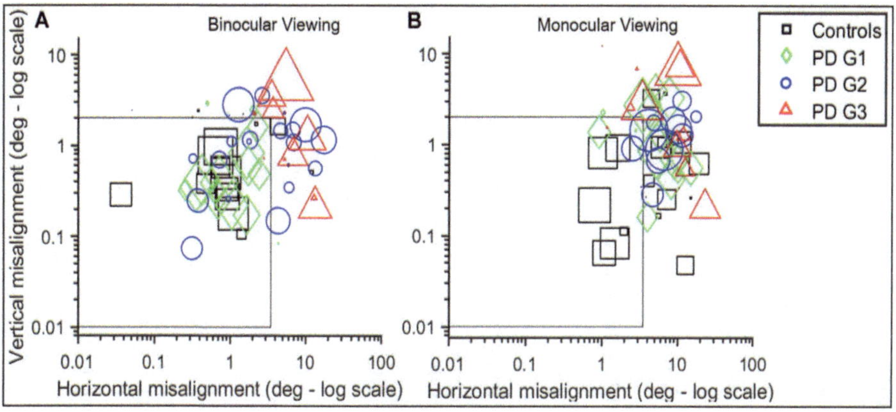

Fig. 26.4 Scatter plot of the log of cluster means generated using the DBSCAN algorithm for each participant in binocular viewing (**a**) and monocular viewing (**b**) in horizontal (x-axis) and vertical (y-axis) planes. Marker size is indicative of the time spent in that region. Controls (black squares) are very well-controlled in binocular viewing with the majority of points within the threshold window in monocular viewing (i.e., within or near gray box depicting threshold for well-aligned clusters—3.5° horizontal; 2° vertical) than PD subjects. PDG1 subjects (green diamonds) are comparable to controls in binocular viewing but show significant deviation in monocular viewing. PDG2 subjects (blue circles) show some increase in eye deviation in binocular viewing with a large increase in eye deviation in monocular viewing. PDG3 subjects (red triangles) show significantly worse eye alignment with extremely large eye deviation reflected as the majority of red symbols being outside the threshold window, particularly in monocular viewing

3 with increased eye misalignment of approximately 7° in binocular viewing and around 18° in monocular viewing, indicating significant exotropia. Histograms reflected constant and extensive exodeviation in both binocular viewing and monocular viewing compared to controls.

We further used the DBSCAN clustering algorithm [14] to further validate these findings. Figures 26.4a, b present mean cluster locations and time spent in each region for binocular viewing and monocular viewing. Control subjects mainly clustered within the normative threshold window, while PD group 1 responses were comparable to controls in binocular viewing but impaired in monocular viewing. PD group 2 showed clusters both inside and outside the window in binocular viewing but were consistently outside in monocular viewing, and PD group 3 clusters were predominantly outside the window in both conditions.

Composite eye deviation was computed for each subject under binocular viewing, right-eye viewing, and left-eye viewing conditions, and the pooled values across all groups showed less deviation in binocular viewing compared to right- and left-eye viewing. PD group 1 had comparable eye alignment to controls in binocular viewing but increased deviation in monocular viewing. PD group 2 exhibited increased deviation in both conditions, while PD group 3 had significantly worse alignment with markedly increased deviation in both binocular and monocular viewing.

Table 26.1 Eye deviation under binocular and monocular viewing across 10 healthy controls and 33 PD participants

Viewing condition		Parameters	Control ($n = 10$)	PDG1 ($n = 15$)	PDG2 ($n = 10$)	PDG3 ($n = 8$)	Kruskal-Wallis test χ^2(df) = chi − sq, p-value
Both eye viewing	Eye deviation	Composite lower bound (deg)	0.05 ± 0.09	0.2 ± 0.35	0.18 ± 0.22	4.84 ± 2.56	$\chi^3(3) = 21.08$, $p < 0.01^{c, d, f}$
		Composite upper bound (deg)	2.05 ± 0.76	2.63 ± 0.93	8.15 ± 5.05	10.19 ± 3.7	$\chi^3(3) = 17.89$, $p < 0.01^{b, c, e}$
		Composite span (deg)	1.99 ± 0.79	2.43 ± 0.9	7.98 ± 5.13	5.36 ± 3.93	$\chi^3(3) = 27.55$, $p < 0.01^{b, c, d, e}$
	Time-based control eye deviation	Good time (%)	98.37 ± 2.29	96.39 ± 5.32	39.12 ± 19.33	10.22 ± 30.04	$\chi^3(3) = 30.76$, $p < 0.01^{b, c, d, e}$
		$\text{Mean}_{\text{Weighted}}$ (deg^2s)	2.88 ± 1.0	1.68 ± 1.79	4.14 ± 8.12	8.74 ± 3.06	$\chi^3(3) = 27.48$, $p < 0.01^{b, c, d, e}$
Left eye viewing	Eye deviation	Composite lower bound (deg)	2.64 ± 3.07	4.91 ± 5.15	4.7 ± 3.39	6.51 ± 3.65	$\chi^3(3) = 6.35$, $p < 0.09$
		Composite upper bound (deg)	4.86 ± 3.49	9.28 ± 5.08	8.08 ± 3.35	10.48 ± 4.75	$\chi^3(3) = 7.67$, $p < 005$
		Composite span (deg)	2.58 ± 1.25	4.57 ± 1.19	3.68 ± 1.39	4.42 ± 1.52	$\chi^3(3) = 12.23$, $p < 0.01^{c}$
	Time-based control eye deviation	Good time (%)	48.92 ± 49.35	12.85 ± 26.08	13.97 ± 34.97	0 ± 0	$\chi^3(3) = 9.23$, $p < 0.023^{c}$
		$\text{Mean}_{\text{Weighted}}$ (deg^2s)	3.70 ± 3.62	8.34 ± 5.04	7.98 ± 2.33	9.74 ± 4.38	$\chi^3(3) = 10.80$, $p < 0.013^{c}$
Right eye viewing	Eye deviation	Composite lower bound (deg)	2.59 ± 3.29	4.46 ± 2.87	6.09 ± 4.8	5.66 ± 2.86	$\chi^3(3) = 5.98$, $p < 0.11$
		Composite upper bound (deg)	5.16 ± 4.21	8.89 ± 4.07	9.51 ± 5.37	11.15 ± 4.47	$\chi^3(3) = 8.96$, $p < 0.03^{c}$
		Composite span (deg)	2.97 ± 1.16	4.6 ± 3.28	3.83 ± 2.29	6.22 ± 4.74	$\chi^3(3) = 3.06$, $p < 0.38$

	Time-based control eye deviation	Good time (%)	50.63 ± 49.53	8.25 ± 20.38	7.96 ± 21.09	0.1 ± 0.29	$\chi^3(3) = 9.17$, $p < 0.03^c$
		$Mean_{Weighted}$ (deg^2s)	3.87 ± 4.35	7.64 ± 3.39	8.57 ± 4.77	10.71 ± 5.41	$\chi^3(3) = 10.01$, $p < 0.02^c$

(Mean ± SD); Composite lower bound: Composite minimum value of spread; Composite upper bound: Composite maximum value of spread; Average Good Time %: total percentage of time when both eyes were well-aligned; $Mean_{Weighted}$: Time-based composite mean of bad clusters of difference of eye positions
[a]Post-hoc Dunn's test significance between controls and PDG1
[b]Post-hoc Dunn's test significance between controls and PDG2
[c]Post-hoc Dunn's test significance between controls and PDG3
[d]Post-hoc Dunn's test significance between PDG1 and PDG3
[e]Post-hoc Dunn's test significance between PDG1 and PDG2
[f]Post-hoc Dunn's test significance between PDG2 and PDG3

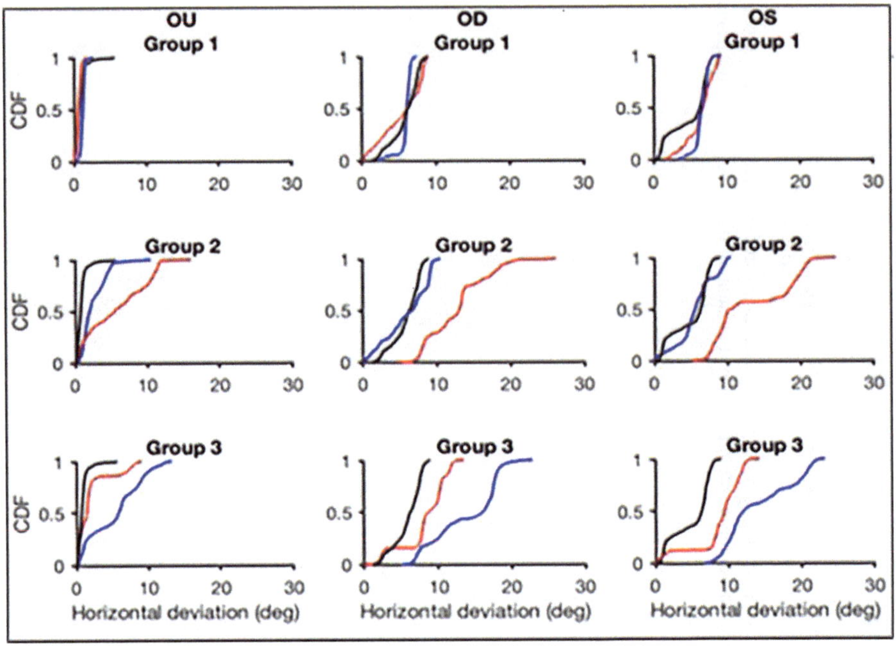

Fig. 26.5 ECDF summary plots of the horizontal deviation in convergent fixation in 21 PD patients and 11 controls. Controls shown in black, DBS off shown in red, and DBS on in blue. Overall, the controls performed better than PD at baseline (off condition) in both binocular (OU) and monocular (OD & OS) viewing fixation. Group 1 had good eye alignment comparable to controls in both binocular and monocular viewing conditions. Group 2 showed severe misalignment which improved with application of DBS across all viewing conditions. Group 3 had mild to moderate misalignment, which worsened with the application of DBS

Table 26.1 summarizes the amount of time spent with eyes aligned within the threshold window for each group, showing that control subjects had longer durations of alignment compared to PD patients. Eyes were aligned for greater periods under binocular viewing than under monocular viewing across all groups. $\text{Mean}_{\text{Weighted}}$ values for misaligned clusters were the lowest in controls and increased progressively from PD group 1 to PD group 3, indicating a worsening of alignment. In conclusion, eye misalignment is prevalent in PD patients, with greater abnormalities observed under monocular viewing compared to binocular viewing.

Effect of DBS on Convergent (near Target) Fixation We studied the response of 22 PD patients to DBS on convergent fixation stability. Based on these responses under binocular viewing, we grouped the PD patients into three main groups (excluding one subject as a case study): Group 1, baseline performance showed a nonsignificant difference compared to controls with no adverse effects with application of DBS ($n = 3$); group 2, baseline performance showed impaired eye alignment which improved with application of DBS ($n = 12$); and group 3, where baseline performance showed mild to moderate misalignment but application of DBS caused an increase in eye deviation ($n = 6$). Fig. 26.5 depicts the groupwise

ECDF curves of horizontal deviation showing the influence of PD on convergent fixation and its modulation with DBS.

26.4 Discussion

Our study aimed to investigate whether ocular alignment is impaired during convergence eye movements that are fundamental to tasks such as reading, looking at the phone, or communicating with objects that are nearby. The task of convergence is known to be impaired in individuals with neurodegenerative disorders, such as PD. In addition to objectively establishing near-vision-associated impairment in binocular alignment, we also asked whether such abnormality is mediated through the STN, and whether DBS can modulate binocular alignment in PD. Consistent with our hypothesis, we found that all PD participants had ocular misalignment at least in the monocular viewing condition, and there was an improvement when DBS was applied, especially in cases where deficits were initially present. As reported previously, our results further validated the claim that PD patients exhibited a significantly higher degree of strabismus at baseline compared to age-matched controls [14, 22]. Spatial and temporal monocular deviation was significantly worse across both populations compared to binocular viewing. This corresponds with previous studies showing decreased foveation and convergence capabilities [13, 14, 17]. Across all cohorts, horizontal deviation was more pronounced than vertical deviation. The baseline performance of PD subjects compared to controls governed the grouping criteria. Group 1 (PDG1) demonstrated performance comparable to that of the control group in both binocular and monocular viewing. Group 2 (PDG2) exhibited performance similar to controls in binocular viewing but displayed pronounced deficits in monocular viewing. Group 3 (PDG3) showed a substantial degree of eye misalignment in both binocular and monocular conditions. It was found that 75% of the participants experienced double vision, whether in binocular or monocular viewing, and had varying degrees of strabismus at the baseline (with DBS off), which then improved with the application of STN-DBS. These results are consistent with previously published qualitative studies with self-reported visual deficits [4, 23, 37].

PD is a progressive neurodegenerative disorder resulting from the degeneration of dopaminergic projections between the substantia nigra pars compacta (SNc) and the rest of the basal ganglia circuitry. Mounting evidence suggests that the STN, a region known for abnormal neural activity in PD, contains visuo-oculomotor neurons [26]. These eye movement cells are predominantly located in the ventral portion of the STN and constitute approximately 20% of its neuronal population [10, 11, 18, 26, 36]. They receive inputs from various regions, including the [28], the frontal eye field (FEF), and the supplementary eye field (SEF) [19, 38]. Both human and nonhuman primate studies provide physiological and anatomical evidence supporting the involvement of the STN in processing information related to eye movements.

In the midbrain, there are neurons crucial for controlling the position and velocity of vergence, located in the mesencephalic reticular formation within a region known as the supra-oculomotor area (SOA)[5]. In normal animals, SOA neurons encode vergence responses, while in strabismic monkeys, they encode horizontal misalignment [8, 9, 44]. Importantly, the SOA receives inputs from the deep cerebellar nuclei, including the fastigial nucleus and the interpositus nucleus, which, in turn, project to the ocular motor neurons responsible for innervating slow extraocular muscle fibers, as described in previous studies [21, 27, 38].

The two potential pathways through which the STN may influence vergence and strabismus control have been identified. The first pathway involves the substantia nigra pars reticulata (SNr), an output region of the basal ganglia, which receives input from the STN and projects to the strabismus angle neurons (SOA) via the superior colliculus [35]. Nonhuman primate (NHP) studies have yielded valuable insights into the neural factors associated with strabismus. When the superior colliculus in strabismic NHPs is electrically microstimulated, it has been observed to elicit disconjugate saccades, differing both in direction and magnitude [12]. Furthermore, cells within the rostral superior colliculus have been found to convey signals linked to horizontal eye misalignment and fixation preference in strabismic NHPs and modulate vergence [6, 7, 20, 24, 30, 39, 40, 42]. Consequently, disruptions in SNr output, as observed in Parkinson's disease, could impact vergence control and eye alignment.

The second pathway directly involves the cerebellum, with subthalamo-ponto-cerebellar projections leading to the deep cerebellar nuclei through precerebellar pontine nuclei. Further downstream, cerebello-pontine fibers connect the deep cerebellar nuclei to neurons sensitive to the strabismus angle within the SOA. Additionally, in strabismic monkeys, neurons within the supra-oculomotor area, typically responsible for encoding vergence responses in normal animals, have been found to encode horizontal misalignment [8, 9, 44]. This intricate network of connections highlights the role of the subthalamic nucleus in regulating strabismus and vergence.

Our study has quantified fixation eye alignment in the context of DBS. This is a critical step in providing further insight into the mechanism behind STN-DBS and adding to our knowledge of the interplay between the established PD pathways and ocular motor regions. We are using a state-of-the-art video-oculography setup and a robust data analysis algorithm to study eye deviation in binocular and monocular viewing that allows for graded evaluation of eye misalignment, giving more precise measurements than a standard clinical exam. All analysis of eye movement-data was blinded to stimulation parameters, disease states, and any demographic information, preventing inherent biasing in the interpretation of our results. Our study is also limited by a small sample size. Furthermore, this study focuses on describing strabismus and eye alignment at near (inherent convergence response). Future studies exploring effects of STN-DBS on eye misalignment at far will aid in developing a more thorough understanding of these complex interactions, and aid in describing the response of divergence controlling neurons to modulation.

References

1. Almer Z, Klein KS, Marsh L, Gerstenhaber M, Repka MX. Ocular motor and sensory function in Parkinson's disease. Ophthalmology. 2012;119:178–82.
2. Armstrong RA. Oculo-visual dysfunction in Parkinson's disease. J Parkinsons Dis. 2015;5: 715–26.
3. Biousse V, Skibell BC, Watts RL, Loupe DN, Drews-Botsch C, Newman NJ. Ophthalmologic features of Parkinson's disease. Neurology. 2004;62:177–80.
4. Borm C, Visser F, Werkmann M, de Graaf D, Putz D, Seppi K, Poewe W, Vlaar AMM, Hoyng C, Bloem BR, Theelen T, de Vries NM. Seeing ophthalmologic problems in Parkinson disease: results of a visual impairment questionnaire. Neurology. 2020;94:e1539–e47.
5. Buttner-Ennever JA. The extraocular motor nuclei: organization and functional neuroanatomy. Prog Brain Res. 2006;151:95–125.
6. Chaturvedi V, van Gisbergen JA. Perturbation of combined saccade-vergence movements by microstimulation in monkey superior colliculus. J Neurophysiol. 1999;81:2279–96.
7. Chaturvedi V, van Gisbergen JA. Stimulation in the rostral pole of monkey superior colliculus: effects on vergence eye movements. Exp Brain Res. 2000;132:72–8.
8. Das VE. Cells in the supraoculomotor area in monkeys with strabismus show activity related to the strabismus angle. Ann N Y Acad Sci. 2011;1233:85–90.
9. Das VE. Responses of cells in the midbrain near-response area in monkeys with strabismus. Invest Ophthalmol Vis Sci. 2012;53:3858–64.
10. Fawcett AP, Cunic D, Hamani C, Hodaie M, Lozano AM, et al. Saccade- related potentials recorded from human subthalamic nucleus. Clin Neurophysiol. 2007;118:155–63.
11. Fawcett AP, Dostrovsky JO, Lozano AM, Hutchison WD. Eye movement- related responses of neurons in human subthalamic nucleus. Exp Brain Res. 2005;162:357–65.
12. Fleuriet J, Walton MM, Ono S, Mustari MJ. Electrical microstimulation of the superior colliculus in Strabismic monkeys. Invest Ophthalmol Vis Sci. 2016;57:3168–80.
13. Gupta P, Beylergil S, Murray J, Jacobs J, Kilbane C, Shaikh AG, Ghasia FF. Effects of Parkinson disease on blur-driven and disparity-driven Vergence eye movements. J Neuroophthalmol. 2021;41:442–51.
14. Gupta P, Murray JM, Beylergil SB, Jacobs J, Kilbane CW, Shaikh AG, Ghasia FF. Objective assessment of eye alignment and disparity-driven vergence in Parkinson's disease. Front Aging Neurosci 2023;15.
15. Hamedani AG, Maguire MG, Marras C, Willis AW. Prevalence and risk factors for double vision in Parkinson disease. Mov Disord Clin Pract. 2021;8:709–12.
16. Hannaway N, Zarkali A, Leyland LA, Bremner F, Nicholas JM, Wagner SK, Roig M, Keane PA, Toosy A, Chataway J, Weil RS. Visual dysfunction is a better predictor than retinal thickness for dementia in Parkinson's disease. J Neurol Neurosurg Psychiatry. 2023;94:742–50.
17. Hanuska J, Bonnet C, Rusz J, Sieger T, Jech R, Rivaud-Pechoux S, Vidailhet M, Gaymard B, Ruzicka E. Fast vergence eye movements are disrupted in Parkinson's disease: a video-oculography study. Parkinsonism Relat Disord. 2015;21:797–9.
18. Hikosaka O, Takikawa Y, Kawagoe R. 'Role of the basal ganglia in the control of purposive saccadic eye movements. Physiol Rev. 2000:953–78.
19. Huerta MF, Kaas JH. Supplementary eye field as defined by intracortical microstimulation: connections in macaques. J Comp Neurol. 1990;293:299–330.
20. Jiang H. Near-response-related neural activity in the rostral superior colliculus of the cat. Soc Neurosci Abstr. 1996:662.
21. Judge SJ, Cumming BG. Neurons in the monkey midbrain with activity related to vergence eye movement and accommodation. J Neurophysiol. 1986;55:915–30.
22. Kang SL, Shaikh AG, Ghasia FF. Vergence and Strabismus in Neurodegenerative Disorders. Front Neurol. 2018;9:299.
23. Kang SL, Beylergil SB, Otero-Millan J, Shaikh AG, Ghasia FF. Fixational eye movement waveforms in amblyopia : characteristics of fast and slow eye movements. JEMR. 2019; (in press).

24. Krauzlis RJ. Neuronal activity in the rostral superior colliculus related to the initiation of pursuit and saccadic eye movements. J Neurosci. 2003;23:4333–44.
25. Lepore FE. Parkinson's disease and diplopia. Neuro-Ophthalmology. 2006;30:37–40.
26. Matsumura M, Kojima J, Gardiner TW, Hikosaka O. Visual and oculomotor functions of monkey subthalamic nucleus. J Neurophysiol. 1992;67:1615–32.
27. Mays LE. Neural control of vergence eye movements: convergence and divergence neurons in midbrain. J Neurophysiol. 1984;51:1091–108.
28. Monakow K, Hartmann-von KA, Künzle H. Projections of the precentral motor cortex and other cortical areas of the frontal lobe to the subthalamic nucleus in the monkey. Exp Brain Res. 1978;33:395–403.
29. Naumann W, Gogarten J, Schönfeld S, Klostermann F, Marzinzik F, Schindlbeck KA. Diplopia in Parkinson's disease: indication of a cortical phenotype with cognitive dysfunction? Acta Neurol Scand. 2021;144:440–9.
30. Pallus AC, Walton MMG, Mustari MJ. Response of supraoculomotor area neurons during combined saccade-vergence movements. J Neurophysiol. 2018;119:585–96.
31. Racette BA, Gokden MS, Tychsen LS, Perlmutter JS. Convergence insufficiency in idiopathic Parkinson's disease responsive to levodopa. Strabismus. 1999;7:169–74.
32. Repka MX, Claro MC, Loupe DN, Reich SG. Ocular motility in Parkinson's disease. J Pediatr Ophthalmol Strabismus. 1996;33:144–7.
33. Schindlbeck KA, Schönfeld S, Naumann W, Friedrich DJ, Maier A, Rewitzer C, Klostermann F, Marzinzik F. Characterization of diplopia in non-demented patients with Parkinson's disease. Parkinsonism Relat Disord. 2017;45:1–6.
34. Searle A, Rowe FJ. Vergence neural pathways: a systematic narrative literature review. Neuroophthalmology. 2016;40:209–18.
35. Shaikh AG, Meng H, Angelaki DE. Multiple reference frames for motion in the primate cerebellum. J Neurosci. 2004;24:4491–7.
36. Sieger T, Bonnet C, Serranová T, Wild J, Novák D, Růžička F, Urgošík D, Růžička E, Gaymard B, Jech R. Basal ganglia neuronal activity during scanning eye movements in Parkinson's disease. PLoS One 2013;8.
37. Smilowska K, Wowra B, Slawek J. Double vision in Parkinson's disease: a systematic review. Neurol Neurochir Pol. 2020;54:502–7.
38. Stanton GB, Goldberg ME, Bruce CJ. Frontal eye field efferents in the macaque monkey: II. Topography of terminal fields in midbrain and pons. J Comp Neurol. 1988;271:493–506.
39. Upadhyaya S, Das VE. Response properties of cells within the rostral superior colliculus of Strabismic monkeys. Invest Ophthalmol Vis Sci. 2019;60:4292–302.
40. Upadhyaya S, Meng H, Das VE. Electrical stimulation of superior colliculus affects strabismus angle in monkey models for strabismus. J Neurophysiol. 2017;117:1281–92.
41. Urwyler P, Nef T, Killen A, Collerton D, Thomas A, Burn D, McKeith I, Mosimann UP. Visual complaints and visual hallucinations in Parkinson's disease. Parkinsonism Relat Disord. 2014;20:318–22.
42. Van Horn MR, Waitzman DM, Cullen KE. Vergence neurons identified in the rostral superior colliculus code smooth eye movements in 3D space. J Neurosci. 2013;33:7274–84.
43. Visser F, Vlaar AMM, Cdjm Borm V, Apostolov YX, Lee IC, Notting HC, Weinstein, Berendse HW. Diplopia in Parkinson's disease: visual illusion or oculomotor impairment? J Neurol. 2019;266:2457–64.
44. Walton MMG, Pallus A, Fleuriet J, Mustari MJ, Tarczy-Hornoch K. Neural mechanisms of oculomotor abnormalities in the infantile strabismus syndrome. J Neurophysiol. 2017;118:280–99.
45. Weil RS, Schrag AE, Warren JD, Crutch SJ, Lees AJ, Morris HR. Visual dysfunction in Parkinson's disease. Brain. 2016;139:2827–43.

Chapter 27
Color Vision Deficits and Eye Movement Abnormalities in Parkinson's Disease

Julia Mleczek, Anim Forjindam, Aasef G. Shaikh, and Fatema Ghasia

Keywords Parkinson's disease · Color vision · Eye movement · Vergence · Deep brain stimulation

27.1 Introduction

Parkinson's disease (PD) is one of the most common neurodegenerative disorders, affecting over ten million people worldwide. It is primarily characterized by progressive dopaminergic neuron loss in the substantia nigra pars compacta (SNpc), leading to progressive dopamine (DA) depletion and classic motor symptoms such as bradykinesia, tremor, and rigidity [34].

Beyond its classic symptom presentation, PD is associated with several other nonmotor symptoms; these include hyposmia, sleep dysfunction (insomnia, somnolence, REM sleep behavior disorder), and psychiatric dysfunction (depression, anxiety, hallucinations) [34]. Among these, visual dysfunction has emerged as a significant and often underrecognized nonmotor symptom of PD [14].

Vision impairments can affect various aspects of visuosensory dysfunction, with effects of PD-related visual impairments being seen in association with both the afferent and efferent visual pathways. Up to 82% of PD patients report at least one

J. Mleczek · A. Forjindam
Case Western Reserve University School of Medicine, Cleveland, OH, USA

A. G. Shaikh
Neurological Institute, University Hospitals Neurology Service, Louis Stokes Cleveland, VA Medical Center, Department of Neurology, Case Western Reserve University, Cleveland, OH, USA

F. Ghasia (✉)
Daroff Dell'Osso Ocular Motilit Laboratory, Cleveland VA Medical Center, Cleveland, OH, USA

Ocular Motility & Vision Neurosciences Laboratory, Cleveland Clinic, Cleveland, OH, USA
e-mail: ghasiaf@ccf.org

P. Ravat et al. (eds.), *Advances in Translational Neuroscience of Vestibular and Eye Movement Disorders*, Contemporary Clinical Neuroscience,
https://doi.org/10.1007/978-3-032-07539-0_27

ophthalmologic symptom, including double vision, blurry vision, misjudging objects and distances, watery eyes, and visual hallucinations [7, 46]. These symptoms can lead to markedly impaired quality of life for these patients, with up to 68% reporting that these visual disturbances interfered with their activities of daily living [41]. Visual dysfunction in PD manifests as increased risk for decreased distance and near best-corrected visual acuity, reduced color vision, contrast sensitivity, and eye movement abnormalities such as vergence insufficiency and strabismus [30]. Importantly, these visual symptoms can be found in the prodromal stages of PD, suggesting its possible utility as a diagnostic tool in PD [29].

In our examination of PD pathology and progression, two interconnected domains of symptomatic visual impairments are specifically of interest: color vision deficits (afferent pathways) and eye movement abnormalities (efferent pathways). In this chapter, we will examine the current knowledge on these domains, explore possible correlations between these visual impairments and neurological severity, and assess the impact of the current treatment options on these symptoms.

27.2 Visual Dysfunction in PD

27.2.1 Color Vision Deficits

Afferent visual impairments in PD include decreased contrast sensitivity and deficits in color vision. Color vision deficiency (dyschromatopsia) refers to the inability of patients to differentiate along the red-green axis (protanomaly/deuteranomaly) and blue-yellow axis (tritanomaly). Poor color discrimination has been suggested as an early symptom of dopaminergic dysfunction in PD [32].

27.2.1.1 Mechanisms and NeuroAnatomy

Color vision disruption in PD can broadly be attributed to subcortical and cortical dysfunction. Subcortical dysfunction, characteristic of PD, disrupts dopaminergic pathways extending from the basal ganglia to visual structures, including the retina.

Changes in retinal structure, including thinning of outer retina [44], the retinal nerve fiber layer (RNFL), and the ganglion cell layer [33] have been described in PD patients. RNFL and ganglion cell layer thinning have been further associated with color vision deficits and overall vision dysfunction in PD [33], corroborating this damage to photoreceptors and their communication pathways in PD as a possible etiology of color vision dysfunction.

More specifically, the retina contains dopaminergic amacrine cells in the retina's inner nuclear layer (INL), with neuronal processes extending into the inner plexiform layer (IPL). Dopamine release from these neurons is essential for photoreceptor signal transmission and synaptic activity, influencing visual acuity, contrast sensitivity, and color perception. Loss of dopamine and dopaminergic neurons in PD, including the IPL cells and amacrine cells of the retina, is therefore suggested to be a cause of the abnormal photoreceptor signaling and subsequent visual deficits. PD patients have been found to have significantly reduced dopaminergic amacrine cells in the retina, further suggesting dopaminergic retinal structural changes as the etiology of this deficiency [31], supporting current theories that color vision impairments in PD are closely related to dopaminergic-amacrine-cell degeneration and these structural changes in the retina.

Normal color vision further involves higher-order processing in the primary visual cortex. In the late stages of PD, cortical dysfunction occurs involving posterior brain structures, including the occipital cortex [18]. This could lead to other disruptions and impairments in higher-order visual processing, such as impaired color and contrast discrimination.

27.2.1.2 Color Vision Dysfunction

Farnsworth-Munsell Test (FMT) Shows Color Vision Abnormalities in PD Patients

Color vision can be measured in several ways. In clinical practice, Ishihara plates (pseudoisochromatic plates) are frequently used as a method for color vision screening, particularly for red-green color vision abnormalities. The ease of use, quick administration, and minimal equipment requirements make this test highly suitable for routine clinical practice. However, it is less attractive for PD color vision assessments, given its limited ability to assess blue-yellow color deficiency.

Frequently used in more extensive color vision workups is the Farnsworth-Munsell 100 Hue Test (FMT), which asks subjects to organize colored tiles into a range of hues, assessing their ability to distinguish color and hue changes. A high error rate is indicative of potential color vision deficiency [21]. The FMT can be divided into subtests of components of Red, Green, Blue, and Purple vision based on the tile or cap color [43], allowing for assessment of color vision across protan, deutan, and tritan axes.

Using the FMT as a measure for color discrimination, PD patients exhibit impaired color discrimination, indicated by higher test error when compared to age-matched controls [4, 36, 44]. This higher error rate also correlates with disease severity [4] and duration [36]. Using a variation of the FMT, the Farnsworth Dichotomous Test D-15, color vision abnormalities were also seen between PD patients and controls (17.53% vs. 5.70%, $p = 0.001$) [30]. More detailed studies show higher error rates among PD patients in the overall test score and among every test subtype, with the highest difference in error rates in the green and purple groups [43].

However, this test has garnered criticism from some, stating that the need for initiation of voluntary movements in testing, a process that is impaired in PD, confounds these results. High error rates and amelioration with medication administration may be due to changes in motor skills rather than changes in color discrimination [39].

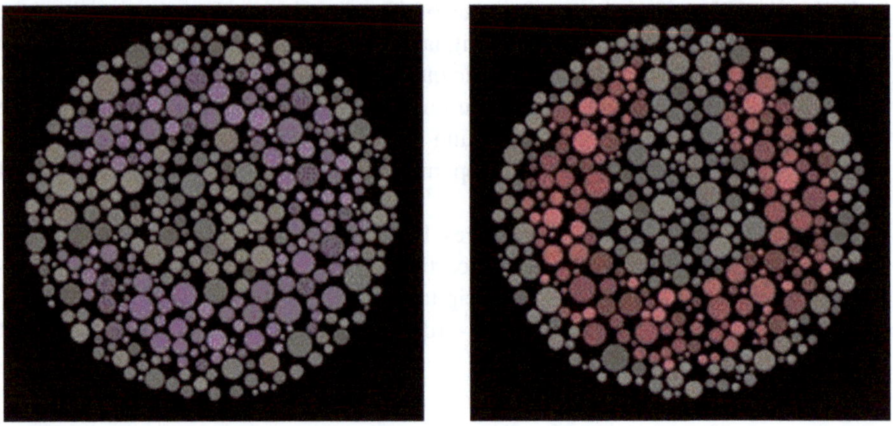

Fig. 27.1 Two examples of CCT pseudoisochromatic Landolt "C" ring stimuli in the left (left panel) and upward (right panel) directions [27]

Studies with Cambridge Color Test (CCT) Show Contrasting Results
The potential confounding factors of the FMT have made finding tests with lower motor demands valuable.

The Cambridge Color Test (CCT) is a potential solution to this dilemma. This computerized test employs pseudoisochromatic stimuli, similar to an Ishihara plate, with disks of varying luminescence creating a background and a target, a Landolt "C" ring. Varying luminescence prevents the subject from using luminescence differences to distinguish the target, therefore improving test validity. The CCT presents the subject with the target-oriented in one of four directions, having the subject select the corresponding directional key [27] (Fig. 27.1).

The CCT has different test versions, each examining different aspects of color vision. The "Trivector" version of the test color discrimination along the protan, deutan, and tritan confusion lines, testing the sensitivity of the L-, M-, and S-cones, respectively. Staircases of each color line are mixed to establish maximum sensitivity to each. Thresholds are established as the smallest perceivable color difference, and are reported in the Commission Internationale de l'Éclairage (CIE) u′,v′ diagram. These limits are scaled by a factor of 10^3 to give normal limits of 10 (protan), 10 (deutan), and 15 (tritan) [27] (Fig. 27.2).

The "Ellipses" version is an extension of this test that reports color discrimination across a range of vectors, producing a "discrimination ellipse". The ellipse's long axis will typically indicate the type of color vision loss for commonly seen color deficiencies [27]. These can be performed across three different background chromaticities to yield three different ellipses.

Typical findings include an axis ratio of less than 2.0. A larger ellipse long axis length, a larger ellipse area, or a higher axis ratio indicates a reduced ability to discriminate color differences [27].

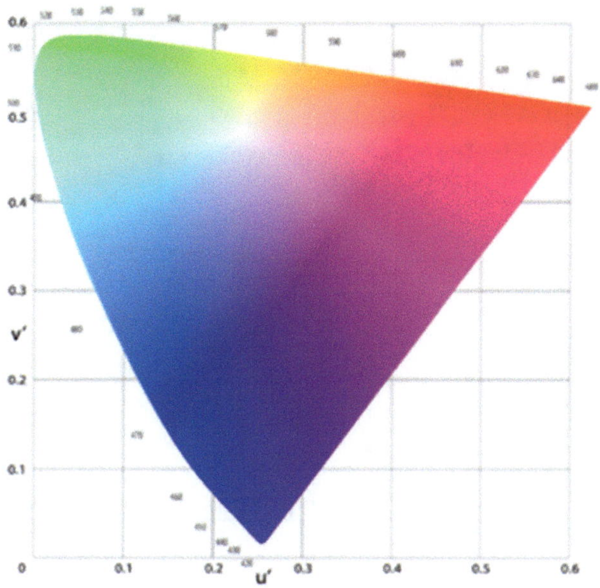

Fig. 27.2 CIE u′,v′ diagram [27].

The angle of the ellipse is also significant in interpreting the test. The direction of the ellipse angle can indicate what discrimination vector the patient has the most difficulty; an axis rotated ~ −90 degrees would show poor discrimination along the tritan axis, suggesting blue-yellow color vision deficiency (Fig. 27.3).

The default test protocol will show the stimuli on screen for 3 seconds or until the subject responds, whichever occurs first. If the subject takes longer than 3 seconds to respond, the test marks this as an incorrect answer. It should be noted that this timing presents challenges in different populations, with slower motor reactions possibly confounding results in mature (> 60 yo) normal trichromats, as evidenced by lower Trivector thresholds at 5 and 8 seconds compared to 3 seconds [10]. This is an essential factor to consider when employing this test with PD patients, as advancing age is a large risk factor for PD.

Current research in PD using this computerized technique is limited. One study of 30 PD patients using this technique found heterogeneous patterns of color vision change with variability in the long axis length and ellipse angle, and significantly increased long axis length compared to controls. They also saw independent impairment along the protan and deutan axes in PD [42].

However, a 1998 study by Regan et al. using this technique found no significant difference in color vision between the PD and the control groups, though they noted some individual PD patients with significantly different thresholds compared to controls.

However, there is a gap in research studies using this test or other similar objective and quantifiable measures of color dysfunction.

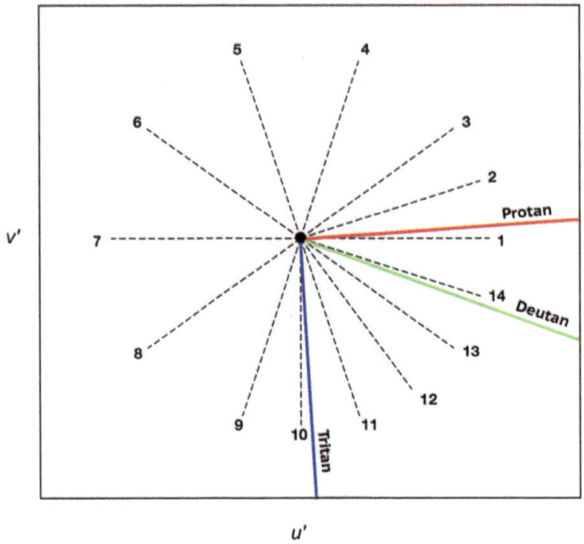

Vector Number	Azimuth (Rad)
1	0
2	0.3142
3	0.6283
4	1.2566
5	1.885
6	2.5133
7	3.1416
8	3.7699
9	4.3982
10	4.7124
11	5.0265
12	5.3407
13	5.6549
14	5.969

Fig. 27.3 Left: Approximate positions of protan, deutan, and tritan Color confusion lines on the u'v' diagram. Numbers indicate the vector number, corresponding to contrast vector locations in a 14-vector test. (Modified from *Metropsis Manual*) Right: Vector number to Azimuth measurements (Rad)

27.2.1.3 Observational Study

We gathered 21 PD subjects and 13 controls to investigate color vision in PD using the Cambridge Color Test-Ellipse Test. We utilized thresholds (mean and SD) obtained along 14 contrast vectors to fit a discrimination ellipse. We compared CCT findings in our PD patients to disease severity (UPDRS), age ($n = 26$; 52–82 y.o), and other measures of ocular motility, including visual acuity, convergence, eye alignment, and stereoacuity.

27.2.1.4 Results

In the PD population, we found that increased age, regardless of PD status, was correlated with larger ellipse area ($p = 0.0193$), and increased length of both long axis ($p = 0.0315$) and short axis ($p = 0.0222$) of the ellipse. No correlation was observed between measures of visual performance on the CCT and UPDRS scores.

The ellipse area of PD subjects was notably increased and neared statistical significance, but ultimately did not reach significance ($p = 0.0534$, Fig. 27.4). PD subjects had significantly increased discrimination vector lengths at vectors 3 ($p = 0.0357$), 5 ($p = 0.0408$), 6 ($p = 0.0264$), and 14 ($p = 0.0174$) (Fig. 27.5). PD patients also showed higher variability in the length of discrimination vectors,

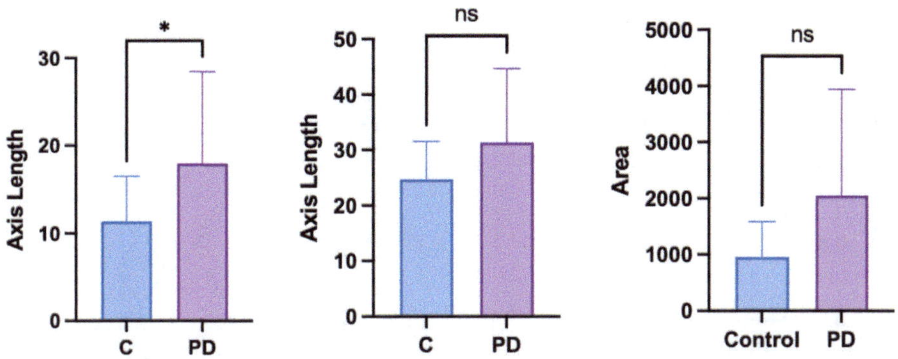

a. Length of the Short Axis of the Ellipse

b. Length of the Long Axis of the Ellipse

c. Area of the Ellipse

Fig. 27.4 (**a**–**c**): Length of short axis of the ellipse (**a**), long axis of the ellipse (**b**), and ellipse area (**c**) in controls vs. PD patients. Statistical significance was established using Mann-Whitney *T* Test (* $p < 0.05$, ** $p < 0.01$, ***$p < 0.001$)

indicated by increased standard deviations in vectors 5 ($p = 0.0043$), 7 ($p = 0.0080$), and 14 ($p = 0.0116$) (Fig. 27.6). PD subjects also had a larger average angular orientation of the ellipse ($p = 0.0442$) and showed a higher range of values (Fig. 27.7).

Notably, control subjects were found to have a larger axial ratio than PD participants (mean control = 2.398, mean PD = 1.994; $p = 0.0442$). However, this may be explained by significantly larger short axis lengths in PD participants ($p = 0.0154$) and nonsignificant increases in long axis lengths ($p = 0.0712$), driving the ratio lower.

The increased length and variability of our discrimination vectors and the larger range of ellipse angles seen in our PD cohort demonstrate an increased variability in color vision deficits in this population. These results suggest increased color discrimination deficits in PD, with longer discrimination vectors and larger ellipse areas, though short of statistical significance, indicating poorer color discrimination.

The variability that is seen in our subjects aligns with previous color vision threshold variability findings using the CCT in Regan et al. Likewise, results for the vector length and ellipse angle range aligned with Silva et al., who found increased vector length along protan, deutan, and tritan confusion lines. However, while nearing significance, our results did not support a significant change in the ellipse's long axis length, as has been previously reported [42]. Our findings of significantly increased axial ratios among controls directly contrast with previous studies, which say significant increases among PD subjects [42].

Seeing consistency across these studies is promising and emphasizes the utility of the CCT in identifying color vision deficiency in this patient population. Further, our findings suggest that color vision deficits occur independently of disease progression, stressing the need for comprehensive eye exams in the clinical management of PD (Fig. 27.8).

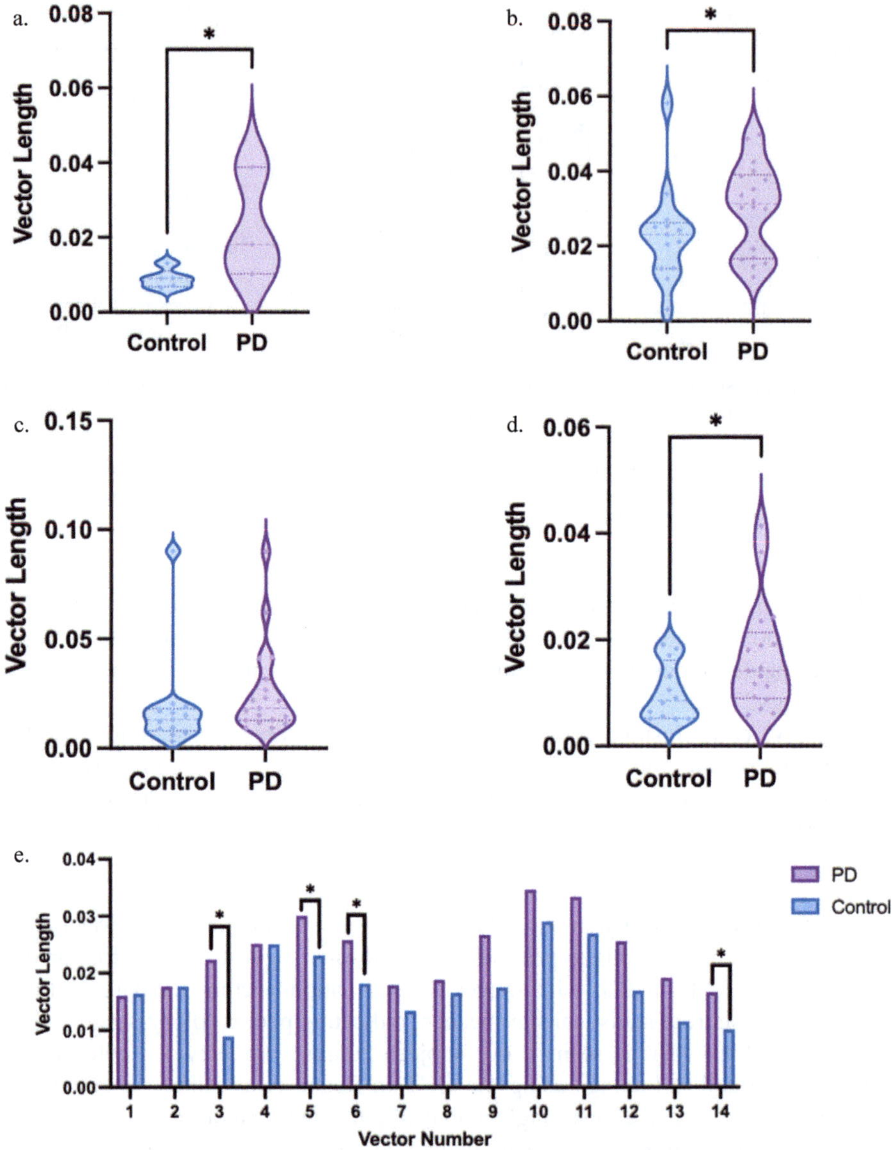

Fig. 27.5 (a–e): Violin plots of the length of threshold vectors in controls vs. PD patients. Statistical significance was established using Mann-Whitney *T* Test (* $p < 0.05$, ** $p < 0.01$, ***$p < 0.001$). (**a**) distribution of threshold length values for controls ($n = 5$) and PD ($n = 3$) at discrimination vector 3 ($p = 0.0357$). (**b**) distribution of threshold length values for controls ($n = 13$) and PD ($n = 18$) at discrimination vector 5 ($p = 0.0408$). (**c**) distribution of threshold length values for controls ($n = 13$) and PD ($n = 19$) at discrimination vector 6 ($p = 0.0264$). (**d**) distribution of threshold length values for controls ($n = 12$) and PD ($n = 17$) at discrimination vector 14 ($p = 0.0174$). (**e**) Plot of mean threshold length of vectors 1–14 for controls vs. PD groups

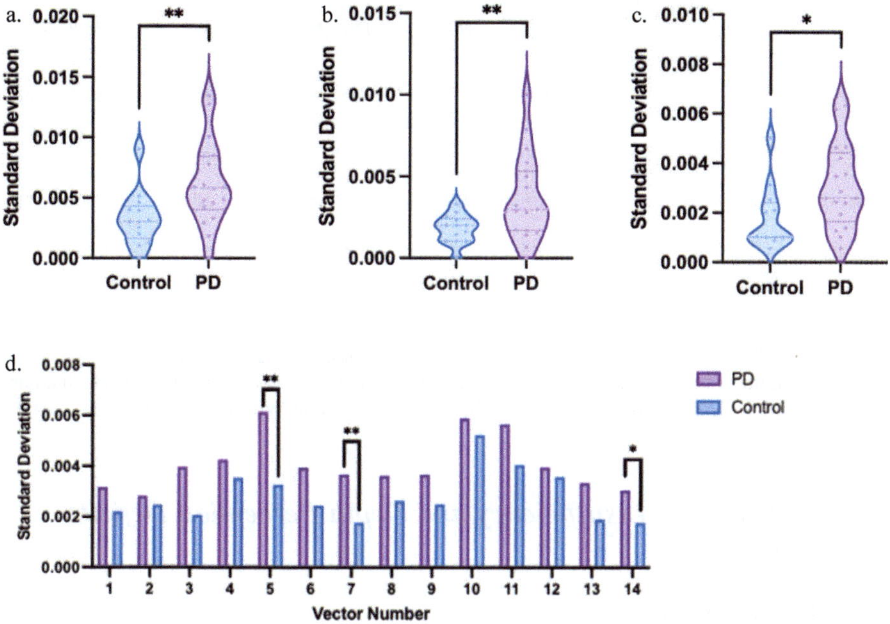

Fig. 27.6 **(a–d):** Standard deviation of threshold vectors in controls vs. PD patients. Statistical significance was established using Mann-Whitney *T* Test (* $p < 0.05$, ** $p < 0.01$, ***$p < 0.001$). (**a**) Distribution of threshold SD values for controls ($n = 11$) and PD ($n = 18$) at discrimination vector 5 ($p = 0.0089$). (**b**) Distribution of threshold SD values for controls ($n = 11$) and PD ($n = 19$) at discrimination vector 7 ($p = 0.0270$). (**c**) Distribution of threshold SD values for controls ($n = 10$) and PD ($n = 17$) at discrimination vector 14 ($p = 0.330$). (**d**) Plot of mean threshold SD of vectors 1–14 for controls vs. PD groups

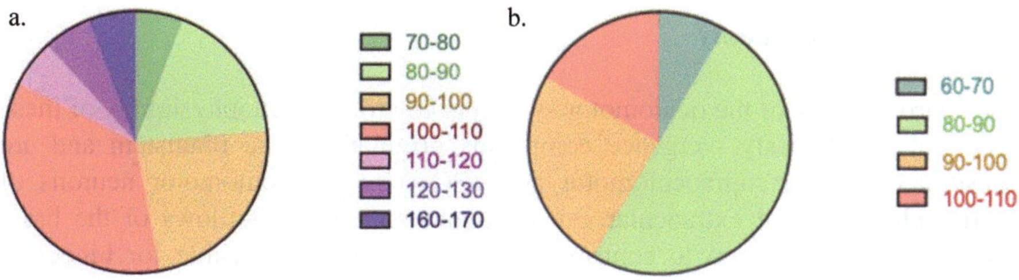

Fig. 27.7 **(a, b):** Distributions of ellipse angles (degrees) in (**a**) PD group ($n = 17$) and (**b**) controls ($n = 12$)

	Parkinson's Disease (Mean, SD)	N	Control (Mean, SD)	N	P-Value	
Ellipse Area	2046 (1896)	17	950.3 (634.6)	12	0.0534	ns
Length of Long Axis	31.29 (13.35)	17	24.67 (6.840)	12	0.0712	ns
Length of Short Axis	17.91 (10.51)	17	11.33 (5.133)	12	0.0154	*
Axial Ratio	1.994 (0.9075)	17	2.398 (0.7629)	12	0.0442	*
Ellipse Angle	101.6 (19.99)	17	90.79 (10.45)	12	0.0442	*

Fig. 27.8 CCT-Ellipse test data of study participants: summary of the ellipse area, length of the long axis, length of the short axis, axial ratio, and ellipse angle. Data presented as mean (Standard Deviation). Statistical significance was established using Mann-Whitney *T* Test (* $p < 0.05$, ** $p < 0.01$, ***$p < 0.001$)

27.2.2 *Vergence Insufficiency and Depth Perception Deficits in PD*

27.2.2.1 Introduction

Eye movement abnormalities are well-established in association PD. Abnormalities include gaze-shifting dysfunction, abnormal smooth pursuits, and abnormal saccades. Here, we explore types of binocular visual dysfunction associated with PD: strabismus, vergence dysfunction, and stereopsis. These abnormalities in ocular processing impair the coordination of fine ocular movement to maintain single binocular vision. The impairment of the perception of binocular input may lead to many of the visual complaints of PD patients, including diplopia.

27.2.2.2 Neuroanatomy

The neuroanatomy of the oculomotor system is key to the pathophysiology of these abnormalities. Grossly, vergence commands originate in the brainstem and are transmitted via the supraoculomotor area (SOA) to the oculomotor neurons of CNIII, which control extraocular eye movements (EOM). Outflows of the basal ganglia are hypothesized to contribute to the network responsible for binocular control. Therefore, impairment of the basal ganglia due to degeneration in PD is likely to lead to abnormal control, as seen in vergence abnormalities [19]. Increasing evidence suggests that it is the subthalamic nucleus (STN), a part of the basal ganglia system, that is connected with visual-oculomotor function. Hypothesized pathways that connect these include (1) a SNr-mediated pathway, with the SNr receiving signals from the STN and subsequently transmitting signals to the SOA via the superior colliculus or (2) a cerebellar pathway, involving precerebellar pontine nuclei connecting the deep cerebellar nuclei to the neurons in SOA [13]. These networks provide a reasonable framework for understanding these changes. However, much is still not understood about the specifics of these pathways.

27.2.2.3 Strabismus and Vergence

Strabismus

Clinically, strabismus is the persistent misalignment of the primary lines of sight of 1 prism diopter (PD) or more; meaning while one eye is focused on a point, the other eye is unable to fixate on the same point. This misalignment leads to diplopia or the suppression of one side of the ocular input to preserve a single input. Convergent (inward) misalignment is defined as esotropia, divergent (outward) misalignment as exotropia, upward and downward misalignment as hypertropia and hypotropia, respectively. Strabismus is a common oculomotor symptom associated with PD, hypothesized to be in relation to the role of dopamine and the basal ganglia in EOM.

Strabismus is commonly measured using the cover test or the alternate cover test, which involves asking the participant to focus on a distant object using binocular vision, occluding one eye, and observing the movement of the contralateral eye. A cover test is positive and indicative of manifest strabismus when the contralateral eye moves after the occlusion of the other eye, as it indicates that the eye was not fixed prior to the introduction of the occluder. An alternate cover test may be used to identify latent strabismus and/or establish the magnitude of strabismus. Here, the occluder is alternated between eyes without permitting binocular vision of the stimulus; movement here indicates shifting fixation [16].

Vergence

Vergence refers to the coordinated disconjugate movement of the eyes to maintain ocular alignment with a target that is changing distance; this is either convergence (inward movement) or divergence (outward movement). Convergence, in particular, is a crucial reflexive visual response that helps us maintain focus on a near target [19, 37]. When a target is moved closer to the eyes, the medial rectus muscles mediate the convergence of the eyes inward to maintain focus on the target. Abnormal convergence, or convergence insufficiency (CI) is associated with symptoms like diplopia, a frequent complaint of PD patients [29], and is attributed to innervation dysfunction [12].

These movements, vergence and ocular alignment, are mediated by signaling from the oculomotor nucleus, which receives signals from the supraoculomotor area (SOA) of the midbrain [19]. Studies have demonstrated that the superior colliculus of the basal ganglia outflow potentially has projections to the SOA, providing a link between pathologic changes seen in basal ganglia outflow tracts in PD and innervation dysfunction leading to CI and strabismus [25].

A commonly used method to identify CI is measuring the near point of convergence (NPC). In this exam, an object, such as a finger or penlight, is moved toward the patient's nose to simulate a near visual target. The "break point" is recorded as the distance of the object from the nose when the examiner notices deviation of one or more eyes (objective version) or the subject reports diplopia (subjective version) [26]. The suggested lower limit for break point varies between studies, and is reported to increase with age [11]. The ease of clinical testing makes it an attractive examination method for CI in an office setting. There are other, more intensive methods of testing CI. For example, fusional vergence amplitude testing measures the compensatory ability of the visual system with gradual increases in base-out prism power until the patient reports diplopia, indicating the break point [40].

More objective methods for testing CI have emerged in the past few years, aiming to produce more reproducible and reliable measurements of fusional vergence amplitudes. Recording eye movements using an EyeLink 1000 plus eye tracker in a haploscopic setup allows more precise tracking of a patient's ocular movements, as it does not rely on the patient's subjective assessment of their vision. This setup utilizes two identical monitors for stimuli presentation and fixes the subject's head in place so the eyelink can calibrate to each patient's baseline convergence at 40 cm. Vergence eye movements were initiated by synchronized movement of the stimuli on the two monitors in either a smooth or stepwise manner. Eye movements are recorded by the EyeLink, and participant data is analyzed via a customized algorithm to determine the break point [40].

Convergence insufficiency is a common oculomotor symptom of PD, with research showing that PD patients are more likely to have CI as compared to controls [17, 24, 30]. This is true regardless of the "on" and "off" states of PD dopaminergic treatment [3].

Current studies further indicate abnormal vergence in PD patients regardless of patients reporting and/or experiencing visual symptoms. A 2021 meta-analysis of 1547 patients with PD shows increased prevalence of both objective and subjective CI compared to controls [2]. Another study of 42 patients found that even in their nonsymptomatic CI PD control group ($n = 9$), 100% of these patients had objective CI when measured [24]. Authors have noted that PD patients may not offer a complaint of diplopia spontaneously, but report it when specifically asked about it [45], so a comprehensive visual history is still a critical aspect of the PD examination.

27.2.2.4 Stereopsis

Stereopsis, the processing of disparities in binocular input into a three-dimensional image, is a critical component of depth perception. This is primarily mediated by the cerebral extrastriatal cortex [20, 22], with sensory fusion occurring within Panum's fusional area [19]. This process is further assisted by extraocular movements, including orthotropia and vergence eye movements that maintain retinal alignment and maintain the eyes within the limits of Panum's area. As seen in CI and strabismus, changes to these movements impact the patient's ability to process this input, leading to stereopsis impairment.

Given the essential nature of intact depth perception for efficient mobility, changes in stereopic ability that we see in PD may lead to worsening of existing PD motor symptoms, including navigation and gait, as well as in neurological processing of spatial orientation. As such, understanding stereopic changes in PD patients is especially meaningful in clinical planning and management, ensuring safe daily function and reducing fall risk.

Methods

Stereopsis and binocular vision are commonly evaluated using stereoscopic plates and polarized glasses. Polarized glasses distort the image so that each eye receives a slightly different image of the same stimulus, testing the patient's stereotypic processing ability, measured in arc seconds. The Titmus Stereo Test utilizes stereoscopic plates to test gross to fine stereoacuity, increasing test precision from the Fly test (~3000 arcsec) to Wirt Circle Plates (< 60 arcsec). Similarly, the Random-dot test methodology includes stereoscopic plates and polarized glasses; however, it asks patients to identify shapes and figures from panels of random dots ("Random-dot stereograms"). This test requires more stereopic ability and is better at detecting fine stereopsis deficiencies. These tests all have high clinical utility due to their timeliness and ease of testing, but are limited in examining depth perception beyond binocular disparity. While binocular input is the main pathway of establishing depth, depth perception additionally relies on monocular cues, which may also negatively impact stereopsis if any one is abnormal.

Link to PD

Stereopsis impairment has been demonstrated to be at a higher frequency in PD patients, even when controlling for age-related decline in stereopsis [20, 43]. PD patients with stereopsis impairment (PDSI) demonstrated abnormal visual perception/constructive function [20], and had worse motor function performance measured by UPDRS than PD patients with normal stereopsis (PDNS). This suggests a possible correlation between stereopsis impairment and PD severity [20, 43]. However, these findings were contradicted in a different study, which showed no differences between PD groups. However, this study notably found significant gray matter reduction in the right extrastriate visual cortex in VBM scans of PDSI compared to PDNS. This reduction was also negatively correlated to the log seconds of arc of the Titmus test, directly linking this gray matter reduction to impaired depth perception [22]. More focused studies on stereopsis in PD are needed to more firmly define the relationship between PD, stereopsis impairments, and motor symptoms.

Observational Study

We looked to further establish the ocular motility changes reported in PD. We again compared our 21 PD subjects and our 13 controls, now comparing in various measures of ocular motility, including eye alignment, convergence, and stereoacuity (Fig. 27.9).

We obtained NPC measurements using the Pen Test and the Royal Air Force (RAF) Rule. In both tests, we found evidence of significant CI in PD patients, evidenced by extension of NPC in both tests (NPC-Pen: $p = 0.0027$; RAF: $p = 0.004$), a change consistent with prior findings discussed above. Eye alignment and magnitude of deviation (prism diopters (Δ)) were measured by the alternate cover test; PD subjects had an increased mean angle of deviation ($p = 0.0006$). Stereoacuity was assessed by Random-dot stereopsis test and log-transformed stereoacuity values were used for statistical comparisons. Stereoacuity significantly decreased in PD patients compared to controls ($p = 0.0293$), again consistent with prior literature.

Overall, our findings demonstrate significant oculomotor changes in PD patients, including increased exotropic ocular deviation, significantly increased convergent insufficiency, and decreased stereoacuity in PD. These findings are consistent with prior literature and offer increased clarity in the context of conflicting study results in prior literature (Fig. 27.10).

	Parkinson's Disease (Mean, SD)	*N*	Control (Mean, SD)	*N*	P-Value	
Age (Years)	67.53 (9.098)	*17*	66.78 (6.924)	9	-	
UPDRS	17.19 (10.432)	*16*	-	-	-	
Strabismus Angle of Deviation	7.722 (6.415)	*18*	2.300 (2.359)	*10*	0.0006	***
Stereoacuity	2.364 (0.7750)	*19*	1.852 (0.2910)		0.0293	*
RAF Near Point of Convergence	16.55 (7.174)	*19*	9.444 (4.447)	9	0.004	**
Pen Test Near Point of Convergence	18.25 (10.80)	*19*	8.633 (5.935)	9	0.0027	**

Fig. 27.9 Demographics and eye movement testing of study participants: summary of age, UPDRS scores, and eye movement testing. Data presented as mean (standard deviation). Statistical significance was established using Mann-Whitney T Test (* $p < 0.05$, ** $p < 0.01$, ***$p < 0.001$)

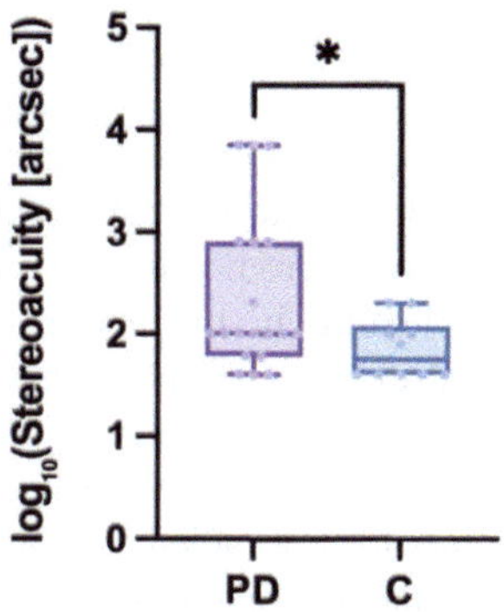

a. Stereoacuity (Arcsec) in PD (N=19) vs. Control (N=10) groups (p=0.0293)

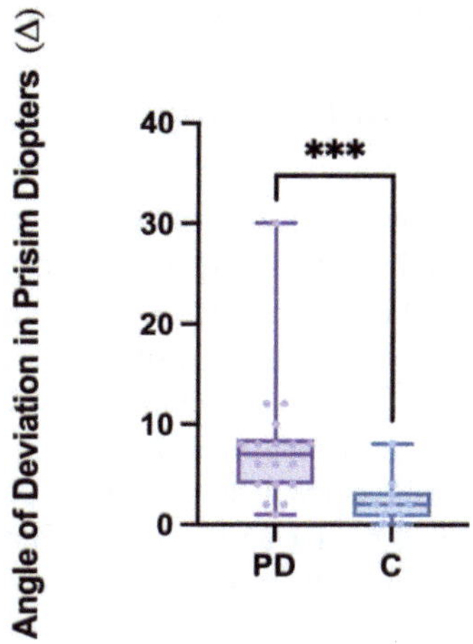

b. Strabismus Angle (Prism Diopters) in PD (N=18) vs. Control (N= 10) groups (p=0.0006)

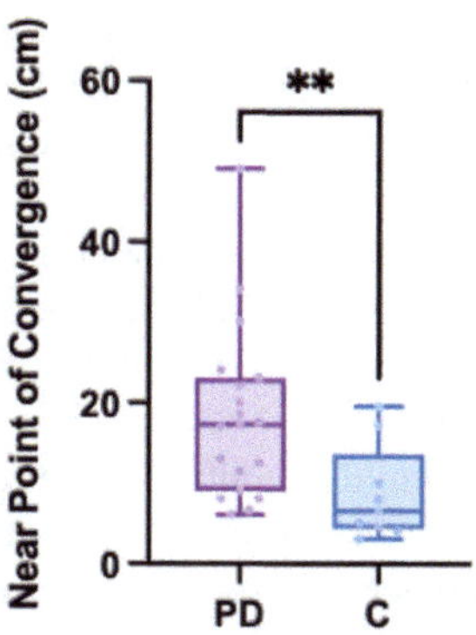

c. Near Point of convergence measured by Pen Test in PD (N=19) vs. Control (N=9) groups (p=0.0027)

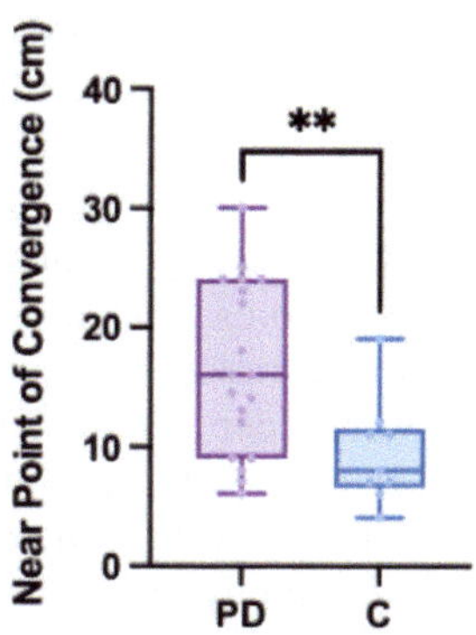

d. Near Point of convergence measured by RAF Test in PD (N=19) vs. Control (N=9) groups (p=0.0044)

Fig. 27.10 (a–d): Box and whiskers plots representing the median, interquartile range, and range in each group. Statistical significance was established using Mann-Whitney T Test (* $p < 0.05$, ** $p < 0.01$, ***$p < 0.001$). (**a**) Stereoacuity (arcsec) in PD ($n = 19$) vs. control ($n = 10$) groups ($p = 0.0293$). (**b**) Strabismus angle (prism diopters) in PD ($n = 18$) vs. control ($n = 10$) groups ($p = 0.0006$). (**c**) Near point of convergence measured by Pen Test in PD ($n = 19$) vs. control ($n = 9$) groups ($p = 0.0027$). (**d**) Near point of convergence measured by RAF Test in PD ($n = 19$) vs. control ($n = 9$) groups ($p = 0.0044$)

27.3 Correlation with Neurological Severity

27.3.1 Introduction

With these symptoms established, observing their correlations with disease severity is crucial in better understanding their pathophysiology and exploring the possibility of using these symptoms to improve treatment outcomes and predict disease course. Given the ease and feasibility of repeated clinical testing in the above cases, understanding the correlations between these tests and disease progression is very attractive in managing PD patients.

Overall, current studies have demonstrated a positive correlation between the number of ocular and visual impairments or complaints from a patient and disease progression [29], and correlations between PD severity and decreased quality of life due to visual complaints [1]. As discussed earlier, these visual symptoms have been found in the prodromal stages of PD, suggesting their possible utility as a diagnostic tool in PD [29].

27.3.1.1 Color Vision

Color vision, measured by the FMT, has been studied as a correlate of neurologic severity in PD patients, with positive correlations between total error score on the FMT (indicating worse color discrimination), disease duration [36], and disease severity [4].

This was supported by Ishihara Color Vision plate testing, which demonstrated worse UPDRS scores with decreased color discrimination. While not a direct measurement of duration, that cohort also saw that their patients with color vision impairment were older and had a later age of PD diagnosis [9].

Notably, studies of PD patients with and without mild cognitive impairment (MCI) showed significantly increased discrimination error scores in PD patients with MCI. Interestingly, there was no significant difference in error scores between the PD group without MCI and controls, and no correlation with test scores and Hoehn and Yahr stage or UPDRS score. Studies of PD patients also found color discrimination errors associated with more severe axial motor impairments in PD, even after controlling for variables such as cognitive function, age, and disease duration [6]. More objective physiological markers of retinal impairment have shown associations between RNFL thinning and UPDRS scores, Hoehn and Yahr stages, and PD duration [23]. Poor FMT performance was also associated with significant changes in the white matter of posterior brain structures, indicating that deterioration of right posterior cortical structures could be responsible for color vision changes [5].

Utilizing the CCT, we compared these discrimination results to PD participant demographics and UPDRS scores. We found no significant difference between disease severity (measured by UPDRS; $n = 16$; range 6–43) and severity of color vision deficits as measured by ellipse area ($p = 0.3504$), axial ratio ($p = 0.2093$), or long axis length ($p = 0.4962$). These findings are consistent with the findings of Silva et al., who also found no correlations with UPDRS scores. Additionally, no significant correlation between disease duration and chromatic color discrimination was noted. However, we did see a correlation between PD patient age ($n = 18$; range 52—82) and ellipse area ($p = 0.0193$), long axis ($p = 0.0315$), and short axis ($p = 0.0222$), but not axial ratio ($p = 0.3181$); our axial ratio findings contrasts previous findings using CCT [42].

Together, these findings support the notion of high variability of color vision deficits in PD, and emphasize the need for comprehensive visual exams regardless of neurological severity. Given that these color vision deficits appear to arise independently of disease progression or motor symptoms severity, restricting extensive color vision examinations to later disease stages may miss early color vision changes. Identifying these nonmotor PD manifestations early could lead to significantly improved patient care and quality of life.

27.3.1.2 Vergence and Strabismus

The connection between convergence insufficiency and PD severity is debated and under-examined. A previous study of 27 PD patients (disease duration of 10.9 ± 6.84 years) found no correlation between duration or severity of PD—measured by UPDRS and Hoehn and Yahr—and the size of convergence insufficiency [3]. However, a 2025 study of 71 PD Patients (14 stage I, 42 stage II, 15 stage III) found that odds ratios for CI prevalence were much higher in stage III than in previous stages, as well as noting a general increase in CI prevalence with advancing Hoehn and Yahr stages [15].

We described above that stereopsis impairment has been demonstrated to have worse motor function performance measured by UPDRS than PD patients with no stereopsis impairment, suggesting a possible correlation between stereopsis impairment and PD severity [20, 42]. While these findings were contradicted in a different study, they showed connections between right extrastriate visual cortex gray matter reduction and impaired depth perception [22]. While initial results appear to favor a connection between disease severity and stereopsis impairment, definitive relationships between stereopsis and disease severity are unable to be drawn from these studies.

We examined our findings of oculomotor dysfunction in our PD population and compared these to UPDRS scores (range: 6–43) and patient age.

Utilizing UPDRS scores, we found no correlation between test results and PD severity for vergence testing (NPC-Pen ($p = 0.2725$), RAF ($p = 0.4167$)). Angle of deviation on the alternating cover test approached statistical significance when correlated with UPDRS scores, but was outside the conventional threshold ($p = 0.0586$). Age was associated with increasing angle of deviation ($p = 0.0252$); but not with any tests of vergence ability (NPC-Pen ($p = 0.3929$),

RAF ($p = 0.1457$)). We found no correlation between stereopsis and UPDRS scores ($p = 0.2945$) or age ($p = 0.2240$).

Our findings here show that while there may be some potential associations between PD severity and both strabismus and vergence insufficiencies, these are not generalizable, and these oculomotor changes are still seen in the earlier PD stages. Again, restricting extensive visual examinations to later disease stages may miss early oculomotor changes. These variable findings indicate that while there is some pathologic connection, the symptomology that we see here in PD is more multifactorial and requires further examination.

27.4 Impact of Neurologic Treatments (Medications & DBS)

Beyond understanding the ocular changes associated with PD, it is necessary to understand how these symptoms may change with clinical PD management. Understanding their reactions to traditional PD treatments, including dopamine agonists (DA) and deep brain stimulation (DBS), provides clinicians with valuable tools in optimizing treatment and improving patient quality of life. These may also demonstrate added utility in evaluating patient response to treatment.

27.4.1 Dopaminergic Medications

Dopaminergic medications include dopamine precursors (levodopa) or dopamine agonists (ropinirole, pramipexole, apomorphine, rotigotine), both predominantly controlling the motor symptoms associated with PD. While these treatments provide symptomatic relief of bradykinesia, tremor, and rigidity, the impact of dopaminergic medications on visual function is less understood and debated.

Treatment with dopaminergic medications generally shows improvements in color vision. Contrast sensitivity deficits have been partially reversed with levodopa, and apomorphine has improved contrast perception at all spatial frequencies [8]. Individual studies have shown improvement in the green-cap portion of the FMT-100 for patients taking medications vs. off medication ($n = 16$) [43]. However, other studies show that intraday oral dosing of L-DOPA does not affect test-retest measures of retinal function and structure following partial washout conditions [44].

Studies on the effect of dopaminergic treatment on eye movement abnormalities are inconsistent [35]. While few studies have investigated treatment effects on stereopsis, presently, studies find no difference in stereopsis [43]. However, convergence insufficiency has been shown to improve upon the administration of levodopa, providing symptomatic relief from symptoms such as diplopia [3, 38]. The severity of convergence insufficiency has also been shown to fluctuate with the dosing schedule of medications, seeing significant improvement in CI in subjects with known "wearing-off" symptoms after a dose of dopamine [3]. However, recent studies have found that medication may adversely affect other eye movements, increasing saccade latency and decreasing saccade peak velocity [28]. The potential adverse effects of medication must be considered in clinical practice, further highlighting the importance of monitoring ocular symptoms in the management of PD patients.

Seeing that dopaminergic treatment has the potential to recover some of the color vision sensitivity and extraocular movements, it implies that the neuroanatomy that governs vergence is grossly intact. The symptoms exhibited in PD patients are a result of the DA depletion and dysfunction. However, varying results indicate a multifactorial pathologic mechanism. More research in these areas is needed to confirm and solidify these results.

27.4.2 Deep Brain Stimulation (DBS)

Subthalamic Nucleus deep brain stimulation (STN-DBS) has been demonstrated to be an effective treatment for PD, improving both motor and nonmotor symptoms through modulation of neural circuits [9]. Given the promising results of studies with dopaminergic medications, DBS has also emerged as a potential mediator of visual dysfunction in PD.

Examining color vision, removal of STN-DBS stimulation negatively impacts color vision assessments, implying a role of DBS in ameliorating the deficit [9]. The limited current research in ocular movements shows that DBS positively affects smooth pursuit and ocular saccade performance [9], with bilateral STN-DBS decreasing saccade latency and increasing saccade peak velocity [28]. Convergence insufficiency has also been shown to improve when used in conjunction with PD medications [9]; however, case studies have reported incidences of worsening visual function when DBS is turned to "ON" state, noting the onset of a new convergence insufficiency that was not present prior [1]. The direct effects of DBS on other forms of ocular movements, including stereopsis, are largely unstudied. While stereopsis would theoretically also be improved with enhanced eye movements, studies directly studying stereopsis are extremely limited.

Seeing as the effects of DBS are not fully understood, more research is needed to examine the therapeutic or adverse effects that occur as a result of treatment (Fig. 27.11).

a.

Discrimination Vector Length	Parkinson's Disease (Mean, SD)	*N*	Control (Mean, SD)	*N*	P-Value	
Vector 1	0.0160	*19*	0.0164	*13*	0.1061	ns
Vector 2	0.0176	*18*	0.0176	*13*	0.1187	ns
Vector 3	0.0223	*3*	0.0089	*5*	0.0357	*
Vector 4	0.0251	*18*	0.0251	*13*	0.1734	ns
Vector 5	0.0300	*18*	0.0230	*13*	0.0408	*
Vector 6	0.0257	*19*	0.0181	*13*	0.0264	*
Vector 7	0.0179	*19*	0.0134	*13*	0.0774	ns
Vector 8	0.0188	*18*	0.0165	*13*	0.1208	ns
Vector 9	0.0267	*3*	0.0175	*5*	0.125	ns
Vector 10	0.0346	*19*	0.0291	*13*	0.2446	ns
Vector 11	0.0334	*19*	0.0269	*13*	0.119	ns
Vector 12	0.0255	*19*	0.0169	*12*	0.1642	ns
Vector 13	0.0192	*18*	0.0115	*12*	0.1038	ns
Vector 14	0.0166	*17*	0.0101	*12*	0.0174	*

Fig. 27.11 (a, b): (**a**) Length of threshold vectors in Parkinson's disease patients vs. controls. Statistical significance was established using Mann-Whitney T Test (* $p < 0.05$, ** $p < 0.01$, ***$p < 0.001$). (**b**) Standard deviation of threshold vectors in Parkinson's disease patients vs. controls. Statistical significance was established using Mann-Whitney *T* Test (* $p < 0.05$, ** $p < 0.01$, ***$p < 0.001$)

b.

Discrimination Vector Standard Deviation	Parkinson's Disease (Mean, SD)	*N*	Control (Mean, SD)	*N*	P-Value	
Vector 1	0.003	*19*	0.002	*13*	0.2444	ns
Vector 2	0.003	*18*	0.002	*13*	0.1489	ns
Vector 3	0.004	*3*	0.002	*5*	0.25	ns
Vector 4	0.004	*18*	0.004	*13*	0.2114	ns
Vector 5	0.006	*18*	0.003	*13*	0.0043	**
Vector 6	0.004	*19*	0.002	*13*	0.0957	ns
Vector 7	0.004	*19*	0.002	*13*	0.008	**
Vector 8	0.004	*18*	0.003	*13*	0.1147	ns
Vector 9	0.004	*3*	0.003	*5*	0.4643	ns
Vector 10	0.006	*19*	0.005	*13*	0.5	ns
Vector 11	0.006	*19*	0.004	*13*	0.2948	ns
Vector 12	0.004	*19*	0.004	*12*	0.3194	ns
Vector 13	0.003	*18*	0.002	*12*	0.0855	ns
Vector 14	0.003	*17*	0.002	*12*	0.0116	*

Fig. 27.11 (continued)

References

1. Abou Khzam R, El Jalbout ND, Seif R, Sadaka A. An unusual presentation of convergence insufficiency in a patient with Parkinson's disease stimulated by deep brain stimulation. Amer J Ophthalmol Case Rep. 2022;26:101531. https://doi.org/10.1016/j.ajoc.2022.101531.
2. Al-Namaeh M. Parkinson's disease and convergence insufficiency: a meta-analysis. Invest Ophthalmol Vis Sci. 2021;62(8):2387.
3. Almer Z, Klein KS, Marsh L, Gerstenhaber M, Repka MX. Ocular motor and sensory function in Parkinson's disease. Ophthalmology. 2012;119(1):178–82. https://doi.org/10.1016/j.ophtha.2011.06.040.
4. Armstrong RA. Oculo-visual dysfunction in Parkinson's disease. J Parkinsons Dis. 2015;5(4): 715–26. https://doi.org/10.3233/JPD-150686.

5. Bertrand JA, Bedetti C, Postuma RB, et al. Color discrimination deficits in Parkinson's disease are related to cognitive impairment and white-matter alterations. Mov Disord. 2012;27(14): 1781–8. https://doi.org/10.1002/mds.25272.
6. Bohnen NI, Haugen J, Ridder A, et al. Color discrimination errors associate with axial motor impairments in Parkinson's disease. Mov Disord Clin Pract. 2017;4(6):864–9. https://doi.org/10.1002/mdc3.12527.
7. Borm CDJM, Visser F, Werkmann M, et al. Seeing ophthalmologic problems in Parkinson disease: results of a visual impairment questionnaire. Neurology. 2020;94(14):e1539–47. https://doi.org/10.1212/WNL.0000000000009214.
8. Büttner T, Müller T, Kuhn W. Effects of apomorphine on visual functions in Parkinson's disease. J Neural Transm. 2000;107(1):87–94. https://doi.org/10.1007/s007020050007.
9. Dumitrascu OM, Kamiński J, Rutishauser U, Tagliati M. Subthalamic nuclei deep brain stimulation improves color vision in patients with Parkinson's disease. Brain Stimul Basic Trans Clin Res Neuromodul. 2016;9(6):948–9. https://doi.org/10.1016/j.brs.2016.08.006.
10. Fars J, Fernandes TP, Huchzermeyer C, Kremers J, Paramei GV. Chromatic discrimination measures in mature observers depend on the response window. Sci Rep. 2022;12(1):9072. https://doi.org/10.1038/s41598-022-13129-w.
11. Gantz L, Stiebel-Kalish H. Convergence insufficiency: review of clinical diagnostic signs. J Opt. 2022;15(4):256–70. https://doi.org/10.1016/j.optom.2021.11.002.
12. Goering M, Drennan KB, Moshirfar M. Convergence insufficiency. In: StatPearls. StatPearls Publishing; 2025. Accessed 24 April 2025. http://www.ncbi.nlm.nih.gov/books/NBK554390/.
13. Gupta P, Murray JM, Beylergil SB, et al. Objective assessment of eye alignment and disparity-driven vergence in Parkinson's disease. Front Aging Neurosci. 2023:15. https://doi.org/10.3389/fnagi.2023.1217765.
14. Hamedani AG, Willis AW. Self-reported visual dysfunction in Parkinson disease: the survey of health, ageing and retirement in Europe. Eur J Neurol. 2020;27(3):484–9. https://doi.org/10.1111/ene.14092.
15. Herrero-Gracia A, Hernández-Andrés R, Luque MJ, Ciuffreda KJ, Díez-Ajenjo MA. Convergence insufficiency and Parkinson's disease progression. Parkinsonism Relat Disord. 2025;133:107341. https://doi.org/10.1016/j.parkreldis.2025.107341.
16. Hull S, Tailor V, Balduzzi S, et al. Tests for detecting strabismus in children aged 1 to 6 years in the community. Cochrane Database Syst Rev. 2017;11 https://doi.org/10.1002/14651858.cd011221.pub2.
17. Irving EL, Chriqui E, Law C, et al. Prevalence of convergence insufficiency in Parkinson's disease. Movement Disorder Clin Pract. 2017;4(3):424–9. https://doi.org/10.1002/mdc3.12453.
18. Jubault T, Gagnon JF, Karama S, et al. Patterns of cortical thickness and surface area in early Parkinson's disease. NeuroImage. 2011;55(2):462–7. https://doi.org/10.1016/j.neuroimage.2010.12.043.
19. Kang SL, Shaikh AG, Ghasia FF. Vergence and Strabismus in Neurodegenerative Disorders. Front Neurol. 2018:9. https://doi.org/10.3389/fneur.2018.00299.
20. Kim SH, Park JH, Kim YH, Koh SB. Stereopsis in drug naïve Parkinson's disease patients. Can J Neurol Sci. 2011;38(2):299–302. https://doi.org/10.1017/S0317167100011501.
21. Kinnear PR, Sahraie A. New Farnsworth-Munsell 100 hue test norms of normal observers for each year of age 5–22 and for age decades 30–70. Br J Ophthalmol. 2002;86(12):1408–11. https://doi.org/10.1136/bjo.86.12.1408.
22. Koh SB, Suh S-I, Kim SH, Kim JH. Stereopsis and extrastriate cortical atrophy in Parkinson's disease: a voxel-based morphometric study. Neuroreport. 2013;24(5):229. https://doi.org/10.1097/WNR.0b013e32835edbc5.
23. Lee JY, Ahn J, Kim TW, Jeon BS. Optical coherence tomography in Parkinson's disease: is the retina a biomarker? J Parkinsons Dis. 2014;4(2):197–204. https://doi.org/10.3233/JPD-130306.
24. Lepore MDFE. Parkinson's disease and diplopia. Neuro-Ophthalmology. 2006;30(2–3):37–40. https://doi.org/10.1080/01658100600742838.

25. May PJ, Bohlen MO, Perkins E, Wang N, Warren S. Superior colliculus projections to target populations in the supraoculomotor area of the macaque monkey. Vis Neurosci. 2021;38:E017. https://doi.org/10.1017/S095252382100016X.
26. Mestre C, Bedell HE, Díaz-Doutón F, Pujol J, Gautier J. Characteristics of saccades during the near point of convergence test. Vis Res. 2021;187:27–40. https://doi.org/10.1016/j.visres.2021.06.001.
27. Started MRG. Version R20. Kent: Cambridge Research Systems Ltd; 2019.
28. Munoz MJ, Arora R, Rivera YM, et al. Medication only improves limb movements while deep brain stimulation improves eye and limb movements during visually-guided reaching in Parkinson's disease. Front Hum Neurosci. 2023:17. https://doi.org/10.3389/fnhum.2023.1224611.
29. Nieto-Escamez F, Obrero-Gaitán E, Cortés-Pérez I. Visual dysfunction in Parkinson's disease. Brain Sci. 2023;13(8):1173. https://doi.org/10.3390/brainsci13081173.
30. Nowacka B, Lubinski W, Honczarenko K, Potemkowski A, Safranow K. Ophthalmological features of Parkinson disease. Med Sci Monit. 2014;20:2243–9. https://doi.org/10.12659/MSM.890861.
31. Ortuño-Lizarán I, Sánchez-Sáez X, Lax P, et al. Dopaminergic retinal cell loss and visual dysfunction in Parkinson disease. Ann Neurol. 2020;88(5):893–906. https://doi.org/10.1002/ana.25897.
32. Piro A, Tagarelli A, Nicoletti G, Fletcher R, Quattrone A. Color vision impairment in Parkinson's disease. J Parkinsons Dis. 2014;4(3):317–9. https://doi.org/10.3233/JPD-140359.
33. Polo V, Satue M, Rodrigo MJ, et al. Visual dysfunction and its correlation with retinal changes in patients with Parkinson's disease: an observational cross-sectional study. BMJ Open. 2016;6(5):e009658. https://doi.org/10.1136/bmjopen-2015-009658.
34. Postuma RB, Berg D, Stern M, et al. MDS clinical diagnostic criteria for Parkinson's disease. Mov Disord. 2015;30(12):1591–601. https://doi.org/10.1002/mds.26424.
35. Pretegiani E, Optican LM. Eye movements in Parkinson's disease and inherited parkinsonian syndromes. Front Neurol. 2017;8:592. https://doi.org/10.3389/fneur.2017.00592.
36. Price MJ, Feldman RG, Adelberg D, Kayne H. Abnormalities in color vision and contrast sensitivity in Parkinson's disease. Neurology. 1992;42(4):887. https://doi.org/10.1212/WNL.42.4.887.
37. Purves D, Augustine GJ, Fitzpatrick D, et al. Types of eye movements and their functions. In: Neuroscience. 2nd ed. Sinauer Associates; 2001. Accessed 24 Apr 2025. https://www.ncbi.nlm.nih.gov/books/NBK10991/.
38. Racette BA, Gokden M, Tychsen L, Perlmutter JS. Convergence insufficiency in idiopathic Parkinson's disease responsive to levodopa. Strabismus. 1999;7(3):169–74. https://doi.org/10.1076/stra.7.3.169.636.
39. Regan BC, Freudenthaler N, Kolle R, Mollon JD, Paulus W. Colour discrimination thresholds in Parkinson's disease: results obtained with a rapid computer-controlled colour vision test. Vis Res. 1998;38(21):3427–31. https://doi.org/10.1016/S0042-6989(97)00402-1.
40. Rovira-Gay C, Mestre C, Argiles M, Vinuela-Navarro V, Pujol J. Feasibility of measuring fusional vergence amplitudes objectively. PLoS One. 2023;18(5):e0284552. https://doi.org/10.1371/journal.pone.0284552.
41. Savitt J, Aouchiche R. Management of Visual Dysfunction in patients with Parkinson's disease. J Parkinsons Dis. 2020;10(s1):S49–56. https://doi.org/10.3233/JPD-202103.
42. Silva MF, Faria P, Regateiro FS, et al. Independent patterns of damage within magno-, parvo- and koniocellular pathways in Parkinson's disease. Brain. 2005;128(10):2260–71. https://doi.org/10.1093/brain/awh581.
43. Sun L, Zhang H, Gu Z, Cao M, Li D, Chan P. Stereopsis impairment is associated with decreased color perception and worse motor performance in Parkinson's disease. Eur J Med Res. 2014;19(1):29. https://doi.org/10.1186/2047-783X-19-29.

44. Tran KKN, Lee PY, Finkelstein DI, et al. Altered outer retinal structure, electrophysiology and visual perception in Parkinson's disease. J Parkinsons Dis. 2024;14(1):167–80. https://doi.org/10.3233/JPD-230293.
45. Ungureanu L, Irincu L, Diaconu S, Oprițoiu B, Chaudhuri KR, Falup-Pecurariu C. Diplopia in movement disorders: a systematic review of the literature. J Pers Med. 2024;14(3):270. https://doi.org/10.3390/jpm14030270.
46. Weil RS, Schrag AE, Warren JD, Crutch SJ, Lees AJ, Morris HR. Visual dysfunction in Parkinson's disease. Brain. 2016;139(11):2827–43. https://doi.org/10.1093/brain/aww175.

Chapter 28
Binocular Visual Acuity as a Key Metric for Evaluating Amblyopia and Strabismus and Fixation Instability

Yulia Haraguchi, Gokce Busra Cakir, Aasef G. Shaikh, and Fatema Ghasia

Keywords Binocular/monocular visual acuity · Nystagmus · Amblyopia · Strabismus · Fixation eye movements

28.1 Introduction and Background

Amblyopia is a neurodevelopmental disorder caused by abnormal visual input during early life, often due to anisometropia, strabismus, or visual deprivation [1]. It results in reduced grating, vernier, and optotype acuities, along with decreased contrast sensitivity in the amblyopic eye [2]. Psychophysical studies indicate that the fellow eye is also affected, exhibiting impairments in contrast sensitivity, grating acuity, motion perception, and visual search performance, though to a lesser degree [3–6].

Under natural binocular conditions, amblyopia impacts stereopsis [7, 8], interocular suppression [8–11], fixation instability [9, 12, 13], and contrast sensitivity [14–17]. It also affects daily activities that require binocular coordination, including visual search efficiency and reading speed [18, 19]. Given its ability to reflect real-world visual function, binocular visual acuity has emerged as a key metric for assessing amblyopia treatment outcomes [20]. Both patching and binocular therapies improve binocular visual acuity, with the latter also enhancing fellow-eye function [21, 22].

Y. Haraguchi
Case Western Reserve University School of Medicine, Cleveland, OH, USA

G. B. Cakir · F. Ghasia (✉)
Ocular Motility & Vision Neurosciences Laboratory, Cleveland Clinic, Cleveland, OH, USA

Cole Eye Institute, Cleveland Clinic, Cleveland, OH, USA
e-mail: ghasiaf@ccf.org

A. G. Shaikh
Neurological Institute, University Hospitals Neurology Service, Louis Stokes Cleveland, VA Medical Center, Department of Neurology, Case Western Reserve University, Cleveland, OH, USA

P. Ravat et al. (eds.), *Advances in Translational Neuroscience of Vestibular and Eye Movement Disorders*, Contemporary Clinical Neuroscience,
https://doi.org/10.1007/978-3-032-07539-0_28

Binocular visual acuity in amblyopia and strabismus is likely influenced by a variety of ocular sensory and motor factors. Sensory factors include fellow-eye acuity deficits, stereopsis impairments, anisometropia severity, and degree of amblyopia [2, 23]. Motor factors involve strabismus and fixation instability, which may stem from nystagmus or altered fixation eye movements (FEM), such as increased saccadic amplitude and intersaccadic drifts [15, 24, 25]. Greater FEM abnormalities correlate with poor stereo-acuity and severe amblyopic eye deficits [12, 23, 25–28].

Binocular summation, which enhances visual performance by integrating input from both eyes, is diminished in amblyopia. This summation is typically observed in visual acuity, contrast sensitivity, and fixation stability in normal subjects [16, 29]. However, amblyopic individuals have reduced binocular contrast sensitivity and diminished binocular summation of contrast sensitivity which has been linked to the depth of amblyopia [14–17]. However, the interplay between sensory and motor deficits in binocular summation in visual acuity remains poorly understood. In this chapter, we will describe visual acuity deficits in amblyopia and strabismus, their influencing factors, and their correlations with fixation eye movement abnormalities.

28.2 Binocular and Monocular Visual Acuity in Amblyopia and Strabismus by Clinical Subtype

We examined the visual acuity of 68 subjects with amblyopia, 17 subjects with strabismus without amblyopia, and 24 control subjects (Fig. 28.1). Amblyopia was attributable to anisometropia ($n = 28$), strabismus ($n = 20$), or a combination of both (mixed amblyopia, $n = 20$), as defined by previous Pediatric Eye Disease Investigator Group (PEDIG) studies. [30] This study employed a psychophysical staircase procedure to quantify visual acuities and assess binocular performance, with visual acuity calculated in arcmin and converted to logMAR for analysis with lower logMAR values indicating better visual acuity [8]. In the control group and in subjects with strabismus without amblyopia, the better monocular acuity eye was designated the fellow/dominant eye, and the worse eye was designated as the amblyopic/nondominant eye. A two-way repeated-measures Analysis of Variance (ANOVA) was conducted to determine the effect of the clinical subtype across viewing conditions on visual acuity.

Binocular Visual Acuity Deficits Binocular visual acuity, measured in logMAR was significantly affected by the presence of amblyopia ($F(4, 104) = 6.14$, $p < 0.001$, $\eta^2_p = 0.19$). The values observed across different groups were:

- **Controls:** -0.10 ± 0.10
- **Anisometropic amblyopia:** 0.02 ± 0.12
- **Strabismic amblyopia:** 0.04 ± 0.12

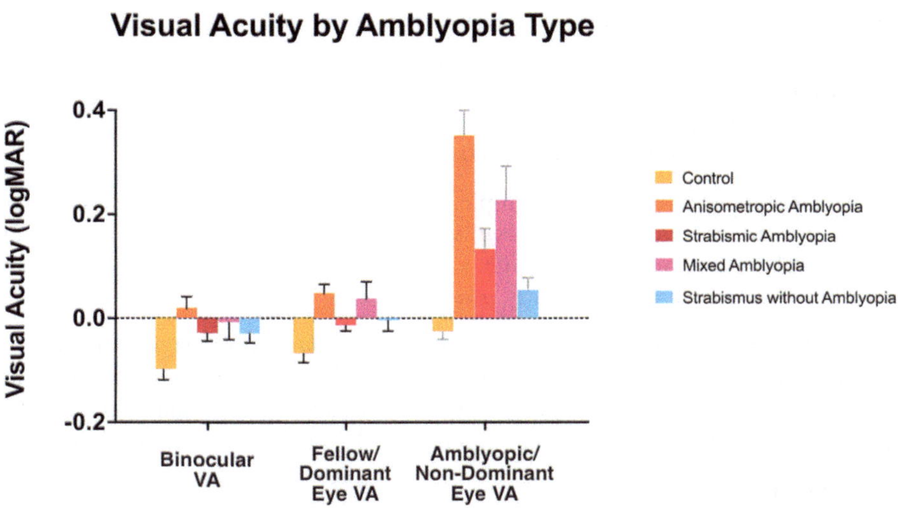

Fig. 28.1 Mean visual acuity by clinical subtype of amblyopia. (Note: Mean ± standard error mean binocular, fellow/dominant eye and amblyopic/nondominant eye visual acuity (logMAR) per clinical subtype of amblyopia)

- **Mixed amblyopia:** 0.00 ± 0.08
- **Strabismus without amblyopia:** −0.03 ± 0.08

Subjects with anisometropic ($p = 0.001$), strabismic ($p < 0.001$), and mixed ($p = 0.02$) amblyopia demonstrated significantly worse binocular visual acuity than the controls. However, subjects with strabismus without amblyopia had binocular visual acuity comparable to controls ($p = 0.46$).

Fellow (Dominant) Eye Visual Acuity Deficits The visual acuity of the fellow or dominant eye was also significantly affected by amblyopia ($F(4, 104) = 6.83$, $p < 0.001$, $\eta^2_p = 0.21$). The mean values for each group were:

- **Controls:** −0.07 ± 0.08
- **Anisometropic amblyopia:** 0.05 ± 0.09
- **Strabismic amblyopia:** 0.07 ± 0.10
- **Mixed amblyopia:** 0.03 ± 0.14
- **Strabismus without amblyopia:** −0.01 ± 0.08

Similar to binocular acuity findings, subjects with anisometropic ($p = 0.001$), strabismic ($p < 0.001$), and mixed ($p = 0.02$) amblyopia demonstrated significantly reduced fellow-eye visual acuity compared to the controls. Subjects with strabismus without amblyopia exhibited similar visual acuity to controls ($p = 0.84$).

Amblyopic (Nondominant) Eye Visual Acuity Deficits As expected, visual acuity deficits in the amblyopic/nondominant eye were significantly affected by the presence of amblyopia ($F(4, 104) = 21.89$, $p < 0.001$, $\eta^2_p = 0.46$). The values recorded were:

- **Controls:** -0.03 ± 0.08
- **Anisometropic amblyopia:** 0.35 ± 0.25
- **Strabismic amblyopia:** 0.25 ± 0.19
- **Mixed amblyopia:** 0.50 ± 0.33
- **Strabismus without amblyopia:** 0.04 ± 0.10

All amblyopic groups (anisometropic, strabismic, and mixed) demonstrated significantly reduced nondominant eye visual acuity compared to controls (all $p < 0.001$). However, subjects with strabismus without amblyopia showed no significant difference from controls ($p = 1.00$).

Binocular Summation of Visual Acuity in Amblyopia Binocular performance of visual acuity was assessed by calculating the ratio of fellow/dominant eye acuity to binocular acuity, where a ratio greater than 1 indicates binocular summation, and a ratio less than 1 indicates binocular inhibition. [31] The observed values were:

- **Controls:** 1.08 ± 0.10
- **Anisometropic amblyopia:** 1.10 ± 0.31
- **Strabismic amblyopia:** 1.11 ± 0.18
- **Mixed amblyopia:** 1.09 ± 0.28
- **Strabismus without amblyopia:** 1.05 ± 0.11

Binocular performance ratios were computed and compared across clinical type using a one-way ANOVA. There were no significant differences in binocular performance ratios among groups ($F(4, 104) = 0.17$, $p = 0.95$, $\eta^2_p = 0.01$). Despite reduced binocular visual acuity, amblyopic and strabismic subjects without amblyopia exhibited levels of binocular summation comparable to controls.

One of the key takeaways from this study is that visual acuity deficits, particularly in the fellow eye and under binocular viewing, are largely attributable to amblyopia rather than strabismus alone. We observed significant reductions in both binocular and fellow-eye visual acuity in individuals with amblyopia, regardless of the type, underscoring the pivotal role of amblyopia in these deficits. In contrast, individuals with strabismus but no amblyopia exhibited smaller, nonsignificant reductions in binocular and dominant eye visual acuity when compared to control subjects. These findings suggest that strabismus, though potentially disruptive to ocular alignment, does not substantially impair visual acuity in the absence of amblyopia.

Another striking observation was the similarity in binocular performance ratios between amblyopic, strabismic, and control subjects. Despite reductions in visual acuity, all groups showed near-1 ratios of fellow-eye visual acuity to binocular visual acuity, indicating the phenomenon of binocular summation. This suggests that binocular vision can partially compensate for visual deficits in monocular viewing, leading to improved overall visual performance. These results are consistent with prior research suggesting that amblyopic individuals can still exhibit some degree of binocular summation, albeit with reduced visual performance compared to controls. [15, 31, 32]

28.3 Binocular and Monocular Fixation Instability in Amblyopia and Strabismus

Fixation instability (FI) is a key characteristic of both amblyopia and strabismus, often manifesting as increased variability in eye position during visual fixation. Our lab uses a high-resolution eye tracker (EyeLink 1000) to record both horizontal and vertical eye positions under binocular and monocular viewing conditions and to assess fixational eye movements. Previous studies, including those conducted in our lab and by others, have consistently shown increased FI in individuals with amblyopia and strabismus. [12, 23, 25, 33] This instability is particularly evident when comparing these groups to normal control subjects. Fixation stability was quantified with the bivariate contour ellipse area (BCEA), encompassing 68% of all the x- and y-coordinates of fixation data points [34] with smaller BCEA values indicating more stable fixation (Fig. 28.2). It was calculated using the equation:

$$BCEA = \pi X^2 \sigma_x \sigma_y \sqrt{1 - p^2}$$

where 2.291 is the X^2 value for a 0.68 probability, σ_x and σ_y are horizontal and vertical eye position standard deviations, and p is the correlation of two position components. [26, 33, 35] A log10 transformation was applied for statistical analysis.

Figure 28.2 illustrates this concept, showing FI of dominant and nondominant eye for a normal subject and comparing it across various subtypes of amblyopia and strabismus without amblyopia under different viewing conditions. The BCEA plots for the nondominant (or amblyopic) eye and the dominant (or fellow) eye are presented in separate boxes, with the top row (2A) showing a control subject. For each condition, three distinct viewing modes were tested: binocular viewing (BEV), fellow-eye/dominant eye viewing, and amblyopic eye/nondominant eye viewing. The plots reveal that the size of the BCEA oval is consistently larger in the eyes of amblyopic subjects, signifying a higher degree of FI compared with the control group. Also, the BCEA values are greater in the AE during AEV than FE during FEV in amblyopic subjects. In strabismic subjects without amblyopia, the size of the BCEA oval is larger in the nonfixing (deviated eye)- however no differences were observed in the fixing eye under monocular viewing.

The Impact of Viewing Condition on Fixation Instability Fixation instability was analyzed using a three-way repeated-measures ANOVA with two within-subject factors—(a) viewing condition: binocular viewing, fellow/dominant eye viewing and amblyopic/nondominant eye viewing; (b) eye: fellow/dominant eye and amblyopic/nondominant eye—and one between subjects factor, clinical subgroups). Mauchly's test of sphericity indicated that the assumption of sphericity had been violated when comparing fixation instability across clinical groups ($\chi^2(2) = 6.82$, $p = 0.03$) and the Greenhouse-Geisser correction was applied.

The viewing condition plays a critical role in determining the extent of FI. There was a significant difference in FI across the three viewing conditions, with the

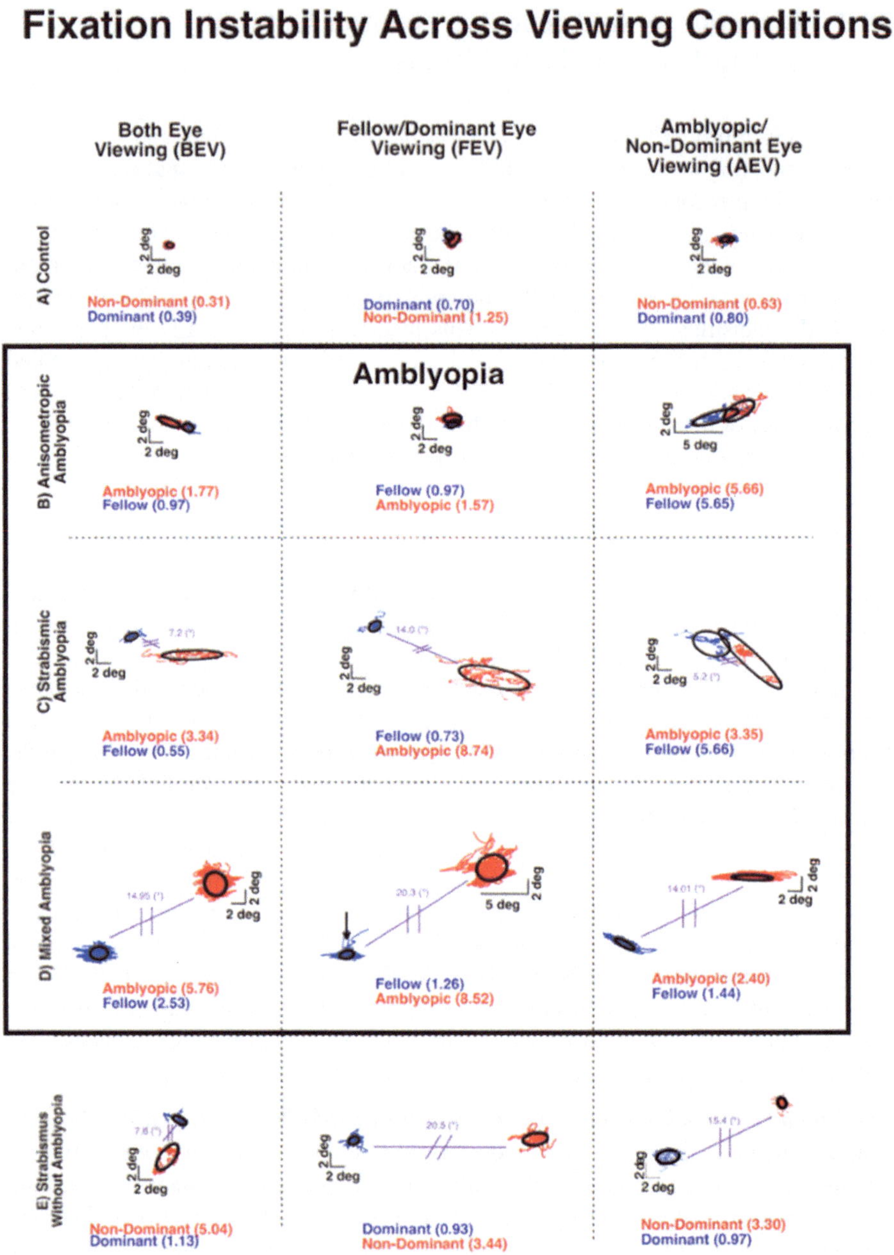

Fig. 28.2 Fixation instability across viewing conditions in amblyopia and control subjects. Fixation instability under binocular (BEV), fellow/dominant eye (FEV), and amblyopic/nondominant eye (AEV) viewing for a control (**A**), anisometropic amblyopia (**B**), strabismic amblyopia (**C**), mixed amblyopia (**D**), and strabismus without amblyopia (**E**). Fixation points: amblyopic/nondominant eye (red), fellow/dominant eye (blue). Black circles indicate logBCEA (logarithm of the bivariate contour ellipse area) encompassing 68% of fixation points. BCEA values (deg^2) plots are provided for each eye in all conditions. Composite strabismic angle shown in purple

Table 28.1 Fixation stability across viewing conditions by amblyopia type

	Control	Anisometropic amblyopia	Strabismic amblyopia	Mixed amblyopia	Strabismus without amblyopia
Binocular viewing					
FE	0.04 ± 0.29	0.05 ± 0.35	0.23 ± 0.41	0.12 ± 0.26	−0.02 ± 0.42
AE	0.03 ± 0.28	0.10 ± 0.38	0.48 ± 0.59	0.49 ± 0.39	0.23 ± 0.36
Fellow/dominant eye viewing					
FE	0.01 ± 0.27	0.08 ± 0.48	0.17 ± 0.46	0.01 ± 0.34	−0.02 ± 0.30
AE	0.16 ± 0.32	0.12 ± 0.57	0.66 ± 0.46	0.43 ± 0.32	0.53 ± 0.44
Amblyopic/nondominant eye viewing					
FE	0.19 ± 0.23	0.28 ± 0.41	0.77 ± 0.51	0.51 ± 0.53	0.77 ± 0.51
AE	0.03 ± 0.31	0.08 ± 0.40	0.39 ± 0.65	0.28 ± 0.40	0.02 ± 0.36

Note: Mean (± standard deviation) fixation instability of the fellow/dominant eye (FE) and amblyopic/nondominant eye (AE) under binocular (BEV), fellow/dominant eye (FEV), and amblyopic/nondominant eye (AEV) viewing conditions for each amblyopia subtype

greatest instability observed during amblyopic eye/nondominant eye viewing ($F(2, 208) = 6.18$, $p = 0.002$). The main effect of eye revealed greater FI in the amblyopic eye compared to the fellow eye ($F(1, 104) = 18.82$, $p < 0.001$). During **binocular viewing (BEV)**, the amblyopic eye demonstrated greater instability than the fellow eye in all forms of amblyopia, with subjects exhibiting more pronounced instability in the amblyopic/nondominant eye. This pattern of increased FI in the amblyopic eye during binocular viewing was consistently observed across all three subtypes of amblyopia ($F(4, 104) = 5.87$, $p < 0.001$), but was particularly noticeable in individuals with coexisting strabismus (i.e., mixed amblyopia) ($p < 0.001$). In contrast, **monocular viewing (MV)** tended to increase FI in all groups, particularly in the occluded eye. This phenomenon is likely attributable to the absence of binocular summation and the loss of feedback from binocular disparity, which normally aid in stabilizing eye position. Interestingly, in monocular viewing, the degree of instability was more pronounced in the amblyopic eye across all amblyopic subjects compared to nondominant eye of controls and strabismus without amblyopia subjects (Table 28.1).

28.4 Nystagmus as an Element of Fixation Instability in Amblyopia and Strabismus

Despite the usefulness of fixation instability (FI) measurements, BCEA alone does not capture the dynamic characteristics of fixation-eye movements (FEMs) in some amblyopes, particularly those with fixational nystagmus. Clinical examinations may not always reliably detect nystagmus, as accurate identification requires careful analysis of eye movement recordings, specifically focusing on the waveforms, direction, and frequency of fast- and slow-phase oscillations (Fig. 28.3).

Classification of Nystagmus in Amblyopia and Strabismus We systematically evaluated fixation eye positions obtained under monocular and binocular viewing conditions for all amblyopic/strabismic subjects and grouped them based on FEM waveforms into (a) controls, (b) those without nystagmus = 49 (Anisometropic Amblyopia = 22, Strabismic Amblyopia = 5, Mixed Amblyopia = 9, Strabismus without Amblyopia = 13), c) those with Fusion Maldevelopment Nystagmus (FMN) = 9 (Anisometropic = 0, Strabismic = 6, Mixed = 2, Strabismus without Amblyopia = 1) and d) those with other nystagmus = 27 (Anisometropic Amblyopia = 6, Strabismic Amblyopia = 9, Mixed Amblyopia = 9, Strabismus without Amblyopia = 3). Figure 28.3 illustrates the different types of eye movement patterns observed in the study.

- **No Nystagmus:** Fig. 28.3a depicts the fixation eye movements (FEMs) of a control subject, whereas Fig. 28.3b presents an example of an amblyopic subject without nystagmus under BEV, fellow-eye viewing (FEV), and amblyopic-eye viewing (AEV) conditions. Both tracings show some fixational saccades interspersed with intersaccadic drifts but lack repetitive oscillations. However, compared with the control subject, the amblyopic subject without nystagmus exhibited increased fixation instability and greater drift (black and gray arrows) in the amblyopic eye across all viewing conditions.
- **Fusion Maldevelopment Nystagmus (FMN):** Fig. 28.3c presents an example of a subject with conjugate horizontal jerk nystagmus during BEV. When viewing with the fellow eye (right eye), the subject exhibited leftward slow-phase nystagmus, whereas viewing with the amblyopic eye (left eye) resulted in an inversion of the nystagmus, with slow phases directed rightward in both eyes at a higher velocity. FMN, which is common in individuals with early-onset strabismus, was identified by a nasalward slow-phase drift with respect to the fixating eye, along with an instantaneous reversal of the fast and slow phases when fixation alternated between the two eyes. [36, 37]
- **Other Nystagmus:** Fig. 28.3d depicts an example of an amblyopic subject with nystagmus that was most evident during amblyopic-eye viewing. Nystagmus was defined as repetitive cycles of a fast-phase saccade followed by a slower,

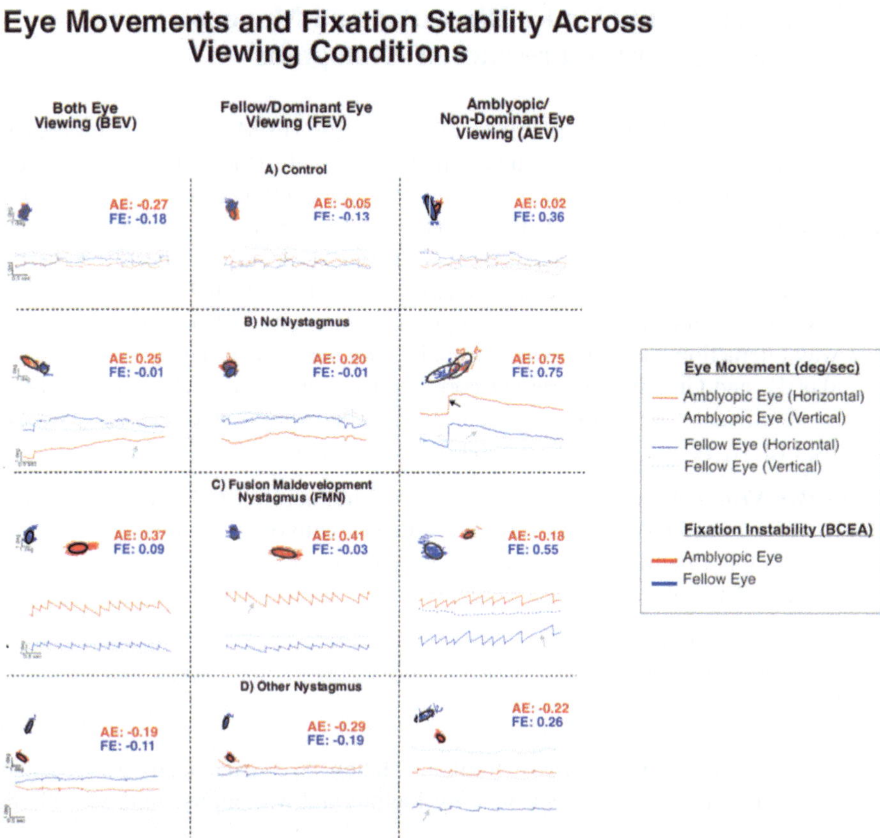

Fig. 28.3 Comparison of fixation eye movements across different viewing conditions in subjects with and without nystagmus. Examples of fixation eye movements in both eyes viewing (BEV), fellow-eye viewing (FEV), and amblyopic eye viewing (AEV) conditions for four different subject groups: (**a**) control subject; (**b**) subject without nystagmus; (**c**) subject with fusion maldevelopment nystagmus; and (**d**) subject with other nystagmus. The x-axis represents time, while the y-axis shows the horizontal (solid line: dark blue for fellow/dominant eye, red for amblyopic/nondominant eye) and vertical (dotted line: pink for fellow eye, light blue for amblyopic eye) positions. Positive values on the vertical axis indicate rightward and upward movements. Each plot includes a bivariate contour ellipse area (BCEA) scatterplot, with black circles representing fixation points. The logarithmic BCEA (logBCEA) values are shown for the amblyopic/nondominant eye (red) and the fellow/dominant eye (blue), with the ellipses encompassing 68% of the fixation points

prolonged slow-phase linear or decreasing-velocity drift of eye position during attempted steady fixation. When the subject switched to FEV, the repetitive nystagmus cycles disappeared. This subject did not meet the criteria for FMN or congenital motor nystagmus and was classified under "other nystagmus."

28.5 Visual Acuity Under Binocular and Monocular Viewing per the Presence of Nystagmus

We then examined how the presence of nystagmus (FEM subgroup) influenced binocular and fellow-eye visual acuity deficits (Fig. 28.4). Due to the limited sample size, for the purpose of subsequent analysis, we have combined subjects with fusion maldevelopment nystagmus and other nystagmus together into the nystagmus group. A two-way repeated-measures ANOVA was run to determine the effect of the clinical subtype of amblyopia across viewing conditions on visual acuity. Mauchly's test of sphericity indicated that the assumption of sphericity had been violated for the two-way interaction when comparing visual acuity per FEM group ($\chi^2(2) = 138.18$, $p < 0.001$), and Greenhouse-Geisser correction was also used. There was a statistically significant interaction between viewing condition and FEM waveform groups, $F(2.31, 122.42) = 7.98$, $p < 0.001$, $\eta^2_p = 0.13$.

Binocular Visual Acuity Deficits Binocular visual acuity (logMAR) was significantly worse in amblyopic and strabismic subjects, with or without nystagmus, with the greatest abnormalities seen in those with nystagmus ($F(2,106) = 7.80$, $p < 0.001$, $\eta^2_p = 0.16$). The values recorded were:

- Controls: -0.10 ± 0.10
- No nystagmus: 0.00 ± 0.10
- Nystagmus: 0.02 ± 0.11

Fellow-Eye Visual Acuity Deficits Similarly, fellow-eye visual acuity was worse in amblyopic and strabismic subjects with nystagmus and without nystagmus, with the

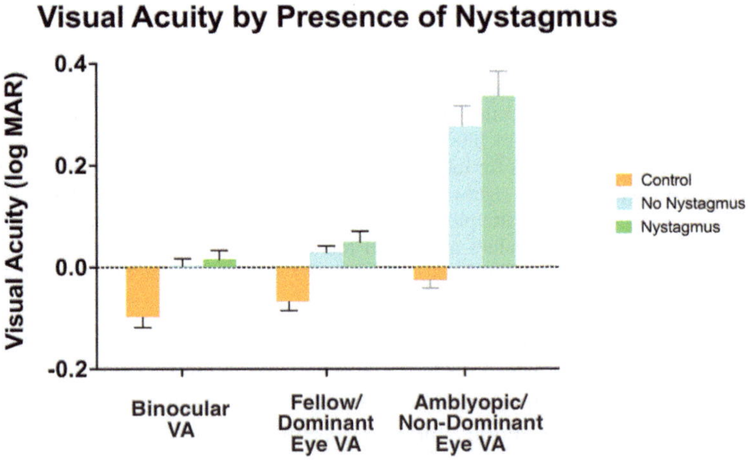

Fig. 28.4 Mean visual acuity by the presence of nystagmus. (Note: Mean ± standard error mean binocular, fellow/dominant eye and amblyopic/nondominant eye visual acuity (logMAR) per FEM subtype)

most pronounced deficits observed in those with nystagmus, ($F(2,106) = 10.09$, $p < 0.001$, $\eta 2p = 0.16$). The values recorded were:

- Controls: −0.07 ± 0.08
- No nystagmus: 0.03 ± 0.09
- Nystagmus: 0.05 ± 0.13

Amblyopic/Nondominant Eye Visual Acuity Deficits Amblyopic/nondominant eye visual acuity was also significantly worse in amblyopic and strabismic subjects with or without nystagmus compared to controls, with the worst outcomes occurring in subjects with nystagmus, ($F(2,106) = 16.25$, $p < 0.001$, $\eta^2_p = 0.24$). The values recorded were:

- Controls: −0.03 ± 0.08
- No nystagmus: 0.28 ± 0.28
- Nystagmus: 0.34 ± 0.29

Binocular Summation of Visual Acuity in Amblyopia and Strabismus Amblyopic subjects, both with and without nystagmus, demonstrated significant binocular summation. Binocular visual acuity (0.02 ± 0.11 logMAR) of subjects with nystagmus significantly improved compared to monocular fellow eye acuity (0.05 ± 0.13) and amblyopic eye acuity (0.34 ± 0.29), $F(2, 105) = 31.61$, $p < 0.001$, $\eta^2_p = 0.38$. Similarly, subjects without nystagmus exhibited improved binocular visual acuity (0.00 ± 0.10 logMAR) compared to amblyopic eye acuity (0.28 ± 0.28), though the improvement in fellow eye acuity was not statistically significant.

A comparison of binocular performance ratios (fellow/dominant eye logMAR VA divided by binocular logMAR VA) showed no significant differences among controls, amblyopic, and strabismic subjects (controls: 1.08 ± 0.10, no nystagmus: 1.08 ± 0.24, nystagmus: 1.10 ± 0.25, $F(2,106) = 0.12$, $p = 0.89$, $\eta^2_p = 0.002$). Thus, amblyopic and strabismic subjects, with or without nystagmus, have worse binocular and fellow-eye visual acuity compared to controls. However, significant binocular summation suggests that binocular viewing improves visual acuity in these populations.

28.6 Impact of Fixation Instability, Nystagmus, and Visual Sensory Factors on Visual Acuity

To evaluate the factors influencing binocular and fellow-eye visual acuity deficits, hierarchical multiple regression analyses were performed. These models assessed binocular and fellow-eye visual acuity (dependent variables) in relation to continuous independent variables such as visual acuity deficits in the amblyopic and fellow eye (while assessing binocular visual acuity), binocular performance ratios, clinically measured strabismus angle, anisometropia, age, stereoacuity deficits (measured by Titmus test), fixation instability, and FEM waveform characteristics. Changes in model performance were evaluated by assessing changes in R^2 and F statistics, with statistical significance set at an alpha value of 0.05.

The Influence of Nystagmus and Clinical Parameters on Binocular Visual Acuity To evaluate how the presence of nystagmus, fixation instability (FI), binocular performance ratios, and other clinical sensory parameters affect binocular visual acuity, we conducted a hierarchical multiple regression analysis. The first model included the following variables: visual acuity of the amblyopic eye (AE) and fellow eye (FE), binocular performance ratio (the ratio of FE and binocular visual acuity), clinically measured strabismus angle, the extent of anisometropia (interocular difference), age, stereoacuity deficits, fixation instability of AE and FE under binocular viewing, and multicategorical waveform characteristics (categorized as with and without nystagmus). In contrast, the second model excluded the multicategorical waveform variable (Table 28.2).

In both models, the following factors were statistically significant predictors of binocular visual acuity deficits: increasing visual acuity deficit of the FE, reduced binocular summation, and increased FI of the AE in binocular viewing. However, the removal of waveform information in the second model did not result in a statistically significant R^2 and F change ($\Delta R^2 = 0.00$, $\Delta F = 0.41$). This indicates that the inclusion of multicategorical waveform characteristics in the first model did not enhance the ability to predict the extent of binocular visual acuity deficits. These

Table 28.2 Evaluation of factors impacting binocular visual acuity deficit using a hierarchical regression model

	Model 1 with waveform	Model 2 without waveform
Constant	0.332	0.333
FE visual acuity	1.071^{***}	1.073^{***}
AE visual acuity	−0.014	−0.013
Binocular ratio (fellow-eye VA/binocular VA)	-0.681^{***}	-0.682^{***}
Strabismus angle	−0.01	−0.008
Refractive error difference	0.03	0.029
Age	0.00	−0.001
Stereopsis	−0.015	−0.014
FI of FE in BEV	−0.025	−0.025
FI of AE in BEV	0.042^{*}	0.043^{*}
No nystagmus	0.007	NA
Nnystagmus	0.007	NA
R^2 F	$0.98, 412.26^{***}$	$0.98, 514.14^{***}$ ΔR^2 $\Delta F = 0, 0.41$

Note: Standardized coefficients from the hierarchical multiple regression model are reported. Significance levels are indicated as follows: $p < 0.05$ (*), $p \leq 0.01$ (**), $p \leq 0.001$ (***). Abbreviations: *FI* fixation instability, *AE* amblyopic/non-dominant eye, *FE* fellow/dominant eye, *BEV* both-eye viewing, *NA* not applicable findings suggest that while FI and binocular summation play a crucial role in determining binocular acuity, waveform characteristics related to nystagmus may not have an independent effect in this context.

The Influence of Nystagmus and Clinical Parameters on Fellow-Eye Visual Acuity A notable finding of our study was the strong correlation between fellow-eye visual acuity deficits and increased fixation instability, particularly in subjects with more significant anisometropia and greater visual acuity deficits in the amblyopic eye. A hierarchical multiple regression analysis was also conducted to evaluate how the presence of nystagmus and other clinical parameters affect fellow eye (FE) visual acuity (Table 28.3). In the first model, the following predictors were included: visual acuity of the AE, clinically measured strabismus angle, the extent of anisometropia (interocular difference), age, stereoacuity deficits, FI of the AE and FE under fellow eye viewing, and multicategorical waveform characteristics (categorized as with and without nystagmus). In contrast, the second model excluded the multicategorical waveform variable.

Results from both models indicated that increasing visual acuity deficit of the AE, greater magnitude of anisometropia, and FI of the FE during fellow eye viewing were significant predictors of FE visual acuity deficits. Additionally, waveform characteristics were statistically significant predictors in the first model. Notably, the removal of waveform information in the second model resulted in a statistically significant R^2 and F change ($\Delta R^2 = -0.05$, $\Delta F = 3.83$, $p = 0.025$). This suggests that the inclusion of waveform characteristics improved the model's ability to predict FE visual acuity deficits, highlighting the potential influence of nystagmus waveform properties on monocular visual function.

Table 28.3 Evaluation of factors impacting fellow-eye visual acuity deficit using a hierarchical regression model

	Model 1 with waveform	Model 2 without waveform
Constant	−0.044	−0.034
AE visual acuity	0.556	0.619^{***}
Strabismus angle	−0.032	0.051
Refractive error difference	-0.377^{***}	-0.311^{**}
Age	−0.096	−0.091
Stereopsis	0.002	0.086
FI of FE in FEV	0.265^{*}	0.235^{*}
FI of AE in FEV	−0.085	−0.035
No nystagmus	0.293^{*}	NA
Nnystagmus	0.378^{**}	NA
R^2 F	$0.40, 7.213^{***}$	$0.36, 7.734^{***}$ ΔR^2 $\Delta F = -0.048, 3.827^{***}$

Note: Standardized coefficients from the hierarchical multiple regression model are reported. Significance levels are indicated as follows: $p < 0.05$ (*), $p \leq 0.01$ (**), $p \leq 0.001$ (***). Abbreviations: *FI* fixation instability, *AE* amblyopic/nondominant eye, *FE* fellow/dominant eye, *FEV* fellow/dominant eye viewing, *NA* not applicable

These findings underscore the differential impact of nystagmus-related factors on binocular versus monocular visual acuity. While fixation instability and binocular summation appear to be the dominant factors in binocular visual acuity deficits, nystagmus waveform characteristics contribute more significantly to explaining fellow-eye visual acuity deficits. This suggests that visual acuity deficits in the fellow eye are often compounded by abnormal eye movements, which, in turn, may negatively influence binocular visual performance. These results emphasize the need for a holistic approach in the assessment and treatment of amblyopia and strabismus, considering both visual acuity and eye-movement abnormalities.

A central factor influencing visual acuity deficits in both amblyopia and strabismus is fixation instability, quantified in this study using the bivariate contour ellipse area (BCEA). Our results indicate that fixation instability, particularly in the amblyopic or nondominant eye, is significantly increased under both binocular and fellow eye viewing conditions. This increased instability was strongly correlated with visual acuity deficits, especially in binocular visual acuity. These findings align with previous studies demonstrating that abnormal eye movements, including fixational saccades and drift, contribute to poorer visual performance in amblyopic eyes. [23, 25]

28.7 Discussion and Conclusion

This chapter set out to investigate the visual acuity profiles of individuals with amblyopia and those with strabismus but without amblyopia, focusing on binocular and fellow-eye visual acuity as well as the factors influencing these measures. Our findings offer important insights into the intricate relationships between amblyopia, strabismus, fixation instability, and visual acuity, contributing to the growing body of literature on childhood visual disorders.

Our study also revealed that nystagmus exerts a unique influence on fellow-eye visual acuity. Nystagmus typically worsens under monocular viewing, which may exacerbate fellow-eye visual acuity deficits. However, under binocular viewing the impact of nystagmus appears to be dampened, suggesting that binocular summation helps mitigate its influence on visual acuity. These findings shed light on why binocular visual acuity may remain relatively intact despite monocular deficits in amblyopic and strabismic individuals with nystagmus. Moreover, our regression analysis showed that the presence of nystagmus significantly improved the prediction of fellow-eye visual acuity deficits, further highlighting the influence of nystagmus on visual performance under monocular viewing. [25, 29, 38–40]

While our study provides valuable insights into the relationship between visual acuity, fixation instability, and nystagmus in amblyopia and strabismus, several limitations must be acknowledged. The sample size was relatively small, limiting the generalizability of our findings and preventing a more nuanced exploration of how fixation eye-movement abnormalities influence binocular and fellow-eye visual acuity across different clinical subtypes of amblyopia. Furthermore, the study did not include subjects with deprivation amblyopia, a form that may present unique challenges in visual acuity and eye-movement dynamics. Future studies with larger, more diverse cohorts are needed to investigate these factors and expand on our findings.

In conclusion, this study highlights the complex relationships between amblyopia, strabismus, visual acuity deficits, fixation instability, and nystagmus. We found that binocular visual acuity deficits were primarily driven by amblyopia rather than strabismus, and that both groups exhibited binocular summation despite reductions in visual acuity. Fixation instability, especially in the amblyopic or nondominant eye, was identified as a crucial factor contributing to these deficits. Moreover, nystagmus was found to exacerbate fellow-eye visual acuity deficits, particularly under monocular viewing conditions. These findings underscore the importance of considering both ocular alignment and fixation eye movements in the assessment and treatment of amblyopia and strabismus. Future research that incorporates larger sample sizes and a more detailed examination of fixation abnormalities across different clinical subtypes will further our understanding of these complex conditions and ultimately contribute to improving visual outcomes for affected individuals.

References

1. Cruz OA, Repka MX, Hercinovic A, Cotter SA, Lambert SR, Hutchinson AK, Sprunger DT, Morse CL, Wallace DK. American Academy of ophthalmology preferred practice pattern pediatric ophthalmology/strabismus, P Amblyopia preferred practice pattern. Ophthalmol. 2023;130:P136–78. https://doi.org/10.1016/j.ophtha.2022.11.003.
2. Wong AM. New concepts concerning the neural mechanisms of amblyopia and their clinical implications. Can J Ophthalmol. 2012;47:399–409. https://doi.org/10.1016/j.jcjo.2012.05.002.
3. Birch EE, Kelly KR, Giaschi DE. Fellow eye deficits in amblyopia. J Binocul Vis Ocul Motil. 2019;69:116–25. https://doi.org/10.1080/2576117X.2019.1624440.
4. Murray J, Garg K, Ghasia F. Monocular and binocular visual function deficits in amblyopic patients with and without fusion Maldevelopment nystagmus. Eye Brain. 2021;13:99–109. https://doi.org/10.2147/EB.S300454.
5. Chen D, Otero-Millan J, Kumar P, Shaikh AG, Ghasia FF. Visual search in amblyopia: abnormal Fixational eye movements and suboptimal sampling strategies. Invest Ophthalmol Vis Sci. 2018;59:4506–17. https://doi.org/10.1167/iovs.18-24794.
6. Ghasia F, Wang J. Amblyopia and fixation eye movements. J Neurol Sci. 2022;441:120373. https://doi.org/10.1016/j.jns.2022.120373.
7. Scaramuzzi M, Murray J, Nucci P, Shaikh AG, Ghasia FF. Fixational eye movements abnormalities and rate of visual acuity and stereoacuity improvement with part time patching. Sci Rep. 2021;11:1217. https://doi.org/10.1038/s41598-020-79077-5.
8. Cakir GB, Murray J, Dulaney C, Ghasia F. Multifaceted interactions of Stereoacuity, interocular suppression, and fixation eye movement abnormalities in amblyopia and strabismus. Invest Ophthalmol Vis Sci. 2024;65:19. https://doi.org/10.1167/iovs.65.3.19.
9. Li J, Thompson B, Lam CS, Deng D, Chan LY, Maehara G, Woo GC, Yu M, Hess RF. The role of suppression in amblyopia. Invest Ophthalmol Vis Sci. 2011;52:4169–76. https://doi.org/10.1167/iovs.11-7233.

10. Farivar R, Thompson B, Mansouri B, Hess RF. Interocular suppression in strabismic amblyopia results in an attenuated and delayed hemodynamic response function in early visual cortex. J Vis. 2011;11 https://doi.org/10.1167/11.14.16.
11. Economides JR, Dilbeck MD, Adams DL, Horton JC. Interocular suppression in primary visual cortex in strabismus: impact of staggering the presentation of stimuli to the eyes. J Neurophysiol. 2021;126:1101–11. https://doi.org/10.1152/jn.00275.2021.
12. Ghasia FF, Otero-Millan J, Shaikh AG. Abnormal fixational eye movements in strabismus. Br J Ophthalmol. 2018;102:253–9. https://doi.org/10.1136/bjophthalmol-2017-310346.
13. Levi DM, Knill DC, Bavelier D. Stereopsis and amblyopia: a mini-review. Vis Res. 2015;114: 17–30. https://doi.org/10.1016/j.visres.2015.01.002.
14. Zlatkova MB, Anderson RS, Ennis FA. Binocular summation for grating detection and resolution in foveal and peripheral vision. Vis Res. 2001;41:3093–100. https://doi.org/10.1016/s0042-6989(01)00191-2.
15. Dulaney CS, Murray J, Ghasia F. Contrast sensitivity, optotype acuity and fixation eye movement abnormalities in amblyopia under binocular viewing. J Neurol Sci. 2023;451: 120721. https://doi.org/10.1016/j.jns.2023.120721.
16. Campbell FW, Green DG. Monocular versus binocular visual acuity. Nature. 1965;208:191–2. https://doi.org/10.1038/208191a0.
17. Wygnanski-Jaffe T, Moshkovitz A, Kushner BJ, Belkin M, Yehezkel O, CureSight Pivotal Trial G. Binocular home treatment for amblyopia: gains stable for one year. Am J Ophthalmol. 2024;262:199–205. https://doi.org/10.1016/j.ajo.2024.02.004.
18. Bhutada I, Skelly P, Jacobs J, Murray J, Shaikh AG, Ghasia FF. Reading difficulties in amblyopia: consequence of visual sensory and oculomotor dysfunction. J Neurol Sci. 2022;442:120438. https://doi.org/10.1016/j.jns.2022.120438.
19. Kelly KR, Jost RM, De La Cruz A, Birch EE. Amblyopic children read more slowly than controls under natural, binocular reading conditions. J AAPOS. 2015;19:515–20. https://doi.org/10.1016/j.jaapos.2015.09.002.
20. Wygnanski-Jaffe T, Kushner BJ, Moshkovitz A, Belkin M, Yehezkel O, Trial CSP, G. An eye-tracking-based dichoptic home treatment for amblyopia: a multicenter randomized clinical Trial. Ophthalmology. 2023;130:274–85. https://doi.org/10.1016/j.ophtha.2022.10.020.
21. Birch EE, Jost RM, Wang YZ, Kelly KR, Giaschi DE. Impaired fellow eye motion perception and abnormal binocular function. Invest Ophthalmol Vis Sci. 2019;60:3374–80. https://doi.org/10.1167/iovs.19-26885.
22. Birch EE, Duffy KR. Leveraging neural plasticity for the treatment of amblyopia. Surv Ophthalmol. 2024;69:818–32. https://doi.org/10.1016/j.survophthal.2024.04.006.
23. Shaikh AG, Otero-Millan J, Kumar P, Ghasia FF. Abnormal Fixational eye movements in amblyopia. PLoS One. 2016;11:e0149953. https://doi.org/10.1371/journal.pone.0149953.
24. Scaramuzzi M, Murray J, Otero-Millan J, Nucci P, Shaikh AG, Ghasia FF. Part time patching treatment outcomes in children with amblyopia with and without fusion maldevelopment nystagmus: an eye movement study. PLoS One. 2020;15:e0237346. https://doi.org/10.1371/journal.pone.0237346.
25. Kang SL, Beylergil SB, Otero-Millan J, Shaikh AG, Ghasia FF. Fixational eye movement waveforms in amblyopia: characteristics of fast and slow eye movements. J Eye Mov Res. 2019;12 https://doi.org/10.16910/jemr.12.6.9.
26. Gonzalez EG, Wong AM, Niechwiej-Szwedo E, Tarita-Nistor L, Steinbach MJ. Eye position stability in amblyopia and in normal binocular vision. Invest Ophthalmol Vis Sci. 2012;53: 5386–94. https://doi.org/10.1167/iovs.12-9941.
27. Chung ST, Kumar G, Li RW, Levi DM. Characteristics of fixational eye movements in amblyopia: limitations on fixation stability and acuity? Vis Res. 2015;114:87–99. https://doi.org/10.1016/j.visres.2015.01.016.
28. Ciuffreda KJ, Kenyon RV, Stark L. Fixational eye movements in amblyopia and strabismus. J Am Optom Assoc. 1979;50:1251–8.

29. Yue SC, Cakir GB, Shaikh A, Ghasia FF. Assessing inter-ocular fixational eye movements throughout the lifespan. Exp Brain Res. 2024;242:2749–63. https://doi.org/10.1007/s00221-024-06936-2.
30. Manh VM, Holmes JM, Lazar EL, Kraker RT, Wallace DK, Kulp MT, Galvin JA, Shah BK, Davis PL, Pediatric Eye Disease Investigator G. A randomized Trial of a binocular iPad game versus part-time patching in children aged 13 to 16 years with amblyopia. Am J Ophthalmol. 2018;186:104–15. https://doi.org/10.1016/j.ajo.2017.11.017.
31. Pardhan S, Gilchrist J. Binocular contrast summation and inhibition in amblyopia. The influence of the interocular difference on binocular contrast sensitivity. Doc Ophthalmol. 1992;82: 239–48. https://doi.org/10.1007/BF00160771.
32. Dorr M, Kwon M, Lesmes LA, Miller A, Kazlas M, Chan K, Hunter DG, Lu ZL, Bex PJ. Binocular summation and suppression of contrast sensitivity in strabismus, fusion and Amblyopia. Front Hum Neurosci. 2019;13:234. https://doi.org/10.3389/fnhum.2019.00234.
33. Subramanian V, Jost RM, Birch EE. A quantitative study of fixation stability in amblyopia. Invest Ophthalmol Vis Sci. 2013;54:1998–2003. https://doi.org/10.1167/iovs.12-11054.
34. Ghasia F, Tychsen L. Inter-ocular fixation instability of amblyopia: relationship to visual acuity, strabismus, nystagmus, stereopsis, Vergence, and age. Am J Ophthalmol. 2024;267:230–48. https://doi.org/10.1016/j.ajo.2024.06.021.
35. Steinman RM, Cushman WB, Martins AJ. The precision of gaze. A review. Hum Neurobiol. 1982;1:97–109.
36. Tychsen L, Richards M, Wong A, Foeller P, Bradley D, Burkhalter A. The neural mechanism for latent (fusion maldevelopment) nystagmus. J Neuroophthalmol. 2010;30:276–83. https://doi.org/10.1097/WNO.0b013e3181dfa9ca.
37. Dell'Osso LF, Daroff RB. Congenital nystagmus waveforms and foveation strategy. Doc Ophthalmol. 1975;39:155–82. https://doi.org/10.1007/BF00578761.
38. Murray J, Gupta P, Dulaney C, Garg K, Shaikh AG, Ghasia FF. Effect of viewing conditions on fixation eye movements and eye alignment in amblyopia. Invest Ophthalmol Vis Sci. 2022;63: 33. https://doi.org/10.1167/iovs.63.2.33.
39. Dorman K. Binocular versus monocular acuity in a patient with latent nystagmus. J Am Optom Assoc. 1982;53:485–6.
40. Evans DE, Biglan AW, Troost BT. Measurement of visual acuity in latent nystagmus. Ophthalmology. 1981;88:134–8. https://doi.org/10.1016/s0161-6420(81)35062-3.

Chapter 29
Ocular Motor Neural Integrator: Mechanistic and Structural Underpinnings and Consequences of Its Lesions

Parthvi Ravat, Anirban Biswas, and Aasef G. Shaikh

Keywords Neural integration · Oculomotor system · Nucleus prepositus hypoglossi (NPH) · Gaze-evoked nystagmus · Sensorimotor control

29.1 Introduction

In 1906, Sir Charles Sherrington introduced the concept of the brain's unity, coining the term "integrative"[41]. This idea, akin to the modern notion of spatial integration, emphasizes how different brain regions collaborate seamlessly. Sherrington's work is particularly evident in the oculomotor and vestibular systems. For instance, the vestibulo-ocular reflex exemplifies this integration by combining signals from the semicircular canals and otoliths to control eye position, showcasing how sensory inputs are unified to produce coordinated motor outputs. His illustrations vividly depicted the brain's capacity to integrate information across both space and time, with a primary focus on spatial integration.

Fast forward nearly seventy years, and David Robinson introduced the concept of the neural integrator in the temporal domain [5, 8–10, 36]. Robinson's theories predicted integration across various eye movements, and his experimental evidence highlighted the balance between sensory differentiation and neural integration.

P. Ravat (✉)
Department of Neurology, Flinders Medical Centre, Bedford Park, Adelaide, SA, Australia

A. Biswas
Vertigo and Deafness Clinic, Kolkatta, India

A. G. Shaikh (✉)
Neurological Institute, University Hospitals Neurology Service, Louis Stokes Cleveland, VA Medical Center, Department of Neurology, Case Western Reserve University, Cleveland, OH, USA
e-mail: axs848@case.edu

P. Ravat et al. (eds.), *Advances in Translational Neuroscience of Vestibular and Eye Movement Disorders*, Contemporary Clinical Neuroscience,
https://doi.org/10.1007/978-3-032-07539-0_29

Sensory differentiation allows the brain to detect and adapt to changes by processing only significant stimuli, reducing redundancy. Conversely, neural integration combines these differentiated signals into coherent, sustained responses, enabling the brain to maintain consistent actions and perceptions over time.

The interrelated concepts of sensory differentiation and neural integration ensure that the brain operates as a cohesive system, capable of accomplishing complex functions. Here we will discuss the mechanistic basis of how ultimate neural substrates for the integration operate at the neuron and circuit levels. We will show that neural integration is neither the property of an isolated neuron nor intervening synapses. The neural integration takes place as the group of neurons with peculiar membrane properties and synaptic connectivity are connected in a particular network arrangement. We will first summarize the extrinsic connectivity of the regions responsible for neural integration of horizontal eye movements. We will then outline the intrinsic structure of the nucleus that forms the internal network essential for neural integration. Finally, we will review the fundamental neurobiology of the synapses and cell membranes that empower the cellular network to perform neural integration. We will focus on the widely studied connectivity and physiology of the neural integrators for horizontal eye movements, the nucleus prepositus hypoglossi (NPH) as a prototype for our discussion. Similar conceptual principles also apply to the neural integrators for vertical and torsional eye movements such as the interstitial nucleus of Cajal (INC); their physiology and any distinction can be compared to the horizontal system.

The chapter will conclude by discussing neural integrator failure. We will illustrate this concept using animal lesion models to explain human neural integrator failure in patients with central ocular motor and vestibular dysfunction.

29.2 Basic Principles of Neural Integration

29.2.1 Extrinsic Connectivity

According to the classic Robinsonian concept, the neural integrator remains in the "final common pathway" [5, 9]. What does that mean? That means that all eye movement signals, such as vestibulo-ocular reflex, saccades, pursuits, and optokinetic nystagmus converge into the neural integrator (Fig. 29.1). This concept is based out of the anatomical connectivity of the NPH.

Input. The bilateral medial and lateral vestibular nuclei project to the NPH for the control of vestibulo-ocular reflex. The signal received from the vestibular nuclei carries otolith- and semicircular canal-generated information, which has already undergone one level of integration at the vestibular velocity storage [21, 22]. The NPH also receives input from the contralateral NPH, facilitating bihemispheric coordination. The saccades related signal to NPH comes from the brainstem reticular formation. Contralateral projections from the medullary reticular formation, an

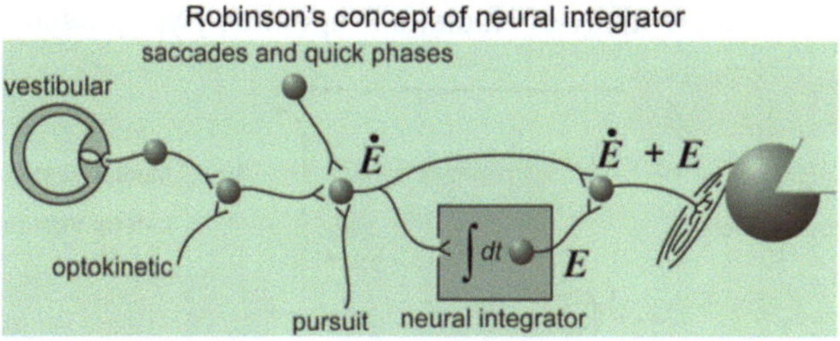

Fig. 29.1 Schematic description of the Robinsonian concept of the neural integrator as a final common pathway. The eye-velocity signal from semicircular canals (partly integrated), optokinetic reflex, pursuit, and saccades converge at the neural integrator. This entity then provides position output to the extraocular muscles, as depicted in the schematic

additional source of signal, emphasize the importance of cross-communication within the brainstem. The Paramedian Pontine Reticular Formation (PPRF) and the Rostral Interstitial Medial Longitudinal Fasciculus (RIMLF) both project to the ipsilateral NPH. These structures are involved in horizontal and vertical eye movements, respectively [43]. Interstitial nucleus of Cajal and mesencephalic reticular formation are additional important sources of NPH input, projecting bilaterally. The oculomotor nuclei and oculomotor internuclear neurons provide additional bilateral NPH inputs, further integrating the signals necessary for the control of eye movements. The visually driven movements, such as pursuits and optokinetic nystagmus also is controlled by the NPH, and it is mediated by the connection of the NPH with the nucleus of the optic tract, and supplementary eye field. NPH, receiving eye movements or vestibular commands to be integrated, further improves its fidelity by the internal cerebellar feedback loop. The cerebellar outflow, the fastigial nuclei, and the vestibular nuclei project bilaterally to the NPH. Figure 29.2 summarizes the inputs to NPH in schematized manner.

Output. The NPH projects to a number of oculomotor nuclei—bilateral abducens and trochlear nerve nuclei, and primarily ipsilateral oculomotor nucleus. In addition to the ocular motor control, the NPH also projects to bilateral vestibular nuclei, with heavy projections sent not only to the medial nucleus but also to the Y-group. The goal of these output signals is to serve gaze stabilization and balance. Figure 29.2 summarizes the outputs of the NPH in schematized manner.

29.2.2 Physiological Properties

To further understand the neural integrator and its contributions to gaze control, we will review the physiological properties of the input and output regions. For conceptual clarity, we will focus on the horizontal system. The paramedian pontine reticular formation (PPRF) provides the saccade signal, which is generated by excitatory burst neurons (EBNs) (Fig. 29.3). These neurons produce bursts of action potentials that generate sufficient force for the high-velocity of saccades (Fig. 29.3),

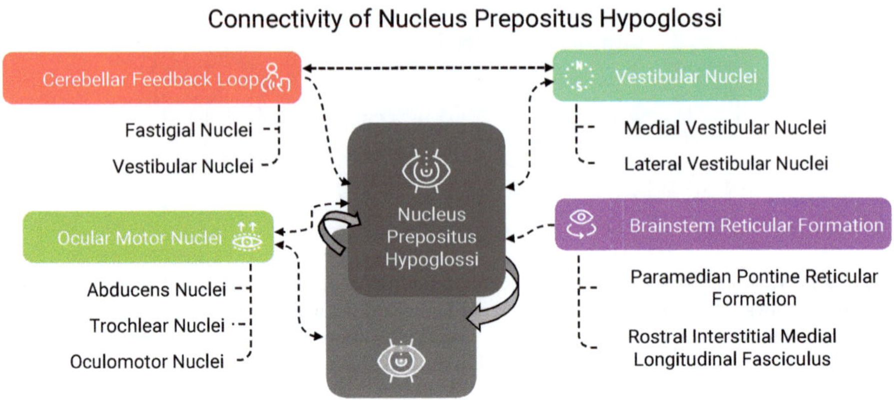

Fig. 29.2 The connectivity of the nucleus prepositus hypoglossi (NPH), the gray structure in the center. The structure is reciprocally connected to its contralateral counterpart. It also has reciprocal connections with oculomotor and vestibular nuclei (shades of green) and receives input from the cerebellum and brainstem reticular formation. The cerebellum receives input from the vestibular nuclei

Fig. 29.3 Paramedian pontine reticular formation (PPRF) projects to the excitatory burst neurons (EBN) providing high-velocity saccade signal. This signal is monosynaptically utilized to generate the "pulse" component of the saccade while multi-synaptically in neural integration to generate the "step" component highlighting steady gaze holding

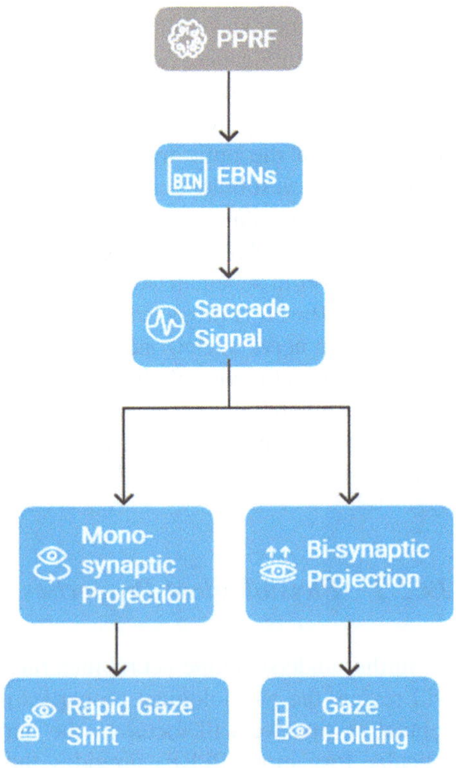

as well as the fast phases of the vestibulo-ocular reflex and optokinetic reflexes [17, 33]. The rapid pulse of neural discharge and the prompt eye movements featuring a high saccade velocity is accomplished by the dual projections of the PPRF neurons to the ocular motor nuclei—bisynaptic projections via the neural integrator, as well as monosynaptic projection directly to the nucleus. The monosynaptic projection provides rapid, high-velocity gaze shift, while multisynaptic projection facilitates gaze holding (Fig. 29.3). Physiological characteristics of the abducens nucleus provide proof of this concept.

Abducens motoneurons fire a burst of action potential, immediately prior-to and during the on-direction saccade. The neurons are quiet during the off-direction saccade. The burst of firing correlates with the saccade velocity; while the tonic firing that follows the burst, determines the correlates of the eye position in space. This property of the motoneurons make them “multistable,” that is stable firing is produced for every eye position, and the firing rate of the neural discharge correlates with the three-dimensional eye position [29].

How does the neural integrator in NPH convert the neuronal burst (input) to a steady-state firing (output)? The discharge rate of the NPH neurons may linearly correlate with the eye position and/or eye velocity [12, 30, 31, 42]. This relationship classifies neurons into three subtypes: position, position-velocity, and velocity-position neurons [15, 16]. Pure position cells monosynaptically project onto abducens motoneurons [16, 17]. The position-velocity NPH neurons encode both eye-position and eye-velocity signals in the horizontal plane; they are activated by eye movements in the ipsilateral direction. The position-velocity neurons project onto abducens motoneurons and the vicinity of the oculomotor complex. The velocity-position neurons, on the contrary, feature an irregular tonic firing, not related to the eye position. These cells present a significant relationship with the eye movements; the velocity-position neurons related to ipsilateral eye movements seem to project to the peri-oculomotor area, while those related to contralateral saccades project mainly to the cerebellum [15]. Such projections putatively participate in the internal feedback necessary to increase the fidelity of the neural integrator.

The velocity-position neurons have highly variable inter-spike intervals depicting irregular spiking; such variability is reduced in case of the position-velocity neurons and even less so in the position sensitive neurons. The abducens motoneurons, receiving the neural integrator input have a very stable firing rate. Intrinsic membrane and synaptic mechanisms justify such variability in firing rate in various subclasses of neurons, with multiple explanations [15, 20]. According to one theory, the stability in firing rate is the algebraic addition, a form of neural integration, from multiple sources of eye-position signals on distal dendrites of ocular motor neurons. Intrinsic membrane properties of the neurons further elaborate on such interactions.

29.2.3 Intrinsic Connectivity and the Cascade Model

The neural connectivity pattern that is fundamental for neural integration was described in the cascade model for the generation of eye-position signals. This model highlights three neuronal types (i.e., velocity-position, position-velocity, and position neurons), receiving comparable inputs from vestibular and reticular (saccade) projections. The position neuron project mono-synaptically onto the abducens motoneurons, and to position-velocity neurons. The position sensitive neurons are housed in the rostrolateral, "marginal" zone of the NPH, abutting the medial vestibular nucleus in primates [27, 43, 45]. The position-velocity neurons project to the position neurons and velocity-position neurons. Velocity-position neurons only project to position-velocity neurons, while receiving the velocity signal from the PPRF. These neurons in more caudal regions of the NPH project to the cerebellum or contralateral NPH [7]. To sum up, the cascade model suggests that eye-velocity signals from EBN primarily target velocity-position neurons. The rest of the cascade, involving position-velocity and position neurons, forms an integrated loop by projecting onto both preceding and following neurons. These chains of neurons overlay the shorter, direct pathways that carry eye-velocity motor commands [16]. This cascade-like, polysynaptic connectivity might account for the high sensitivity of eye-position neuronal systems to various factors such as drugs, anesthetics, conscious state, and attention level. Figure 29.4 summarizes the neural interconnectivity comprising the cascade concept.

The cascade model of the NPH comprises of a network of neurons temporally integrating the velocity signals, showing a rostral-caudal gradient, in a way that rostral neurons encode position signals while caudal neurons present the velocity [15, 32]. As we further elaborate the mechanistic underpinning of the cascade concept in neural integration, it is likely that the NPH neurons generate the eye-position signal by a cholinergic-mediated sustained depolarization [34], as well as by means of the integration of velocity signals from the sustained local networks in NPH [37], and positive feedback pathways [4] that probably include cerebellum [25] [28] and orbital proprioception [18]. In the subsequent section we will review the mechanistic underpinnings of interconnected circuits and how synaptic contribution and cholinergic mechanisms enhance neural integration.

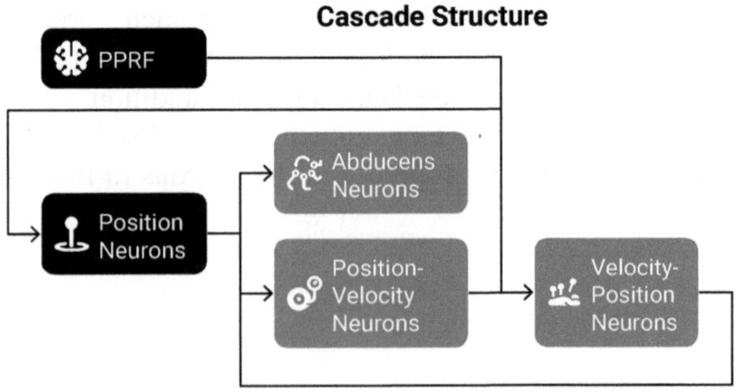

Fig. 29.4 Cascade model for neural integration depicting layered complex connectivity of the position, position-velocity, and velocity-position neurons of the NPH as well as abducens position cells

29.2.4 *Interconnected Network, Membrane Properties, and Neural Integration*

The functioning of the horizontal neural integrator at the interconnected neural circuit level has been elucidated in the goldfish's medullary area 1. This structure is analogous to the horizontal velocity-to-position integrators found in the primate NPH [1]. The research revealed that dendrites in this region extend their branches to the ventral half of the medulla and toward the contralateral position neurons in the contralateral abducens nucleus. This is like how position neurons project to the contralateral NPH and position-velocity neurons. As anticipated in mammalian position neurons, the neurons in the goldfish medullary area 1 exhibit firing rates that rise with eye position [1]. Intracellular recordings from these neurons provided further insights into the membrane mechanism of the neural integration [3]. The researchers observed that sustained alterations in firing rates following saccades, indicative of neural integration, were linked to step changes in the inter-spike membrane potential. This potential was found to correlate with eye position. This raised the question of whether the inter-spike membrane potential alone dictates neural integration behavior. To explore this, Aksay et al. [3] simulated saccades by injecting intracellular current pulses [3]. The hypothesis posited that if neural integration were solely governed by neuronal membrane conductance and properties, then intracellular current pulses would result in a sustained change in firing rate and inter-spike membrane potential. However, their experiment did not support this concept. Instead, the transient changes in the inter-spike interval and firing rate following electrical current injection in goldfish medullary area 1 neurons indicated that other network properties, such as synaptic inputs, had a crucial role in effective neural integrator function [3]. These results provided additional confirmation that neural networks are essential components assuring the fidelity of the neural integration. Subsequent study focusing on cross-correlational analysis of simultaneously recorded neuronal pairs further supported the critical role of synaptic input in neural integration [2]. This study provided the first direct measurement of rate covariation in oculomotor integrator position neurons through simultaneous recordings, revealing patterns influenced by saccade history, termed "pair hysteresis" [2]. Hysteresis refers to a phenomenon where a system's state or behavior is influenced not only by its current conditions but also by its past. Two types of hysteresis were identified: "ON–OFF hysteresis," where neuron pairs showed consistent differences in firing rates between ON and OFF fixations, and "ON2–ON1 hysteresis," where firing rates during later ON fixations were higher than those during earlier ON fixations. These patterns, observed in approximately 60% and 50% of neuron pairs respectively, were linked to the neurons' relative thresholds and persistent firing dynamics. Unlike nonlinearities caused by rate saturation, these hysteresis effects demonstrated a nonuniqueness related to the history of preceding saccades, suggesting a functional role in stabilizing gaze and encoding eye position [2]. The reason for such differences in the firing rate of paired partnering neurons might reflect that synaptic feedback is not the only mechanism for persistent neuronal firing. A burst input in response to saccade also contributes to persistent activity, triggering plateau potentials [23, 26]. Such correlations were not affected by the presence or absence of vision [2].

29.2.5 Role of Synapses and Neurotransmitters in Neural Integration

Aksay and colleagues hypothesized that the oculomotor system integration might be driven by two opposing cell populations: one increasing and the other decreasing their firing rates in response to changes in eye position. Traditional theories highlighted the significance of positive feedback through mutually inhibiting, opposing neuron populations in maintaining persistent neuronal firing activity. To test this hypothesis, they measured the eye position and firing rates of one population of goldfish neurons while pharmacologically inactivating the opposing population [4]. The authors discovered that, contrary to traditional views, mutual inhibition acts as feedforward control. In this model, above-equilibrium firing—neuronal activity surpassing the baseline rate—of one population triggers below-equilibrium activity—where firing rates fall below the baseline—in the opposing population. Persistence mechanisms within each population are active only at above-equilibrium rates, suggesting that integration happens independently across various eye position ranges [4]. These findings shed light on the function of mutual inhibition, backed by anatomical and physiological data, and demonstrated that persistence depends on thresholding mechanisms or dendritic bistability, which reduces positive feedback. Even when inactivation occurred, strong persistence was noted, with firing rate drifts aligning with integrator time constants.

A further follow-up study found variability in firing-rate persistence and topographically organized groups of neurons. Two adjacent neurons have similar firing persistence, which decreases as the distance between recorded neurons increases along the rostrocaudal and dorsoventral axes [32]. This pattern reflects topographically determined circuit architecture heterogeneity; more closely located cells are strongly connected, while distances have weaker communication [32]. Consequently, it is plausible that a single neural integrator circuit encompasses multiple time constants. Supporting this, research using two-photon calcium imaging and serial electron microscopy in larval zebrafish revealed that the oculomotor integrator performs leaky integration across various distributed timescales. This behavior aligns with an inverse model for producing oculomotor commands and suggests that multi-timescale dynamics might be a common characteristic of motor circuits [44].

The populations of neurons in the integrator regions also show topographic variability in their biological properties. For example, the zebra fish neural integrator has three morphological neuronal classes. The ipsilaterally projecting neurons are located medially and are glutamatergic. Contralaterally projecting neurons are situated laterally and GABAergic, while neuronal projections of extreme lateral edge in the hindbrain remain undetermined [44].

The sustained firing of NPH position neurons is crucial for keeping the eye stable within the orbit [10, 33]. Similarly, in goldfish, certain brainstem neurons exhibit persistent firing that generates eye-position signals [2, 3]. These persistent neuronal responses last beyond the transient stimulus and resemble mechanisms seen in working memory and other cognitive functions [29]. A muscarinic cholinergic mechanism supports persistent firing in rat entorhinal cortex slices, while a similar mechanism might partly drive eye-position signal generation. Persistent activity in prepositus neurons likely results from interactions between AMPA/kainate and M1 muscarinic receptors. Postsynaptic M1 receptors were shown to maintain carbachol-induced depolarization despite tetrodotoxin, while pirenzepine blocked both sustained depolarization and carbachol effects.

Based on these findings, prepositus neurons in vitro exhibited similar functional properties (such as sustained firing proportional to train stimulation of the PPRF area and hysteresis) to those observed in alert mammals [15]. To verify the cholinergic nature of the synaptic mechanisms responsible for the persistent activity in prepositus neurons, a pharmacological study was conducted in alert cats. The prepositus nucleus was pinpointed using electrophysiological techniques [17, 34], followed by microinjections of specific drugs during spontaneous eye movements. Noradrenaline (NA) also modulates neural integrators. Strong rostral projections from the locus coeruleus to the prepositus have been identified in monkeys. Intracellular recordings from guinea pig brainstem slices classified prepositus neurons into linear tonic (type A), nonlinear phasic (type B), and oscillatory (type D) subtypes, showing that NA depolarizes types A and B while hyperpolarizing type D. These responses suggest noradrenergic modulation of neural integrator functions, though the specific adrenoceptor subtypes mediating these effects remain unclear [37].

29.3 Lesions of Neural Integrator and Clinical Syndromes

To further understand the dysfunction of the ocular motor neural integrator, we will examine gaze-evoked nystagmus, a condition where the integrator's function is compromised, failing to transform the pulse of neural discharge into a step [5, 9]. This dysfunction results in the eye position decaying toward a neutral position, often straight ahead, referred to as the "null" position (Fig. 29.5a–d). In gaze-evoked nystagmus, when a participant attempts to orient their gaze to the right, the eyes drift to the left, and vice versa. The velocity of the drifting gaze is proportionate to the orientation of the eyes within the orbit [6]. This failure of the neural integrator, a well-documented concept in ocular motor physiology, has also been associated with head movements and dysfunction in head holding, such as in cervical dystonia [38, 39].

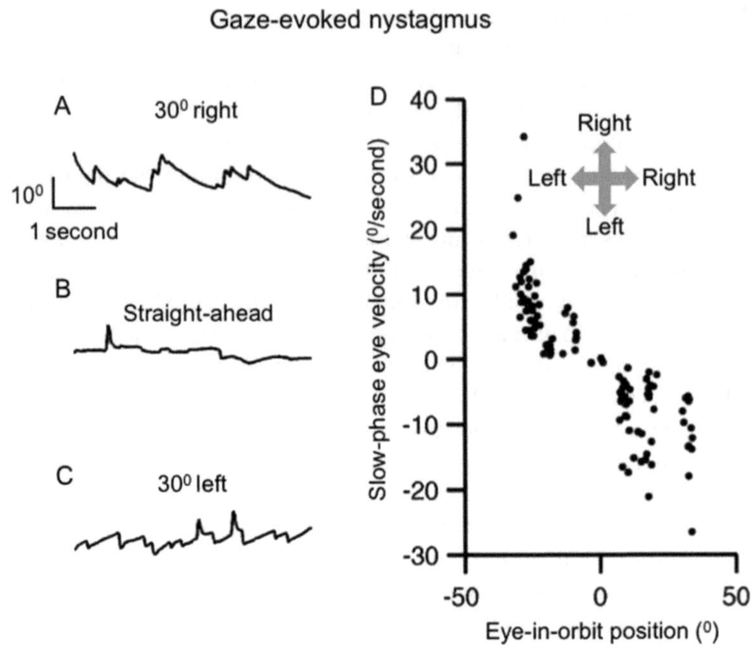

Fig. 29.5 Gaze-evoked nystagmus of the eyes and conceptual network model: (**a**–**c**): This example shows gaze-evoked nystagmus recorded over a 5 second period as a subject with a cerebellar lesion attempts to fixate their gaze to the right (**a**), straight ahead (**b**), or to the left (**c**). The nystagmus is characterized by slow phases and rapid corrective quick phases. During a slow phase, the eyes drift toward the null position (**b**). (**d**): Slow-phase velocity changes systematically with the eye's orientation in the orbit. Each circle in (**d**) represents one cycle of gaze-evoked nystagmus, with slow-phase velocity on the y-axis and the corresponding eye-in-orbit position on the x-axis. Positive values indicate a rightward direction, while negative values indicate a leftward direction

The notion that neural integrator dysfunction is solely reflective within gaze holding is far from accurate. Nearly half a century ago, Daroff and Hoyt demonstrated that unilateral brainstem lesions affecting the PPRF result in the loss of all ipsilateral conjugate eye movements [13]. Their studies suggested damage to the ipsilateral premotor pathways, predicting it as a final common pathway from the integrator [40]. Cohen and Komatsuzaki elicited ipsilateral eye deviation with PPRF electrical stimulation, concluding that the neural integrator was located in the PPRF [11]. However, PPRF lesions abolished ipsilateral saccades and quick phases of nystagmus, while pursuits and the ability to maintain eccentric gaze were preserved [14, 19, 24, 35, 40]. This indicates that electrical stimulation and inactivation (or human lesion) of the PPRF may involve neighboring oculomotor regions such as the Nucleus Prepositus Hypoglossi (NPH) and the medial vestibular nucleus (MVN). Subsequent studies by Cannon and Robinson further examined the effects of NPH and MVN in animal models [9].

The classic Robinsonian concept depicted the neural integrator in the final common pathway, meaning all oculomotor signals, such as those from the vestibulo-ocular reflex, saccades, pursuit, and optokinetic nystagmus, converge into the neural integrator located in the MVN and NPH [5, 9] (Fig. 29.1). Therefore, the Robinsonian concept predicted that a failure of the neural integrator would lead to more global dysfunction. Robinson and colleagues demonstrated in nonhuman primates that acute (yet transient) lesions in areas of the MVN and NPH, induced by an inactivating agent kainic acid, would lead to gaze-evoked nystagmus [6, 9]. If the injections were unilateral, the nystagmus would have a contralateral bias. In these models with inactivation, the Vestibulo Ocular Reflex (VOR) and Optokinetic Nystagmus (OKN) would lack quick phases, and there would be a sustained shift, suggesting a lack of vestibular velocity storage mechanism. The dysfunction of the neural integrator would also result in a paucity of quick phases in neural OKN and in a sustained shift. The lesions and their effects on OKN paralleled pursuit dysfunction; saccades were also slow, if present. The hallmark of kainic acid injection was a robust decline in the time constant of the neural integrator, manifesting as rapid exponentially shaped drifts in eye position [9]. This acute phenomenon is reversible, as seen in Robinson and Cannon's experiment, where the reversal of inhibition over time due to wearing off from the locally injected pharmacological agent, correlated with the monkey regaining function.

The acute and robust, yet transient, failure of the neural integrator is not exclusive to animal models. This phenomenon can be observed in human patients when there is an acute yet transient mechanical compression of the regions of the neural integrator. For example, a mass-generated edema compressing the neighboring structures within the dorsal midbrain or acute intracranial bleeding can lead to complete recovery. Such transient yet acute effects mimic the animal model of Cannon and Robinson in highly unusual cases, such as intracranial hemorrhage or worsening edema surrounding the focal brainstem mass, outlining proof of the principle demonstrated in macaque models, in human patients.

Acknowledgments AS was supported by grants from Department of Veterans Affairs I01CX002086-04. Some work presented in this review was supported by NIH F32DC006540-01 to AS.

References

1. Aksay E, Baker R, Seung HS, Tank DW. Anatomy and discharge properties of pre-motor neurons in the goldfish medulla that have eye-position signals during fixations. J Neurophysiol. 2000;84(2):1035–49.
2. Aksay E, Baker R, Seung HS, Tank DW. Correlated discharge among cell pairs within the oculomotor horizontal velocity-to-position integrator. J Neurosci. 2003;23(34):10852–8.
3. Aksay E, Gamkrelidze G, Seung HS, Baker R, Tank DW. In vivo intracellular recording and perturbation of persistent activity in a neural integrator. Nat Neurosci. 2001;4(2):184–93.
4. Aksay E, Olasagasti I, Mensh BD, Baker R, Goldman MS, Tank DW. Functional dissection of circuitry in a neural integrator. Nat Neurosci. 2007;10(4):494–504.
5. Arnold DB, Robinson DA. The oculomotor integrator: testing of a neural network model. Exp Brain Res. 1997;113(1):57–74.

6. Arnold DB, Robinson DA, Leigh RJ. Nystagmus induced by pharmacological inactivation of the brainstem ocular motor integrator in monkey. Vis Res. 1999;39(25):4286–95.
7. Baizer JS, Baker JF. Neurochemically defined cell columns in the nucleus prepositus hypoglossi of the cat and monkey. Brain Res. 2006;1094(1):127–37.
8. Cannon SC, Robinson DA. An improved neural-network model for the neural integrator of the oculomotor system: more realistic neuron behavior. Biol Cybern. 1985;53(2):93–108.
9. Cannon SC, Robinson DA. Loss of the neural integrator of the oculomotor system from brain stem lesions in monkey. J Neurophysiol. 1987;57(5):1383–409.
10. Cannon SC, Robinson DA, Shamma S. A proposed neural network for the integrator of the oculomotor system. Biol Cybern. 1983;49(2):127–36.
11. Cohen B, Komatsuzaki A. Eye movements induced by stimulation of the pontine reticular formation: evidence for integration in oculomotor pathways. Exp Neurol. 1972;36(1):101–17.
12. Dale A, Cullen KE. The nucleus prepositus predominantly outputs eye movement-related information during passive and active self-motion. J Neurophysiol. 2013;109(7):1900–11.
13. Daroff RB, Hoyt WF. Supranuclear disorders of ocular control systems in man. Clinical, anatomical, and physiological correlations. New York: Academic; 1972.
14. Deleu D, Michotte A, Ebinger G. Impairment of smooth pursuit in pontine lesions: functional topography based on MRI and neuropathologic findings. Acta Neurol Belg. 1997;97(1):28–35.
15. Delgado-Garcia JM, Vidal PP, Gomez C, Berthoz A. A neurophysiological study of prepositus hypoglossi neurons projecting to oculomotor and preoculomotor nuclei in the alert cat. Neuroscience. 1989;29(2):291–307.
16. Escudero M, de la Cruz RR, Delgado-Garcia JM. A physiological study of vestibular and prepositus hypoglossi neurones projecting to the abducens nucleus in the alert cat. J Physiol. 1992;458:539–60.
17. Escudero M, Delgado-Garcia JM. Behavior of reticular, vestibular and prepositus neurons terminating in the abducens nucleus of the alert cat. Exp Brain Res. 1988;71(1):218–22.
18. Ghasia FF, Gulati D, Westbrook EL, Shaikh AG. Viewing condition dependence of the gaze-evoked nystagmus in Arnold Chiari type 1 malformation. J Neurol Sci. 2014;339(1–2):134–9.
19. Goebel HH, Komatsuzaki A, Bender MB, Cohen B. Lesions of the pontine tegmentum and conjugate gaze paralysis. Arch Neurol. 1971;24(5):431–40.
20. Gonzalez-Forero D, Alvarez FJ, de la Cruz RR, Delgado-Garcia JM, Pastor AM. Influence of afferent synaptic innervation on the discharge variability of cat abducens motoneurones. J Physiol. 2002;541(Pt 1):283–99.
21. Green AM, Angelaki DE. Resolution of sensory ambiguities for gaze stabilization requires a second neural integrator. J Neurosci. 2003;23(28):9265–75.
22. Green AM, Angelaki DE. An integrative neural network for detecting inertial motion and head orientation. J Neurophysiol. 2004;92(2):905–25.
23. Hounsgaard J, Hultborn H, Jespersen B, Kiehn O. Bistability of alpha-motoneurones in the decerebrate cat and in the acute spinal cat after intravenous 5-hydroxytryptophan. J Physiol. 1988;405:345–67.
24. Johkura K, Matsumoto S, Komiyama A, Hasegawa O, Kuroiwa Y. Unilateral saccadic pursuit in patients with sensory stroke: sign of a pontine tegmentum lesion. Stroke. 1998;29(11): 2377–80.
25. Kamath BY, Keller EL. A neurological integrator for the oculomotor control system. Math Biosci. 1976;30:341–52.
26. Kiehn O. Plateau potentials and active integration in the 'final common pathway' for motor behaviour. Trends Neurosci. 1991;14(2):68–73.
27. Langer T, Kaneko CR, Scudder CA, Fuchs AF. Afferents to the abducens nucleus in the monkey and cat. J Comp Neurol. 1986;245(3):379–400.
28. Maex R, Steuber V. An integrator circuit in cerebellar cortex. Eur J Neurosci. 2013;38(6): 2917–32.
29. Major G, Tank D. Persistent neural activity: prevalence and mechanisms. Curr Opin Neurobiol. 2004;14(6):675–84.

30. McConville K, Tomlinson RD, King WM, Paige G, Na EQ. Eye position signals in the vestibular nuclei: consequences for models of integrator function. J Vestib Res. 1994;4(5): 391–400.
31. McFarland JL, Fuchs AF. Discharge patterns in nucleus prepositus hypoglossi and adjacent medial vestibular nucleus during horizontal eye movement in behaving macaques. J Neurophysiol. 1992;68(1):319–32.
32. Miri A, Daie K, Arrenberg AB, Baier H, Aksay E, Tank DW. Spatial gradients and multidimensional dynamics in a neural integrator circuit. Nat Neurosci. 2011;14(9):1150–9.
33. Moschovakis AK, Dalezios Y, Petit J, Grantyn AA. New mechanism that accounts for position sensitivity of saccades evoked in response to stimulation of superior colliculus. J Neurophysiol. 1998;80(6):3373–9.
34. Navarro-Lopez JD, Delgado-Garcia JM, Yajeya J. Cooperative glutamatergic and cholinergic mechanisms generate short-term modifications of synaptic effectiveness in prepositus hypoglossi neurons. J Neurosci. 2005;25(43):9902–6.
35. Pierrot-Deseilligny C, Chain F, Serdaru M, Gray F, Lhermitte F. The 'one-and-a-half' syndrome. Electro-oculographic analyses of five cases with deductions about the physiological mechanisms of lateral gaze. Brain. 1981;104(Pt 4):665–99.
36. Robinson DA. The effect of cerebellectomy on the cat's bestibulo-ocular integrator. Brain Res. 1974;71(2–3):195–207.
37. Saito Y, Yanagawa Y. Synaptic mechanism for the sustained activation of oculomotor integrator circuits in the rat prepositus hypoglossi nucleus: contribution of Ca2+-permeable AMPA receptors. J Neurosci. 2010;30(47):15735–46.
38. Shaikh AG, Wong AL, Zee DS, Jinnah HA. Keeping your head on target. J Neurosci. 2013;33 (27):11281–95.
39. Shaikh AG, Zee DS, Crawford JD, Jinnah HA. Cervical dystonia: a neural integrator disorder. Brain. 2016;139(Pt 10):2590–9.
40. Sharpe JA, Rosenberg MA, Hoyt WF, Daroff RB. Paralytic pontine exotropia. A sign of acute unilateral pontine gaze palsy and internuclear ophthalmoplegia. Neurology. 1974;24(11): 1076–81.
41. Sherrington CS. Integrative action of nervous system. New York; 1906.
42. Sylvestre PA, Choi JT, Cullen KE. Discharge dynamics of oculomotor neural integrator neurons during conjugate and disjunctive saccades and fixation. J Neurophysiol. 2003;90(2): 739–54.
43. Ugolini G, Klam F, Doldan DM, Dubayle D, Brandi AM, Buttner-Ennever J, Graf W. Horizontal eye movement networks in primates as revealed by retrograde transneuronal transfer of rabies virus: differences in monosynaptic input to "slow" and "fast" abducens motoneurons. J Comp Neurol. 2006;498(6):762–85.
44. Vishwanathan A, Daie K, Ramirez AD, Lichtman JW, Aksay ERF, Seung HS. Electron microscopic reconstruction of functionally identified cells in a neural integrator. Curr Biol. 2017;27(14):2137–2147 e2133.
45. Wasicky, R., A. K. Horn and J. A. B?tner-Ennever (2004). "Twitch and nontwitch motoneuron subgroups in the oculomotor nucleus of monkeys receive different afferent projections." J Comp Neurol 479(2): 117–129.

Chapter 30
Skew Deviation

David E. Hale, Anand K. Bery, and Daniel R. Gold

Keywords Skew deviation · Ocular tilt reaction · Vestibular pathway · Brainstem localization · Diplopia

30.1 Introduction

Balance, spatial orientation, and steady gaze all depend critically on a well-functioning vestibular system. Unlike the semicircular canals, which detect angular acceleration, the otolith organs detect linear acceleration. The utricle also detects head tilts relative to gravity, leading to a predictable pattern of compensatory cyclovertical eye movements known as the physiologic ocular tilt reaction (OTR, discussed below).

Clinically, the dysfunction of the utricle pathways causes a pathologic OTR, which can be highly localizing. These pathways begin in the inner ear (labyrinth) and traverse the posterior fossa (especially the brainstem) to reach the cyclovertical ocular motor nuclei (i.e., third and fourth nuclei). Along this course, they travel in close proximity to a variety of important neuro-ophthalmic and neuro-otologic structures, thereby adding to their localizing value.

In this chapter, we will review the utriculo-ocular motor pathway (including the important clinical sign known as a *skew deviation*). We will walk through three sequenced illustrative cases. Importantly, each case serves as a "waystation" along the same pathway originating in the right ear and ending in the left midbrain. As you review the cases, note how different anatomic lesions (in some cases, across the midline) can yield similar clinical findings.

D. E. Hale · D. R. Gold (✉)
Department of Neurology, The Johns Hopkins Hospital, Baltimore, MD, USA
e-mail: dhale12@jh.edu; dgold@juh.edu

A. K. Bery
Otolaryngology, Massachusetts Eye and Ear / Harvard Medical School, Boston, MA, USA

P. Ravat et al. (eds.), *Advances in Translational Neuroscience of Vestibular and Eye Movement Disorders*, Contemporary Clinical Neuroscience,
https://doi.org/10.1007/978-3-032-07539-0_30

30.2 What Is the Physiologic Ocular Tilt Reaction (OTR)?

Let us first review the utriculo-ocular motor pathway. Consider a voluntary body tilt to the right (such as a motorcyclist going around a tight curve to the right). The pathway helps the individual reattain gravitational vertical in three ways. First, there will be a (large) compensatory head tilt to the left. Second, there will be a (small) physiologic ocular counter-roll, with the top pole of each eye rotated toward the left ear. Third, there will be a (minute) elevation of the right eye and depression of the left eye, termed a skew deviation. Formally, a skew deviation is a *nonparalytic* vertical ocular misalignment, meaning that it is not caused by a cranial nerve palsy. From a neuroanatomic perspective, the body tilt excites the ipsilateral (right) utricle (Fig. 30.1) which sends information via cranial nerve VIII. The utriculo-ocular motor pathway then proceeds to the ipsilateral vestibular nucleus (VN) before crossing the midline and ascending in the medial longitudinal fasciculus (MLF) to synapse on the cranial nerve subnuclei that have vertical action (cranial nerves III and IV), and finally on the interstitial nucleus of Cajal (INC). In this way, the circuitry can achieve and sustain the necessary eye movements to compensate for voluntary body tilt. In summary, the physiologic OTR has three components: (1) a head tilt; (2) an ocular counter-roll; and (3) a (negligible) vertical skewing of the eyes.

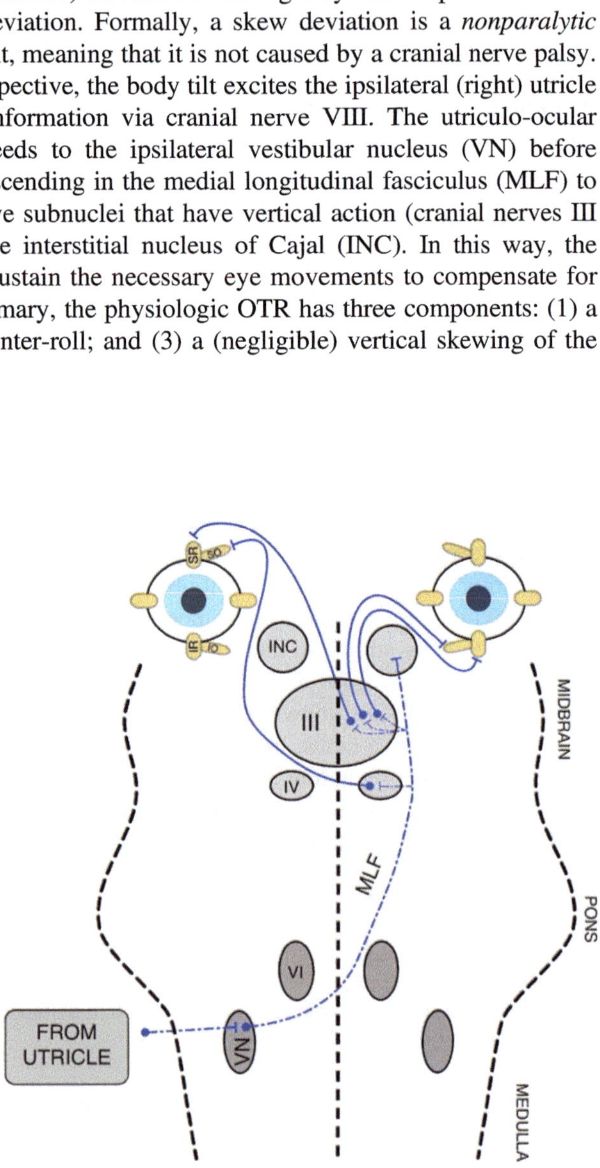

Fig. 30.1 The utriculo-ocular motor pathway begins peripherally in the utricle, synapses in the caudal vestibular nucleus (VN) in the medulla. The pathway decussates at the pontomedullary junction and ascends via the medial longitudinal fasciculus (MLF). In the midbrain, it synapses in the cranial nerve nuclei controlling the cycloverticallly-acting extraocular muscles via CN III and CN IV as well as the interstitial nucleus of Cajal (INC) to maintain this new eye position

30.3 How Do We Evaluate the Utriculo-Ocular Motor Pathway?

The utriculo-ocular motor pathway can be evaluated at the bedside by first evaluating ocular alignment using the cross-cover and cover-uncover tests to look for a vertical misalignment in primary gaze. By convention in neuro-ophthalmology, the higher eye is typically named (i.e., a right hypertropia rather than a left hypotropia) when describing a vertical ocular misalignment; however, that does not indicate the side of the dysfunction or lesion. In most cases, a skew deviation is comitant, meaning that there is the same degree of vertical deviation in each direction of gaze, and typically with right and left head tilt as well. An exception to this is an alternating skew deviation, which typically indicates bilateral and symmetric cerebellar dysfunction (rather than an asymmetric unilateral brainstem disease with a more typical primary gaze skew deviation). An alternating skew deviation is characterized by a right hypertropia in rightward gaze and a left hypertropia in leftward gaze. Fundus torsion caused by an ocular counter-roll can be evaluated subjectively by using double Maddox rod or the bucket test (below) or objectively by measuring the disc-foveal angle on fundus photography, as exemplified in the cases below. The disc-foveal angle is the angle between the center of the optic nerve and fovea relative to the horizon. In a patient with an ocular counter-roll, this will be represented as an excycloduction of one eye and an incycloduction of the other, i.e., the top poles of both eyes will be rotated in the same direction. The subjective visual vertical, the patient's perceived vertical orientation relative to gravity as compared to earth vertical, is altered with the disruption of the utriculo-ocular motor pathway and can be evaluated by using the bucket test. Briefly, the bucket test is performed by asking a patient to look into a bucket with a straight line drawn on the inside bottom of the bucket. The bucket blocks their peripheral visual cues as the examiner rotates the bucket (thereby avoiding proprioceptive clues) and asks the patient to indicate when the line appears vertical. This angle of rotation is measured by a protractor on the outside bottom of the bucket and is considered abnormal when >2–3 degrees off the true (earth) vertical. This can be performed monocularly as well to evaluate for subjective torsion.

30.4 Localizing Lesions Within the Utriculo-Ocular Motor Pathway

As discussed above, the utriculo-ocular motor pathway begins peripherally in the labyrinth, connects to the brainstem via CN VIII, synapses ipsilaterally in the caudal medulla at the vestibular nucleus and decussates at the pontomedullary junction where it ascends via the MLF to reach the sub-nuclei of the cyclovertically acting cranial nerve nuclei (CN III and CN IV) as well as the INC. A disruption can occur anywhere along this pathway, producing similar symptoms and examination

Table 30.1 Summary of expected ocular tilt reaction findings along the utriculo-ocular motor pathway

Localization	Head tilt	Ocular counter-roll	Skew deviation	Key localizing oculao-motor findings
Peripheral (e.g., *right* utricle or vestibular nerve)	Ipsiversive *(rightward)*; minimal or absent	Ipsiversive *(rightward);* common, but subtle	Ipsilateral *(right)* hypodeviation; small and asymptomatic with vestibular neuritis (e.g., 1–2 prism diopters)	Ipsilateral *(right)* abnormal head impulse test
Medulla (e.g., *right* caudal vestibular nucleus)	Ipsiversive *(rightward)*	Ipsiversive *(rightward)*	Ipsilateral *(right)* hypotropia; diplopia present	Saccadic dysmetria, ipsipulsion (lateral medulla), contrapulsion (medial medulla)
Pons (e.g., *left* MLF)	Contraversive *(rightward)*	Contraversive *(rightward)*	Ipsilateral *(left)* hypertropia; diplopia present	Internuclear ophthalmoplegia; spontaneous vertical-torsional nystagmus
Midbrain (e.g., *left* INC)	Contraversive *(rightward)*	Contraversive *(rightward)*	Ipsilateral *(left)* hypertropia; diplopia present	Vertical gaze-evoked nystagmus; spontaneous ipsiversive torsional nystagmus

findings. *The company that it keeps*, as in, the additional findings on the patient's ocular motor and neurologic examination provide clues to differentiate between peripheral, medullary, pontine, or midbrain localizations. This is summarized in the table below. The "high-high" and "low-low" rule is used to remember that if the lesion is "high" (midbrain, pons) then the lesion will be ipsilateral to the hypertropic eye. Conversely, if the lesion is "low" (medulla, vestibular nerve) then the lesion will be ipsilateral to the hypotropic eye (Table 30.1).

30.4.1 Lesions of the Labyrinth, Vestibular Nerve, or Vestibular Nucleus

Consider a lesion of the left vestibular nucleus (Fig. 30.2), such as a stroke. This will cause hypofunction of the pathway originating in the left utricle. There will then be relative *hyperfunction* of the pathway originating in the right utricle. This will cause the same eye movements as a voluntary body and head tilt to the right, similar to the motorcycle example above. The physiologic response to a rightward body/head tilt would be a compensatory head tilt to the left (to gravitational vertical), an ocular counter-roll (with the top pole of each eye rotated to the left), an elevation of the right

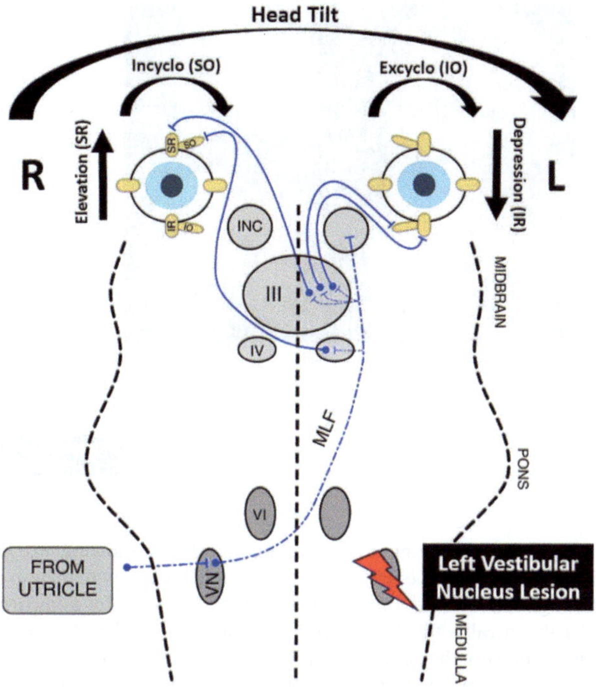

Fig. 30.2 A lesion of the left vestibular nucleus will cause relative hyperfunction of the right utriculo-ocular motor pathway, which would be equivalent to the body and head sensing a tilt to the right. This results in a left (ipsilateral) head tilt, ocular counter-roll with the top pole of both eyes rotated toward the left (ipsilateral) side, and a skew deviation with a right (contralateral) hypertropia

eye, and depression of the left eye. Here, there is a lesion of the utriculo-ocular motor pathway so that this response becomes a pathologic (but still compensatory) OTR. It is important to note that a vertical skewing of the eyes is a normal response in lateral-eyed animals (rabbits), and a skew deviation in frontal-eyed humans can be thought of as the emergence of this primitive reflex. Note that peripheral lesions (such as of the labyrinth or vestibular nerve) typically cause small skew deviations. So, if a clinically significant skew deviation is seen, a central lesion should be suspected until proven otherwise.

Case 1 A 55-year-old male patient presented with acute vestibular syndrome. He had typical findings of an acute unilateral peripheral vestibulopathy on the right side, including spontaneous left-beating nystagmus and corrective saccades on head impulse test to the right. The most common etiologic cause of his presentation is a right vestibular neuritis (Link video: https://collections.lib.utah.edu/ark:/87278/s6ht70fx). Ocular fundus photos were obtained in the acute period—with a normal patient's fundus photos provided as a reference (Fig. 30.3a). The patient's fundus photos demonstrate the expected findings for a right vestibular nerve lesion. There is an ocular counter-roll with the top pole of each eye rotated toward the affected right ear. The patient's perception of the vertical was similarly tilted in the same direction (i.e., to the right) (Fig. 30.3b) when measured using the bucket test. Figure 30.3c reviews the neuroanatomic localization of the case. With a right vestibular nerve localization, the body effectively perceives a (false) tilt of the body and heads to the

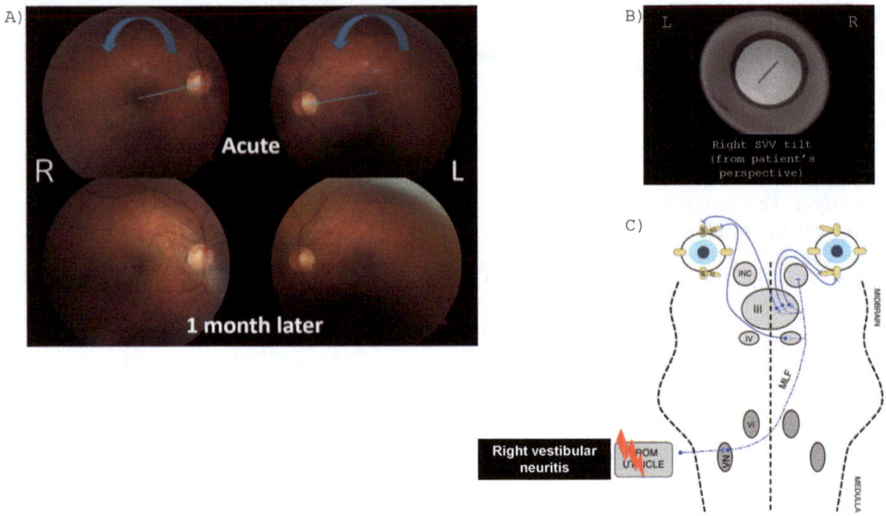

Fig. 30.3 (Illustrating Case 1). Fundus photos showing excycloduction of the right eye and incycloduction of the left eye, an ocular counter-roll with the top pole of both eyes rotating toward the right ear with a normal subject's fundus photos for comparison (**a**). The patient's subjective visual vertical is tilted to the right side, as evidenced by the bucket test. The patient perceives the line to be straight up and down despite being rotated >3 degrees past the true vertical (**b**). The utricle pathway begins peripherally in the inner ear and travels to the brainstem via the vestibulocochlear nerve where the pathway can be interrupted (**c**). When a peripheral lesion causes an ocular tilt reaction, additional brainstem localizing findings would be absent and, instead, an ipsilesional catch up saccade would be expected

(unaffected) left side. Peripheral lesions (of the labyrinth or vestibular nerve) like this can cause small, 1–2 prism diopter skew deviations.

30.4.2 Lesions of the Lower Brainstem

Case 2 A patient had a right lateral medullary stroke. In addition to the usual ocular motor and neuro-ophthalmic findings of the lateral medullary syndrome (e.g., ipsipulsion, ipsilesional hypermetria, contralesional hypometria, or ipsilateral Horner's syndrome), his fundus images showed an ocular counter-roll with the top pole of each eye rotated toward the affected right side. This exemplifies the fact that not all components of the ocular tilt reaction (head tilt, skew deviation, or ocular counter-roll) will be present in each patient. Instead, an *incomplete* ocular tilt reaction can occur.

30.4.3 Lesions of the Upper Brainstem

Case 3 A six-year-old male presented with vertical diplopia on awakening. He held his head slightly tilted to the right. On exam, he had a large (16 prism diopter) left hypertropia in primary gaze and a right head tilt. He also had vertical gaze-evoked nystagmus. Vertical and horizontal saccades appeared normal.

Fundus photos (Fig. 30.4a) showed an ocular counter-roll with the top pole of each eye rotated toward the right ear. Diffusion-weighted MRI demonstrated a lesion of the left interstitial nucleus of Cajal (Fig. 30.4b). The localization is outlined in Fig. 30.4c.

30.5 How Do You Differentiate a Skew Deviation from a Cranial Nerve IV Palsy?

Both a skew deviation and a cranial nerve IV (CN IV) palsy can cause a vertical misalignment of the eyes. Importantly, a skew deviation is typically comitant, meaning that it is the same degree of vertical deviation in each direction of gaze. Conversely, a CN IV palsy follows a typical pattern which can be evaluated using the Parks-Bielschowsky 3 Step Test. First, the side of the CN IV palsy is on the side of the hypertropic eye. Second, the hypertropia will increase in contralateral gaze.

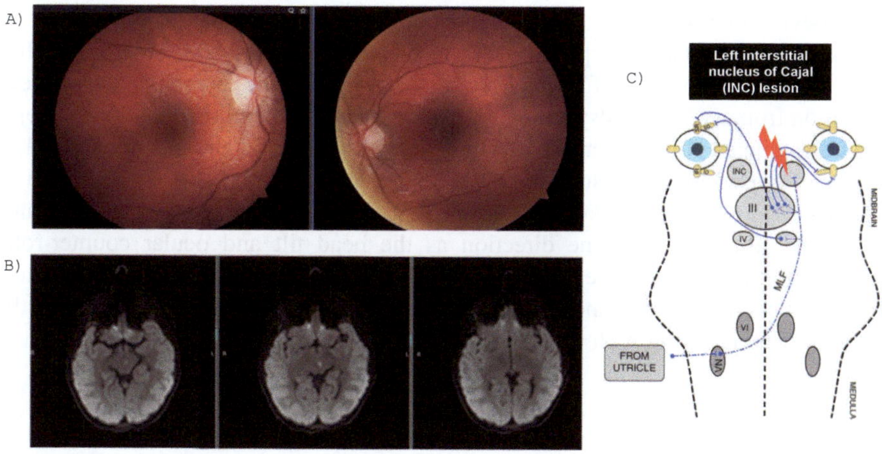

Fig. 30.4 (Illustrating Case 3). Fundus photos showing excycloduction of the right eye and incycloduction of the left eye, an ocular counter-roll with the top pole of both eyes rotating toward the right ear (**a**). In this case, the hypertropic (incycloducted) left eye is ipsilateral to the left interstitial nucleus of Cajal (INC) lesion as seen on the diffusion weighted imaging (DWI) sequences (**b**). The utriculo-ocular motor pathway synapses in the INC in the midbrain, which can cause the features of the ocular tilt reaction as seen in this patient

Third, the hypertropia will increase in ipsilateral head tilt. For example, in a patient with a right CN IV palsy, there would be a right hypertropia that increases in leftward gaze and with rightward head tilt. In addition to the incomitant nature of a CN IV palsy, there are also differences in ocular torsion. In a skew deviation, the hypertropic eye is *incycloducted* while in a CN IV palsy, the hypertropic eye is *excycloducted.* Last, a CN IV palsy typically is due to a peripheral lesion and occurs in isolation, whereas a skew deviation most commonly occurs due to a central lesion and is associated with ocular motor and/or neurologic abnormalities.

30.6 Conclusions

Although patients with a skew deviation typically experience diplopia, poor (or no) vision in one or both eyes can prevent the subjective experience of diplopia. Since small (<1–2 prism diopters) peripheral skew deviations due to vestibular neuritis usually do not cause diplopia, the presence of diplopia (usually with >3–4 prism diopters) is concerning, and should be considered central until proven otherwise. Acute diplopia due to a skew deviation is usually accompanied by symptoms such as imbalance, dizziness, or vertigo, in addition to signs such as central vestibular or gaze-evoked nystagmus. The "company that it keeps" should be evaluated with a comprehensive ocular motor and neurologic examination, which can be highly localizing and also help elucidate etiology (e.g., the combination of skew deviation and internuclear ophthalmoplegia (INO) strongly suggests an MLF localization; consider multiple sclerosis in the young and stroke in older patients). When a skew deviation is suspected, other features of the OTR should be sought.

Expensive ophthalmic equipment is not required to diagnose a skew/OTR; that said, evaluating objective fundus torsion can be helpful in distinguishing a skew deviation from a CN IV palsy, using a fundus camera or an indirect ophthalmoscope through a dilated pupil. Alternatively, the double Maddox rod and monocular bucket test can be used to assess subjective fundus torsion. Think of the SVV—which is also assessed using the bucket test—as the perceptual consequence of the OTR, and it will be tilted in the same direction as the head tilt and ocular counter-roll. Recognition of an acute skew deviation can be potentially life-saving with certain etiologies such as stroke, and the clinical implications (and localization) are vastly different when dealing with a skew versus a CN IV palsy.

Chapter 31
Training the Vestibulo-ocular Reflex with Computer Games

Mark F. Walker, Qi Li, and Michael J. Fu

Keywords Vestibulo-ocular reflex · Rehabilitation · Virtual reality · Exergaming

31.1 Introduction to Vestibular Rehabilitation

Dizziness is one of the most common presenting symptoms to the emergency room and outpatient clinics. Dizziness is a nonspecific symptom with a wide range of causes, but many cases are due to vestibular dysfunction. As a consequence, the prevalence of vestibular dysfunction affects many individuals; a recent study determined the prevalence to be about 3% in the general population [1]. The prevalence of vestibular disorders and other causes of imbalance increases dramatically with age and is a significant cause of morbidity [2].

A classic vestibular presentation to emergency care is the acute peripheral vestibular syndrome or vestibular neuritis (VN). The annual incidence of VN has been estimated to be 3.5–15 per 100,000 [3, 4]. There are no pharmacological treatments to improve vestibular function after VN. Vestibular suppressant medications may reduce the uncomfortable symptoms of acute vestibular hypofunction, but they do not promote functional recovery. In fact, they may impede the physiological

M. F. Walker (✉)
Daroff-Dell'Osso Ocular Motility Laboratory, Department of Neurology, VA Northeast Ohio Healthcare System/Case Western Reserve University (CWRU), Cleveland, OH, USA

Advanced Platform Technology Center, VA Northeast Ohio Healthcare System, Cleveland, OH, USA
e-mail: mark.walker@case.edu

Q. Li
Department of Electrical, Computer, and Systems Engineering, CWRU, Cleveland, OH, USA

M. J. Fu
Advanced Platform Technology Center, VA Northeast Ohio Healthcare System, Cleveland, OH, USA

Department of Electrical, Computer, and Systems Engineering, CWRU, Cleveland, OH, USA

P. Ravat et al. (eds.), *Advances in Translational Neuroscience of Vestibular and Eye Movement Disorders*, Contemporary Clinical Neuroscience,
https://doi.org/10.1007/978-3-032-07539-0_31

learning process by which balance and vision during head motion are restored. Thus, the mainstay of treatment for VN and other causes of vestibular hypofunction is vestibular rehabilitation therapy.

Vestibular therapy consists of a set of office-based and home-based gaze stability and balance exercises that are prescribed by a vestibular therapist. Current recommendations are for a protocol of home exercises, performed several times per day for a total of about 6 weeks for unilateral vestibular dysfunction and longer for bilateral vestibular hypofunction, adjusted on a case-by-case basis at the clinician's discretion based on therapeutic response [5]. Traditional home gaze stability exercises consist of repeated head movements during attempted fixation of a visual target [6]. The target is typically fixed relative to external space (x1 viewing), or the patient may be asked to move a hand-held target in the opposite direction to head rotation (x2 viewing). More recently, Schubert and colleagues introduced a novel approach of incremental VOR training, in which vestibular demand is gradually increased using a head-fixed laser target [7]. The rationale for this approach, reviewed elsewhere in this volume, is that it is easier for the motor learning process to overcome a smaller error, rather than trying to train the reflex all at once.

If a patient engaged in vestibular therapy is not making progress in functional gaze stability, the vestibular therapist must decide whether to suspend treatment. To determine if the treatment is truly ineffective, however, it is important for the therapist to know whether the exercises have been performed consistently according to the prescribed protocol, and whether they have been done correctly. This is difficult to achieve with current home-based exercises, in which assessment of patient adherence is based on self-report. There are also several other limitations of standard gaze stability exercises. In general, they are not interactive and engaging, so that the patient may not be motivated to continue them. It is difficult to control the exact difficulty of the exercises; for example, head rotations that are too slow or too fast may be suboptimal for learning. The characteristics of the head movements may also be age dependent: a recent study showed that older individuals rotate their heads at lower frequencies during gaze stabilization exercises [8]; they may, especially, need feedback related to their head movements during training. These limitations of standard therapy motivated our effort to develop a computer game to implement vestibular exercises, with the ultimate goal of deploying it to a virtual reality platform for home use.

31.2 VOR Adaptation as a Foundation for Vestibular Rehabilitation

The effectiveness of exercise-based rehabilitation therapy depends on the ability of those exercises to induce motor learning in the brain that restores neurological function. In the case of vestibular hypofunction, the desired outcomes are improved balance and postural stability and restoration of gaze stability during movement. The rationale for vestibular gaze stability exercises is based on a solid foundation of research establishing the ability of exposure to visual-vestibular mismatch to induce learning in the vestibulo-ocular reflex (VOR), so-called VOR adaptation. For example, prolonged or repeated rotation of the head during viewing of a target or visual scene that moves with the head leads to a reduction in the gain of the VOR (ratio of evoked eye velocity to head velocity), concordant with the reduced vestibular requirements of this artificial visual environment.

Prior work has shown that VOR gain can be adapted both up and down by visual-vestibular mismatch during continuous sinusoidal motion [9], brief high-acceleration head rotations [7], and position errors [10, 11]. For rapid rotations, Schubert and colleagues found that stepwise adaptation is more effective [7] and demonstrated that it is possible to adapt the yaw VOR to different gains for leftward and rightward rotations [12]. With the appropriate combination of head rotation and visual motion, the direction [13, 14] and phase [15] of the VOR can also be altered.

31.3 Custom VOR Training with a Computer Game

Considering the limitations of conventional vestibular rehabilitation therapy, recent studies have investigated the use of virtual reality and computer games to enhance the exercises and make them more engaging [16–24]. Our approach was to develop a computer game that can be customized to each patient's degree of vestibular impairment; that can implement the incremental learning approach; and that produces a record of exercise time (adherence to prescribed protocol), intensity of exercise (total dose of head rotation), quality of head rotations (e.g., profile of head speeds), and performance (by incorporating a task that requires gaze stability to be performed accurately).

In the game scenario we chose, the player assumes the role of a soccer goalkeeper (Fig. 31.1), whose objective is to prevent a goal by guessing correctly the direction that the ball will move toward the goal. The direction of travel is cued by a Landolt-C optotype that flashes briefly while the head rotates, thus creating a dynamic acuity task. The game is structured as a series of head rotation trials, similar to clinical head impulses. Each trial begins with a cue indicating whether the head should be rotated to the right or left. Once a head speed of 100 °/s is reached, the optotype flashes on the ball for 100 ms and then disappears. At the end of the head rotation, the player uses a joystick to report the perceived optotype orientation. If the correct orientation is selected, the goal is prevented, and a save is recorded. An incorrect choice leads to a goal scored against the player. The player is given the running score as feedback for performance and is alerted if the peak speed of the head rotation for that trial was slower or faster than the target range.

We tested the ability of this game to induce VOR adaptation in 18 adults with normal vestibular function. Full details of the methods and the complete results of this study have been published elsewhere [25]. To simulate exercises for vestibular hypofunction, in which the goal is to enhance the response toward normal, we

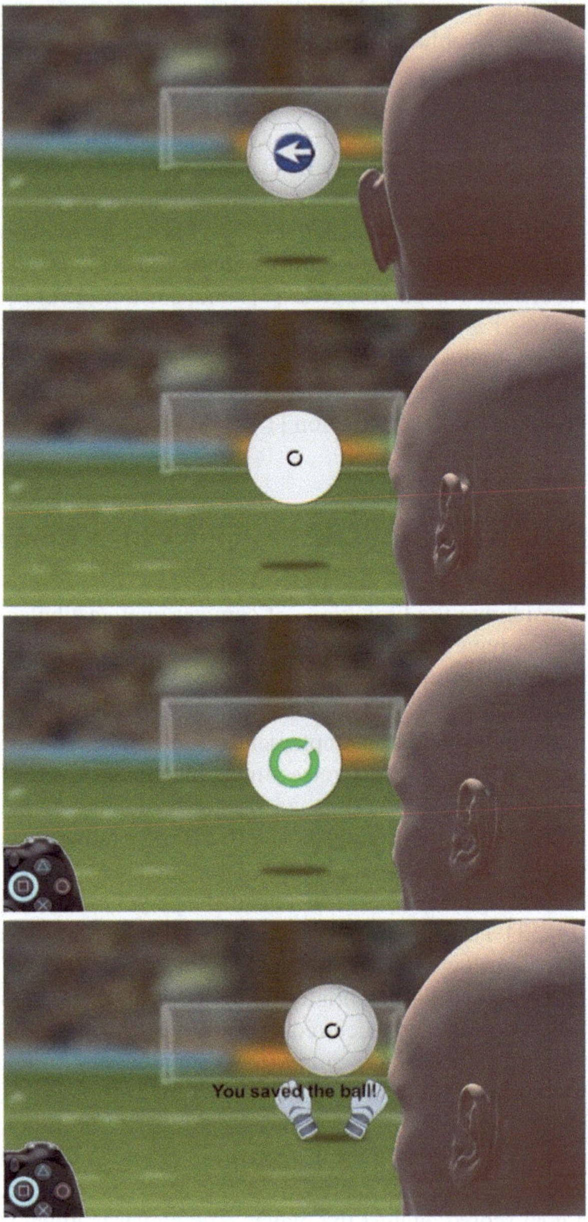

Fig. 31.1 Screenshots from VOR training game depicting game trial sequence (see text for description)

designed a protocol that would increase VOR gain above its baseline. To accomplish this, we moved the game scene on the monitor in the direction opposite to head rotation, creating a visual-vestibular mismatch that would require the eyes to move faster than the head to maintain fixation. Each playing session consisted of 300 head rotation trials. We used two different learning paradigms: (1) fixed mismatch, in

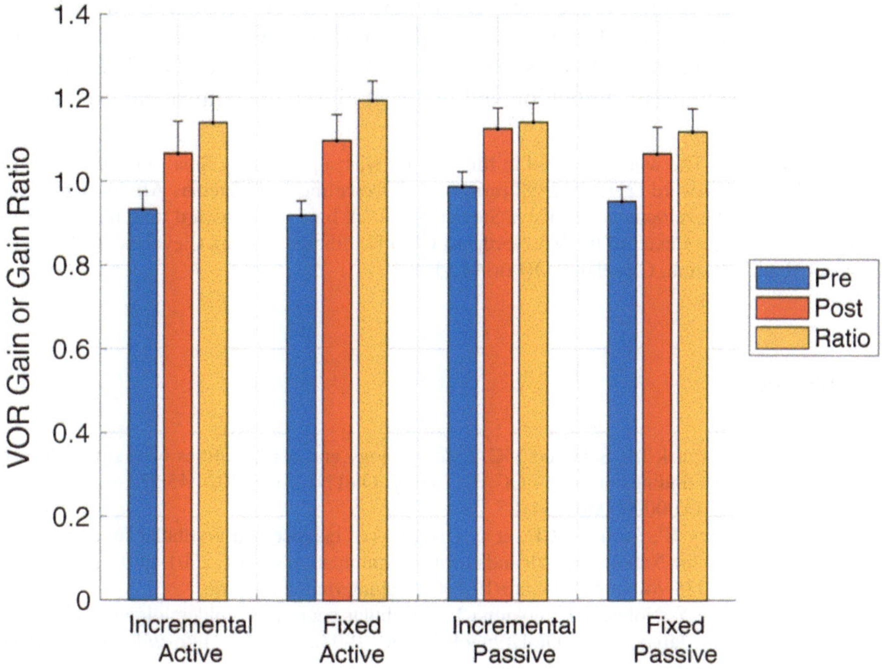

Fig. 31.2 VOR gains before (pre) and after (post) training, VOR gain ratio (post/pre), and mean ± 95% confidence interval. (Adapted from Fig. 5 in Li, et al. [25])

which the instantaneous speed of scene motion was half that of yaw head rotation (x1.5 viewing condition) for all 300 trials, and (2) incremental mismatch, in which playing was divided into three equal blocks of 100 trials of increasing mismatch (x1.17, x1.33, x1.5 viewing conditions). The two training paradigms occurred in separate sessions on different days. To quantify learning, we determined the ratio of the VOR gains measured in the dark before and after game playing, in response to both passive (delivered by the examiner) and active (self-generated) head rotations.

Our game produced robust and similar VOR learning for both training approaches (Fig. 31.2). Combining all conditions, the VOR gain ratio was 1.15 ± 0.047 (mean ± 95% CI, $p < 0.001$). Additionally, there was no effect of training strategy on gaze stability during the head rotations, nor was there a difference in dynamic acuity in the final block of training [25].

31.4 Conclusion

The use of game-based gaze stabilization exercises and virtual reality in vestibular rehabilitation is a promising tool to provide engaging, customized therapy to patients with vestibular dysfunction. Its primary advantages include the ability to provide an interactive training experience, to titrate exercises to each patient's level of dysfunction, to track adherence to the prescribed protocol, and to measure functional progress over time. More study is needed to compare the efficacy of this approach to standard vestibular rehabilitation and its feasibility in the clinical environment.

Acknowledgment The work reviewed in this chapter was supported by Small Projects in Rehabilitation Research Award I21RX002892 from the US Department of Veterans Affairs Rehabilitation Research and Development Service; a Steven Garverick Innovation Award from the Advanced Platform Technology Center of the VA Northeast Ohio Healthcare System, Cleveland, OH; and by the MetroHealth System, Cleveland, OH (to M.J.F.).

References

1. Thorman IB, Schrack JA, Schubert MC. Epidemiology and comorbidities of vestibular disorders: preliminary findings of the AVOCADO study. Otol Neurotol. 2024;45:572–9. https://doi.org/10.1097/MAO.0000000000004185.
2. Agrawal Y, Carey JP, Santina CCD, et al. Disorders of balance and vestibular function in US adults: data from the National Health and nutrition examination survey, 2001–2004. Arch Intern Med. 2009;169:938–44. https://doi.org/10.1001/archinternmed.2009.66.
3. Sekitani T, Imate Y, Noguchi T, Inokuma T. Vestibular neuronitis: epidemiological survey by questionnaire in Japan. Acta Otolaryngol Suppl. 1993;503:9–12. https://doi.org/10.3109/00016489309128061.
4. Adamec I, Krbot Skorić M, Handžić J, Habek M. Incidence, seasonality and comorbidity in vestibular neuritis. Neurol Sci. 2015;36:91–5. https://doi.org/10.1007/s10072-014-1912-4.
5. Hall CD, Herdman SJ, Whitney SL, et al. Vestibular rehabilitation for peripheral vestibular Hypofunction: an updated clinical practice guideline from the academy of neurologic physical therapy of the American Physical Therapy Association. J Neurol Phys Ther. 2022;46:118–77. https://doi.org/10.1097/NPT.0000000000000382.
6. Meldrum D, Jahn K. Gaze stabilisation exercises in vestibular rehabilitation: review of the evidence and recent clinical advances. J Neurol. 2019;266:11–8. https://doi.org/10.1007/s00415-019-09459-x.
7. Schubert MC, Della Santina CC, Shelhamer M. Incremental angular vestibulo-ocular reflex adaptation to active head rotation. Exp Brain Res. 2008;191:435–46. https://doi.org/10.1007/s00221-008-1537-z.
8. Souza WH, Zobeiri OA, Millar JL, et al. Aging delays completion of head rotation cycles in continuous gaze stabilization exercises despite putative healthy vestibular function. Gerontology. 2024;70:1113–27. https://doi.org/10.1159/000540230.
9. Gonshor A, Jones GM. Short-term adaptive changes in the human vestibulo-ocular reflex arc. J Physiol. 1976;256:361–79. https://doi.org/10.1113/jphysiol.1976.sp011329.
10. Eggers SDZ, De Pennington N, Walker MF, et al. Short-term adaptation of the VOR: non-retinal-slip error signals and saccade substitution. Ann N Y Acad Sci. 2003;1004:94–110.
11. Scherer M, Schubert MC. High-velocity angular Vestibulo-ocular reflex adaptation to position error signals. J Neurol Phys Ther. 2010;34:82–6. https://doi.org/10.1097/NPT.0b013e3181dde7bc.
12. Migliaccio AA, Schubert MC. Unilateral adaptation of the human angular Vestibulo-ocular reflex. JARO-J Assoc Res Otolaryngol. 2013;14:29–36. https://doi.org/10.1007/s10162-012-0359-7.

13. Walker MF, Zee DS. Rectified cross-axis adaptation of the vestibulo-ocular reflex in rhesus monkey. Ann N Y Acad Sci. 2002;956:543–5. https://doi.org/10.1111/j.1749-6632.2002.tb02877.x.
14. Schultheis LW, Robinson DA. Directional plasticity of the vestibuloocular reflex in the cat. Ann N Y Acad Sci. 1981;374:504–12. https://doi.org/10.1111/j.1749-6632.1981.tb30895.x.
15. Kramer PD, Shelhamer M, Zee DS. Short-term adaptation of the phase of the vestibulo-ocular reflex (VOR) in normal human subjects. Exp Brain Res. 1995;106:318–26. https://doi.org/10.1007/BF00241127.
16. Mempouo E, Lau K, Green F, et al. Customised vestibular rehabilitation with the addition of virtual reality based therapy in the management of persistent postural-perceptual dizziness. J Laryngol Otol. 2021;135:887–91. https://doi.org/10.1017/S0022215121002127.
17. Kanyılmaz T, Topuz O, Ardıç FN, et al. Effectiveness of conventional versus virtual reality-based vestibular rehabilitation exercises in elderly patients with dizziness: a randomized controlled study with 6-month follow-up. Braz J Otorhinolaryngol. 2021;88:S41–9. https://doi.org/10.1016/j.bjorl.2021.08.010.
18. Meldrum D, Herdman S, Moloney R, et al. Effectiveness of conventional versus virtual reality based vestibular rehabilitation in the treatment of dizziness, gait and balance impairment in adults with unilateral peripheral vestibular loss: a randomised controlled trial. BMC Ear Nose Throat Disord. 2012;12:3. https://doi.org/10.1186/1472-6815-12-3.
19. Xie M, Zhou K, Patro N, et al. Virtual reality for vestibular rehabilitation: a systematic review. Otol Neurotol. 2021;42:967–77. https://doi.org/10.1097/MAO.0000000000003155.
20. Micarelli A, Viziano A, Micarelli B, et al. Vestibular rehabilitation in older adults with and without mild cognitive impairment: effects of virtual reality using a head-mounted display. Arch Gerontol Geriatr. 2019;83:246–56. https://doi.org/10.1016/j.archger.2019.05.008.
21. Heffernan A, Abdelmalek M, Nunez DA. Virtual and augmented reality in the vestibular rehabilitation of peripheral vestibular disorders: systematic review and meta-analysis. Sci Rep. 2021;11:17843. https://doi.org/10.1038/s41598-021-97370-9.
22. Chu H-Y, Song N, Zhou Z-R, et al. Can virtual reality-assisted therapy offer additional benefits to patients with vestibular disorders compared with conventional vestibular physical therapy? A meta-analysis. Arch Phys Med Rehabil. 2023;104:490–501. https://doi.org/10.1016/j.apmr.2022.08.972.
23. Rosiak O, Krajewski K, Woszczak M, Jozefowicz-Korczynska M. Evaluation of the effectiveness of a virtual reality-based exercise program for unilateral peripheral vestibular deficit. J Vestib Res. 2019;28:409–15. https://doi.org/10.3233/VES-180647.
24. Lee JW, Yoon CY, Kim JH, et al. Virtual reality-based vestibular rehabilitation therapy in patients with acute unilateral vestibulopathy: a randomized controlled trial. Front Neurol. 2025;16:1519470. https://doi.org/10.3389/fneur.2025.1519470.
25. Li Q, Xu H, Chen W, et al. Short-term learning of the vestibulo-ocular reflex induced by a custom interactive computer game. J Neurophysiol. 2024;131:16–27. https://doi.org/10.1152/jn.00130.2023.

Index

P. Ravat et al. (eds.), *Advances in Translational Neuroscience of Vestibular and Eye Movement Disorders*, Contemporary Clinical Neuroscience,
https://doi.org/10.1007/978-3-032-07539-0

W

MIX
Papier aus verantwortungsvollen Quellen
Paper from responsible sources
FSC® C105338
FSC
www.fsc.org

If you have any concerns about our products,
you can contact us on
ProductSafety@springernature.com

In case Publisher is established outside the EU,
the EU authorized representative is:
Springer Nature Customer Service Center GmbH
Europaplatz 3, 69115 Heidelberg, Germany

Printed by Libri Plureos GmbH
in Hamburg, Germany